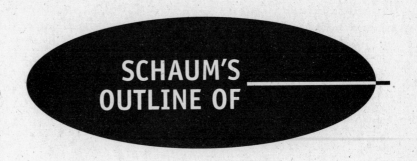

SCHAUM'S
OUTLINE OF

Discrete
Mathematics

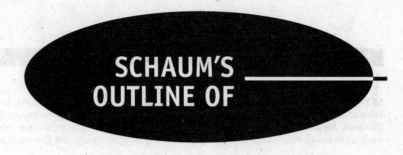

SCHAUM'S
OUTLINE OF

Discrete
Mathematics

Third Edition

Seymour Lipschutz, Ph.D.
Temple University

Marc Lars Lipson, Ph.D.
University of Virginia

Schaum's Outline Series

New York   Chicago   San Francisco   Lisbon
London   Madrid   Mexico City   Milan   New Delhi
San Juan   Seoul   Singapore   Sydney   Toronto

The McGraw·Hill Companies

SEYMOUR LIPSCHUTZ is on the mathematics faculty of Temple University and formerly taught at the Polytechnic Institute of Brooklyn. He received his Ph.D. in 1960 at Courant Institute of Mathematical Sciences of New York University. He is one of Schaum's most prolific authors, and has also written *Probability; Finite Mathematics*, 2nd edition; *Linear Algebra*, 3rd edition; *Beginning Linear Algebra; Set Theory;* and *Essential Computer Mathematics.*

MARC LARS LIPSON is on the faculty of the University of Virginia and previously was on the faculty of the University of Georgia. He received his Ph.D. in finance in 1994 from the University of Michigan. He is also the coauthor of *Linear Algebra*, 3rd edition, and *2000 Solved Problems in Discrete Mathematics* with Seymour Lipschutz.

Schaum's Outline of Theory and Problems of
DISCRETE MATHEMATICS

4 5 6 7 8 9 0   CUS/CUS   0 1 4 3 2

ISBN   978-0-07-161586-0
MHID   0-07-161586-5

Library of Congress Cataloguing-in-Publication Data is on file with the Library of Congress.

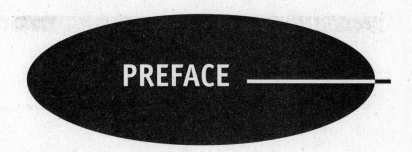

PREFACE

Discrete mathematics, the study of finite systems, has become increasingly important as the computer age has advanced. The digital computer is basically a finite structure, and many of its properties can be understood and interpreted within the framework of finite mathematical systems. This book, in presenting the more essential material, may be used as a textbook for a formal course in discrete mathematics or as a supplement to all current texts.

The first three chapters cover the standard material on sets, relations, and functions and algorithms. Next come chapters on logic, counting, and probability. We then have three chapters on graph theory: graphs, directed graphs, and binary trees. Finally there are individual chapters on properties of the integers, languages, machines, ordered sets and lattices, and Boolean algebra, and appendices on vectors and matrices, and algebraic systems. The chapter on functions and algorithms includes a discussion of cardinality and countable sets, and complexity. The chapters on graph theory include discussions on planarity, traversability, minimal paths, and Warshall's and Huffman's algorithms. We emphasize that the chapters have been written so that the order can be changed without difficulty and without loss of continuity.

Each chapter begins with a clear statement of pertinent definitions, principles, and theorems with illustrative and other descriptive material. This is followed by sets of solved and supplementary problems. The solved problems serve to illustrate and amplify the material, and also include proofs of theorems. The supplementary problems furnish a complete review of the material in the chapter. More material has been included than can be covered in most first courses. This has been done to make the book more flexible, to provide a more useful book of reference, and to stimulate further interest in the topics.

We wish to thank the staff of the McGraw-Hill Schaum Outline Series for their invaluable suggestions and helpful cooperation. We also wish to thank Michel Falus for his careful review of the manuscript. Lastly, Professor Lipschutz wants to express his gratitude to his teacher, advisor, and friend, William Magnus, who introduced him to the beauty of mathematics.

SEYMOUR LIPSCHUTZ
MARC LARS LIPSON

CONTENTS

SCHAUM'S OUTLINE OF

Theory and Problems of

DISCRETE MATHEMATICS

CHAPTER 1

Set Theory

1.1 INTRODUCTION

The concept of a *set* appears in all mathematics. This chapter introduces the notation and terminology of set theory which is basic and used throughout the text. The chapter closes with the formal definition of mathematical induction, with examples.

1.2 SETS AND ELEMENTS, SUBSETS

A *set* may be viewed as any well-defined collection of objects, called the *elements* or *members* of the set. One usually uses capital letters, $A, B, X, Y, \ldots$, to denote sets, and lowercase letters, $a, b, x, y, \ldots$, to denote elements of sets. Synonyms for "set" are "class," "collection," and "family."

Membership in a set is denoted as follows:

$a \in S$ denotes that a belongs to a set S

$a, b \in S$ denotes that a and b belong to a set S

Here $\in$ is the symbol meaning "is an element of." We use $\notin$ to mean "is not an element of."

Specifying Sets

There are essentially two ways to specify a particular set. One way, if possible, is to list its members separated by commas and contained in braces { }. A second way is to state those properties which characterized the elements in the set. Examples illustrating these two ways are:

$$A = \{1, 3, 5, 7, 9\} \quad \text{and} \quad B = \{x \mid x \text{ is an even integer, } x > 0\}$$

That is, A consists of the numbers 1, 3, 5, 7, 9. The second set, which reads:

B is the set of x such that x is an even integer and x is greater than 0,

denotes the set B whose elements are the positive even integers. Note that a letter, usually x, is used to denote a typical member of the set; and the vertical line | is read as "such that" and the comma as "and."

EXAMPLE 1.1

(a) The set A above can also be written as $A = \{x \mid x \text{ is an odd positive integer, } x < 10\}$.

(b) We cannot list all the elements of the above set B although frequently we specify the set by

$$B = \{2, 4, 6, \ldots\}$$

where we assume that everyone knows what we mean. Observe that $8 \in B$, but $3 \notin B$.

(c) Let $E = \{x \mid x^2 - 3x + 2 = 0\}$, $F = \{2, 1\}$ and $G = \{1, 2, 2, 1\}$.   Then $E = F = G$.

We emphasize that a set does not depend on the way in which its elements are displayed. A set remains the same if its elements are repeated or rearranged.

Even if we can list the elements of a set, it may not be practical to do so. That is, we describe a set by listing its elements only if the set contains a few elements; otherwise we describe a set by the property which characterizes its elements.

Subsets

Suppose every element in a set A is also an element of a set B, that is, suppose $a \in A$ implies $a \in B$. Then A is called a *subset* of B. We also say that A is *contained* in B or that B *contains* A. This relationship is written

$$A \subseteq B \quad \text{or} \quad B \supseteq A$$

Two sets are equal if they both have the same elements or, equivalently, if each is contained in the other. That is:

$$\boxed{A = B \text{ if and only if } A \subseteq B \text{ and } B \subseteq A}$$

If A is not a subset of B, that is, if at least one element of A does not belong to B, we write $A \nsubseteq B$.

The empty set is a subset of ALL sets

EXAMPLE 1.2 Consider the sets:

$$A = \{1, 3, 4, 7, 8, 9\}, \quad B = \{1, 2, 3, 4, 5\}, \quad C = \{1, 3\}.$$

Then $C \subseteq A$ and $C \subseteq B$ since 1 and 3, the elements of C, are also members of A and B. But $B \nsubseteq A$ since some of the elements of B, e.g., 2 and 5, do not belong to A. Similarly, $A \nsubseteq B$.

Property 1: It is common practice in mathematics to put a vertical line "|" or slanted line "/" through a symbol to indicate the opposite or negative meaning of a symbol.

Property 2: The statement $A \subseteq B$ does not exclude the possibility that $A = B$. In fact, for every set A we have $A \subseteq A$ since, trivially, every element in A belongs to A. However, if $A \subseteq B$ and $A \neq B$, then we say A is a proper subset of B (sometimes written $A \subset B$).

Property 3: Suppose every element of a set A belongs to a set B and every element of B belongs to a set C. Then clearly every element of A also belongs to C. In other words, if $A \subseteq B$ and $B \subseteq C$, then $A \subseteq C$.

The above remarks yield the following theorem.

Theorem 1.1: Let A, B, C be any sets. Then:

  (i) $A \subseteq A$

 (ii) If $A \subseteq B$ and $B \subseteq A$, then $A = B$

(iii) If $A \subseteq B$ and $B \subseteq C$, then $A \subseteq C$

Special symbols

Some sets will occur very often in the text, and so we use special symbols for them. Some such symbols are:
 N = the set of *natural numbers* or positive integers: $1, 2, 3, \ldots$
 Z = the set of all integers: $\ldots, -2, -1, 0, 1, 2, \ldots$
 Q = the set of rational numbers
 R = the set of real numbers
 C = the set of complex numbers
Observe that $\mathbf{N} \subseteq \mathbf{Z} \subseteq \mathbf{Q} \subseteq \mathbf{R} \subseteq \mathbf{C}$.

Universal Set, Empty Set

All sets under investigation in any application of set theory are assumed to belong to some fixed large set called the *universal set* which we denote by

$$U$$

unless otherwise stated or implied.

Given a universal set U and a property P, there may not be any elements of U which have property P. For example, the following set has no elements:

$$S = \{x \mid x \text{ is a positive integer, } x^2 = 3\}$$

Such a set with no elements is called the *empty set* or *null set* and is denoted by

$$\emptyset$$

There is only one empty set. That is, if S and T are both empty, then $S = T$, since they have exactly the same elements, namely, none.

The empty set $\emptyset$ is also regarded as a subset of every other set. Thus we have the following simple result which we state formally.

Theorem 1.2: For any set A, we have $\emptyset \subseteq A \subseteq U$.

Disjoint Sets

Two sets A and B are said to be *disjoint* if they have no elements in common. For example, suppose

$$A = \{1, 2\}, \quad B = \{4, 5, 6\}, \quad \text{and} \quad C = \{5, 6, 7, 8\}$$

Then A and B are disjoint, and A and C are disjoint. But B and C are not disjoint since B and C have elements in common, e.g., 5 and 6. We note that if A and B are disjoint, then neither is a subset of the other (unless one is the empty set).

1.3 VENN DIAGRAMS

A Venn diagram is a pictorial representation of sets in which sets are represented by enclosed areas in the plane. The universal set U is represented by the interior of a rectangle, and the other sets are represented by disks lying within the rectangle. If $A \subseteq B$, then the disk representing A will be entirely within the disk representing B as in Fig. 1-1(a). If A and B are disjoint, then the disk representing A will be separated from the disk representing B as in Fig. 1-1(b).

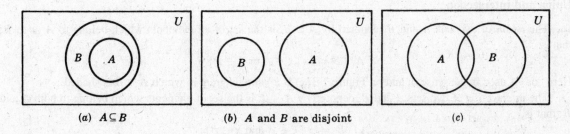

(a) $A \subseteq B$ (b) A and B are disjoint   (c)

Fig. 1-1

However, if A and B are two arbitrary sets, it is possible that some objects are in A but not in B, some are in B but not in A, some are in both A and B, and some are in neither A nor B; hence in general we represent A and B as in Fig. 1-1(c).

Arguments and Venn Diagrams

Many verbal statements are essentially statements about sets and can therefore be described by Venn diagrams. Hence Venn diagrams can sometimes be used to determine whether or not an argument is valid.

EXAMPLE 1.3 Show that the following argument (adapted from a book on logic by Lewis Carroll, the author of *Alice in Wonderland*) is valid:

S_1: All my tin objects are saucepans.
S_2: I find all your presents very useful.
S_3: None of my saucepans is of the slightest use.

S : Your presents to me are not made of tin.

The statements S_1, S_2, and S_3 above the horizontal line denote the assumptions, and the statement S below the line denotes the conclusion. The argument is valid if the conclusion S follows logically from the assumptions S_1, S_2, and S_3.

By S_1 the tin objects are contained in the set of saucepans, and by S_3 the set of saucepans and the set of useful things are disjoint. Furthermore, by S_2 the set of "your presents" is a subset of the set of useful things. Accordingly, we can draw the Venn diagram in Fig. 1-2.

The conclusion is clearly valid by the Venn diagram because the set of "your presents" is disjoint from the set of tin objects.

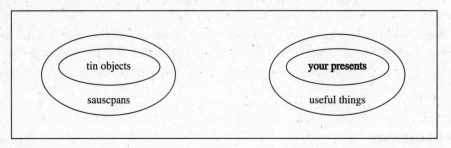

Fig. 1-2

1.4 SET OPERATIONS

This section introduces a number of set operations, including the basic operations of union, intersection, and complement.

Union and Intersection

The *union* of two sets A and B, denoted by $A \cup B$, is the set of all elements which belong to A or to B; that is,

$$A \cup B = \{x \mid x \in A \text{ or } x \in B\}$$

Here "or" is used in the sense of and/or. Figure 1-3(a) is a Venn diagram in which $A \cup B$ is shaded.

The *intersection* of two sets A and B, denoted by $A \cap B$, is the set of elements which belong to both A and B; that is,

$$A \cap B = \{x \mid x \in A \text{ and } x \in B\}$$

Figure 1-3(b) is a Venn diagram in which $A \cap B$ is shaded.

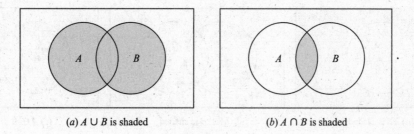

(a) $A \cup B$ is shaded   (b) $A \cap B$ is shaded

Fig. 1-3

Recall that sets A and B are said to be *disjoint* or *nonintersecting* if they have no elements in common or, using the definition of intersection, if $A \cap B = \emptyset$, the empty set. Suppose

$$S = A \cup B \quad \text{and} \quad A \cap B = \emptyset$$

Then S is called the *disjoint union* of A and B.

EXAMPLE 1.4

(a) Let $A = \{1, 2, 3, 4\}$, $B = \{3, 4, 5, 6, 7\}$, $C = \{2, 3, 8, 9\}$. Then

$$A \cup B = \{1, 2, 3, 4, 5, 6, 7\}, \quad A \cup C = \{1, 2, 3, 4, 8, 9\}, \quad B \cup C = \{2, 3, 4, 5, 6, 7, 8, 9\},$$
$$A \cap B = \{3, 4\}, \quad\quad\quad\quad A \cap C = \{2, 3\}, \quad\quad\quad\quad B \cap C = \{3\}.$$

(b) Let **U** be the set of students at a university, and let M denote the set of male students and let F denote the set of female students. The **U** is the disjoint union of M of F; that is,

$$\mathbf{U} = M \cup F \quad \text{and} \quad M \cap F = \emptyset$$

This comes from the fact that every student in **U** is either in M or in F, and clearly no student belongs to both M and F, that is, M and F are disjoint.

The following properties of union and intersection should be noted.

Property 1: Every element x in $A \cap B$ belongs to both A and B; hence x belongs to A and x belongs to B. Thus $A \cap B$ is a subset of A and of B; namely

$$A \cap B \subseteq A \quad \text{and} \quad A \cap B \subseteq B$$

Property 2: An element x belongs to the union $A \cup B$ if x belongs to A or x belongs to B; hence every element in A belongs to $A \cup B$, and every element in B belongs to $A \cup B$. That is,

$$A \subseteq A \cup B \quad \text{and} \quad B \subseteq A \cup B$$

We state the above results formally:

Theorem 1.3: For any sets A and B, we have:

(i) $A \cap B \subseteq A \subseteq A \cup B$ and (ii) $A \cap B \subseteq B \subseteq A \cup B$.

The operation of set inclusion is closely related to the operations of union and intersection, as shown by the following theorem.

Theorem 1.4: The following are equivalent: $A \subseteq B$, $A \cap B = A$, $A \cup B = B$.

This theorem is proved in Problem 1.8. Other equivalent conditions to $A \subseteq B$ are given in Problem 1.31.

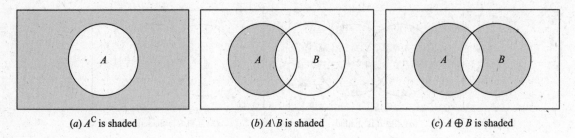

(a) A^C is shaded   (b) $A \backslash B$ is shaded   (c) $A \oplus B$ is shaded

Fig. 1-4

Complements, Differences, Symmetric Differences

Recall that all sets under consideration at a particular time are subsets of a fixed universal set **U**. The *absolute complement* or, simply, *complement* of a set A, denoted by A^C, is the set of elements which belong to **U** but which do not belong to A. That is,

$$A^C = \{x \mid x \in \mathbf{U}, x \notin A\}$$

Some texts denote the complement of A by A' or $\bar{A}$. Fig. 1-4(a) is a Venn diagram in which A^C is shaded.

The *relative complement* of a set B with respect to a set A or, simply, the *difference* of A and B, denoted by $A \backslash B$, is the set of elements which belong to A but which do not belong to B; that is

$$A \backslash B = \{x \mid x \in A, x \notin B\} \qquad \text{Also } A\text{-}B$$

The set $A \backslash B$ is read "A minus B." Many texts denote $A \backslash B$ by $A - B$ or $A \sim B$. Fig. 1-4(b) is a Venn diagram in which $A \backslash B$ is shaded.

The *symmetric difference* of sets A and B, denoted by $A \oplus B$, consists of those elements which belong to A or B but not to both. That is, *Everything in A or B, but excludes $A \cap B$*

$$A \oplus B = (A \cup B) \backslash (A \cap B) \quad \text{or} \quad A \oplus B = (A \backslash B) \cup (B \backslash A)$$

Figure 1-4(c) is a Venn diagram in which $A \oplus B$ is shaded.

EXAMPLE 1.5 Suppose $\mathbf{U} = \mathbf{N} = \{1, 2, 3, \ldots\}$ is the universal set. Let

$$A = \{1, 2, 3, 4\}, \quad B = \{3, 4, 5, 6, 7\}, \quad C = \{2, 3, 8, 9\}, \quad E = \{2, 4, 6, \ldots\}$$

(Here E is the set of positive even integers.) Then:

$$A^C = \{5, 6, 7, \ldots\}, \quad B^C = \{1, 2, 8, 9, 10, \ldots\}, \quad E^C = \{1, 3, 5, 7, \ldots\}$$

That is, E^C is the set of odd positive integers. Also:

$$A \backslash B = \{1, 2\}, \qquad A \backslash C = \{1, 4\}, \qquad B \backslash C = \{4, 5, 6, 7\}, \qquad A \backslash E = \{1, 3\},$$
$$B \backslash A = \{5, 6, 7\}, \qquad C \backslash A = \{8, 9\}, \qquad C \backslash B = \{2, 8, 9\}, \qquad E \backslash A = \{6, 8, 10, 12, \ldots\}.$$

Furthermore:

$$A \oplus B = (A \backslash B) \cup (B \backslash A) = \{1, 2, 5, 6, 7\}, \quad B \oplus C = \{2, 4, 5, 6, 7, 8, 9\},$$
$$A \oplus C = (A \backslash C) \cup (C \backslash A) = \{1, 4, 8, 9\}, \qquad A \oplus E = \{1, 3, 6, 8, 10, \ldots\}.$$

Fundamental Products

Consider n distinct sets $A_1, A_2, \ldots, A_n$. A *fundamental product* of the sets is a set of the form

$$A_1^* \cap A_2^* \cap \ldots \cap A_n^* \quad \text{where} \quad A_i^* = A \quad \text{or} \quad A_i^* = A^C$$

We note that:

   (i) There are $m = 2^n$ such fundamental products.
  (ii) Any two such fundamental products are disjoint.
 (iii) The universal set **U** is the union of all fundamental products.

Thus **U** is the disjoint union of the fundamental products (Problem 1.60). There is a geometrical description of these sets which is illustrated below.

EXAMPLE 1.6 Figure 1-5(a) is the Venn diagram of three sets A, B, C. The following lists the $m = 2^3 = 8$ fundamental products of the sets A, B, C:

$$P_1 = A \cap B \cap C, \quad P_3 = A \cap B^C \cap C, \quad P_5 = A^C \cap B \cap C, \quad P_7 = A^C \cap B^C \cap C,$$
$$P_2 = A \cap B \cap C^C, \quad P_4 = A \cap B^C \cap C^C, \quad P_6 = A^C \cap B \cap C^C, \quad P_8 = A^C \cap B^C \cap C^C.$$

The eight products correspond precisely to the eight disjoint regions in the Venn diagram of sets A, B, C as indicated by the labeling of the regions in Fig. 1-5(b).

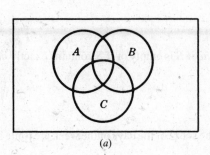
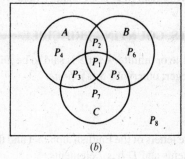

(a) (b)

Fig. 1-5

1.5 ALGEBRA OF SETS, DUALITY

Sets under the operations of union, intersection, and complement satisfy various laws (identities) which are listed in Table 1-1. In fact, we formally state this as:

Theorem 1.5: Sets satisfy the laws in Table 1-1.

Table 1-1 Laws of the algebra of sets

Idempotent laws:	(1a) $A \cup A = A$	(1b) $A \cap A = A$
Associative laws:	(2a) $(A \cup B) \cup C = A \cup (B \cup C)$	(2b) $(A \cap B) \cap C = A \cap (B \cap C)$
Commutative laws:	(3a) $A \cup B = B \cup A$	(3b) $A \cap B = B \cap A$
Distributive laws:	(4a) $A \cup (B \cap C) = (A \cup B) \cap (A \cup C)$	(4b) $A \cap (B \cup C) = (A \cap B) \cup (A \cap C)$
Identity laws:	(5a) $A \cup \emptyset = A$	(5b) $A \cap \mathbf{U} = A$
	(6a) $A \cup \mathbf{U} = \mathbf{U}$	(6b) $A \cap \emptyset = \emptyset$
Involution laws:	(7) $(A^C)^C = A$	
Complement laws:	(8a) $A \cup A^C = \mathbf{U}$	(8b) $A \cap A^C = \emptyset$
	(9a) $\mathbf{U}^C = \emptyset$	(9b) $\emptyset^C = \mathbf{U}$
DeMorgan's laws:	(10a) $(A \cup B)^C = A^C \cap B^C$	(10b) $(A \cap B)^C = A^C \cup B^C$

Remark: Each law in Table 1-1 follows from an equivalent logical law. Consider, for example, the proof of DeMorgan's Law 10(a):

$$(A \cup B)^C = \{x \mid x \notin (A \text{ or } B)\} = \{x \mid x \notin A \text{ and } x \notin B\} = A^C \cap B^C$$

Here we use the equivalent (DeMorgan's) logical law:

$$\neg(p \vee q) = \neg p \wedge \neg q$$

where $\neg$ means "not," $\vee$ means "or," and $\wedge$ means "and." (Sometimes Venn diagrams are used to illustrate the laws in Table 1-1 as in Problem 1.17.)

Duality

The identities in Table 1-1 are arranged in pairs, as, for example, (2a) and (2b). We now consider the principle behind this arrangement. Suppose E is an equation of set algebra. The dual E^* of E is the equation obtained by replacing each occurrence of $\cup, \cap, \mathbf{U}$ and $\emptyset$ in E by $\cap, \cup, \emptyset$, and $\mathbf{U}$, respectively. For example, the dual of

$$(\mathbf{U} \cap A) \cup (B \cap A) = A \quad \text{is} \quad (\emptyset \cup A) \cap (B \cup A) = A$$

Observe that the pairs of laws in Table 1-1 are duals of each other. It is a fact of set algebra, called the *principle of duality*, that if any equation E is an identity then its dual E^* is also an identity.

1.6 FINITE SETS, COUNTING PRINCIPLE

Sets can be finite or infinite. A set S is said to be *finite* if S is empty or if S contains exactly m elements where m is a positive integer; otherwise S is *infinite*.

EXAMPLE 1.7

(a) The set A of the letters of the English alphabet and the set D of the days of the week are finite sets. Specifically, A has 26 elements and D has 7 elements.

(b) Let E be the set of even positive integers, and let $\mathbf{I}$ be the *unit interval*, that is,

$$E = \{2, 4, 6, \ldots\} \quad \text{and} \quad \mathbf{I} = [0, 1] = \{x \mid 0 \le x \le 1\}$$

Then both E and $\mathbf{I}$ are infinite.

A set S is *countable* if S is finite or if the elements of S can be arranged as a sequence, in which case S is said to be *countably infinite*; otherwise S is said to be *uncountable*. The above set E of even integers is countably infinite, whereas one can prove that the unit interval $\mathbf{I} = [0, 1]$ is uncountable.

Counting Elements in Finite Sets

The notation $n(S)$ or $|S|$ will denote the number of elements in a set S. (Some texts use $\#(S)$ or $\text{card}(S)$ instead of $n(S)$.) Thus $n(A) = 26$, where A is the letters in the English alphabet, and $n(D) = 7$, where D is the days of the week. Also $n(\emptyset) = 0$ since the empty set has no elements.

The following lemma applies.

Lemma 1.6: Suppose A and B are finite disjoint sets. Then $A \cup B$ is finite and

$$n(A \cup B) = n(A) + n(B)$$

This lemma may be restated as follows:

Lemma 1.6: Suppose S is the disjoint union of finite sets A and B. Then S is finite and

$$n(S) = n(A) + n(B)$$

Proof. In counting the elements of $A \cup B$, first count those that are in A. There are $n(A)$ of these. The only other elements of $A \cup B$ are those that are in B but not in A. But since A and B are disjoint, no element of B is in A, so there are $n(B)$ elements that are in B but not in A. Therefore, $n(A \cup B) = n(A) + n(B)$.

For any sets A and B, the set A is the disjoint union of $A \backslash B$ and $A \cap B$. Thus Lemma 1.6 gives us the following useful result.

Corollary 1.7: Let A and B be finite sets. Then

$$n(A \backslash B) = n(A) - n(A \cap B)$$

For example, suppose an art class A has 25 students and 10 of them are taking a biology class B. Then the number of students in class A which are not in class B is:

$$n(A \backslash B) = n(A) - n(A \cap B) = 25 - 10 = 15$$

Given any set A, recall that the universal set $\mathbf{U}$ is the disjoint union of A and A^C. Accordingly, Lemma 1.6 also gives the following result.

Corollary 1.8: Let A be a subset of a finite universal set $\mathbf{U}$. Then

$$n(A^C) = n(\mathbf{U}) - n(A)$$

For example, suppose a class $\mathbf{U}$ with 30 students has 18 full-time students. Then there are $30 - 18 = 12$ part-time students in the class $\mathbf{U}$.

Inclusion–Exclusion Principle

There is a formula for $n(A \cup B)$ even when they are not disjoint, called the Inclusion–Exclusion Principle. Namely:

Theorem (Inclusion–Exclusion Principle) 1.9: Suppose A and B are finite sets. Then $A \cup B$ and $A \cap B$ are finite and

$$n(A \cup B) = n(A) + n(B) - n(A \cap B)$$

That is, we find the number of elements in A or B (or both) by first adding $n(A)$ and $n(B)$ (inclusion) and then subtracting $n(A \cap B)$ (exclusion) since its elements were counted twice.

We can apply this result to obtain a similar formula for three sets:

Corollary 1.10: Suppose A, B, C are finite sets. Then $A \cup B \cup C$ is finite and

$$n(A \cup B \cup C) = n(A) + n(B) + n(C) - n(A \cap B) - n(A \cap C) - n(B \cap C) + n(A \cap B \cap C)$$

Mathematical induction (Section 1.8) may be used to further generalize this result to any number of finite sets.

EXAMPLE 1.8 Suppose a list A contains the 30 students in a mathematics class, and a list B contains the 35 students in an English class, and suppose there are 20 names on both lists. Find the number of students: (a) only on list A, (b) only on list B, (c) on list A or B (or both), (d) on exactly one list.

(a) List A has 30 names and 20 are on list B; hence $30 - 20 = 10$ names are only on list A.

(b) Similarly, $35 - 20 = 15$ are only on list B.

(c) We seek $n(A \cup B)$. By inclusion–exclusion,

$$n(A \cup B) = n(A) + n(B) - n(A \cap B) = 30 + 35 - 20 = 45.$$

In other words, we combine the two lists and then cross out the 20 names which appear twice.

(d) By (a) and (b), $10 + 15 = 25$ names are only on one list; that is, $n(A \oplus B) = 25$.

1.7 CLASSES OF SETS, POWER SETS, PARTITIONS

Given a set S, we might wish to talk about some of its subsets. Thus we would be considering a *set of sets*. Whenever such a situation occurs, to avoid confusion, we will speak of a *class* of sets or *collection* of sets rather than a *set* of sets. If we wish to consider some of the sets in a given class of sets, then we speak of *subclass* or *subcollection*.

EXAMPLE 1.9 Suppose $S = \{1, 2, 3, 4\}$.

(a) Let A be the class of subsets of S which contain exactly three elements of S. Then

$$A = [\{1, 2, 3\}, \{1, 2, 4\}, \{1, 3, 4\}, \{2, 3, 4\}]$$

That is, the elements of A are the sets $\{1, 2, 3\}$, $\{1, 2, 4\}$, $\{1, 3, 4\}$, and $\{2, 3, 4\}$.

(b) Let B be the class of subsets of S, each which contains 2 and two other elements of S. Then

$$B = [\{1, 2, 3\}, \{1, 2, 4\}, \{2, 3, 4\}]$$

The elements of B are the sets $\{1, 2, 3\}$, $\{1, 2, 4\}$, and $\{2, 3, 4\}$. Thus B is a subclass of A, since every element of B is also an element of A. (To avoid confusion, we will sometimes enclose the sets of a class in brackets instead of braces.)

Power Sets

For a given set S, we may speak of the class of all subsets of S. This class is called the *power set* of S, and will be denoted by $P(S)$. If S is finite, then so is $P(S)$. In fact, the number of elements in $P(S)$ is 2 raised to the power $n(S)$. That is,

$$n(P(S)) = 2^{n(S)}$$

(For this reason, the power set of S is sometimes denoted by 2^S.)

EXAMPLE 1.10 Suppose $S = \{1, 2, 3\}$. Then

$$P(S) = [\emptyset, \{1\}, \{2\}, \{3\}, \{1, 2\}, \{1, 3\}, \{2, 3\}, S]$$

Note that the empty set $\emptyset$ belongs to $P(S)$ since $\emptyset$ is a subset of S. Similarly, S belongs to $P(S)$. As expected from the above remark, $P(S)$ has $2^3 = 8$ elements.

Partitions

Let S be a nonempty set. A *partition* of S is a subdivision of S into nonoverlapping, nonempty subsets. Precisely, a *partition* of S is a collection $\{A_i\}$ of nonempty subsets of S such that:

(i) Each a in S belongs to one of the A_i.
(ii) The sets of $\{A_i\}$ are mutually disjoint; that is, if

$$A_j \neq A_k \quad \text{then} \quad A_j \cap A_k = \emptyset$$

The subsets in a partition are called *cells*. Figure 1-6 is a Venn diagram of a partition of the rectangular set S of points into five cells, A_1, A_2, A_3, A_4, A_5.

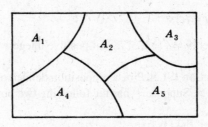

Fig. 1-6

EXAMPLE 1.11 Consider the following collections of subsets of $S = \{1, 2, \ldots, 8, 9\}$:

   (i) $[\{1, 3, 5\}, \{2, 6\}, \{4, 8, 9\}]$

   (ii) $[\{1, 3, 5\}, \{2, 4, 6, 8\}, \{5, 7, 9\}]$

   (iii) $[\{1, 3, 5\}, \{2, 4, 6, 8\}, \{7, 9\}]$

Then (i) is not a partition of S since 7 in S does not belong to any of the subsets. Furthermore, (ii) is not a partition of S since $\{1, 3, 5\}$ and $\{5, 7, 9\}$ are not disjoint. On the other hand, (iii) is a partition of S.

Generalized Set Operations   SKIP

The set operations of union and intersection were defined above for two sets. These operations can be extended to any number of sets, finite or infinite, as follows.

Consider first a finite number of sets, say, $A_1, A_2, \ldots, A_m$. The union and intersection of these sets are denoted and defined, respectively, by

$$A_1 \cup A_2 \cup \ldots \cup A_m = \bigcup_{i=1}^{m} A_i = \{x \mid x \in A_i \text{ for some } A_i\}$$
$$A_1 \cap A_2 \cap \ldots \cap A_m = \bigcap_{i=1}^{m} A_i = \{x \mid x \in A_i \text{ for every } A_i\}$$

That is, the union consists of those elements which belong to at least one of the sets, and the intersection consists of those elements which belong to all the sets.

Now let $\mathscr{A}$ be any collection of sets. The union and the intersection of the sets in the collection $\mathscr{A}$ is denoted and defined, respectively, by

$$\bigcup (A \mid A \in \mathscr{A}) = \{x \mid x \in A_i \text{ for some } A_i \in \mathscr{A}\}$$
$$\bigcap (A \mid A \in \mathscr{A}) = \{x \mid x \in A_i \text{ for every } A_i \in \mathscr{A}\}$$

That is, the union consists of those elements which belong to at least one of the sets in the collection $\mathscr{A}$ and the intersection consists of those elements which belong to every set in the collection A.

EXAMPLE 1.12 Consider the sets

$$A_1 = \{1, 2, 3, \ldots\} = \mathbf{N}, \quad A_2 = \{2, 3, 4, \ldots\}, \quad A_3 = \{3, 4, 5, \ldots\}, \quad A_n = \{n, n+1, n+2, \ldots\}.$$

Then the union and intersection of the sets are as follows:

$$\bigcup (A_k \mid k \in \mathbf{N}) = \mathbf{N} \quad \text{and} \quad \bigcap (A_k \mid k \in \mathbf{N}) = \emptyset$$

DeMorgan's laws also hold for the above generalized operations. That is:

Theorem 1.11: Let $\mathscr{A}$ be a collection of sets. Then:

   (i) $\left[\bigcup (A \mid A \in \mathscr{A}) \right]^C = \bigcap (A^C \mid A \in \mathscr{A})$

   (ii) $\left[\bigcap (A \mid A \in \mathscr{A}) \right]^C = \bigcup (A^C \mid A \in \mathscr{A})$

1.8 MATHEMATICAL INDUCTION

SAVE FORLATER B.t still skip Braos

An essential property of the set $N = \{1, 2, 3, \ldots\}$ of positive integers follows:

Principle of Mathematical Induction I: Let P be a proposition defined on the positive integers N; that is, $P(n)$ is either true or false for each $n \in N$. Suppose P has the following two properties:

   (i) $P(1)$ is true.

  (ii) $P(k + 1)$ is true whenever $P(k)$ is true.

Then P is true for every positive integer $n \in N$.

We shall not prove this principle. In fact, this principle is usually given as one of the axioms when N is developed axiomatically.

EXAMPLE 1.13 Let P be the proposition that the sum of the first n odd numbers is n^2; that is,

$$P(n) : 1 + 3 + 5 + \cdots + (2n - 1) = n^2$$

(The kth odd number is $2k - 1$, and the next odd number is $2k + 1$.) Observe that $P(n)$ is true for $n = 1$; namely,

$$P(1) : 1 = 1^2$$

Assuming $P(k)$ is true, we add $2k + 1$ to both sides of $P(k)$, obtaining

$$1 + 3 + 5 + \cdots + (2k - 1) + (2k + 1) = k^2 + (2k + 1) = (k + 1)^2$$

which is $P(k + 1)$. In other words, $P(k + 1)$ is true whenever $P(k)$ is true. By the principle of mathematical induction, P is true for all n.

There is a form of the principle of mathematical induction which is sometimes more convenient to use. Although it appears different, it is really equivalent to the above principle of induction.

Principle of Mathematical Induction II: Let P be a proposition defined on the positive integers N such that:

   (i) $P(1)$ is true.

  (ii) $P(k)$ is true whenever $P(j)$ is true for all $1 \leq j < k$.

Then P is true for every positive integer $n \in N$.

Remark: Sometimes one wants to prove that a proposition P is true for the set of integers

$$\{a, a + 1, a + 2, a + 3, \ldots\}$$

where a is any integer, possibly zero. This can be done by simply replacing 1 by a in either of the above Principles of Mathematical Induction.

Solved Problems

SETS AND SUBSETS

1.1 Which of these sets are equal: $\{x, y, z\}$, $\{z, y, z, x\}$, $\{y, x, y, z\}$, $\{y, z, x, y\}$?

They are all equal. Order and repetition do not change a set.

1.2 List the elements of each set where $N = \{1, 2, 3, \ldots\}$.

   (a) $A = \{x \in N \mid 3 < x < 9\}$

   (b) $B = \{x \in N \mid x$ is even, $x < 11\}$

(c) $C = \{x \in \mathbf{N} \mid 4 + x = 3\}$

    (a) A consists of the positive integers between 3 and 9; hence $A = \{4, 5, 6, 7, 8\}$.

    (b) B consists of the even positive integers less than 11; hence $B = \{2, 4, 6, 8, 10\}$.

    (c) No positive integer satisfies $4 + x = 3$; hence $C = \emptyset$, the empty set.

1.3 Let $A = \{2, 3, 4, 5\}$.

    (a) Show that A is not a subset of $B = \{x \in \mathbf{N} \mid x \text{ is even}\}$.

    (b) Show that A is a proper subset of $C = \{1, 2, 3, \ldots, 8, 9\}$.

    (a) It is necessary to show that at least one element in A does not belong to B. Now $3 \in A$ and, since B consists of even numbers, $3 \notin B$; hence A is not a subset of B.

    (b) Each element of A belongs to C so $A \subseteq C$. On the other hand, $1 \in C$ but $1 \notin A$. Hence $A \neq C$. Therefore A is a proper subset of C.

SET OPERATIONS

1.4 Let $\mathbf{U} = \{1, 2, \ldots, 9\}$ be the universal set, and let

$$A = \{1, 2, 3, 4, 5\}, \quad C = \{5, 6, 7, 8, 9\}, \quad E = \{2, 4, 6, 8\},$$
$$B = \{4, 5, 6, 7\}, \quad D = \{1, 3, 5, 7, 9\}, \quad F = \{1, 5, 9\}.$$

Find: (a) $A \cup B$ and $A \cap B$; (b) $A \cup C$ and $A \cap C$; (c) $D \cup F$ and $D \cap F$.

Recall that the union $X \cup Y$ consists of those elements in either X or Y (or both), and that the intersection $X \cap Y$ consists of those elements in both X and Y.

    (a) $A \cup B = \{1, 2, 3, 4, 5, 6, 7\}$ and $A \cap B = \{4, 5\}$

    (b) $A \cup C = \{1, 2, 3, 4, 5, 6, 7, 8, 9\} = \mathbf{U}$ and $A \cap C = \{5\}$

    (c) $D \cup F = \{1, 3, 5, 7, 9\} = D$ and $D \cap F = (1, 5, 9) = F$

        Observe that $F \subseteq D$, so by Theorem 1.4 we must have $D \cup F = D$ and $D \cap F = F$.

1.5 Consider the sets in the preceding Problem 1.4. Find:

    (a) A^C, B^C, D^C, E^C; (b) $A \backslash B, B \backslash A, D \backslash E, F \backslash D$; (c) $A \oplus B, C \oplus D, E \oplus F$.

Recall that:

    (1) The complements X^C consists of those elements in $\mathbf{U}$ which do not belong to X.

    (2) The difference $X \backslash Y$ consists of the elements in X which do not belong to Y.

    (3) The symmetric difference $X \oplus Y$ consists of the elements in X or in Y but not in both.

Therefore:

    (a) $A^C = \{6, 7, 8, 9\}$; $B^C = \{1, 2, 3, 8, 9\}$; $D^C = \{2, 4, 6, 8\} = E$; $E^C = \{1, 3, 5, 7, 9\} = D$.

    (b) $A \backslash B = \{1, 2, 3\}$; $B \backslash A = \{6, 7\}$; $D \backslash E = \{1, 3, 5, 7, 9\} = D$; $F \backslash D = \emptyset$.

    (c) $A \oplus B = \{1, 2, 3, 6, 7\}$; $C \oplus D = \{1, 3, 6, 8\}$; $E \oplus F = \{2, 4, 6, 8, 1, 5, 9\} = E \cup F$.

1.6 Show that we can have: (a) $A \cap B = A \cap C$ without $B = C$; (b) $A \cup B = A \cup C$ without $B = C$.

    (a) Let $A = \{1, 2\}$, $B = \{2, 3\}$, $C = \{2, 4\}$. Then $A \cap B = \{2\}$ and $A \cap C = \{2\}$; but $B \neq C$.

    (b) Let $A = \{1, 2\}$, $B = \{1, 3\}$, $C = \{2, 3\}$. Then $A \cup B = \{1, 2, 3\}$ and $A \cup C = \{1, 2, 3\}$ but $B \neq C$.

1.7 Prove: $B \backslash A = B \cap A^C$. Thus, the set operation of difference can be written in terms of the operations of intersection and complement.

$$B \backslash A = \{x \mid x \in B, \ x \notin A\} = \{x \mid x \in B, \ x \in A^C\} = B \cap A^C.$$

1.8 Prove Theorem 1.4. The following are equivalent: $A \subseteq B$, $A \cap B = A$, $A \cup B = B$.

Suppose $A \subseteq B$ and let $x \in A$. Then $x \in B$, hence $x \in A \cap B$ and $A \subseteq A \cap B$. By Theorem 1.3, $(A \cap B) \subseteq A$. Therefore $A \cap B = A$. On the other hand, suppose $A \cap B = A$ and let $x \in A$. Then $x \in (A \cap B)$; hence $x \in A$ and $x \in B$. Therefore, $A \subseteq B$. Both results show that $A \subseteq B$ is equivalent to $A \cap B = A$.

Suppose again that $A \subseteq B$. Let $x \in (A \cup B)$. Then $x \in A$ or $x \in B$. If $x \in A$, then $x \in B$ because $A \subseteq B$. In either case, $x \in B$. Therefore $A \cup B \subseteq B$. By Theorem 1.3, $B \subseteq A \cup B$. Therefore $A \cup B = B$. Now suppose $A \cup B = B$ and let $x \in A$. Then $x \in A \cup B$ by definition of the union of sets. Hence $x \in B = A \cup B$. Therefore $A \subseteq B$. Both results show that $A \subseteq B$ is equivalent to $A \cup B = B$.

Thus $A \subseteq B$, $A \cup B = A$ and $A \cup B = B$ are equivalent.

VENN DIAGRAMS, ALGEBRA OF SETS, DUALITY

1.9 Illustrate DeMorgan's Law $(A \cup B)^C = A^C \cap B^C$ using Venn diagrams.

Shade the area outside $A \cup B$ in a Venn diagram of sets A and B. This is shown in Fig. 1-7(a); hence the shaded area represents $(A \cup B)^C$. Now shade the area outside A in a Venn diagram of A and B with strokes in one direction (////), and then shade the area outside B with strokes in another direction (\\\\). This is shown in Fig. 1-7(b); hence the cross-hatched area (area where both lines are present) represents $A^C \cap B^C$. Both $(A \cup B)^C$ and $A^C \cap B^C$ are represented by the same area; thus the Venn diagram indicates $(A \cup B)^C = A^C \cap B^C$. (We emphasize that a Venn diagram is not a formal proof, but it can indicate relationships between sets.)

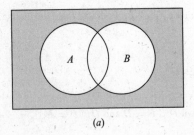   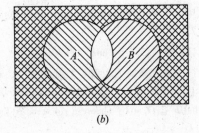

(a) (b)

Fig. 1-7

1.10 Prove the Distributive Law: $A \cap (B \cup C) = (A \cap B) \cup (A \cap C)$.

$$A \cap (B \cup C) = \{x \mid x \in A, x \in (B \cup C)\}$$
$$= \{x \mid x \in A, x \in B \text{ or } x \in A, x \in C\} = (A \cap B) \cup (A \cap C)$$

Here we use the analogous logical law $p \wedge (q \vee r) \equiv (p \wedge q) \vee (p \wedge r)$ where $\wedge$ denotes "and" and $\vee$ denotes "or."

1.11 Write the dual of: (a) $(\mathbf{U} \cap A) \cup (B \cap A) = A$;   ($b$) $(A \cap \mathbf{U}) \cap (\emptyset \cup A^C) = \emptyset$.

Interchange $\cup$ and $\cap$ and also $\mathbf{U}$ and $\emptyset$ in each set equation:

(a) $(\emptyset \cup A) \cap (B \cup A) = A$;   ($b$) $(A \cup \emptyset) \cup (\mathbf{U} \cap A^C) = \mathbf{U}$.

1.12 Prove: $(A \cup B) \backslash (A \cap B) = (A \backslash B) \cup (B \backslash A)$. (Thus either one may be used to define $A \oplus B$.)

Using $X \backslash Y = X \cap Y^C$ and the laws in Table 1.1, including DeMorgan's Law, we obtain:

$$(A \cup B) \backslash (A \cap B) = (A \cup B) \cap (A \cap B)^C = (A \cup B) \cap (A^C \cup B^C)$$
$$= (A \cup A^C) \cup (A \cap B^C) \cup (B \cap A^C) \cup (B \cap B^C)$$
$$= \emptyset \cup (A \cap B^C) \cup (B \cap A^C) \cup \emptyset$$
$$= (A \cap B^C) \cup (B \cap A^C) = (A \backslash B) \cup (B \backslash A)$$

1.13 Determine the validity of the following argument:

S_1: All my friends are musicians.

S_2: John is my friend.

S_3: None of my neighbors are musicians.

S : John is not my neighbor.

The premises S_1 and S_3 lead to the Venn diagram in Fig. 1-8(a). By S_2, John belongs to the set of friends which is disjoint from the set of neighbors. Thus S is a valid conclusion and so the argument is valid.

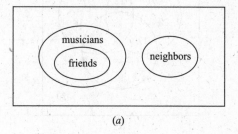   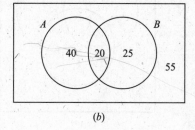

(a)   (b)

Fig. 1-8

FINITE SETS AND THE COUNTING PRINCIPLE

1.14 Each student in Liberal Arts at some college has a mathematics requirement A and a science requirement B. A poll of 140 sophomore students shows that:

60 completed A,   45 completed B,   20 completed both A and B.

Use a Venn diagram to find the number of students who have completed:

(a) At least one of A and B; (b) exactly one of A or B; (c) neither A nor B.

Translating the above data into set notation yields:

$n(A) = 60$, $n(B) = 45$, $n(A \cap B) = 20$, $n(\mathbf{U}) = 140$

Draw a Venn diagram of sets A and B as in Fig. 1-1(c). Then, as in Fig. 1-8(b), assign numbers to the four regions as follows:

20 completed both A and B, so $n(A \cap B) = 20$.

$60 - 20 = 40$ completed A but not B, so $n(A \backslash B) = 40$.

$45 - 20 = 25$ completed B but not A, so $n(B \backslash A) = 25$.

$140 - 20 - 40 - 25 = 55$ completed neither A nor B.

By the Venn diagram:

(a) $20 + 40 + 25 = 85$ completed A or B. Alternately, by the Inclusion–Exclusion Principle:
$$n(A \cup B) = n(A) + n(B) - n(A \cap B) = 60 + 45 - 20 = 85$$
(b) $40 + 25 = 65$ completed exactly one requirement. That is, $n(A \oplus B) = 65$.

(c) 55 completed neither requirement, i.e. $n(A^C \cap B^C) = n[(A \cup B)^C] = 140 - 85 = 55$.

1.15 In a survey of 120 people, it was found that:

65 read *Newsweek* magazine,   20 read both *Newsweek* and *Time*,

45 read *Time*,   25 read both *Newsweek* and *Fortune*,

42 read *Fortune*,   15 read both *Time* and *Fortune*,

8 read all three magazines.

(*a*) Find the number of people who read at least one of the three magazines.

(*b*) Fill in the correct number of people in each of the eight regions of the Venn diagram in Fig. 1-9(*a*) where *N*, *T*, and *F* denote the set of people who read *Newsweek*, *Time*, and *Fortune*, respectively.

(*c*) Find the number of people who read exactly one magazine.

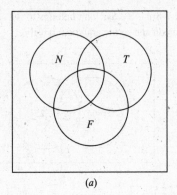
(a)

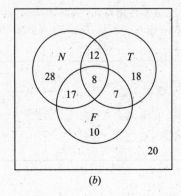
(b)

Fig. 1-9

(*a*) We want to find $n(N \cup T \cup F)$. By Corollary 1.10 (Inclusion–Exclusion Principle),

$$n(N \cup T \cup F) = n(N) + n(T) + n(F) - n(N \cap T) - n(N \cap F) - n(T \cap F) + n(N \cap T \cap F)$$
$$= 65 + 45 + 42 - 20 - 25 - 15 + 8 = 100$$

(*b*) The required Venn diagram in Fig. 1-9(*b*) is obtained as follows:

  8 read all three magazines,

  $20 - 8 = 12$ read *Newsweek* and *Time* but not all three magazines,

  $25 - 8 = 17$ read *Newsweek* and *Fortune* but not all three magazines,

  $15 - 8 = 7$ read *Time* and *Fortune* but not all three magazines,

  $65 - 12 - 8 - 17 = 28$ read only *Newsweek*,

  $45 - 12 - 8 - 7 = 18$ read only *Time*,

  $42 - 17 - 8 - 7 = 10$ read only *Fortune*,

  $120 - 100 = 20$ read no magazine at all.

(*c*) $28 + 18 + 10 = 56$ read exactly one of the magazines.

1.16 Prove Theorem 1.9. Suppose *A* and *B* are finite sets. Then $A \cup B$ and $A \cap B$ are finite and

$$n(A \cup B) = n(A) + n(B) - n(A \cap B)$$

If *A* and *B* are finite then, clearly, $A \cup B$ and $A \cap B$ are finite.

Suppose we count the elements in *A* and then count the elements in *B*.

Then every element in $A \cap B$ would be counted twice, once in *A* and once in *B*. Thus

$$n(A \cup B) = n(A) + n(B) - n(A \cap B)$$

CLASSES OF SETS

1.17 Let $A = [\{1, 2, 3\}, \{4, 5\}, \{6, 7, 8\}]$. (a) List the elements of A; (b) Find $n(A)$.

(a) A has three elements, the sets $\{1, 2, 3\}$, $\{4, 5\}$, and $\{6, 7, 8\}$.

(b) $n(A) = 3$.

1.18 Determine the power set $P(A)$ of $A = \{a, b, c, d\}$.

The elements of $P(A)$ are the subsets of A. Hence

$$P(A) = [A, \{a, b, c\}, \{a, b, d\}, \{a, c, d\}, \{b, c, d\}, \{a, b\}, \{a, c\}, \{a, d\}, \{b, c\}, \{b, d\},$$
$$\{c, d\}, \{a\}, \{b\}, \{c\}, \{d\}, \emptyset]$$

As expected, $P(A)$ has $2^4 = 16$ elements.

1.19 Let $S = \{a, b, c, d, e, f, g\}$. Determine which of the following are partitions of S:

(a) $P_1 = [\{a, c, e\}, \{b\}, \{d, g\}]$,   (c) $P_3 = [\{a, b, e, g\}, \{c\}, \{d, f\}]$,

(b) $P_2 = [\{a, e, g\}, \{c, d\}, \{b, e, f\}]$,   (d) $P_4 = [\{a, b, c, d, e, f, g\}]$.

(a) P_1 is not a partition of S since $f \in S$ does not belong to any of the cells.

(b) P_2 is not a partition of S since $e \in S$ belongs to two of the cells.

(c) P_3 is a partition of S since each element in S belongs to exactly one cell.

(d) P_4 is a partition of S into one cell, S itself.

1.20 Find all partitions of $S = \{a, b, c, d\}$.

Note first that each partition of S contains either 1, 2, 3, or 4 distinct cells. The partitions are as follows:

(1) $[\{a, b, c, d\}]$

(2) $[\{a\}, \{b, c, d\}], [\{b\}, \{a, c, d\}], [\{c\}, \{a, b, d\}], [\{d\}, \{a, b, c\}],$
$[\{a, b\}, \{c, d\}], [\{a, c\}, \{b, d\}], [\{a, d\}, \{b, c\}]$

(3) $[\{a\}, \{b\}, \{c, d\}], [\{a\}, \{c\}, \{b,d\}], [\{a\}, \{d\}, \{b, c\}],$
$[\{b\}, \{c\}, \{a, d\}], [\{b\}, \{d\}, \{a, c\}], [\{c\}, \{d\}, \{a, b\}]$

(4) $[\{a\}, \{b\}, \{c\}, \{d\}]$

There are 15 different partitions of S.

1.21 Let $\mathbf{N} = \{1, 2, 3,...\}$ and, for each $n \in \mathbf{N}$, Let $A_n = \{n, 2n, 3n,...\}$. Find:

(a) $A_3 \cap A_5$; (b) $A_4 \cap A_5$; (c) $\bigcup_{i \in Q} A_i$ where $Q = \{2, 3, 5, 7, 11, ...\}$ is the set of prime numbers.

(a) Those numbers which are multiples of both 3 and 5 are the multiples of 15; hence $A_3 \cap A_5 = A_{15}$.

(b) The multiples of 12 and no other numbers belong to both A_4 and A_6, hence $A_4 \cap A_6 = A_{12}$.

(c) Every positive integer except 1 is a multiple of at least one prime number; hence

$$\bigcup_{i \in Q} A_i = \{2, 3, 4, ...\} = \mathbf{N} \setminus \{1\}$$

1.22 Let $\{A_i \mid i \in I\}$ be an indexed class of sets and let $i_0 \in I$. Prove

$$\bigcap_{i \in I} A_i \subseteq A_{i_0} \subseteq \bigcup_{i \in I} A_i.$$

Let $x \in \bigcap_{i \in I} A_i$ then $x \in A_i$ for every $i \in I$. In particular, $x \in A_{i_0}$. Hence $\bigcap_{i \in I} A_i \subseteq A_{i_0}$. Now let $y \in A_{i_0}$. Since $i_0 \in I$, $y \in \bigcap_{i \in I} A_i$. Hence $A_{i_0} \subseteq \bigcup_{i \in I} A_i$.

1.23 Prove (De Morgan's law): For any indexed class $\{A_i \mid i \in I\}$, we have $\left(\bigcup_i A_i\right)^C = \bigcap_i A_i^C$.

Using the definitions of union and intersection of indexed classes of sets:

$$\left(\bigcup_i A_i\right)^C = \{x \mid x \notin \bigcup_i A_i\} = \{x \mid x \notin A_i \text{ for every } i\}$$
$$= \{x \mid x \in A_i^C \text{ for every } i\} = \bigcap_i A_i^C$$

MATHEMATICAL INDUCTION

1.24 Prove the proposition $P(n)$ that the sum of the first n positive integers is $\frac{1}{2}n(n+1)$; that is,

$$P(n) = 1 + 2 + 3 + \cdots + n = \tfrac{1}{2}n(n+1)$$

The proposition holds for $n = 1$ since:

$$P(1) : 1 = \tfrac{1}{2}(1)(1+1)$$

Assuming $P(k)$ is true, we add $k+1$ to both sides of $P(k)$, obtaining

$$1 + 2 + 3 + \cdots + k + (k+1) = \tfrac{1}{2}k(k+1) + (k+1)$$
$$= \tfrac{1}{2}[k(k+1) + 2(k+1)]$$
$$= \tfrac{1}{2}[(k+1)(k+2)]$$

which is $P(k+1)$. That is, $P(k+1)$ is true whenever $P(k)$ is true. By the Principle of Induction, P is true for all n.

1.25 Prove the following proposition (for $n \geq 0$):

$$P(n) : 1 + 2 + 2^2 + 2^3 + \cdots + 2^n = 2^{n+1} - 1$$

$P(0)$ is true since $1 = 2^1 - 1$. Assuming $P(k)$ is true, we add 2^{k+1} to both sides of $P(k)$, obtaining

$$1 + 2 + 2^2 + 2^3 + \cdots + 2^k + 2^{k+1} = 2^{k+1} - 1 + 2^{k+1} = 2(2^{k+1}) - 1 = 2^{k+2} - 1$$

which is $P(k+1)$. That is, $P(k+1)$ is true whenever $P(k)$ is true. By the principle of induction, $P(n)$ is true for all n.

Supplementary Problems

SETS AND SUBSETS

1.26 Which of the following sets are equal?

$$A = \{x \mid x^2 - 4x + 3 = 0\}, \quad C = \{x \mid x \in \mathbf{N}, x < 3\}, \qquad E = \{1, 2\}, \quad G = \{3, 1\},$$
$$B = \{x \mid x^2 - 3x + 2 = 0\}, \quad D = \{x \mid x \in \mathbf{N}, x \text{ is odd}, x < 5\}, \quad F = \{1, 2, 1\}, \quad H = \{1, 1, 3\}.$$

1.27 List the elements of the following sets if the universal set is $\mathbf{U} = \{a, b, c, \ldots, y, z\}$.

Furthermore, identify which of the sets, if any, are equal.

$$A = \{x \mid x \text{ is a vowel}\}, \qquad\qquad C = \{x \mid x \text{ precedes } f \text{ in the alphabet}\},$$
$$B = \{x \mid x \text{ is a letter in the word "little"}\}, \quad D = \{x \mid x \text{ is a letter in the word "title"}\}.$$

1.28 Let $A = \{1, 2, \ldots, 8, 9\}$, $B = \{2, 4, 6, 8\}$, $C = \{1, 3, 5, 7, 9\}$, $D = \{3, 4, 5\}$, $E = \{3, 5\}$.
Which of the these sets can equal a set X under each of the following conditions?

(a) X and B are disjoint.   (c) $X \subseteq A$ but $X \not\subseteq C$.

(b) $X \subseteq D$ but $X \not\subseteq B$.   (d) $X \subseteq C$ but $X \not\subseteq A$.

SET OPERATIONS

1.29 Consider the universal set $\mathbf{U} = \{1, 2, 3, \ldots, 8, 9\}$ and sets $A = \{1, 2, 5, 6\}$, $B = \{2, 5, 7\}$, $C = \{1, 3, 5, 7, 9\}$. Find:

(a) $A \cap B$ and $A \cap C$ (c) A^C and C^C (e) $A \oplus B$ and $A \oplus C$

(b) $A \cup B$ and $B \cup C$ (d) $A \backslash B$ and $A \backslash C$ (f) $(A \cup C) \backslash B$ and $(B \oplus C) \backslash A$

1.30 Let A and B be any sets. Prove:

(a) A is the disjoint union of $A \backslash B$ and $A \cap B$.

(b) $A \cup B$ is the disjoint union of $A \backslash B$, $A \cap B$, and $B \backslash A$.

1.31 Prove the following:

(a) $A \subseteq B$ if and only if $A \cap B^C = \emptyset$ (c) $A \subseteq B$ if and only if $B^C \subseteq A^C$

(b) $A \subseteq B$ if and only if $A^C \cup B = \mathbf{U}$ (d) $A \subseteq B$ if and only if $A \backslash B = \emptyset$

(Compare the results with Theorem 1.4.)

1.32 Prove the Absorption Laws: (a) $A \cup (A \cap B) = A$;   (b) $A \cap (A \cup B) = A$.

1.33 The formula $A \backslash B = A \cap B^C$ defines the difference operation in terms of the operations of intersection and complement. Find a formula that defines the union $A \cup B$ in terms of the operations of intersection and complement.

VENN DIAGRAMS

1.34 The Venn diagram in Fig. 1-5(a) shows sets A, B, C. Shade the following sets:

(a) $A \backslash (B \cup C)$;   (b) $A^C \cap (B \cup C)$;   (c) $A^C \cap (C \backslash B)$.

1.35 Use the Venn diagram in Fig. 1-5(b) to write each set as the (disjoint) union of fundamental products:

(a) $A \cap (B \cup C)$;   (b) $A^C \cap (B \cup C)$;   (c) $A \cup (B \backslash C)$.

1.36 Consider the following assumptions:

S_1: All dictionaries are useful.

S_2: Mary owns only romance novels.

S_3: No romance novel is useful.

Use a Venn diagram to determine the validity of each of the following conclusions:

(a) Romance novels are not dictionaries.

(b) Mary does not own a dictionary.

(c) All useful books are dictionaries.

ALGEBRA OF SETS AND DUALITY

1.37 Write the dual of each equation:

(a) $A = (B^C \cap A) \cup (A \cap B)$

(b) $(A \cap B) \cup (A^C \cap B) \cup (A \cap B^C) \cup (A^C \cap B^C) = \mathbf{U}$

1.38 Use the laws in Table 1-1 to prove each set identity:

(a) $(A \cap B) \cup (A \cap B^C) = A$

(b) $A \cup B = (A \cap B^C) \cup (A^C \cap B) \cup (A \cap B)$

FINITE SETS AND THE COUNTING PRINCIPLE

1.39 Determine which of the following sets are finite:

    (a) Lines parallel to the x axis.   (c) Integers which are multiples of 5.

    (b) Letters in the English alphabet.   (d) Animals living on the earth.

1.40 Use Theorem 1.9 to prove Corollary 1.10: Suppose A, B, C are finite sets. Then $A \cup B \cup C$ is finite and

$$n(A \cup B \cup C) = n(A) + n(B) + n(C) - n(A \cap B) - n(A \cap C) - n(B \cap C) + n(A \cap B \cap C)$$

1.41 A survey on a sample of 25 new cars being sold at a local auto dealer was conducted to see which of three popular options, air-conditioning (A), radio (R), and power windows (W), were already installed. The survey found:

               15 had air-conditioning (A),   5 had A and P,

               12 had radio (R),   9 had A and R,   3 had all three options.

               11 had power windows (W),   4 had R and W,

    Find the number of cars that had: (a) only W;   (b) only A;   (c) only R;   (d) R and W but not A;   (e) A and R but not W;   (f) only one of the options;   (g) at least one option;   (h) none of the options.

CLASSES OF SETS

1.42 Find the power set $P(A)$ of $A = \{1, 2, 3, 4, 5\}$.

1.43 Given $A = [\{a, b\}, \{c\}, \{d, e, f\}]$.

    (a) List the elements of A.   (b) Find $n(A)$.   (c) Find the power set of A.

1.44 Suppose A is finite and $n(A) = m$. Prove the power set $P(A)$ has 2^m elements.

PARTITIONS

1.45 Let $S = \{1, 2, \ldots, 8, 9\}$. Determine whether or not each of the following is a partition of S:

    (a) $[\{1, 3, 6\}, \{2, 8\}, \{5, 7, 9\}]$ (c) $[\{2, 4, 5, 8\}, \{1, 9\}, \{3, 6, 7\}]$

    (b) $[\{1, 5, 7\}, \{2, 4, 8, 9\}, \{3, 5, 6\}]$ (d) $[\{1, 2, 7\}, \{3, 5\}, \{4, 6, 8, 9\}, \{3, 5\}]$

1.46 Let $S = \{1, 2, 3, 4, 5, 6\}$. Determine whether or not each of the following is a partition of S:

    (a) $P_1 = [\{1, 2, 3\}, \{1, 4, 5, 6\}]$ (c) $P_3 = [\{1, 3, 5\}, \{2, 4\}, \{6\}]$

    (b) $P_2 = [\{1, 2\}, \{3, 5, 6\}]$ (d) $P_4 = [\{1, 3, 5\}, \{2, 4, 6, 7\}]$

1.47 Determine whether or not each of the following is a partition of the set **N** of positive integers:

    (a) $[\{n \mid n > 5\}, \{n \mid n < 5\}]$;   (b) $[\{n \mid n > 6\}, \{1, 3, 5\}, \{2, 4\}]$;

    (c) $[\{n \mid n^2 > 11\}, \{n \mid n^2 < 11\}]$.

1.48 Let $[A_1, A_2, \ldots, A_m]$ and $[B_1, B_2, \ldots, B_n]$ be partitions of a set S.

    Show that the following collection of sets is also a partition (called the *cross partition*) of S:

$$P = [A_i \cap B_j \mid i = 1, \ldots, m, \ j = 1, \ldots, n] \backslash \emptyset$$

    Observe that we deleted the empty set $\emptyset$.

1.49 Let $S = \{1, 2, 3, \ldots, 8, 9\}$. Find the cross partition P of the following partitions of S:

$$P_1 = [\{1, 3, 5, 7, 9\}, \{2, 4, 6, 8\}] \quad \text{and} \quad P_2 = [\{1, 2, 3, 4\}, \{5, 7\}, \{6, 8, 9\}]$$

INDUCTION

1.50 Prove: $2 + 4 + 6 + \cdots + 2n = n(n+1)$

1.51 Prove: $1 + 4 + 7 + \cdots + 3n - 2 = \frac{n(3n-1)}{2}$

1.52 Prove: $1^2 + 2^2 + 3^2 + \cdots + n^2 = \frac{n(n+1)(2n+1)}{6}$

1.53 Prove: $\frac{1}{1\cdot3} + \frac{1}{3\cdot5} + \frac{1}{5\cdot7} + \cdots + \frac{1}{(2n-1)(2n+1)} = \frac{n}{2n+1}$

1.54 Prove: $\frac{1}{1\cdot5} + \frac{1}{5\cdot9} + \frac{1}{9\cdot13} + \cdots + \frac{1}{(4n-3)(4n+1)} = \frac{n}{4n+1}$

1.55 Prove $7^n - 2^n$ is divisible by 5 for all $n \in \mathbf{N}$

1.56 Prove $n^3 - 4n + 6$ is divisible by 3 for all $n \in \mathbf{N}$

1.57 Use the identity $1 + 2 + 3 + \cdots + n = n(n+1)/2$ to prove that

$$1^3 + 2^3 + 3^3 + \cdots + n^3 = (1 + 2 + 3 + \cdots + n)^2$$

Miscellaneous Problems

1.58 Suppose $\mathbf{N} = \{1, 2, 3, \ldots\}$ is the universal set, and

$$A = \{n \mid n \le 6\}, \quad B = \{n \mid 4 \le n \le 9\}, \quad C = \{1, 3, 5, 7, 9\}, \quad D = \{2, 3, 5, 7, 8\}.$$

Find: (a) $A \oplus B$;   (b) $B \oplus C$;   (c) $A \cap (B \oplus D)$;   (d) $(A \cap B) \oplus (A \cap D)$.

1.59 Prove the following properties of the symmetric difference:

(a) $(A \oplus B) \oplus C = A \oplus (B \oplus C)$ (Associative Law).

(b) $A \oplus B = B \oplus A$ (Commutative Law).

(c) If $A \oplus B = A \oplus C$, then $B = C$ (Cancellation Law).

(d) $A \cap (B \oplus C) = (A \cap B) \oplus (A \cap C)$ (Distributive Law).

1.60 Consider m nonempty distinct sets $A_1, A_2, \ldots, A_m$ in a universal set $\mathbf{U}$. Prove:

(a) There are 2^m fundamental products of the m sets.

(b) Any two fundamental products are disjoint.

(c) $\mathbf{U}$ is the union of all the fundamental products.

Answers to Supplementary Problems

1.26 $B = C = E = F, A = D = G = H.$

1.27 $A = \{a, e, i, o, u\}$, $B = D = \{l, i, t, e\}$, $C = \{a, b, c, d, e\}.$

1.28 (a) C and E; (b) D and E; (c) A, B, and D; (d) None.

1.29 (a) $A \cap B = \{2, 5\}$, $A \cap C = \{1, 5\}$;
(b) $A \cup B = \{1, 2, 5, 6, 7\}$, $B \cup C = \{1, 2, 3, 5, 7, 9\}$;
(c) $A^C = \{3, 4, 7, 8, 9\}$, $C^C = \{2, 4, 6, 8\}$;
(d) $A \backslash B = \{1, 6\}$, $A \backslash C = \{2, 6\}$;
(e) $A \oplus B = \{1, 6, 7\}$, $A \oplus C = \{2, 3, 6, 7, 9\}$;
(f) $(A \cup C) \backslash B = \{1, 3, 6, 9\}$, $(B \oplus C) \backslash A = \{3, 9\}$.

1.33 $A \cup B = (A^C \cap B^C)^C.$

1.34 See Fig. 1-10.

1.35 (a) $(A \cap B \cap C) \cup (A \cap B \cap C^C) \cup (A \cap B^C \cap C)$

(b) $(A^C \cap B \cap C^C) \cup (A^C \cap B \cap C) \cup (A^C \cap B^C \cap C)$

(c) $(A \cap B \cap C) \cup (A \cap B \cap C^C) \cup (A \cap B^C \cap C)$ $\cup (A^C \cap B \cap C^C) \cup (A \cap B^C \cap C^C)$

1.36 The three premises yield the Venn diagram in Fig. 1-11(a). (a) and (b) are valid, but (c) is not valid.

1.37 (a) $A = (B^C \cup A) \cap (A \cup B)$

(b) $(A \cup B) \cap (A^C \cup B) \cap (A \cup B^C) \cap (A^C \cup B^C) = \emptyset$

1.39 (a) Infinite; (b) finite; (c) infinite; (d) finite.

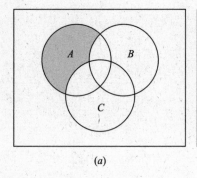
(a)

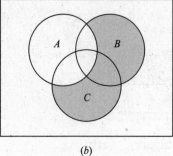
(b)

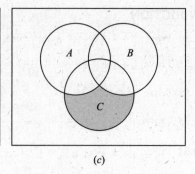
(c)

Fig. 1-10

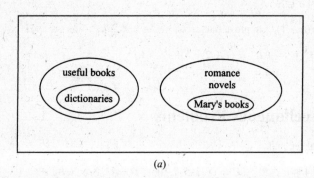
(a)

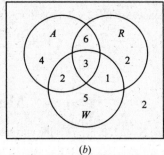
(b)

Fig. 1-11

1.41 Use the data to fill in the Venn diagram in Fig. 1-11(b). Then:

(a) 5; (b) 4; (c) 2; (d) 1; (e) 6; (f) 11; (g) 23; (h) 2.

1.42 $P(A)$ has $2^5 = 32$ elements as follows:

[∅, {1}, {2}, {3}, {4}, {5}, {1, 2}, {1, 3}, {1, 4}, {1, 5}, {2, 3}, {2, 4}, {2, 5}, {3, 4}, {3, 5}, {4, 5}, {1, 2, 3}, {1, 2, 4}, {1, 2, 5}, {2, 3, 4}, {2, 3, 5}, {3, 4, 5}, {1, 3, 4}, {1, 3, 5}, {1, 4, 5}, {2, 4, 5}, {1, 2, 3, 4}, {1, 2, 3, 5}, {1, 2, 4, 5}, {1, 3, 4, 5}, {2, 3, 4, 5}, A]

1.43 (a) Three elements: [a, b], (c), and {d, e, f}. (b) 3. (c) $P(A)$ has $2^3 = 8$ elements as follows:

$P(A) = \{A, [\{a, b\}, \{c\}], [\{a, b\}, \{d, e, f\}],$

$[\{c\}, \{d, e, f\}], [\{a, b\}], [\{c\}], [\{d, e, f\}], \emptyset\}$

1.44 Let X be an element in $P(A)$. For each $a \in A$, either $a \in X$ or $a \notin X$. Since $n(A) = m$, there are 2^m different sets X. That is $|P(A)| = 2^m$.

1.45 (a) No, (b) no, (c) yes, (d) yes.

1.46 (a) No, (b) no, (c) yes, (d) no.

1.47 (a) No, (b) no, (c) yes.

1.49 [{1,3}, {2,4}, {5,7}, {9}, {6,8}]

1.55 Hint: $7^{k+1} - 2^{k+1} = 7^{k+1} - 7(2^k) + 7(2^k) - 2^{k+1} = 7(7^k - 2^k) + (7 - 2)2^k$

1.58 (a) {1, 2, 3, 7, 8, 9}; (b) {1, 3, 4, 6, 8}; (c) and (d) {2, 3, 4, 6}.

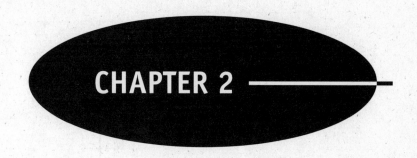

CHAPTER 2

Relations

2.1 INTRODUCTION

The reader is familiar with many relations such as "less than," "is parallel to," "is a subset of," and so on. In a certain sense, these relations consider the existence or nonexistence of a certain connection between pairs of objects taken in a definite order. Formally, we define a relation in terms of these "ordered pairs."

An *ordered pair* of elements a and b, where a is designated as the first element and b as the second element, is denoted by (a, b). In particular,

$$(a, b) = (c, d)$$

if and only if $a = c$ and $b = d$. Thus $(a, b) \neq (b, a)$ unless $a = b$. This contrasts with sets where the order of elements is irrelevant; for example, $\{3, 5\} = \{5, 3\}$.

2.2 PRODUCT SETS

Consider two arbitrary sets A and B. The set of all ordered pairs (a, b) where $a \in A$ and $b \in B$ is called the *product*, or *Cartesian product*, of A and B. A short designation of this product is $A \times B$, which is read "A cross B." By definition,

$$A \times B = \{(a, b) \mid a \in A \text{ and } b \in B\}$$

One frequently writes A^2 instead of $A \times A$.

EXAMPLE 2.1 $\mathbf{R}$ denotes the set of real numbers and so $\mathbf{R}^2 = \mathbf{R} \times \mathbf{R}$ is the set of ordered pairs of real numbers. The reader is familiar with the geometrical representation of $\mathbf{R}^2$ as points in the plane as in Fig. 2-1. Here each point P represents an ordered pair (a, b) of real numbers and vice versa; the vertical line through P meets the x-axis at a, and the horizontal line through P meets the y-axis at b. $\mathbf{R}^2$ is frequently called the *Cartesian plane*.

EXAMPLE 2.2 Let $A = \{1, 2\}$ and $B = \{a, b, c\}$. Then

$$A \times B = \{(1, a), \ (1, b), \ (1, c), \ (2, a), \ (2, b), \ (2, c)\}$$
$$B \times A = \{(a, 1), \ (b, 1), \ (c, 1), \ (a, 2), \ (b, 2), \ (c, 2)\}$$

Also, $A \times A = \{(1, 1), (1, 2), (2, 1), (2, 2)\}$

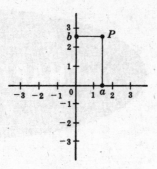

Fig. 2-1

There are two things worth noting in the above examples. First of all $A \times B \neq B \times A$. The Cartesian product deals with ordered pairs, so naturally the order in which the sets are considered is important. Secondly, using $n(S)$ for the number of elements in a set S, we have:

$$n(A \times B) = 6 = 2(3) = n(A)n(B)$$

In fact, $n(A \times B) = n(A)n(B)$ for any finite sets A and B. This follows from the observation that, for an ordered pair (a, b) in $A \times B$, there are $n(A)$ possibilities for a, and for each of these there are $n(B)$ possibilities for b.

The idea of a product of sets can be extended to any finite number of sets. For any sets $A_1, A_2, \ldots, A_n$, the set of all ordered n-tuples $(a_1, a_2, \ldots, a_n)$ where $a_1 \in A_1, a_2 \in A_2, \ldots, a_n \in A_n$ is called the *product* of the sets $A_1, \ldots, A_n$ and is denoted by

$$A_1 \times A_2 \times \cdots \times A_n \quad \text{or} \quad \prod_{i=1}^{n} A_i$$

Just as we write A^2 instead of $A \times A$, so we write A^n instead of $A \times A \times \cdots \times A$, where there are n factors all equal to A. For example, $\mathbf{R}^3 = \mathbf{R} \times \mathbf{R} \times \mathbf{R}$ denotes the usual three-dimensional space.

2.3 RELATIONS

We begin with a definition.

Definition 2.1: Let A and B be sets. A *binary relation* or, simply, *relation* from A to B is a subset of $A \times B$.

Suppose R is a relation from A to B. Then R is a set of ordered pairs where each first element comes from A and each second element comes from B. That is, for each pair $a \in A$ and $b \in B$, exactly one of the following is true:

  (i) $(a, b) \in R$; we then say "a is R-related to b", written aRb.
  (ii) $(a, b) \notin R$; we then say "a is not R-related to b", written $a\cancel{R}b$.

If R is a relation from a set A to itself, that is, if R is a subset of $A^2 = A \times A$, then we say that R is a relation *on A*.

The *domain* of a relation R is the set of all first elements of the ordered pairs which belong to R, and the *range* is the set of second elements.

Although n-ary relations, which involve ordered n-tuples, are introduced in Section 2.10, the term relation shall then mean binary relation unless otherwise stated or implied.

EXAMPLE 2.3

(a) $A = (1, 2, 3)$ and $B = \{x, y, z\}$, and let $R = \{(1, y), (1, z), (3, y)\}$. Then R is a relation from A to B since R is a subset of $A \times B$. With respect to this relation,

$$1Ry, 1Rz, 3Ry, \quad \text{but} \quad 1\not{R}x, 2\not{R}x, 2\not{R}y, 2\not{R}z, 3\not{R}x, 3\not{R}z$$

The domain of R is $\{1, 3\}$ and the range is $\{y, z\}$.

(b) Set inclusion $\subseteq$ is a relation on any collection of sets. For, given any pair of set A and B, either $A \subseteq B$ or $A \nsubseteq B$.

(c) A familiar relation on the set $\mathbf{Z}$ of integers is "m divides n." A common notation for this relation is to write $m \mid n$ when m divides n. Thus $6 \mid 30$ but $7 \nmid 25$.

(d) Consider the set L of lines in the plane. Perpendicularity, written "$\perp$," is a relation on L. That is, given any pair of lines a and b, either $a \perp b$ or $a \not\perp b$. Similarly, "is parallel to," written "$\parallel$," is a relation on L since either $a \parallel b$ or $a \nparallel b$.

(e) Let A be any set. An important relation on A is that of *equality*,

$$\{(a, a) \mid a \in A\}$$

which is usually denoted by "$=$." This relation is also called the *identity* or *diagonal* relation on A and it will also be denoted by Δ_A or simply Δ.

(f) Let A be any set. Then $A \times A$ and $\emptyset$ are subsets of $A \times A$ and hence are relations on A called the *universal relation* and *empty relation*, respectively.

Inverse Relation

Let R be any relation from a set A to a set B. The *inverse* of R, denoted by R^{-1}, is the relation from B to A which consists of those ordered pairs which, when reversed, belong to R; that is,

$$R^{-1} = \{(b, a) \mid (a, b) \in R\}$$

For example, let $A = \{1, 2, 3\}$ and $B = \{x, y, z\}$. Then the inverse of

$$R = \{(1, y), (1, z), (3, y)\} \quad \text{is} \quad R^{-1} = \{(y, 1), (z, 1), (y, 3)\}$$

Clearly, if R is any relation, then $(R^{-1})^{-1} = R$. Also, the domain and range of R^{-1} are equal, respectively, to the range and domain of R. Moreover, if R is a relation on A, then R^{-1} is also a relation on A.

2.4 PICTORIAL REPRESENTATIVES OF RELATIONS

There are various ways of picturing relations.

Relations on R

Let S be a relation on the set $\mathbf{R}$ of real numbers; that is, S is a subset of $\mathbf{R}^2 = \mathbf{R} \times \mathbf{R}$. Frequently, S consists of all ordered pairs of real numbers which satisfy some given equation $E(x, y) = 0$ (such as $x^2 + y^2 = 25$).

Since $\mathbf{R}^2$ can be represented by the set of points in the plane, we can picture S by emphasizing those points in the plane which belong to S. The pictorial representation of the relation is sometimes called the *graph* of the relation. For example, the graph of the relation $x^2 + y^2 = 25$ is a circle having its center at the origin and radius 5. See Fig. 2-2(a).

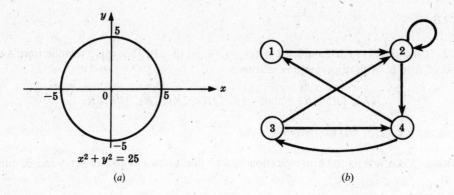

Fig. 2-2

Directed Graphs of Relations on Sets

There is an important way of picturing a relation R on a finite set. First we write down the elements of the set, and then we draw an arrow from each element x to each element y whenever x is related to y. This diagram is called the *directed graph* of the relation. Figure 2-2(b), for example, shows the directed graph of the following relation R on the set $A = \{1, 2, 3, 4\}$:

$$R = \{(1, 2), (2, 2), (2, 4), (3, 2), (3, 4), (4, 1), (4, 3)\}$$

Observe that there is an arrow from 2 to itself, since 2 is related to 2 under R.

These directed graphs will be studied in detail as a separate subject in Chapter 8. We mention it here mainly for completeness.

Pictures of Relations on Finite Sets

Suppose A and B are finite sets. There are two ways of picturing a relation R from A to B.

(i) Form a rectangular array (matrix) whose rows are labeled by the elements of A and whose columns are labeled by the elements of B. Put a 1 or 0 in each position of the array according as $a \in A$ is or is not related to $b \in B$. This array is called the *matrix of the relation*.

(ii) Write down the elements of A and the elements of B in two disjoint disks, and then draw an arrow from $a \in A$ to $b \in B$ whenever a is related to b. This picture will be called the *arrow diagram* of the relation.

Figure 2-3 pictures the relation R in Example 2.3(a) by the above two ways.

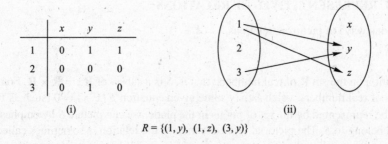

	x	y	z
1	0	1	1
2	0	0	0
3	0	1	0

(i)   (ii)

$$R = \{(1, y),\ (1, z),\ (3, y)\}$$

Fig. 2-3

2.5 COMPOSITION OF RELATIONS

Let A, B and C be sets, and let R be a relation from A to B and let S be a relation from B to C. That is, R is a subset of $A \times B$ and S is a subset of $B \times C$. Then R and S give rise to a relation from A to C denoted by $R \circ S$ and defined by:

$$a(R \circ S)c \text{ if for some } b \in B \text{ we have } aRb \text{ and } bSc.$$

That is,

$$R \circ S = \{(a, c) \mid \text{there exists } b \in B \text{ for which } (a, b) \in R \text{ and } (b, c) \in S\}$$

The relation $R \circ S$ is called the *composition* of R and S; it is sometimes denoted simply by RS.

Suppose R is a relation on a set A, that is, R is a relation from a set A to itself. Then $R \circ R$, the composition of R with itself, is always defined. Also, $R \circ R$ is sometimes denoted by R^2. Similarly, $R^3 = R^2 \circ R = R \circ R \circ R$, and so on. Thus R^n is defined for all positive n.

Warning: Many texts denote the composition of relations R and S by $S \circ R$ rather than $R \circ S$. This is done in order to conform with the usual use of $g \circ f$ to denote the composition of f and g where f and g are functions. Thus the reader may have to adjust this notation when using this text as a supplement with another text. However, when a relation R is composed with itself, then the meaning of $R \circ R$ is unambiguous.

EXAMPLE 2.4 Let $A = \{1, 2, 3, 4\}$, $B = \{a, b, c, d\}$, $C = \{x, y, z\}$ and let

$$R = \{(1, a), (2, d), (3, a), (3, b), (3, d)\} \quad \text{and} \quad S = \{(b, x), (b, z), (c, y), (d, z)\}$$

Consider the arrow diagrams of R and S as in Fig. 2-4. Observe that there is an arrow from 2 to d which is followed by an arrow from d to z. We can view these two arrows as a "path" which "connects" the element $2 \in A$ to the element $z \in C$. Thus:

$$2(R \circ S)z \quad \text{since } 2Rd \text{ and } dSz$$

Similarly there is a path from 3 to x and a path from 3 to z. Hence

$$3(R \circ S)x \quad \text{and} \quad 3(R \circ S)z$$

No other element of A is connected to an element of C. Accordingly,

$$R \circ S = \{(2, z), (3, x), (3, z)\}$$

Our first theorem tells us that composition of relations is associative.

Theorem 2.1: Let A, B, C and D be sets. Suppose R is a relation from A to B, S is a relation from B to C, and T is a relation from C to D. Then

$$(R \circ S) \circ T = R \circ (S \circ T)$$

We prove this theorem in Problem 2.8.

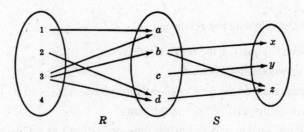

Fig. 2-4

Composition of Relations and Matrices

There is another way of finding $R \circ S$. Let M_R and M_S denote respectively the matrix representations of the relations R and S. Then

$$
M_R = \begin{array}{c} \\ 1 \\ 2 \\ 3 \\ 4 \end{array}
\begin{array}{cccc} a & b & c & d \\ \end{array}
\left[\begin{array}{cccc} 1 & 0 & 0 & 0 \\ 0 & 0 & 0 & 1 \\ 1 & 1 & 0 & 1 \\ 0 & 0 & 0 & 0 \end{array} \right]
\quad \text{and} \quad
M_S = \begin{array}{c} \\ a \\ b \\ c \\ d \end{array}
\begin{array}{ccc} x & y & z \\ \end{array}
\left[\begin{array}{ccc} 0 & 0 & 0 \\ 1 & 0 & 1 \\ 0 & 1 & 0 \\ 0 & 0 & 1 \end{array} \right]
$$

Multiplying M_R and M_S we obtain the matrix

$$
M = M_R M_S = \begin{array}{c} \\ 1 \\ 2 \\ 3 \\ 4 \end{array}
\begin{array}{ccc} x & y & z \\ \end{array}
\left[\begin{array}{ccc} 0 & 0 & 0 \\ 0 & 0 & 1 \\ 1 & 0 & 2 \\ 0 & 0 & 0 \end{array} \right]
$$

The nonzero entries in this matrix tell us which elements are related by $R \circ S$. Thus $M = M_R M_S$ and $M_{R \circ S}$ have the same nonzero entries.

2.6 TYPES OF RELATIONS

This section discusses a number of important types of relations defined on a set A.

Reflexive Relations

A relation R on a set A is *reflexive* if aRa for every $a \in A$, that is, if $(a, a) \in R$ for every $a \in A$. Thus R is not reflexive if there exists $a \in A$ such that $(a, a) \notin R$.

EXAMPLE 2.5 Consider the following five relations on the set $A = \{1, 2, 3, 4\}$:

$$R_1 = \{(1, 1), (1, 2), (2, 3), (1, 3), (4, 4)\}$$
$$R_2 = \{(1, 1)(1, 2), (2, 1), (2, 2), (3, 3), (4, 4)\}$$
$$R_3 = \{(1, 3), (2, 1)\}$$
$$R_4 = \emptyset, \text{ the empty relation}$$
$$R_5 = A \times A, \text{ the universal relation}$$

Determine which of the relations are reflexive.

Since A contains the four elements 1, 2, 3, and 4, a relation R on A is reflexive if it contains the four pairs $(1, 1)$, $(2, 2)$, $(3, 3)$, and $(4, 4)$. Thus only R_2 and the universal relation $R_5 = A \times A$ are reflexive. Note that R_1, R_3, and R_4 are not reflexive since, for example, $(2, 2)$ does not belong to any of them.

EXAMPLE 2.6 Consider the following five relations:

(1) Relation $\leq$ (less than or equal) on the set $\mathbf{Z}$ of integers.

(2) Set inclusion $\subseteq$ on a collection C of sets.

(3) Relation $\perp$ (perpendicular) on the set L of lines in the plane.

(4) Relation $\parallel$ (parallel) on the set L of lines in the plane.

(5) Relation $|$ of divisibility on the set $\mathbf{N}$ of positive integers. (Recall $x \,|\, y$ if there exists z such that $xz = y$.)

Determine which of the relations are reflexive.

The relation (3) is not reflexive since no line is perpendicular to itself. Also (4) is not reflexive since no line is parallel to itself. The other relations are reflexive; that is, $x \leq x$ for every $x \in \mathbf{Z}$, $A \subseteq A$ for any set $A \in C$, and $n \mid n$ for every positive integer $n \in \mathbf{N}$.

Symmetric and Antisymmetric Relations

A relation R on a set A is *symmetric* if whenever aRb then bRa, that is, if whenever $(a, b) \in R$ then $(b, a) \in R$. Thus R is not symmetric if there exists $a, b \in A$ such that $(a, b) \in R$ but $(b, a) \notin R$.

EXAMPLE 2.7

(a) Determine which of the relations in Example 2.5 are symmetric.

R_1 is not symmetric since $(1, 2) \in R_1$ but $(2, 1) \notin R_1$. R_3 is not symmetric since $(1, 3) \in R_3$ but $(3, 1) \notin R_3$. The other relations are symmetric.

(b) Determine which of the relations in Example 2.6 are symmetric.

The relation $\perp$ is symmetric since if line a is perpendicular to line b then b is perpendicular to a. Also, $\parallel$ is symmetric since if line a is parallel to line b then b is parallel to line a. The other relations are not symmetric. For example:

$$3 \leq 4 \text{ but } 4 \nleq 3; \quad \{1, 2\} \subseteq \{1, 2, 3\} \text{ but } \{1, 2, 3\} \nsubseteq \{1, 2\}; \quad \text{and} \quad 2 \mid 6 \text{ but } 6 \nmid 2.$$

A relation R on a set A is *antisymmetric* if whenever aRb and bRa then $a = b$, that is, if $a \neq b$ and aRb then $b\cancel{R}a$. Thus R is not antisymmetric if there exist distinct elements a and b in A such that aRb and bRa.

EXAMPLE 2.8

(a) Determine which of the relations in Example 2.5 are antisymmetric.

R_2 is not antisymmetric since $(1, 2)$ and $(2, 1)$ belong to R_2, but $1 \neq 2$. Similarly, the universal relation R_5 is not antisymmetric. All the other relations are antisymmetric.

(b) Determine which of the relations in Example 2.6 are antisymmetric.

The relation $\leq$ is antisymmetric since whenever $a \leq b$ and $b \leq a$ then $a = b$. Set inclusion $\subseteq$ is antisymmetric since whenever $A \subseteq B$ and $B \subseteq A$ then $A = B$. Also, divisibility on $\mathbf{N}$ is antisymmetric since whenever $m \mid n$ and $n \mid m$ then $m = n$. (Note that divisibility on $\mathbf{Z}$ is not antisymmetric since $3 \mid -3$ and $-3 \mid 3$ but $3 \neq -3$.) The relations $\perp$ and $\parallel$ are not antisymmetric.

Remark: The properties of being symmetric and being antisymmetric are not negatives of each other. For example, the relation $R = \{(1, 3), (3, 1), (2, 3)\}$ is neither symmetric nor antisymmetric. On the other hand, the relation $R' = \{(1, 1), (2, 2)\}$ is both symmetric and antisymmetric.

Transitive Relations

A relation R on a set A is *transitive* if whenever aRb and bRc then aRc, that is, if whenever $(a, b), (b, c) \in R$ then $(a, c) \in R$. Thus R is not transitive if there exist $a, b, c \in R$ such that $(a, b), (b, c) \in R$ but $(a, c) \notin R$.

EXAMPLE 2.9

(a) Determine which of the relations in Example 2.5 are transitive.

The relation R_3 is not transitive since $(2, 1), (1, 3) \in R_3$ but $(2, 3) \notin R_3$. All the other relations are transitive.

(b) Determine which of the relations in Example 2.6 are transitive.

The relations $\leq$, $\subseteq$, and $|$ are transitive, but certainly not $\perp$. Also, since no line is parallel to itself, we can have $a \parallel b$ and $b \parallel a$, but $a \not\parallel a$. Thus $\parallel$ is not transitive. (We note that the relation "is parallel or equal to" is a transitive relation on the set L of lines in the plane.)

The property of transitivity can also be expressed in terms of the composition of relations. For a relation R on A we did define $R^2 = R \circ R$ and, more generally, $R^n = R^{n-1} \circ R$. Then we have the following result:

Theorem 2.2: A relation R is transitive if and only if, for every $n \geq 1$, we have $R^n \subseteq R$.

2.7 CLOSURE PROPERTIES

Consider a given set A and the collection of all relations on A. Let P be a property of such relations, such as being symmetric or being transitive. A relation with property P will be called a P-relation. The P-closure of an arbitrary relation R on A, written $P(R)$, is a P-relation such that

$$R \subseteq P(R) \subseteq S$$

for every P-relation S containing R. We will write

$$\text{reflexive}(R), \quad \text{symmetric}(R), \quad \text{and} \quad \text{transitive}(R)$$

for the reflexive, symmetric, and transitive closures of R.

Generally speaking, $P(R)$ need not exist. However, there is a general situation where $P(R)$ will always exist. Suppose P is a property such that there is at least one P-relation containing R and that the intersection of any P-relations is again a P-relation. Then one can prove (Problem 2.12) that

$$P(R) = \cap(S \mid S \text{ is a } P\text{-relation and } R \subseteq S)$$

Thus one can obtain $P(R)$ from the "top-down," that is, as the intersection of relations. However, one usually wants to find $P(R)$ from the "bottom-up," that is, by adjoining elements to R to obtain $P(R)$. This we do below.

Reflexive and Symmetric Closures

The next theorem tells us how to obtain easily the reflexive and symmetric closures of a relation. Here $\Delta_A = \{(a, a) \mid a \in A\}$ is the diagonal or equality relation on A.

Theorem 2.3: Let R be a relation on a set A. Then:

       (i) $R \cup \Delta_A$ is the reflexive closure of R.

      (ii) $R \cup R^{-1}$ is the symmetric closure of R.

In other words, reflexive(R) is obtained by simply adding to R those elements (a, a) in the diagonal which do not already belong to R, and symmetric(R) is obtained by adding to R all pairs (b, a) whenever (a, b) belongs to R.

EXAMPLE 2.10 Consider the relation $R = \{(1, 1), (1, 3), (2, 4), (3, 1), (3, 3), (4, 3)\}$ on the set $A = \{1, 2, 3, 4\}$. Then

$$\text{reflexive}(R) = R \cup \{(2, 2), (4, 4)\} \quad \text{and} \quad \text{symmetric}(R) = R \cup \{(4, 2), (3, 4)\}$$

Transitive Closure

Let R be a relation on a set A. Recall that $R^2 = R \circ R$ and $R^n = R^{n-1} \circ R$. We define

$$R^* = \bigcup_{i=1}^{\infty} R^i$$

The following theorem applies:

Theorem 2.4: R^* is the transitive closure of R.

Suppose A is a finite set with n elements. We show in Chapter 8 on graphs that

$$R^* = R \cup R^2 \cup \ldots \cup R^n$$

This gives us the following theorem:

Theorem 2.5: Let R be a relation on a set A with n elements. Then

$$\text{transitive } (R) = R \cup R^2 \cup \ldots \cup R^n$$

EXAMPLE 2.11   Consider the relation $R = \{(1, 2), (2, 3), (3, 3)\}$ on $A = \{1, 2, 3\}$. Then:

$$R^2 = R \circ R = \{(1, 3), (2, 3), (3, 3)\} \quad \text{and} \quad R^3 = R^2 \circ R = \{(1, 3), (2, 3), (3, 3)\}$$

Accordingly,

$$\text{transitive } (R) = \{(1, 2), (2, 3), (3, 3), (1, 3)\}$$

2.8   EQUIVALENCE RELATIONS

Consider a nonempty set S. A relation R on S is an *equivalence relation* if R is reflexive, symmetric, and transitive. That is, R is an equivalence relation on S if it has the following three properties:

   (1) For every $a \in S$, aRa.   (2) If aRb, then bRa.   (3) If aRb and bRc, then aRc.

The general idea behind an equivalence relation is that it is a classification of objects which are in some way "alike." In fact, the relation "$=$" of equality on any set S is an equivalence relation; that is:

   (1) $a = a$ for every $a \in S$.   (2) If $a = b$, then $b = a$.   (3) If $a = b$, $b = c$, then $a = c$.

Other equivalence relations follow.

EXAMPLE 2.12

(a) Let L be the set of lines and let T be the set of triangles in the Euclidean plane.

    (i) The relation "is parallel to or identical to" is an equivalence relation on L.

    (ii) The relations of congruence and similarity are equivalence relations on T.

(b) The relation $\subseteq$ of set inclusion is not an equivalence relation. It is reflexive and transitive, but it is not symmetric since $A \subseteq B$ does not imply $B \subseteq A$.

(c) Let m be a fixed positive integer. Two integers a and b are said to be *congruent modulo m*, written

$$a \equiv b \ (\text{mod } m)$$

if m divides $a - b$. For example, for the modulus $m = 4$, we have

$$11 \equiv 3 \ (\text{mod } 4) \quad \text{and} \quad 22 \equiv 6 \ (\text{mod } 4)$$

since 4 divides $11 - 3 = 8$ and 4 divides $22 - 6 = 16$. This relation of congruence modulo m is an important equivalence relation.

Equivalence Relations and Partitions

This subsection explores the relationship between equivalence relations and partitions on a non-empty set S. Recall first that a partition P of S is a collection $\{A_i\}$ of nonempty subsets of S with the following two properties:

(1) Each $a \in S$ belongs to some A_i.
(2) If $A_i \neq A_j$ then $A_i \cap A_j = \emptyset$.

In other words, a partition P of S is a subdivision of S into disjoint nonempty sets. (See Section 1.7.)

Suppose R is an equivalence relation on a set S. For each $a \in S$, let [a] denote the set of elements of S to which a is related under R; that is:

$$[a] = \{x \mid (a, x) \in R\}$$

We call [a] the *equivalence class* of a in S; any $b \in [a]$ is called a *representative* of the equivalence class.

The collection of all equivalence classes of elements of S under an equivalence relation R is denoted by S/R, that is,

$$S/R = \{[a] \mid a \in S\}$$

It is called the *quotient set* of S by R. The fundamental property of a quotient set is contained in the following theorem.

Theorem 2.6: Let R be an equivalence relation on a set S. Then S/R is a partition of S. Specifically:

(i)   For each a in S, we have $a \in [a]$.
(ii) $[a] = [b]$ if and only if $(a, b) \in R$.
(iii) If $[a] \neq [b]$, then [a] and [b] are disjoint.

Conversely, given a partition $\{A_i\}$ of the set S, there is an equivalence relation R on S such that the sets A_i are the equivalence classes.

This important theorem will be proved in Problem 2.17.

EXAMPLE 2.13

(a) Consider the relation $R = \{(1, 1), (1, 2), (2, 1), (2, 2), (3, 3)\}$ on $S = \{1, 2, 3\}$.

One can show that R is reflexive, symmetric, and transitive, that is, that R is an equivalence relation. Also:

$$[1] = \{1, 2\}, [2] = \{1, 2\}, [3] = \{3\}$$

Observe that $[1] = [2]$ and that $S/R = \{[1], [3]\}$ is a partition of S. One can choose either $\{1, 3\}$ or $\{2, 3\}$ as a set of representatives of the equivalence classes.

(b) Let R_5 be the relation of congruence modulo 5 on the set $\mathbf{Z}$ of integers denoted by

$$x \equiv y \pmod 5$$

This means that the difference $x - y$ is divisible by 5. Then R_5 is an equivalence relation on $\mathbf{Z}$. The quotient set $\mathbf{Z}/R_5$ contains the following five equivalence classes:

$$A_0 = \{\ldots, -10, -5, 0, 5, 10, \ldots\}$$
$$A_1 = \{\ldots, -9, -4, 1, 6, 11, \ldots\}$$
$$A_2 = \{\ldots, -8, -3, 2, 7, 12, \ldots\}$$
$$A_3 = \{\ldots, -7, -2, 3, 8, 13, \ldots\}$$
$$A_4 = \{\ldots, -6, -1, 4, 9, 14, \ldots\}$$

Any integer x, uniquely expressed in the form $x = 5q + r$ where $0 \le r < 5$, is a member of the equivalence class A_r, where r is the remainder. As expected, $\mathbf{Z}$ is the disjoint union of equivalence classes A_0, A_1, A_2, A_3, A_4. Usually one chooses $\{0, 1, 2, 3, 4\}$ or $\{-2, -1, 0, 1, 2\}$ as a set of representatives of the equivalence classes.

2.9 PARTIAL ORDERING RELATIONS

A relation R on a set S is called a *partial ordering* or a *partial order* of S if R is reflexive, antisymmetric, and transitive. A set S together with a partial ordering R is called a *partially ordered set* or *poset*. Partially ordered sets will be studied in more detail in Chapter 14, so here we simply give some examples.

EXAMPLE 2.14

(a) The relation $\subseteq$ of set inclusion is a partial ordering on any collection of sets since set inclusion has the three desired properties. That is,

   (1) $A \subseteq A$ for any set A.
   (2) If $A \subseteq B$ and $B \subseteq A$, then $A = B$.
   (3) If $A \subseteq B$ and $B \subseteq C$, then $A \subseteq C$.

(b) The relation $\le$ on the set $\mathbf{R}$ of real numbers is reflexive, antisymmetric, and transitive. Thus $\le$ is a partial ordering on $\mathbf{R}$.

(c) The relation "a divides b," written $a \mid b$, is a partial ordering on the set $\mathbf{N}$ of positive integers. However, "a divides b" is not a partial ordering on the set $\mathbf{Z}$ of integers since $a \mid b$ and $b \mid a$ need not imply $a = b$. For example, $3 \mid -3$ and $-3 \mid 3$ but $3 \ne -3$.

2.10 *n*-ARY RELATIONS

All the relations discussed above were binary relations. By an *n-ary relation*, we mean a set of ordered *n*-tuples. For any set S, a subset of the product set S^n is called an *n*-ary relation on S. In particular, a subset of S^3 is called a *ternary relation* on S.

EXAMPLE 2.15

(a) Let L be a line in the plane. Then "betweenness" is a ternary relation R on the points of L; that is, $(a, b, c) \in R$ if b lies between a and c on L.

(b) The equation $x^2 + y^2 + z^2 = 1$ determines a ternary relation T on the set $\mathbf{R}$ of real numbers. That is, a triple (x, y, z) belongs to T if (x, y, z) satisfies the equation, which means (x, y, z) is the coordinates of a point in $\mathbf{R}^3$ on the sphere S with radius 1 and center at the origin $O = (0, 0, 0)$.

Solved Problems

PRODUCT SETS

2.1. Given: $A = \{1, 2\}$, $B = \{x, y, z\}$, and $C = \{3, 4\}$. Find: $A \times B \times C$.

    $A \times B \times C$ consists of all ordered triplets (a, b, c) where $a \in A$, $b \in B$, $c \in C$. These elements of $A \times B \times C$ can be systematically obtained by a so-called tree diagram (Fig. 2-5). The elements of $A \times B \times C$ are precisely the 12 ordered triplets to the right of the tree diagram.

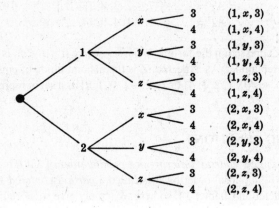

Fig. 2-5

    Observe that $n(A) = 2$, $n(B) = 3$, and $n(C) = 2$ and, as expected,

$$n(A \times B \times C) = 12 = n(A) \cdot n(B) \cdot n(C)$$

2.2. Find x and y given $(2x, x + y) = (6, 2)$.

    Two ordered pairs are equal if and only if the corresponding components are equal. Hence we obtain the equations

$$2x = 6 \quad \text{and} \quad x + y = 2$$

from which we derive the answers $x = 3$ and $y = -1$.

RELATIONS AND THEIR GRAPHS

2.3. Find the number of relations from $A = \{a, b, c\}$ to $B = \{1, 2\}$.

    There are $3(2) = 6$ elements in $A \times B$, and hence there are $m = 2^6 = 64$ subsets of $A \times B$. Thus there are $m = 64$ relations from A to B.

2.4. Given $A = \{1, 2, 3, 4\}$ and $B = \{x, y, z\}$. Let R be the following relation from A to B:

$$R = \{(1, y), (1, z), (3, y), (4, x), (4, z)\}$$

    (a) Determine the matrix of the relation.

    (b) Draw the arrow diagram of R.

    (c) Find the inverse relation R^{-1} of R.

    (d) Determine the domain and range of R.

    (a) See Fig. 2-6(a) Observe that the rows of the matrix are labeled by the elements of A and the columns by the elements of B. Also observe that the entry in the matrix corresponding to $a \in A$ and $b \in B$ is 1 if a is related to b and 0 otherwise.

    (b) See Fig. 2.6(b) Observe that there is an arrow from $a \in A$ to $b \in B$ iff a is related to b, i.e., iff $(a, b) \in R$.

$$\begin{array}{c} \quad x \quad y \quad z \\ \begin{array}{c} 1 \\ 2 \\ 3 \\ 4 \end{array} \left[\begin{array}{ccc} 0 & 1 & 1 \\ 0 & 0 & 0 \\ 0 & 1 & 0 \\ 1 & 0 & 1 \end{array} \right] \end{array}$$

(a)

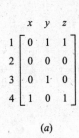 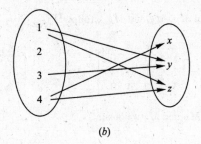

(b)

Fig. 2-6

(c) Reverse the ordered pairs of R to obtain R^{-1}:

$$R^{-1} = \{(y, 1), (z, 1), (y, 3), (x, 4), (z, 4)\}$$

Observe that by reversing the arrows in Fig. 2.6(b), we obtain the arrow diagram of R^{-1}.

(d) The domain of R, $\mathrm{Dom}(R)$, consists of the first elements of the ordered pairs of R, and the range of R, $\mathrm{Ran}(R)$, consists of the second elements. Thus,

$$\mathrm{Dom}(R) = \{1, 3, 4\} \quad \text{and} \quad \mathrm{Ran}(R) = \{x, y, z\}$$

2.5. Let $A = \{1, 2, 3\}$, $B = \{a, b, c\}$, and $C = \{x, y, z\}$. Consider the following relations R and S from A to B and from B to C, respectively.

$$R = \{(1, b), (2, a), (2, c)\} \quad \text{and} \quad S = \{(a, y), (b, x), (c, y), (c, z)\}$$

(a) Find the composition relation $R \circ S$.

(b) Find the matrices M_R, M_S, and $M_{R \circ S}$ of the respective relations R, S, and $R \circ S$, and compare $M_{R \circ S}$ to the product $M_R M_S$.

(a) Draw the arrow diagram of the relations R and S as in Fig. 2-7(a). Observe that 1 in A is "connected" to x in C by the path $1 \rightarrow b \rightarrow x$; hence $(1, x)$ belongs to $R \circ S$. Similarly, $(2, y)$ and $(2, z)$ belong to $R \circ S$. We have

$$R \circ S = \{(1, x), (2, y), (2, z)\}$$

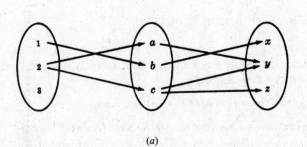 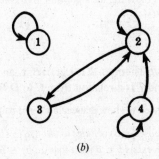

(a)   (b)

Fig. 2-7

(b) The matrices of M_R, M_S, and $M_{R \circ S}$ follow:

$$M_R = \begin{array}{c} \\ 1 \\ 2 \\ 3 \end{array} \begin{array}{ccc} a & b & c \\ \left[\begin{array}{ccc} 0 & 1 & 0 \\ 1 & 0 & 1 \\ 0 & 0 & 0 \end{array} \right] \end{array} \qquad M_S = \begin{array}{c} \\ a \\ b \\ c \end{array} \begin{array}{ccc} x & y & z \\ \left[\begin{array}{ccc} 0 & 1 & 0 \\ 1 & 0 & 0 \\ 0 & 1 & 1 \end{array} \right] \end{array} \qquad M_{R \circ S} = \begin{array}{c} \\ 1 \\ 2 \\ 3 \end{array} \begin{array}{ccc} x & y & z \\ \left[\begin{array}{ccc} 1 & 0 & 0 \\ 0 & 1 & 1 \\ 0 & 0 & 0 \end{array} \right] \end{array}$$

Multiplying M_R and M_S we obtain

$$M_R M_S = \left[\begin{array}{ccc} 1 & 0 & 0 \\ 0 & 2 & 1 \\ 0 & 0 & 0 \end{array} \right]$$

Observe that $M_{R \circ S}$ and $M_R M_S$ have the same zero entries.

2.6. Consider the relation $R = \{(1, 1), (2, 2), (2, 3), (3, 2), (4, 2), (4, 4)\}$ on $A = \{1, 2, 3, 4\}$.

(a) Draw its directed graph. (b) Find $R^2 = R \circ R$.

(a) For each $(a, b) \in R$, draw an arrow from a to b as in Fig. 2-7(b).

(b) For each pair $(a, b) \in R$, find all $(b, c) \in R$. Then $(a, c) \in R^2$. Thus

$$R^2 = \{(1, 1), (2, 2), (2, 3), (3, 2), (3, 3), (4, 2), (4, 3), (4, 4)\}$$

2.7. Let R and S be the following relations on $A = \{1, 2, 3\}$:

$$R = \{(1, 1), (1, 2), (2, 3), (3, 1), (3, 3)\}, \quad S = \{(1, 2), (1, 3), (2, 1), (3, 3)\}$$

Find (a) $R \cup S$, $R \cap S$, R^C; (b) $R \circ S$; (c) $S^2 = S \circ S$.

(a) Treat R and S simply as sets, and take the usual intersection and union. For R^C, use the fact that $A \times A$ is the universal relation on A.

$$R \cap S = \{(1, 2), (3, 3)\}$$
$$R \cup S = \{(1, 1), (1, 2), (1, 3), (2, 1), (2, 3), (3, 1), (3, 3)\}$$
$$R^C = \{(1, 3), (2, 1), (2, 2), (3, 2)\}$$

(b) For each pair $(a, b) \in R$, find all pairs $(b, c) \in S$. Then $(a, c) \in R \circ S$. For example, $(1, 1) \in R$ and $(1, 2)$, $(1, 3) \in S$; hence $(1, 2)$ and $(1, 3)$ belong to $R \circ S$. Thus,

$$R \circ S = \{(1, 2), (1, 3), (1, 1), (2, 3), (3, 2), (3, 3)\}$$

(c) Following the algorithm in (b), we get

$$S^2 = S \circ S = \{(1, 1), (1, 3), (2, 2), (2, 3), (3, 3)\}$$

2.8. Prove Theorem 2.1: Let A, B, C and D be sets. Suppose R is a relation from A to B, S is a relation from B to C and T is a relation from C to D. Then $(R \circ S) \circ T = R \circ (S \circ T)$.

We need to show that each ordered pair in $(R \circ S) \circ T$ belongs to $R \circ (S \circ T)$, and vice versa.

Suppose (a, d) belongs to $(R \circ S) \circ T$. Then there exists $c \in C$ such that $(a, c) \in R \circ S$ and $(c, d) \in T$. Since $(a, c) \in R \circ S$, there exists $b \in B$ such that $(a, b) \in R$ and $(b, c) \in S$. Since $(b, c) \in S$ and $(c, d) \in T$, we have $(b, d) \in S \circ T$; and since $(a, b) \in R$ and $(b, d) \in S \circ T$, we have $(a, d) \in R \circ (S \circ T)$. Therefore, $(R \circ S) \circ T \subseteq R \circ (S \circ T)$. Similarly $R \circ (S \circ T) \subseteq (R \circ S) \circ T$. Both inclusion relations prove $(R \circ S) \circ T = R \circ (S \circ T)$.

TYPES OF RELATIONS AND CLOSURE PROPERTIES

2.9. Consider the following five relations on the set $A = \{1, 2, 3\}$:

$$R = \{(1, 1), (1, 2), (1, 3), (3, 3)\}, \qquad \emptyset = \text{empty relation}$$
$$S = \{(1, 1)(1, 2), (2, 1)(2, 2), (3, 3)\}, \qquad A \times A = \text{universal relation}$$
$$T = \{(1, 1), (1, 2), (2, 2), (2, 3)\}$$

Determine whether or not each of the above relations on A is: (a) reflexive; (b) symmetric; (c) transitive; (d) antisymmetric.

(a) R is not reflexive since $2 \in A$ but $(2, 2) \notin R$. T is not reflexive since $(3, 3) \notin T$ and, similarly, $\emptyset$ is not reflexive. S and $A \times A$ are reflexive.

(b) R is not symmetric since $(1, 2) \in R$ but $(2, 1) \notin R$, and similarly T is not symmetric. S, $\emptyset$, and $A \times A$ are symmetric.

(c) T is not transitive since $(1, 2)$ and $(2, 3)$ belong to T, but $(1, 3)$ does not belong to T. The other four relations are transitive.

(d) S is not antisymmetric since $1 \neq 2$, and $(1, 2)$ and $(2, 1)$ both belong to S. Similarly, $A \times A$ is not antisymmetric. The other three relations are antisymmetric.

2.10. Give an example of a relation R on $A = \{1, 2, 3\}$ such that:

(a) R is both symmetric and antisymmetric.

(b) R is neither symmetric nor antisymmetric.

(c) R is transitive but $R \cup R^{-1}$ is not transitive.

There are several such examples. One possible set of examples follows:

(a) $R = \{(1, 1), (2, 2)\}$;   (b) $R = \{(1, 2), (2, 1), (2, 3)\}$;   (c) $R = \{(1, 2)\}$.

2.11. Suppose C is a collection of relations S on a set A, and let T be the intersection of the relations S in C, that is, $T = \cap(S \mid S \in C)$ Prove:

(a) If every S is symmetric, then T is symmetric.

(b) If every S is transitive, then T is transitive.

(a) Suppose $(a, b) \in T$. Then $(a, b) \in S$ for every S. Since each S is symmetric, $(b, a) \in S$ for every S. Hence $(b, a) \in T$ and T is symmetric.

(b) Suppose (a, b) and (b, c) belong to T. Then (a, b) and (b, c) belong to S for every S. Since each S is transitive, (a, c) belongs to S for every S. Hence, $(a, c) \in T$ and T is transitive.

2.12. Let R be a relation on a set A, and let P be a property of relations, such as symmetry and transitivity. Then P will be called R-closable if P satisfies the following two conditions:

(1) There is a P-relation S containing R.

(2) The intersection of P-relations is a P-relation.

(a) Show that symmetry and transitivity are R-closable for any relation R.

(b) Suppose P is R-closable. Then $P(R)$, the P-closure of R, is the intersection of all P-relations S containing R, that is,

$$P(R) = \cap(S \mid S \text{ is a } P\text{-relation and } R \subseteq S)$$

(a) The universal relation $A \times A$ is symmetric and transitive and $A \times A$ contains any relation R on A. Thus (1) is satisfied. By Problem 2.11, symmetry and transitivity satisfy (2). Thus symmetry and transitivity are R-closable for any relation R.

(b) Let $T = \cap(S \mid S$ is a P-relation and $R \subseteq S)$. Since P is R-closable, T is nonempty by (1) and T is a P-relation by (2). Since each relation S contains R, the intersection T contains R. Thus, T is a P-relation containing R. By definition, $P(R)$ is the smallest P-relation containing R; hence $P(R) \subseteq T$. On the other hand, $P(R)$ is one of the sets S defining T, that is, $P(R)$ is a P-relation and if $R \subseteq P(R)$. Therefore, $T \subseteq P(R)$. Accordingly, $P(R) = T$.

2.13. Consider the relation $R = \{(a, a), (a, b), (b, c), (c, c)\}$ on the set $A = \{a, b, c\}$. Find: (a) reflexive(R); (b) symmetric(R); (c) transitive(R).

(a) The reflexive closure on R is obtained by adding all diagonal pairs of $A \times A$ to R which are not currently in R. Hence,

$$\text{reflexive}(R) = R \cup \{(b, b)\} = \{(a, a), (a, b), (b, b), (b, c), (c, c)\}$$

(b) The symmetric closure on R is obtained by adding all the pairs in R^{-1} to R which are not currently in R. Hence,

$$\text{symmetric}(R) = R \cup \{(b, a), (c, b)\} = \{(a, a), (a, b), (b, a), (b, c), (c, b), (c, c)\}$$

(c) The transitive closure on R, since A has three elements, is obtained by taking the union of R with $R^2 = R \circ R$ and $R^3 = R \circ R \circ R$. Note that

$$R^2 = R \circ R = \{(a, a), (a, b), (a, c), (b, c), (c, c)\}$$
$$R^3 = R \circ R \circ R = \{(a, a), (a, b), (a, c), (b, c), (c, c)\}$$

Hence

$$\text{transitive}(R) = R \cup R^2 \cup R^3 = \{(a, a), (a, b), (a, c), (b, c), (c, c)\}$$

EQUIVALENCE RELATIONS AND PARTITIONS

2.14. Consider the **Z** of integers and an integer $m > 1$. We say that x is congruent to y modulo m, written

$$x \equiv y \pmod{m}$$

if $x - y$ is divisible by m. Show that this defines an equivalence relation on **Z**.

We must show that the relation is reflexive, symmetric, and transitive.

(i) For any x in Z we have $x \equiv x \pmod{m}$ because $x - x = 0$ is divisible by m. Hence the relation is reflexive.

(ii) Suppose $x \equiv y \pmod{m}$, so $x - y$ is divisible by m. Then $-(x - y) = y - x$ is also divisible by m, so $y \equiv x \pmod{m}$. Thus the relation is symmetric.

(iii) Now suppose $x \equiv y \pmod{m}$ and $y \equiv z \pmod{m}$, so $x - y$ and $y - z$ are each divisible by m. Then the sum

$$(x - y) + (y - z) = x - z$$

is also divisible by m; hence $x \equiv z \pmod{m}$. Thus the relation is transitive.

Accordingly, the relation of congruence modulo m on **Z** is an equivalence relation.

2.15. Let A be a set of nonzero integers and let $\approx$ be the relation on $A \times A$ defined by

$$(a, b) \approx (c, d) \quad \text{whenever} \quad ad = bc$$

Prove that $\approx$ is an equivalence relation.

We must show that $\approx$ is reflexive, symmetric, and transitive.

(i) *Reflexivity*: We have $(a, b) \approx (a, b)$ since $ab = ba$. Hence $\approx$ is reflexive.

(ii) *Symmetry*: Suppose $(a, b) \approx (c, d)$. Then $ad = bc$. Accordingly, $cb = da$ and hence $(c, d) \approx (a, b)$. Thus, $\approx$ is symmetric.

(iii) *Transitivity*: Suppose $(a, b) \approx (c, d)$ and $(c, d) \approx (e, f)$. Then $ad = bc$ and $cf = de$. Multiplying corresponding terms of the equations gives $(ad)(cf) = (bc)(de)$. Canceling $c \neq 0$ and $d \neq 0$ from both sides of the equation yields $af = be$, and hence $(a, b) \approx (e, f)$. Thus $\approx$ is transitive. Accordingly, $\approx$ is an equivalence relation.

2.16. Let R be the following equivalence relation on the set $A = \{1, 2, 3, 4, 5, 6\}$:

$$R = \{(1, 1), (1, 5), (2, 2), (2, 3), (2, 6), (3, 2), (3, 3), (3, 6), (4, 4), (5, 1), (5, 5), (6, 2), (6, 3), (6, 6)\}$$

Find the partition of A induced by R, i.e., find the equivalence classes of R.

Those elements related to 1 are 1 and 5 hence

$$[1] = \{1, 5\}$$

We pick an element which does not belong to $[1]$, say 2. Those elements related to 2 are 2, 3, and 6, hence

$$[2] = \{2, 3, 6\}$$

The only element which does not belong to $[1]$ or $[2]$ is 4. The only element related to 4 is 4. Thus

$$[4] = \{4\}$$

Accordingly, the following is the partition of A induced by R:

$$[\{1, 5\}, \{2, 3, 6\}, \{4\}]$$

2.17. Prove Theorem 2.6: Let R be an equivalence relation in a set A. Then the quotient set A/R is a partition of A. Specifically,

   (i) $a \in [a]$, for every $a \in A$.

   (ii) $[a] = [b]$ if and only if $(a, b) \in R$.

   (iii) If $[a] \neq [b]$, then $[a]$ and $[b]$ are disjoint.

   (a) *Proof of (i)*: Since R is reflexive, $(a, a) \in R$ for every $a \in A$ and therefore $a \in [a]$.

   (b) *Proof of (ii)*: Suppose $(a, b) \in R$. We want to show that $[a] = [b]$. Let $x \in [b]$; then $(b, x) \in R$. But by hypothesis $(a, a) \in R$ and so, by transitivity, $(a, x) \in R$. Accordingly $x \in [a]$. Thus $[b] \subseteq [a]$. To prove that $[a] \subseteq [b]$ we observe that $(a, b) \in R$ implies, by symmetry, that $(b, a) \in R$. Then, by a similar argument, we obtain $[a] \subseteq [b]$. Consequently, $[a] = [b]$.

   On the other hand, if $[a] = [b]$, then, by (i), $b \in [b] = [a]$; hence $(a, b) \in R$.

   (c) *Proof of (iii)*: We prove the equivalent contrapositive statement:

$$\text{If } [a] \cap [b] \neq \emptyset \quad \text{then} \quad [a] = [b]$$

   If $[a] \cap [b] \neq \emptyset$, then there exists an element $x \in A$ with $x \in [a] \cap [b]$. Hence $(a, x) \in R$ and $(b, x) \in R$. By symmetry, $(x, b) \in R$ and by transitivity, $(a, b) \in R$. Consequently by (ii), $[a] = [b]$.

PARTIAL ORDERINGS

2.18. Let ℓ be any collection of sets. Is the relation of set inclusion $\subseteq$ a partial order on ℓ?

   Yes, since set inclusion is reflexive, antisymmetric, and transitive. That is, for any sets A, B, C in ℓ we have: (i) $A \subseteq A$; (ii) if $A \subseteq B$ and $B \subseteq A$, then $A = B$; (iii) if $A \subseteq B$ and $B \subseteq C$, then $A \subseteq C$.

2.19. Consider the set $\mathbf{Z}$ of integers. Define aRb by $b = a^r$ for some positive integer r. Show that R is a partial order on $\mathbf{Z}$, that is, show that R is: (a) reflexive; (b) antisymmetric; (c) transitive.

   (a) R is reflexive since $a = a^1$.

   (b) Suppose aRb and bRa, say $b = a^r$ and $a = b^s$. Then $a = (a^r)^s = a^{rs}$. There are three possibilities: (i) $rs = 1$, (ii) $a = 1$, and (iii) $a = -1$. If $rs = 1$ then $r = 1$ and $s = 1$ and so $a = b$. If $a = 1$ then $b = 1^r = 1 = a$, and, similarly, if $b = 1$ then $a = 1$. Lastly, if $a = -1$ then $b = -1$ (since $b \neq 1$) and $a = b$. In all three cases, $a = b$. Thus R is antisymmetric.

   (c) Suppose aRb and bRc say $b = a^r$ and $c = b^s$. Then $c = (a^r)^s = a^{rs}$ and, therefore, aRc. Hence R is transitive.

   Accordingly, R is a partial order on $\mathbf{Z}$.

Supplementary Problems

RELATIONS

2.20. Let $S = \{a, b, c\}$, $T = \{b, c, d\}$, and $W = \{a, d\}$. Find $S \times T \times W$.

2.21. Find x and y where: (a) $(x + 2, 4) = (5, 2x + y)$;   (b) $(y - 2, 2x + 1) = (x - 1, y + 2)$.

2.22. Prove: (a) $A \times (B \cap C) = (A \times B) \cap (A \times C)$;   (b) $A \times (B \cup C) = (A \times B) \cup (A \times C)$.

2.23. Consider the relation $R = \{(1, 3), (1, 4), (3, 2), (3, 3), (3, 4)\}$ on $A = \{1, 2, 3, 4\}$.

    (a) Find the matrix M_R of R.   (d) Draw the directed graph of R.

    (b) Find the domain and range of R.   (e) Find the composition relation $R \circ R$.

    (c) Find R^{-1}.   (f) Find $R \circ R^{-1}$ and $R^{-1} \circ R$.

2.24. Let $A = \{1, 2, 3, 4\}$, $B = \{a, b, c\}$, $C = \{x, y, z\}$. Consider the relations R from A to B and S from B to C as follows:

$$R = \{(1, b), (3, a), (3, b), (4, c)\} \quad \text{and} \quad S = \{(a, y), (c, x), (a, z)\}$$

    (a) Draw the diagrams of R and S.

    (b) Find the matrix of each relation R, S (composition) $R \circ S$.

    (c) Write R^{-1} and the composition $R \circ S$ as sets of ordered pairs.

2.25. Let R and S be the following relations on $B = \{a, b, c, d\}$:

$$R = \{(a, a), (a, c), (c, b), (c, d), (d, b)\} \quad \text{and} \quad S = \{(b, a), (c, c), (c, d), (d, a)\}$$

    Find the following composition relations: (a) $R \circ S$; (b) $S \circ R$; (c) $R \circ R$; (d) $S \circ S$.

2.26. Let R be the relation on **N** defined by $x + 3y = 12$, i.e. $R = \{(x, y) \mid x + 3y = 12\}$.

    (a) Write R as a set of ordered pairs.   (c) Find R^{-1}.

    (b) Find the domain and range of R.   (d) Find the composition relation $R \circ R$.

PROPERTIES OF RELATIONS

2.27. Each of the following defines a relation on the positive integers **N**:

    (1) "x is greater than y."   (3) $x + y = 10$

    (2) "xy is the square of an integer."   (4) $x + 4y = 10$.

    Determine which of the relations are: (a) reflexive; (b) symmetric; (c) antisymmetric; (d) transitive.

2.28. Let R and S be relations on a set A. Assuming A has at least three elements, state whether each of the following statements is true or false. If it is false, give a counterexample on the set $A = \{1, 2, 3\}$:

    (a) If R and S are symmetric then $R \cap S$ is symmetric.

    (b) If R and S are symmetric then $R \cup S$ is symmetric.

    (c) If R and S are reflexive then $R \cap S$ is reflexive.

(d) If R and S are reflexive then $R \cup S$ is reflexive.

(e) If R and S are transitive then $R \cup S$ is transitive.

(f) If R and S are antisymmetric then $R \cup S$ is antisymmetric.

(g) If R is antisymmetric, then R^{-1} is antisymmetric.

(h) If R is reflexive then $R \cap R^{-1}$ is not empty.

(i) If R is symmetric then $R \cap R^{-1}$ is not empty.

2.29. Suppose R and S are relations on a set A, and R is antisymmetric. Prove that $R \cap S$ is antisymmetric.

EQUIVALENCE RELATIONS

2.30. Prove that if R is an equivalence relation on a set A, than R^{-1} is also an equivalence relation on A.

2.31. Let $S = \{1, 2, 3, \ldots, 18, 19\}$. Let R be the relation on S defined by "xy is a square," (a) Prove R is an equivalence relation. (b) Find the equivalence class [1]. (c) List all equivalence classes with more than one element.

2.32. Let $S = \{1, 2, 3, \ldots, 14, 15\}$. Let R be the equivalence relation on S defined by $x \equiv y \pmod 5$, that is, $x - y$ is divisible by 5. Find the partition of S induced by R, i.e. the quotient set S/R.

2.33. Let $S = \{1, 2, 3, \ldots, 9\}$, and let $\sim$ be the relation on $A \times A$ defined by

$$(a, b) \sim (c, d) \quad \text{whenever} \quad a + d = b + c.$$

(a) Prove that $\sim$ is an equivalence relation.

(b) Find $[(2, 5)]$, that is, the equivalence class of $(2, 5)$.

Answers to Supplementary Problems

2.20. $\{(a, b, a), (a, b, d), (a, c, a), (a, c, d),$
$(a, d, a), (a, d, d), (b, b, a), (b, b, d),$
$(b, c, a), (b, c, d), (b, d, a), (b, d, d),$
$(c, b, a), (c, b, d), (c, c, a), (c, c, d),$
$(c, d, a), (c, d, d)\}$

2.21. (a) $x = 3, y = -2$; (b) $x = 2, y = 3$.

2.23. (a) $M_R = [0, 0, 1, 1; 0, 0, 0, 0;$
$0, 1, 1, 1; 0, 0, 0, 0]$;

(b) Domain $= \{1, 3\}$, range $= \{2, 3, 4\}$;

(c) $R^{-1} = \{(3, 1), (4, 1), (2, 3), (3, 3),$
$(4, 3)\}$;

(d) See Fig. 2-8(a);

(e) $R \circ R = \{(1, 2), (1, 3), (1, 4), (3, 2),$
$(3, 3), (3, 4)\}$.

2.24. (a) See Fig. 2-8(b);

(b) $R = [0, 1, 0; 0, 0, 0; 1, 1, 0; 0, 0, 1]$,
$S = [0, 1, 1; 0, 0, 0; 1, 0, 0]$,
$R \circ S = [0, 0, 0; 0, 0, 0; 0, 1, 1; 1, 0, 0]$;

(c) $\{(b, 1), (a, 3), (b, 3), (c, 4)\}, \{(3, y),$
$(3, z), (4, x)\}$.

2.25. (a) $R \circ S = \{(a, c), (a, d), (c, a), (d, a)\}$

(b) $S \circ R = \{(b, a), (b, c), (c, b), (c, d),$
$(d, a), (d, c)\}$

(c) $R \circ R = \{(a, a), (a, b), (a, c), (a, d), (c, b)\}$

(d) $S \circ S = \{(c, c), (c, a), (c, d)\}$

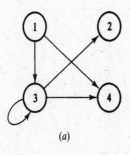

(a)

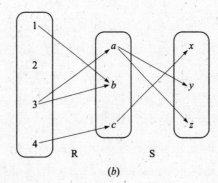

(b)

Fig. 2-8

2.26. (a)   {(9, 1), (6, 2), (3, 3)};   (b)   (i)   {9, 6, 3},
(ii) {1, 2, 3}, (iii) {(1, 9), (2, 6), (3, 3)}; (c) {(3, 3)}.

2.27. (a) None; (b) (2) and (3); (c) (1) and (4); (d) all
except (3).

2.28. All are true except: (e) $R = \{(1, 2)\}$, $S = \{(2, 3)\}$;
(f) $R = \{(1, 2)\}$, $S = \{(2, 1)\}$.

2.31. (b) {1, 4, 9, 16}; (c) {1, 4, 9, 16}, {2, 8, 18}, {3, 12}.

2.32. [{1, 6, 11}, {2, 7, 12}, {3, 8, 133}, {4, 9, 14},
{5, 10, 15}]

2.33. (b) {(1, 4), (2, 5), (3, 6), (4, 7), (5, 8), (6, 9)}.

CHAPTER 3

Functions and Algorithms

3.1 INTRODUCTION

One of the most important concepts in mathematics is that of a function. The terms "map," "mapping," "transformation," and many others mean the same thing; the choice of which word to use in a given situation is usually determined by tradition and the mathematical background of the person using the term.

Related to the notion of a function is that of an algorithm. The notation for presenting an algorithm and a discussion of its complexity is also covered in this chapter.

3.2 FUNCTIONS

Suppose that to each element of a set A we assign a unique element of a set B; the collection of such assignments is called *a function* from A into B. The set A is called the *domain* of the function, and the set B is called the *target set* or *codomain*.

Functions are ordinarily denoted by symbols. For example, let f denote a function from A into B. Then we write

$$f: A \to B$$

which is read: "f is a function from A into B," or "f takes (or maps) A into B." If $a \in A$, then $f(a)$ (read: "f of a") denotes the unique element of B which f assigns to a; it is called the *image* of a under f, or the *value* of f at a. The set of all image values is called the *range* or *image* of f. The image of $f: A \to B$ is denoted by $\text{Ran}(f)$, $\text{Im}(f)$ or $f(A)$.

Frequently, a function can be expressed by means of a mathematical formula. For example, consider the function which sends each real number into its square. We may describe this function by writing

$$f(x) = x^2 \quad \text{or} \quad x \mapsto x^2 \quad \text{or} \quad y = x^2$$

In the first notation, x is called a *variable* and the letter f denotes the function. In the second notation, the barred arrow $\mapsto$ is read "goes into." In the last notation, x is called the *independent variable* and y is called the *dependent variable* since the value of y will depend on the value of x.

Remark: Whenever a function is given by a formula in terms of a variable x, we assume, unless it is otherwise stated, that the domain of the function is **R** (or the largest subset of **R** for which the formula has meaning) and the codomain is **R**.

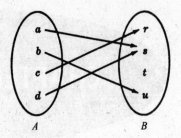

Fig. 3-1

EXAMPLE 3.1

(a) Consider the function $f(x) = x^3$, i.e., f assigns to each real number its cube. Then the image of 2 is 8, and so we may write $f(2) = 8$.

(b) Figure 3-1 defines a function f from $A = \{a, b, c, d\}$ into $B = \{r, s, t, u\}$ in the obvious way. Here

$$f(a) = s, \quad f(b) = u, \quad f(c) = r, \quad f(d) = s$$

The image of f is the set of image values, $\{r, s, u\}$. Note that t does not belong to the image of f because t is not the image of any element under f.

(c) Let A be any set. The function from A into A which assigns to each element in A the element itself is called the *identity function* on A and it is usually denoted by 1_A, or simply 1. In other words, for every $a \in A$,

$$1_A(a) = a.$$

(d) Suppose S is a subset of A, that is, suppose $S \subseteq A$. The *inclusion map* or *embedding* of S into A, denoted by $i: S \hookrightarrow A$ is the function such that, for every $x \in S$,

$$i(x) = x$$

The *restriction* of any function $f: A \to B$, denoted by $f|_S$ is the function from S into B such that, for any $x \in S$,

$$f|_S(x) = f(x)$$

Functions as Relations

There is another point of view from which functions may be considered. First of all, every function $f: A \to B$ gives rise to a relation from A to B called the *graph of f* and defined by

$$\text{Graph of } f = \{(a, b) \mid a \in A, b = f(a)\}$$

Two functions $f: A \to B$ and $g: A \to B$ are defined to be *equal*, written $f = g$, if $f(a) = g(a)$ for every $a \in A$; that is, if they have the same graph. Accordingly, we do not distinguish between a function and its graph. Now, such a graph relation has the property that each a in A belongs to a unique ordered pair (a, b) in the relation. On the other hand, any relation f from A to B that has this property gives rise to a function $f: A \to B$, where $f(a) = b$ for each (a, b) in f. Consequently, one may equivalently define a function as follows:

Definition: A function $f: A \to B$ is a relation from A to B (i.e., a subset of $A \times B$) such that each $a \in A$ belongs to a unique ordered pair (a, b) in f.

Although we do not distinguish between a function and its graph, we will still use the terminology "graph of f" when referring to f as a set of ordered pairs. Moreover, since the graph of f is a relation, we can draw its picture as was done for relations in general, and this pictorial representation is itself sometimes called the graph of f. Also, the defining condition of a function, that each $a \in A$ belongs to a unique pair (a, b) in f, is equivalent to the geometrical condition of each vertical line intersecting the graph in exactly one point.

EXAMPLE 3.2

(a) Let $f: A \rightarrow B$ be the function defined in Example 3.1 (b). Then the graph of f is as follows:

$$\{(a, s), (b, u), (c, r), (d, s)\}$$

(b) Consider the following three relations on the set $A = \{1, 2, 3\}$:

$$f = \{(1, 3), (2, 3), (3, 1)\}, \quad g = \{(1, 2), (3, 1)\}, \quad h = \{(1, 3), (2, 1), (1, 2), (3, 1)\}$$

f is a function from A into A since each member of A appears as the first coordinate in exactly one ordered pair in f; here $f(1) = 3$, $f(2) = 3$, and $f(3) = 1$. g is not a function from A into A since $2 \in A$ is not the first coordinate of any pair in g and so g does not assign any image to 2. Also h is not a function from A into A since $1 \in A$ appears as the first coordinate of two distinct ordered pairs in h, $(1, 3)$ and $(1, 2)$. If h is to be a function it cannot assign both 3 and 2 to the element $1 \in A$.

(c) By a *real polynomial function*, we mean a function $f: \mathbf{R} \rightarrow \mathbf{R}$ of the form

$$f(x) = a_n x^n + a_{n-1} x^{n-1} + \cdots + a_1 x + a_0$$

where the a_i are real numbers. Since $\mathbf{R}$ is an infinite set, it would be impossible to plot each point of the graph. However, the graph of such a function can be approximated by first plotting some of its points and then drawing a smooth curve through these points. The points are usually obtained from a table where various values are assigned to x and the corresponding values of $f(x)$ are computed. Figure 3-2 illustrates this technique using the function $f(x) = x^2 - 2x - 3$.

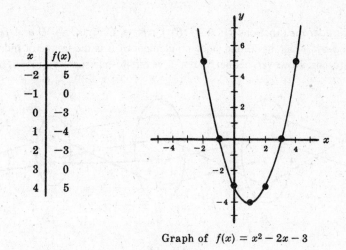

x	$f(x)$
-2	5
-1	0
0	-3
1	-4
2	-3
3	0
4	5

Graph of $f(x) = x^2 - 2x - 3$

Fig. 3-2

Composition Function

Consider functions $f: A \rightarrow B$ and $g: B \rightarrow C$; that is, where the codomain of f is the domain of g. Then we may define a new function from A to C, called the *composition* of f and g and written $g \circ f$, as follows:

$$(g \circ f)(a) \equiv g(f(a))$$

That is, we find the image of a under f and then find the image of $f(a)$ under g. This definition is not really new. If we view f and g as relations, then this function is the same as the composition of f and g as relations (see Section 2.5) except that here we use the functional notation $g \circ f$ for the composition of f and g instead of the notation $f \circ g$ which was used for relations.

Consider any function $f: A \rightarrow B$. Then

$$f \circ 1_A = f \quad \text{and} \quad 1_B \circ f = f$$

where 1_A and 1_B are the identity functions on A and B, respectively.

3.3 ONE-TO-ONE, ONTO, AND INVERTIBLE FUNCTIONS

A function $f: A \rightarrow B$ is said to be *one-to-one* (written 1-1) if different elements in the domain A have distinct images. Another way of saying the same thing is that f is *one-to-one* if $f(a) = f(a')$ implies $a = a'$.

A function $f: A \rightarrow B$ is said to be an *onto* function if each element of B is the image of some element of A. In other words, $f: A \rightarrow B$ is onto if the image of f is the entire codomain, i.e., if $f(A) = B$. In such a case we say that f is a function from A onto B or that f maps A onto B.

A function $f: A \rightarrow B$ is *invertible* if its inverse relation f^{-1} is a function from B to A. In general, the inverse relation f^{-1} may not be a function. The following theorem gives simple criteria which tells us when it is.

Theorem 3.1: A function $f: A \rightarrow B$ is invertible if and only if f is both one-to-one and onto.

If $f: A \rightarrow B$ is one-to-one and onto, then f is called a *one-to-one correspondence* between A and B. This terminology comes from the fact that each element of A will then correspond to a unique element of B and vice versa.

Some texts use the terms *injective* for a one-to-one function, *surjective* for an onto function, and *bijective* for a one-to-one correspondence.

EXAMPLE 3.3 Consider the functions $f_1: A \rightarrow B$, $f_2: B \rightarrow C$, $f_3: C \rightarrow D$ and $f_4: D \rightarrow E$ defined by the diagram of Fig. 3-3. Now f_1 is one-to-one since no element of B is the image of more than one element of A. Similarly, f_2 is one-to-one. However, neither f_3 nor f_4 is one-to-one since $f_3(r) = f_3(u)$ and $f_4(v) = f_4(w)$

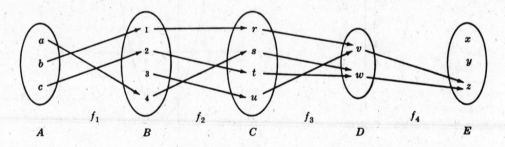

Fig. 3-3

As far as being onto is concerned, f_2 and f_3 are both onto functions since every element of C is the image under f_2 of some element of B and every element of D is the image under f_3 of some element of C, $f_2(B) = C$ and $f_3(C) = D$. On the other hand, f_1 is not onto since $3 \in B$ is not the image under f_1 of any element of A. and f_4 is not onto since $x \in E$ is not the image under f_4 of any element of D.

Thus f_1 is one-to-one but not onto, f_3 is onto but not one-to-one and f_4 is neither one-to-one nor onto. However, f_2 is both one-to-one and onto, i.e., is a one-to-one correspondence between B and C. Hence f_2 is invertible and f_2^{-1} is a function from C to B.

Geometrical Characterization of One-to-One and Onto Functions

Consider now functions of the form $f: \mathbf{R} \rightarrow \mathbf{R}$. Since the graphs of such functions may be plotted in the Cartesian plane $\mathbf{R}^2$ and since functions may be identified with their graphs, we might wonder whether the concepts of being one-to-one and onto have some geometrical meaning. The answer is yes. Specifically:

(1) $f:\mathbf{R} \to \mathbf{R}$ is one-to-one if each horizontal line intersects the graph of f in at most one point.

(2) $f:\mathbf{R} \to \mathbf{R}$ is an onto function if each horizontal line intersects the graph of f at one or more points.

Accordingly, if f is both one-to-one and onto, i.e. invertible, then each horizontal line will intersect the graph of f at exactly one point.

EXAMPLE 3.4 Consider the following four functions from $\mathbf{R}$ into $\mathbf{R}$:

$$f_1(x) = x^2, \quad f_2(x) = 2^x, \quad f_3(x) = x^3 - 2x^2 - 5x + 6, \quad f_4(x) = x^3$$

The graphs of these functions appear in Fig. 3-4. Observe that there are horizontal lines which intersect the graph of f_1 twice and there are horizontal lines which do not intersect the graph of f_1 at all; hence f_1 is neither one-to-one nor onto. Similarly, f_2 is one-to-one but not onto, f_3 is onto but not one-to-one and f_4 is both one-to-one and onto. The inverse of f_4 is the cube root function, i.e., $f_4^{-1}(x) = \sqrt[3]{x}$.

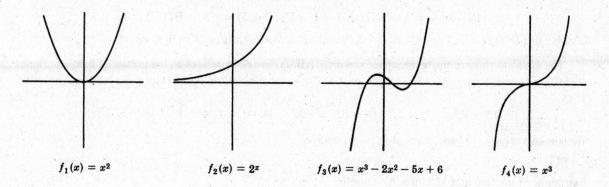

$$f_1(x) = x^2 \qquad f_2(x) = 2^x \qquad f_3(x) = x^3 - 2x^2 - 5x + 6 \qquad f_4(x) = x^3$$

Fig. 3-4

Permutations

An invertible (bijective) function $\sigma: X \to X$ is called a *permutation* on X. The composition and inverses of permutations on X and the identity function on X are also permutations on X.

Suppose $X = \{1, 2, \ldots, n\}$. Then a permutation σ on X is frequently denoted by

$$\sigma = \begin{pmatrix} 1 & 2 & 3 & \cdots & n \\ j_1 & j_2 & j_3 & \cdots & j_n \end{pmatrix}$$

where $j_i = \sigma(i)$. The set of all such permutations is denoted by S_n, and there are $n! = n(n-1)\cdots 3 \cdot 2 \cdot 1$ of them. For example,

$$\sigma = \begin{pmatrix} 1 & 2 & 3 & 4 & 5 & 6 \\ 4 & 6 & 2 & 5 & 1 & 3 \end{pmatrix} \quad \text{and} \quad \tau = \begin{pmatrix} 1 & 2 & 3 & 4 & 5 & 6 \\ 6 & 4 & 3 & 1 & 2 & 5 \end{pmatrix}$$

are permutations in S_6, and there are $6! = 720$ of them. Sometimes, we only write the second line of the permutation, that is, we denote the above permutations by writing $\sigma = 462513$ and $\tau = 643125$.

3.4 MATHEMATICAL FUNCTIONS, EXPONENTIAL AND LOGARITHMIC FUNCTIONS

This section presents various mathematical functions which appear often in the analysis of algorithms, and in computer science in general, together with their notation. We also discuss the exponential and logarithmic functions, and their relationship.

Floor and Ceiling Functions

Let x be any real number. Then x lies between two integers called the floor and the ceiling of x. Specifically,

$\lfloor x \rfloor$, called the *floor* of x, denotes the greatest integer that does not exceed x.

$\lceil x \rceil$, called the *ceiling* of x, denotes the least integer that is not less than x.

If x is itself an integer, then $\lfloor x \rfloor = \lceil x \rceil$; otherwise $\lfloor x \rfloor + 1 = \lceil x \rceil$. For example,

$$\lfloor 3.14 \rfloor = 3, \quad \left\lfloor \sqrt{5} \right\rfloor = 2, \quad \lfloor -8.5 \rfloor = -9, \quad \lfloor 7 \rfloor = 7, \quad \lfloor -4 \rfloor = -4,$$

$$\lceil 3.14 \rceil = 4, \quad \left\lceil \sqrt{5} \right\rceil = 3, \quad \lceil -8.5 \rceil = -8, \quad \lceil 7 \rceil = 7, \quad \lceil -4 \rceil = -4$$

Integer and Absolute Value Functions

Let x be any real number. The *integer value* of x, written $\text{INT}(x)$, converts x into an integer by deleting (truncating) the fractional part of the number. Thus

$$\text{INT}(3.14) = 3, \quad \text{INT}(\sqrt{5}) = 2, \quad \text{INT}(-8.5) = -8, \quad \text{INT}(7) = 7$$

Observe that $\text{INT}(x) = \lfloor x \rfloor$ or $\text{INT}(x) = \lceil x \rceil$ according to whether x is positive or negative.

The *absolute value* of the real number x, written $\text{ABS}(x)$ or $|x|$, is defined as the greater of x or $-x$. Hence $\text{ABS}(0) = 0$, and, for $x \neq 0$, $\text{ABS}(x) = x$ or $\text{ABS}(x) = -x$, depending on whether x is positive or negative. Thus

$$|-15| = 15, \quad |7| = 7, \quad |-3.33| = 3.33, \quad |4.44| = 4.44, \quad |-0.075| = 0.075$$

We note that $|x| = |-x|$ and, for $x \neq 0$, $|x|$ is positive.

Remainder Function and Modular Arithmetic

Let k be any integer and let M be a positive integer. Then

$$k \ (\text{mod } M)$$

(read: k modulo M) will denote the integer remainder when k is divided by M. More exactly, $k \ (\text{mod } M)$ is the unique integer r such that

$$k = Mq + r \quad \text{where} \quad 0 \leq r < M$$

When k is positive, simply divide k by M to obtain the remainder r. Thus

$$25 \ (\text{mod } 7) = 4, \quad 25 \ (\text{mod } 5) = 0, \quad 35 \ (\text{mod } 11) = 2, \quad 3 \ (\text{mod } 8) = 3$$

If k is negative, divide $|k|$ by M to obtain a remainder r'; then $k \ (\text{mod } M) = M - r'$ when $r' \neq 0$. Thus

$$-26 \ (\text{mod } 7) = 7 - 5 = 2, \quad -371 \ (\text{mod } 8) = 8 - 3 = 5, \quad -39 \ (\text{mod } 3) = 0$$

The term "mod" is also used for the mathematical congruence relation, which is denoted and defined as follows:

$$a \equiv b \ (\text{mod } M) \quad \text{if any only if} \quad M \text{ divides } b - a$$

M is called the *modulus*, and $a \equiv b \ (\text{mod } M)$ is read "a is congruent to b modulo M". The following aspects of the congruence relation are frequently useful:

$$0 \equiv M \ (\text{mod } M) \quad \text{and} \quad a \pm M \equiv a \ (\text{mod } M)$$

Arithmetic modulo M refers to the arithmetic operations of addition, multiplication, and subtraction where the arithmetic value is replaced by its equivalent value in the set

$$\{0, 1, 2, \ldots, M - 1\} \quad \text{or in the set} \quad \{1, 2, 3, \ldots, M\}$$

For example, in arithmetic modulo 12, sometimes called "clock" arithmetic,

$$6 + 9 \equiv 3, \quad 7 \times 5 \equiv 11, \quad 1 - 5 \equiv 8, \quad 2 + 10 \equiv 0 \equiv 12$$

(The use of 0 or M depends on the application.)

Exponential Functions

Recall the following definitions for integer exponents (where m is a positive integer):

$$a^m = a \cdot a \cdots a (m \text{ times}), \quad a^0 = 1, \quad a^{-m} = \frac{1}{a^m}$$

Exponents are extended to include all rational numbers by defining, for any rational number m/n,

$$a^{m/n} = \sqrt[n]{a^m} = (\sqrt[n]{a})^m$$

For example,

$$2^4 = 16, \quad 2^{-4} = \frac{1}{2^4} = \frac{1}{16}, \quad 125^{2/3} = 5^2 = 25$$

In fact, exponents are extended to include all real numbers by defining, for any real number x,

$$a^x = \lim_{r \to x} a^r, \quad \text{where } r \text{ is a rational number}$$

Accordingly, the exponential function $f(x) = a^x$ is defined for all real numbers.

Logarithmic Functions

Logarithms are related to exponents as follows. Let b be a positive number. The logarithm of any positive number x to the base b, written

$$\log_b x$$

represents the exponent to which b must be raised to obtain x. That is,

$$y = \log_b x \quad \text{and} \quad b^y = x$$

are equivalent statements. Accordingly,

$$\log_2 8 = 3 \quad \text{since} \quad 2^3 = 8; \quad \log_{10} 100 = 2 \quad \text{since} \quad 10^2 = 100$$
$$\log_2 64 = 6 \quad \text{since} \quad 2^6 = 64; \quad \log_{10} 0.001 = -3 \quad \text{since} \quad 10^{-3} = 0.001$$

Furthermore, for any base b, we have $b^0 = 1$ and $b^1 = b$; hence

$$\log_b 1 = 0 \quad \text{and} \quad \log_b b = 1$$

The logarithm of a negative number and the logarithm of 0 are not defined.

Frequently, logarithms are expressed using approximate values. For example, using tables or calculators, one obtains

$$\log_{10} 300 = 2.4771 \quad \text{and} \quad \log_e 40 = 3.6889$$

as approximate answers. (Here $e = 2.718281\ldots$)

Three classes of logarithms are of special importance: logarithms to base 10, called *common logarithms*; logarithms to base e, called *natural logarithms;* and logarithms to base 2, called *binary logarithms*. Some texts write

$$\ln x \text{ for } \log_e x \quad \text{and} \quad \lg x \text{ or } \log x \text{ for } \log_2 x$$

The term $\log x$, by itself, usually means $\log_{10} x$; but it is also used for $\log_e x$ in advanced mathematical texts and for $\log_2 x$ in computer science texts.

Frequently, we will require only the floor or the ceiling of a binary logarithm. This can be obtained by looking at the powers of 2. For example,

$$\lfloor \log_2 100 \rfloor = 6 \quad \text{since} \quad 2^6 = 64 \quad \text{and} \quad 2^7 = 128$$
$$\lceil \log_2 1000 \rceil = 9 \quad \text{since} \quad 2^9 = 512 \quad \text{and} \quad 2^{10} = 1024$$

and so on.

Relationship between the Exponential and Logarithmic Functions

The basic relationship between the exponential and the logarithmic functions

$$f(x) = b^x \quad \text{and} \quad g(x) = \log_b x$$

is that they are inverses of each other; hence the graphs of these functions are related geometrically. This relationship is illustrated in Fig. 3-5 where the graphs of the exponential function $f(x) = 2^x$, the logarithmic function $g(x) = \log_2 x$, and the linear function $h(x) = x$ appear on the same coordinate axis. Since $f(x) = 2^x$ and $g(x) = \log_2 x$ are inverse functions, they are symmetric with respect to the linear function $h(x) = x$ or, in other words, the line $y = x$.

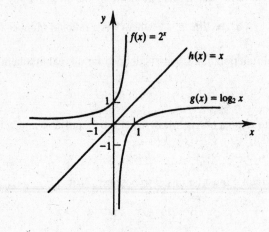

Fig. 3-5

Figure 3-5 also indicates another important property of the exponential and logarithmic functions. Specifically, for any positive c, we have

$$g(c) < h(c) < f(c), \quad \text{that is,} \quad g(c) < c < f(c)$$

In fact, as c increases in value, the vertical distances $h(c) - g(c)$ and $f(c) - g(c)$ increase in value. Moreover, the logarithmic function $g(x)$ grows very slowly compared with the linear function $h(x)$, and the exponential function $f(x)$ grows very quickly compared with $h(x)$.

3.5 SEQUENCES, INDEXED CLASSES OF SETS

Sequences and indexed classes of sets are special types of functions with their own notation. We discuss these objects in this section. We also discuss the summation notation here.

Sequences

A *sequence* is a function from the set $\mathbf{N} = \{1, 2, 3, \ldots\}$ of positive integers into a set A. The notation a_n is used to denote the image of the integer n. Thus a sequence is usually denoted by

$$a_1, a_2, a_3, \ldots \quad \text{or} \quad \{a_n : n \in \mathbf{N}\} \quad \text{or} \quad \text{simply} \quad \{a_n\}$$

Sometimes the domain of a sequence is the set $\{0, 1, 2, \ldots\}$ of nonnegative integers rather than $\mathbf{N}$. In such a case we say n begins with 0 rather than 1.

A *finite sequence* over a set A is a function from $\{1, 2, \ldots, m\}$ into A, and it is usually denoted by

$$a_1, a_2, \ldots, a_m$$

Such a finite sequence is sometimes called a *list* or an *m-tuple*.

EXAMPLE 3.5

(a) The following are two familiar sequences:

    (i) $1, \frac{1}{2}, \frac{1}{3}, \frac{1}{4}, \ldots$ which may be defined by $a_n = \frac{1}{n}$;

    (ii) $1, \frac{1}{2}, \frac{1}{4}, \frac{1}{8}, \ldots$ which may be defined by $b_n = 2^{-n}$

    Note that the first sequence begins with $n = 1$ and the second sequence begins with $n = 0$.

(b) The important sequence $1, -1, 1, -1, \ldots$ may be formally defined by

$$a_n = (-1)^{n+1} \quad \text{or, equivalently, by} \quad b_n = (-1)^n$$

    where the first sequence begins with $n = 1$ and the second sequence begins with $n = 0$.

(c) *Strings*   Suppose a set A is finite and A is viewed as a character set or an alphabet. Then a finite sequence over A is called a *string* or *word*, and it is usually written in the form $a_1 a_2 \ldots a_m$, that is, without parentheses. The number m of characters in the string is called its *length*. One also views the set with zero characters as a string; it is called the *empty string* or *null string*. Strings over an alphabet A and certain operations on these strings will be discussed in detail in Chapter 13.

Summation Symbol, Sums

Here we introduce the summation symbol $\sum$ (the Greek letter sigma). Consider a sequence $a_1, a_2, a_3, \ldots$. Then we define the following:

$$\sum_{J=1}^{n} a_j = a_1 + a_2 + \cdots + a_n \quad \text{and} \quad \sum_{j=m}^{n} a_j = a_m + a_{m+1} + \cdots + a_n$$

The letter j in the above expressions is called a *dummy index* or *dummy variable*. Other letters frequently used as dummy variables are i, k, s, and t.

EXAMPLE 3.6

$$\sum_{i=1}^{n} a_i b_i = a_1 b_1 + a_2 b_2 + \cdots + a_n b_n$$

$$\sum_{j=2}^{5} j^2 = 2^2 + 3^2 + 4^2 + 5^2 = 4 + 9 + 16 + 25 = 54$$

$$\sum_{j=1}^{n} j = 1 + 2 + \cdots + n$$

The last sum appears very often. It has the value $n(n+1)/2$. That is:

$$1 + 2 + 3 + \cdots + n = \frac{n(n+1)}{2}, \quad \text{for example,} \quad 1 + 2 + \cdots + 50 = \frac{50(51)}{2} = 1275$$

Indexed Classes of Sets

Let I be any nonempty set, and let S be a collection of sets. An *indexing function* from I to S is a function $f: I \to S$. For any $i \in I$, we denote the image $f(i)$ by A_i. Thus the indexing function f is usually denoted by

$$\{A_i \mid i \in I\} \quad \text{or} \quad \{A_i\}_{i \in I} \quad \text{or simply} \quad \{A_i\}$$

The set I is called the *indexing set*, and the elements of I are called *indices*. If f is one-to-one and onto, we say that S is indexed by I.

The concepts of union and intersection are defined for indexed classes of sets as follows:

$$\cup_{i \in I} A_i = \{x \mid x \in A_i \text{ for some } i \in I\} \quad \text{and} \quad \cap_{i \in I} A_i = \{x \mid x \in A_i \text{ for all } i \in I\}$$

In the case that I is a finite set, this is just the same as our previous definition of union and intersection. If I is $\mathbf{N}$, we may denote the union and intersection, respectively, as follows:

$$A_1 \cup A_2 \cup A_3 \cup \dots \quad \text{and} \quad A_1 \cap A_2 \cap A_3 \cap \dots$$

EXAMPLE 3.7 Let I be the set $\mathbf{Z}$ of integers. To each $n \in \mathbf{Z}$, we assign the following infinite interval in $\mathbf{R}$:

$$A_n = \{x \mid x \le n\} = (-\infty, n]$$

For any real number a, there exists integers n_1 and n_2 such that $n_1 < a < n_2$; so $a \in A_{n_2}$ but $a \notin A_{n_1}$. Hence

$$a \in \cup_n A_n \quad \text{but} \quad a \notin \cap_n A_n$$

Accordingly,

$$\cup_n A_n = \mathbf{R} \quad \text{but} \quad \cap_n A_n = \emptyset$$

3.6 RECURSIVELY DEFINED FUNCTIONS

A function is said to be *recursively defined* if the function definition refers to itself. In order for the definition not to be circular, the function definition must have the following two properties:

(1) There must be certain arguments, called *base values*, for which the function does not refer to itself.
(2) Each time the function does refer to itself, the argument of the function must be closer to a base value.

A recursive function with these two properties is said to be *well-defined*.

The following examples should help clarify these ideas.

Factorial Function

The product of the positive integers from 1 to n, inclusive, is called "n factorial" and is usually denoted by $n!$. That is,

$$n! = n(n-1)(n-2) \cdots 3 \cdot 2 \cdot 1$$

It is also convenient to define $0! = 1$, so that the function is defined for all nonnegative integers. Thus:

$$0! = 1, \quad 1! = 1, \quad 2! = 2 \cdot 1 = 2, \quad 3! = 3 \cdot 2 \cdot 1 = 6, \quad 4! = 4 \cdot 3 \cdot 2 \cdot 1 = 24$$

$$5! = 5 \cdot 4 \cdot 3 \cdot 2 \cdot 1 = 120, \quad 6! = 6 \cdot 5 \cdot 4 \cdot 3 \cdot 2 \cdot 1 = 720$$

And so on. Observe that

$$5! = 5 \cdot 4! = 5 \cdot 24 = 120 \quad \text{and} \quad 6! = 6 \cdot 5! = 6 \cdot 120 = 720$$

This is true for every positive integer n; that is,

$$n! = n \cdot (n - 1)!$$

Accordingly, the factorial function may also be defined as follows:

Definition 3.1 (Factorial Function):
        (a) If $n = 0$, then $n! = 1$.
        (b) If $n > 0$, then $n! = n \cdot (n - 1)!$

Observe that the above definition of $n!$ is recursive, since it refers to itself when it uses $(n - 1)!$. However:

(1) The value of $n!$ is explicitly given when $n = 0$ (thus 0 is a base value).
(2) The value of $n!$ for arbitrary n is defined in terms of a smaller value of n which is closer to the base value 0.

Accordingly, the definition is not circular, or, in other words, the function is well-defined.

EXAMPLE 3.8 Figure 3-6 shows the nine steps to calculate 4! using the recursive definition. Specifically:

Step 1. This defines 4! in terms of 3!, so we must postpone evaluating 4! until we evaluate 3. This postponement is indicated by indenting the next step.

Step 2. Here 3! is defined in terms of 2!, so we must postpone evaluating 3! until we evaluate 2!.

Step 3. This defines 2! in terms of 1!.

Step 4. This defines 1! in terms of 0!.

Step 5. This step can explicitly evaluate 0!, since 0 is the base value of the recursive definition.

Steps 6 to 9. We backtrack, using 0! to find 1!, using 1! to find 2!, using 2! to find 3!, and finally using 3! to find 4!. This backtracking is indicated by the "reverse" indention.

Observe that we backtrack in the reverse order of the original postponed evaluations.

```
(1)    4! = 4 · 3!
(2)          3! = 3 · 2!
(3)               2! = 2 · 1!
(4)                    1! = 1 · 0!
(5)                         0! = 1
(6)                    1! = 1 · 1 = 1
(7)               2! = 2 · 1 = 2
(8)          3! = 3 · 2 = 6
(9)    4! = 4 · 6 = 24
```

Fig. 3-6

Level Numbers

Let P be a procedure or recursive formula which is used to evaluate $f(X)$ where f is a recursive function and X is the input. We associate a *level number* with each execution of P as follows. The original execution of P is assigned level 1; and each time P is executed because of a recursive call, its level is one more than the level of the execution that made the recursive call. The *depth* of recursion in evaluating $f(X)$ refers to the maximum level number of P during its execution.

Consider, for example, the evaluation of 4! Example 3.8, which uses the recursive formula $n! = n(n-1)!$. Step 1 belongs to level 1 since it is the first execution of the formula. Thus:

$$\text{Step 2 belongs to level 2;} \quad \text{Step 3 to level 3, \ldots;} \quad \text{Step 5 to level 5.}$$

On the other hand, Step 6 belongs to level 4 since it is the result of a return from level 5. In other words, Step 6 and Step 4 belong to the same level of execution. Similarly,

$$\text{Step 7 belongs to level 3;} \quad \text{Step 8 to level 2;} \quad \text{and Step 9 to level 1.}$$

Accordingly, in evaluating 4!, the depth of the recursion is 5.

Fibonacci Sequence

The celebrated Fibonacci sequence (usually denoted by $F_0, F_1, F_2, \ldots$) is as follows:

$$0, \quad 1, \quad 1, \quad 2, \quad 3, \quad 5, \quad 8, \quad 13, \quad 21, \quad 34, \quad 55, \quad \ldots$$

That is, $F_0 = 0$ and $F_1 = 1$ and each succeeding term is the sum of the two preceding terms. For example, the next two terms of the sequence are

$$34 + 55 = 89 \quad \text{and} \quad 55 + 89 = 144$$

A formal definition of this function follows:

Definition 3.2 (Fibonacci Sequence):

    (a)   If $n = 0$, or $n = 1$, then $F_n = n$.
    (b)   If $n > 1$, then $F_n = F_{n-2} + F_{n-1}$.

This is another example of a recursive definition, since the definition refers to itself when it uses F_{n-2} and F_{n-1} However:

(1) The base values are 0 and 1.
(2) The value of F_n is defined in terms of smaller values of n which are closer to the base values.

Accordingly, this function is well-defined.

Ackermann Function

The Ackermann function is a function with two arguments, each of which can be assigned any nonnegative interger, that is, $0, 1, 2, \ldots$. This function is defined as:

Definition 3.3 (Ackermann function):

    (a)   If $m = 0$, then $A(m, n) = n + 1$.
    (b)   If $m \neq 0$ but $n = 0$, then $A(m, n) = A(m - 1, 1)$.
    (c)   If $m \neq 0$ and $n \neq 0$, then $A(m, n) = A(m - 1, A(m, n - 1))$.

Once more, we have a recursive definition, since the definition refers to itself in parts (b) and (c). Observe that $A(m, n)$ is explicitly given only when $m = 0$. The base criteria are the pairs

$$(0, 0), \quad (0, 1), \quad (0, 2), \quad (0, 3), \quad \ldots, (0, n), \quad \ldots$$

Although it is not obvious from the definition, the value of any $A(m, n)$ may eventually be expressed in terms of the value of the function on one or more of the base pairs.

The value of $A(1, 3)$ is calculated in Problem 3.21. Even this simple case requires 15 steps. Generally speaking, the Ackermann function is too complex to evaluate on any but a trivial example. Its importance comes from its use in mathematical logic. The function is stated here mainly to give another example of a classical recursive function and to show that the recursion part of a definition may be complicated.

3.7 CARDINALITY

Two sets A and B are said to be *equipotent*, or to have the *same number of elements* or the *same cardinality*, written $A \simeq B$, if there exists a one-to-one correspondence $f \colon A \to B$. A set A is *finite* if A is empty or if A has the same cardinality as the set $\{1, 2, \ldots, n\}$ for some positive integer n. A set is *infinite* if it is not finite. Familiar examples of infinite sets are the natural numbers $\mathbf{N}$, the integers $\mathbf{Z}$, the rational numbers $\mathbf{Q}$, and the real numbers $\mathbf{R}$.

We now introduce the idea of "cardinal numbers". We will consider cardinal numbers simply as symbols assigned to sets in such a way that two sets are assigned the same symbol if and only if they have the same cardinality. The cardinal number of a set A is commonly denoted by $|A|$, $n(A)$, or card (A). We will use $|A|$.

The obvious symbols are used for the cardinality of finite sets. That is, 0 is assigned to the empty set $\emptyset$, and n is assigned to the set $\{1, 2, \ldots, n\}$. Thus $|A| = n$ if and only if A has n elements. For example,

$$|\{x, y, z\}| = 3 \quad \text{and} \quad |\{1, 3, 5, 7, 9\}| = 5$$

The cardinal number of the infinite set $\mathbf{N}$ of positive integers is $\aleph_0$ ("aleph-naught"). This symbol was introduced by Cantor. Thus $|A| = \aleph_0$ if and only if A has the same cardinality as $\mathbf{N}$.

EXAMPLE 3.9 Let $E = \{2, 4, 6, \ldots\}$, the set of even positive integers. The function $f \colon \mathbf{N} \to E$ defined by $f(n) = 2n$ is a one-to-one correspondence between the positive integers $\mathbf{N}$ and E. Thus E has the same cardinality as $\mathbf{N}$ and so we may write

$$|E| = \aleph_0$$

A set with cardinality $\aleph_0$ is said to be *denumerable* or *countably infinite*. A set which is finite or denumerable is said to be *countable*. One can show that the set $\mathbf{Q}$ of rational numbers is countable. In fact, we have the following theorem (proved in Problem 3.13) which we will use subsequently.

Theorem 3.2: A countable union of countable sets is countable.

That is, if $A_1, A_2, \ldots$ are each countable sets, then the following union is countable:

$$A_1 \cup A_2 \cup A_3 \cup \ldots$$

An important example of an infinite set which is uncountable, i.e., not countable, is given by the following theorem which is proved in Problem 3.14.

Theorem 3.3: The set $\mathbf{I}$ of all real numbers between 0 and 1 is uncountable.

Inequalities and Cardinal Numbers

One also wants to compare the size of two sets. This is done by means of an inequality relation which is defined for cardinal numbers as follows. For any sets A and B, we define $|A| \leq |B|$ if there exists a function $f \colon A \to B$ which is one-to-one. We also write

$$|A| < |B| \quad \text{if} \quad |A| \leq |B| \quad \text{but} \quad |A| \neq |B|$$

For example, $|\mathbf{N}| < |\mathbf{I}|$, where $\mathbf{I} = \{x \colon 0 \leq x \leq 1\}$, since the function $f \colon \mathbf{N} \to \mathbf{I}$ defined by $f(n) = 1/n$ is one-to-one, but $|\mathbf{N}| \neq |\mathbf{I}|$ by Theorem 3.3.

Cantor's Theorem, which follows and which we prove in Problem 3.25, tells us that the cardinal numbers are unbounded.

Theorem 3.4 (Cantor): For any set A, we have $|A| < |\text{Power}(A)|$ (where Power(A) is the power set of A, i.e., the collection of all subsets of A).

The next theorem tells us that the inequality relation for cardinal numbers is antisymmetric.

Theorem 3.5: (Schroeder-Bernstein): Suppose A and B are sets such that

$$|A| \leq |B| \quad \text{and} \quad |B| \leq |A|$$

Then $|A| = |B|$.

We prove an equivalent formulation of this theorem in Problem 3.26.

3.8 ALGORITHMS AND FUNCTIONS

An algorithm M is a finite step-by-step list of well-defined instructions for solving a particular problem, say, to find the output $f(X)$ for a given function f with input X. (Here X may be a list or set of values.) Frequently, there may be more than one way to obtain $f(X)$, as illustrated by the following examples. The particular choice of the algorithm M to obtain $f(X)$ may depend on the "efficiency" or "complexity" of the algorithm; this question of the complexity of an algorithm M is formally discussed in the next section.

EXAMPLE 3.10 (Polynomial Evaluation)   Suppose, for a given polynomial $f(x)$ and value $x = a$, we want to find $f(a)$, say,

$$f(x) = 2x^3 - 7x^2 + 4x - 15 \quad \text{and} \quad a = 5$$

This can be done in the following two ways.

(a) (*Direct Method*):   Here we substitute $a = 5$ directly in the polynomial to obtain

$$f(5) = 2(125) - 7(25) + 4(5) - 7 = 250 - 175 + 20 - 15 = 80$$

Observe that there are $3 + 2 + 1 = 6$ multiplications and 3 additions. In general, evaluating a polynomial of degree n directly would require approximately

$$n + (n - 1) + \cdots + 1 = \frac{n(n + 1)}{2} \text{ multiplications and } n \text{ additions.}$$

(b) (*Horner's Method or Synthetic Division*):   Here we rewrite the polynomial by successively factoring out x (on the right) as follows:

$$f(x) = (2x^2 - 7x + 4)x - 15 = ((2x - 7)x + 4)x - 15$$

Then

$$f(5) = ((3)5 + 4)5 - 15 = (19)5 - 15 = 95 - 15 = 80$$

For those familiar with synthetic division, the above arithmetic is equivalent to the following synthetic division:

$$
\begin{array}{r|rrrr}
5 & 2 & -\ \ 7 & +\ \ 4 & -\ \ 15 \\
  &   & 10 & +\ 15 & +\ 95 \\
\hline
  & 2 & +\ \ 3 & +\ 19 & +\ 80
\end{array}
$$

Observe that here there are 3 multiplications and 3 additions. In general, evaluating a polynomial of degree n by Horner's method would require approximately

$$n \text{ multiplications and } n \text{ additions}$$

Clearly Horner's method (b) is more efficient than the direct method (a).

EXAMPLE 3.11 (Greatest Common Divisor)   Let a and b be positive integers with, say, $b < a$; and suppose we want to find $d = \text{GCD}(a, b)$, the greatest common divisor of a and b. This can be done in the following two ways.

(a) (***Direct Method***):   Here we find all the divisors of a, say by testing all the numbers from 2 to $a/2$, and all the divisors of b. Then we pick the largest common divisor. For example, suppose $a = 258$ and $b = 60$. The divisors of a and b follow:

$$a = 258; \quad \text{divisors:} \quad 1, \quad 2, \quad 3, \quad 6, \quad 86, \quad 129, \quad 258$$
$$b = 60; \quad \text{divisors:} \quad 1, \quad 2, \quad 3, \quad 4, \quad 5, \quad 6, \quad 10, \quad 12, \quad 15, \quad 20, \quad 30, \quad 60$$

Accordingly, $d = \text{GCD}(258, 60) = 6$.

(b) (***Euclidean Algorithm***):   Here we divide a by b to obtain a remainder r_1. (Note $r_1 < b$.) Then we divide b by the remainder r_1 to obtain a second remainder r_2. (Note $r_2 < r_1$.) Next we divide r_1 by r_2 to obtain a third remainder r_3. (Note $r_3 < r_2$.) We continue dividing r_k by r_{k+1} to obtain a remainder r_{k+2}. Since

$$a > b > r_1 > r_2 > r_3 \ldots \tag{*}$$

eventually we obtain a remainder $r_m = 0$. Then $r_{m-1} = \text{GCD}\,(a, b)$. For example, suppose $a = 258$ and $b = 60$. Then:

    (1) Dividing $a = 258$ by $b = 60$ yields the remainder $r_1 = 18$.
    (2) Dividing $b = 60$ by $r_1 = 18$ yields the remainder $r_2 = 6$.
    (3) Dividing $r_1 = 18$ by $r_2 = 6$ yields the remainder $r_3 = 0$.

Thus $r_2 = 6 = \text{GCD}(258, 60)$.

The Euclidean algorithm is a very efficient way to find the greatest common divisor of two positive integers a and b. The fact that the algorithm ends follows from (*). The fact that the algorithm yields $d = \text{GCD}(a, b)$ is not obvious; it is discussed in Section 11.6.

3.9   COMPLEXITY OF ALGORITHMS

The analysis of algorithms is a major task in computer science. In order to compare algorithms, we must have some criteria to measure the efficiency of our algorithms. This section discusses this important topic.

Suppose M is an algorithm, and suppose n is the size of the input data. The time and space used by the algorithm are the two main measures for the efficiency of M. The time is measured by counting the number of "key operations;" for example:

(a) In sorting and searching, one counts the number of comparisons.
(b) In arithmetic, one counts multiplications and neglects additions.

Key operations are so defined when the time for the other operations is much less than or at most proportional to the time for the key operations. The space is measured by counting the maximum of memory needed by the algorithm.

The *complexity* of an algorithm M is the function $f(n)$ which gives the running time and/or storage space requirement of the algorithm in terms of the size n of the input data. Frequently, the storage space required by an algorithm is simply a multiple of the data size. Accordingly, unless otherwise stated or implied, the term "complexity" shall refer to the running time of the algorithm.

The complexity function $f(n)$, which we assume gives the running time of an algorithm, usually depends not only on the size n of the input data but also on the particular data. For example, suppose we want to search through an English short story TEXT for the first occurrence of a given 3-letter word W. Clearly, if W is the 3-letter word "the," then W likely occurs near the beginning of TEXT, so $f(n)$ will be small. On the other hand, if W is the 3-letter word "zoo," then W may not appear in TEXT at all, so $f(n)$ will be large.

The above discussion leads us to the question of finding the complexity function $f(n)$ for certain cases. The two cases one usually investigates in complexity theory are as follows:

(1) *Worst case*: The maximum value of $f(n)$ for any possible input.
(2) *Average case*: The expected value of $f(n)$.

The analysis of the average case assumes a certain probabilistic distribution for the input data; one possible assumption might be that the possible permutations of a data set are equally likely. The average case also uses the following concept in probability theory. Suppose the numbers $n_1, n_2, \ldots, n_k$ occur with respective probabilities $p_1, p_2, \ldots, p_k$. Then the *expectation* or *average value E* is given by

$$E = n_1 p_1 + n_2 p_2 + \cdots + n_k p_k$$

These ideas are illustrated below.

Linear Search

Suppose a linear array DATA contains n elements, and suppose a specific ITEM of information is given. We want either to find the location LOC of ITEM in the array DATA, or to send some message, such as LOC = 0, to indicate that ITEM does not appear in DATA. The linear search algorithm solves this problem by comparing ITEM, one by one, with each element in DATA. That is, we compare ITEM with DATA[1], then DATA[2], and so on, until we find LOC such that ITEM = DATA[LOC].

The complexity of the search algorithm is given by the number C of comparisons between ITEM and DATA[K]. We seek $C(n)$ for the worst case and the average case.

(1) **Worst Case**: Clearly the worst case occurs when ITEM is the last element in the array DATA or is not there at all. In either situation, we have

$$C(n) = n$$

Accordingly, $C(n) = n$ is the worst-case complexity of the linear search algorithm.

(2) **Average Case**: Here we assume that ITEM does appear in DATA, and that it is equally likely to occur at any position in the array. Accordingly, the number of comparisons can be any of the numbers $1, 2, 3, \ldots, n$, and each number occurs with probability $p = 1/n$. Then

$$C(n) = 1 \cdot \frac{1}{n} + 2 \cdot \frac{1}{n} + \cdots + n \cdot \frac{1}{n}$$

$$= (1 + 2 + \cdots + n) \cdot \frac{1}{n}$$

$$= \frac{n(n+1)}{2} \cdot \frac{1}{n} = \frac{n+1}{2}$$

This agrees with our intuitive feeling that the average number of comparisons needed to find the location of ITEM is approximately equal to half the number of elements in the DATA list.

Remark: The complexity of the average case of an algorithm is usually much more complicated to analyze than that of the worst case. Moreover, the probabilistic distribution that one assumes for the average case may not actually apply to real situations. Accordingly, unless otherwise stated or implied, the complexity of an algorithm shall mean the function which gives the running time of the worst case in terms of the input size. This is not too strong an assumption, since the complexity of the average case for many algorithms is proportional to the worst case.

Rate of Growth; Big O Notation

Suppose M is an algorithm, and suppose n is the size of the input data. Clearly the complexity $f(n)$ of M increases as n increases. It is usually the rate of increase of $f(n)$ that we want to examine. This is usually done by comparing $f(n)$ with some standard function, such as

$$\log n, \quad n, \quad n \log n, \quad n^2, \quad n^3, \quad 2^n$$

The rates of growth for these standard functions are indicated in Fig. 3-7, which gives their approximate values for certain values of n. Observe that the functions are listed in the order of their rates of growth: the logarithmic function $\log_2 n$ grows most slowly, the exponential function 2^n grows most rapidly, and the polynomial functions n^c grow according to the exponent c.

n \ $g(n)$	$\log n$	n	$n \log n$	n^2	n^3	2^n
5	3	5	15	25	125	32
10	4	10	40	100	10^3	10^3
100	7	100	700	10^4	10^6	10^{30}
1000	10	10^3	10^4	10^6	10^9	10^{300}

Fig. 3-7   Rate of growth of standard functions

The way we compare our complexity function $f(n)$ with one of the standard functions is to use the functional "big O" notation which we formally define below.

Definition 3.4: Let $f(x)$ and $g(x)$ be arbitrary functions defined on $\mathbf{R}$ or a subset of $\mathbf{R}$. We say "$f(x)$ is of order $g(x)$," written

$$f(x) = O(g(x))$$

if there exists a real number k and a positive constant C such that, for all $x > k$, we have

$$|f(x)| \leq C|g(x)|$$

In other words, $f(x) = 0(g(x))$ if a constant multiple of $|g(x)|$ exceeds $|f(x)|$ for all x greater than some real number k.

We also write:

$$f(x) = h(x) + O(g(x)) \quad \text{when} \quad f(x) - h(x) = O(g(x))$$

(The above is called the "big O" notation since $f(x) = o(g(x))$ has an entirely different meaning.)

Consider now a polynomial $P(x)$ of degree m. We show in Problem 3.24 that $P(x) = O(x^m)$. Thus, for example,

$$7x^2 - 9x + 4 = O(x^2) \quad \text{and} \quad 8x^3 - 576x^2 + 832x - 248 = O(x^3)$$

Complexity of Well-known Algorithms

Assuming $f(n)$ and $g(n)$ are functions defined on the positive integers, then

$$f(n) = O(g(n))$$

means that $f(n)$ is bounded by a constant multiple of $g(n)$ for almost all n.

To indicate the convenience of this notation, we give the complexity of certain well-known searching and sorting algorithms in computer science:

(a) Linear search: $O(n)$ (c) Bubble sort: $O(n^2)$
(b) Binary search: $O(\log n)$ (d) Merge-sort: $O(n \log n)$

Solved Problems

FUNCTIONS

3.1. Let $X = \{1, 2, 3, 4\}$. Determine whether each relation on X is a function from X into X.

    (a) $f = \{(2, 3), (1, 4), (2, 1), (3.2), (4, 4)\}$

    (b) $g = \{(3, 1), (4, 2), (1, 1)\}$

    (c) $h = \{(2, 1), (3, 4), (1, 4), (2, 1), (4, 4)\}$

           Recall that a subset f of $X \times X$ is a function $f : X \to X$ if and only if each $a \in X$ appears as the first coordinate in exactly one ordered pair in f.

    (a) No. Two different ordered pairs $(2, 3)$ and $(2, 1)$ in f have the same number 2 as their first coordinate.

    (b) No. The element $2 \in X$ does not appear as the first coordinate in any ordered pair in g.

    (c) Yes. Although $2 \in X$ appears as the first coordinate in two ordered pairs in h, these two ordered pairs are equal.

3.2. Sketch the graph of: (a) $f(x) = x^2 + x - 6$; (b) $g(x) = x^3 - 3x^2 - x + 3$.

    Set up a table of values for x and then find the corresponding values of the function. Since the functions are polynomials, plot the points in a coordinate diagram and then draw a smooth continuous curve through the points. See Fig. 3-8.

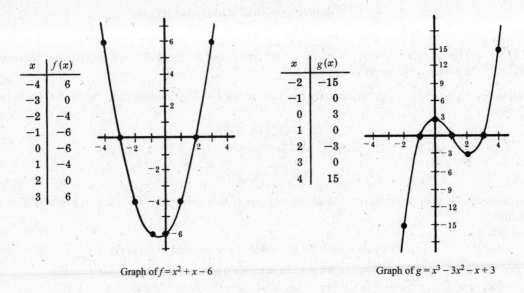

Graph of $f = x^2 + x - 6$ Graph of $g = x^3 - 3x^2 - x + 3$

Fig. 3-8

3.3. Let $A = \{a, b, c\}$, $B = \{x, y, z\}$, $C = \{r, s, t\}$. Let $f : A \to B$ and $g : B \to C$ be defined by:

$$f = \{(a, y)(b, x), (c, y)\} \quad \text{and} \quad g = \{(x, s), (y, t), (z, r)\}.$$

Find: (a) composition function $g \circ f : A \to C$; (b) $\text{Im}(f)$, $\text{Im}(g)$, $\text{Im}(g \circ f)$.

    (a) Use the definition of the composition function to compute:

$$(g \circ f)(a) = g(f(a)) = g(y) = t$$
$$(g \circ f)(b) = g(f(b)) = g(x) = s$$
$$(g \circ f)(c) = g(f(c)) = g(y) = t$$

    That is $g \circ f = \{(a, t), (b, s), (c, t)\}$.

    (b) Find the image points (or second coordinates):

$$\text{Im}(f) = \{x, y\}, \quad \text{Im}(g) = \{r, s, t\}, \quad \text{Im}(g \circ f) = \{s, t\}$$

3.4. Let $f: \mathbf{R} \to \mathbf{R}$ and $g: \mathbf{R} \to \mathbf{R}$ be defined by $f(x) = 2x + 1$ and $g(x) = x^2 - 2$. Find the formula for the composition function $g \circ f$.

Compute $g \circ f$ as follows: $(g \circ f)(x) = g(f(x)) = g(2x + 1) = (2x + 1)^2 - 2 = 4x^2 + 4x - 1$.

Observe that the same answer can be found by writing

$$y = f(x) = 2x + 1 \quad \text{and} \quad z = g(y) = y^2 - 2$$

and then eliminating y from both equations:

$$z = y^2 - 2 = (2x + 1)^2 - 2 = 4x^2 + 4x - 1$$

ONE-TO-ONE, ONTO, AND INVERTIBLE FUNCTIONS

3.5. Let the functions $f: A \to B$, $g: B \to C$, $h: C \to D$ be defined by Fig. 3-9. Determine if each function is: (a) onto, (b) one-to-one, (c) invertible.

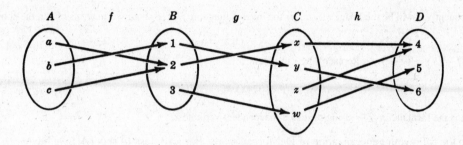

Fig. 3-9

(a) The function $f: A \to B$ is not onto since $3 \in B$ is not the image of any element in A.

The function $g: B \to C$ is not onto since $z \in C$ is not the image of any element in B.

The function $h: C \to D$ is onto since each element in D is the image of some element of C.

(b) The function $f: A \to B$ is not one-to-one since a and c have the same image 2.

The function $g: B \to C$ is one-to-one since 1, 2 and 3 have distinct images.

The function $h: C \to D$ is not one-to-one since x and z have the same image 4.

(c) No function is one-to-one and onto; hence no function is invertible.

3.6. Consider permutations $\sigma = \begin{pmatrix} 1 & 2 & 3 & 4 & 5 & 6 \\ 3 & 6 & 4 & 5 & 1 & 2 \end{pmatrix}$ and $\tau = \begin{pmatrix} 1 & 2 & 3 & 4 & 5 & 6 \\ 2 & 4 & 6 & 5 & 3 & 1 \end{pmatrix}$ in S_6. Find: (a) composition $\tau \circ \sigma$; (b) σ^{-1}.

(a) Note that σ sends 1 into 3 and τ sends 3 into 6. So the composition $\tau \circ \sigma$ sends 1 into 6. I.e. $(\tau \circ \sigma)(1) = 6$. Moreover, $\tau \circ \sigma$ sends 2 into 6 into 1 that is, $(\tau \circ \sigma)(2) = 1$, Similarly,

$$(\tau \circ \sigma)(3) = 5, \quad (\tau \circ \sigma)(4) = 3, \quad (\tau \circ \sigma)(5) = 2, \quad (\tau \circ \sigma)(6) = 4$$

Thus

$$\tau \circ \sigma = \begin{pmatrix} 1 & 2 & 3 & 4 & 5 & 6 \\ 6 & 1 & 5 & 3 & 2 & 4 \end{pmatrix}$$

(b) Look for 1 in the second row of σ. Note σ sends 5 into 1. Hence $\sigma^{-1}(1) = 5$. Look for 2 in the second row of σ. Note σ sends 6 into 2. Hence $\sigma^{-1}(2) = 6$. Similarly, $\sigma^{-1}(3) = 1$, $\sigma^{-1}(4) = 3$, $\sigma^{-1}(5) = 4$, $\sigma^{-1}(6) = 2$. Thus

$$\sigma^{-1} = \begin{pmatrix} 1 & 2 & 3 & 4 & 5 & 6 \\ 5 & 6 & 1 & 3 & 4 & 2 \end{pmatrix}$$

3.7. Consider functions $f: A \to B$ and $g: B \to C$. Prove the following:

    (a) If f and g are one-to-one, then the composition function $g \circ f$ is one-to-one.

    (b) If f and g are onto functions, then $g \circ f$ is an onto function.

    (a) Suppose $(g \circ f)(x) = (g \circ f)(y)$; then $g(f(x)) = g(f(y))$. Hence $f(x) = f(y)$ because g is one-to-one. Furthermore, $x = y$ since f is one-to-one. Accordingly $g \circ f$ is one-to-one.

    (b) Let c be any arbitrary element of C. Since g is onto, there exists a $b \in B$ such that $g(b) = c$. Since f is onto, there exists an $a \in A$ such that $f(a) = b$. But then

$$(g \circ f)(a) = g(f(a)) = g(b) = c$$

Hence each $c \in C$ is the image of some element $a \in A$. Accordingly, $g \circ f$ is an onto function.

3.8. Let $f: \mathbf{R} \to \mathbf{R}$ be defined by $f(x) = 2x - 3$. Now f is one-to-one and onto; hence f has an inverse function f^{-1}. Find a formula for f^{-1}.

Let y be the image of x under the function f:

$$y = f(x) = 2x - 3$$

Consequently, x will be the image of y under the inverse function f^{-1}. Solve for x in terms of y in the above equation:

$$x = (y + 3)/2$$

Then $f^{-1}(y) = (y + 3)/2$. Replace y by x to obtain

$$f^{-1}(x) = \frac{x + 3}{2}$$

which is the formula for f^{-1} using the usual independent variable x.

3.9. Prove the following generalization of DeMorgan's law: For any class of sets $\{A_i\}$ we have

$$(\cup_i A_i)^c = \cap_i A_i^c$$

We have:

$$x \in (\cup_i A_i)^c \quad \text{iff } x \notin \cup_i A_i, \quad \text{iff } \forall_i \in I, x \notin A_i, \quad \text{iff } \forall_i \in I, x \in A_i^c, \quad \text{iff } x \in \cap_i A_i^c$$

Therefore, $(\cup_i A_i)^c = \cap_i A_i^c$. (Here we have used the logical notations iff for "if and only" if and $\forall$ for "for all.")

CARDINALITY

3.10. Find the cardinal number of each set:

    (a) $A = \{a, b, c, \ldots, y, z\}$,   (c) $C = \{10, 20, 30, 40, \ldots\}$,

    (b) $B = \{x \mid x \in \mathbf{N}, x^2 = 5\}$,   (d) $D = \{6, 7, 8, 9, \ldots\}$.

    (a) $|A| = 26$ since there are 26 letters in the English alphabet.

    (b) $|B| = 0$ since there is no positive integer whose square is 5, that is, B is empty.

    (c) $|C| = \aleph_0$ because $f: \mathbf{N} \to C$, defined by $f(n) = 10n$, is a one-to-one correspondence between $\mathbf{N}$ and C.

    (d) $|D| = \aleph_0$ because $g: \mathbf{N} \to D$, defined by $g(n) = n + 5$ is a one-to-one correspondence between $\mathbf{N}$ and D.

3.11. Show that the set $\mathbf{Z}$ of integers has cardinality $\aleph_0$.

The following diagram shows a one-to-one correspondence between $\mathbf{N}$ and $\mathbf{Z}$:

$$
\begin{array}{ccccccccc}
\mathbf{N} = & 1 & 2 & 3 & 4 & 5 & 6 & 7 & 8 & \ldots \\
& \downarrow & \downarrow & \downarrow & \downarrow & \downarrow & \downarrow & \downarrow & \downarrow & \ldots \\
\mathbf{Z} = & 0 & 1 & -1 & 2 & -2 & 3 & -3 & 4 & \ldots
\end{array}
$$

That is, the following function $f: \mathbf{N} \to \mathbf{Z}$ is one-to-one and onto:

$$f(n) = \begin{cases} n/2 & \text{if } n \text{ is even} \\ (1 - n)/2 & \text{if } n \text{ is odd} \end{cases}$$

Accordingly, $|\mathbf{Z}| = |\mathbf{N}| = \aleph_0$.

3.12. Let $A_1, A_2, \ldots$ be a countable number of finite sets. Prove that the union $S = \cup_i A_i$ is countable.

Essentially, we list the elements of A_1, then we list the elements of A_2 which do not belong to A_1, then we list the elements of A_3 which do not belong to A_1 or A_2, i.e., which have not already been listed, and so on. Since the A_i are finite, we can always list the elements of each set. This process is done formally as follows.

First we define sets $B_1, B_2, \ldots$ where B_i contains the elements of A_i which do not belong to preceding sets, i.e., we define

$$B_1 = A_1 \quad \text{and} \quad B_k = A_k \backslash (A_1 \cup A_2 \cup \cdots \cup A_{k-1})$$

Then the B_i are disjoint and $S = \cup_i B_i$. Let $b_{i1}, b_{i2}, \ldots, b_{im}$, be the elements of B_i. Then $S = \{b_{ij}\}$. Let $f: S \to \mathbf{N}$ be defined as follows:

$$f(b_{ij}) = m_1 + m_2 + \cdots + m_{i-1} + j$$

If S is finite, then S is countable. If S is infinite then f is a one-to-one correspondence between S and $\mathbf{N}$. Thus S is countable.

3.13. Prove Theorem 3.2: A countable union of countable sets is countable.

Suppose $A_1, A_2, A_3, \ldots$ are a countable number of countable sets. In particular, suppose $a_{i1}, a_{i2}, a_{i3}, \ldots$ are the elements of A_i. Define sets $B_2, B_3, B_4, \ldots$ as follows:

$$B_k = \{a_{ij} \mid i + j = k\}$$

For example, $B_6 = \{a_{15}, \; a_{24}, \; a_{33}, \; a_{12}, \; a_{51}\}$. Observe that each B_k is finite and

$$S = \cup_i A_i = \cup_k B_k$$

By the preceding problem $\cup_k B_k$ is countable. Hence $S = \cup_i A_i$ is countable and the theorem is proved.

3.14. Prove Theorem 3.3: The set $\mathbf{I}$ of all real numbers between 0 and 1 inclusive is uncountable.

The set $\mathbf{I}$ is clearly infinite, since it contains $1, \frac{1}{2}, \frac{1}{3}, \ldots$. Suppose $\mathbf{I}$ is denumerable. Then there exists a one-to-one correspondence $f: N \to \mathbf{I}$. Let $f(1) = a_1, f(2) = a_2, \ldots$; that is, $\mathbf{I} = \{a_1, a_2, a_3, \ldots\}$. We list the elements $a_1, a_2, \ldots$ in a column and express each in its decimal expansion:

$$a_1 = 0.x_{11}x_{12}x_{13}x_{14}\ldots$$
$$a_2 = 0.x_{21}x_{22}x_{23}x_{24}\ldots$$
$$a_3 = 0.x_{31}x_{32}x_{33}x_{34}\ldots$$
$$a_4 = 0.x_{41}x_{42}x_{43}x_{44}\ldots$$
$$\ldots\ldots\ldots\ldots\ldots\ldots\ldots\ldots\ldots$$

where $x_{ij} \in \{0, 1, 2, \ldots, 9\}$. (For those numbers which can be expressed in two different decimal expansions, e.g., $0.2000000\ldots = 0.1999999\ldots$, we choose the expansion which ends with nines.)

Let $b = 0.y_1 y_2 y_3 y_4 \ldots$ be the real number obtained as follows:

$$y_i = \begin{cases} 1 & \text{if } x_{ii} \neq 1 \\ 2 & \text{if } x_{ii} = 1 \end{cases}$$

Now $b \in \mathbf{I}$. But

$$b \neq a_1 \text{ because } y_1 \neq x_{11}$$
$$b \neq a_2 \text{ because } y_2 \neq x_{22}$$
$$b \neq a_3 \text{ because } y_3 \neq x_{33}$$
$$\ldots\ldots\ldots\ldots\ldots\ldots\ldots\ldots\ldots$$

Therefore b does not belong to $\mathbf{I} = \{a_1, a_2, \ldots\}$. This contradicts the fact that $b \in \mathbf{I}$. Hence the assumption that $\mathbf{I}$ is denumerable must be false, so $\mathbf{I}$ is uncountable.

SPECIAL MATHEMATICAL FUNCTIONS

3.15. Find: (a) $\lfloor 7.5 \rfloor, \lfloor -7.5 \rfloor, \lfloor -18 \rfloor$; (b) $\lceil 7.5 \rceil, \lceil -7.5 \rceil, \lceil -18 \rceil$.

(a) By definition, $\lfloor x \rfloor$ denotes the greatest integer that does not exceed x, hence $\lfloor 7.5 \rfloor = 7$, $\lfloor -7.5 \rfloor = -8$, $\lfloor -18 \rfloor = -18$.

(b) By definition, $\lceil x \rceil$ denotes the least integer that is not less than x, hence $\lceil 7.5 \rceil = 8$, $\lceil -7.5 \rceil = -7$, $\lceil -18 \rceil = -18$.

3.16. Find: (a) 25 (mod7); (b) 25 (mod5); (c) −35 (mod 11); (d) −3 (mod 8).

When k is positive, simply divide k by the modulus M to obtain the remainder r. Then $r = k(\bmod M)$. If k is negative, divide $|k|$ by M to obtain the remainder r'. Then $k(\bmod M) = M − r'$ (when $r' \neq 0$). Thus:

(a) 25 (mod 7) = 4 (c) −35 (mod 11) = 11 − 2 = 9

(b) 25 (mod 5) = 0 (d) −3 (mod 8) = 8 − 3 = 5

3.17. Evaluate modulo $M = 15$: (a) $9 + 13$; (b) $7 + 11$; (c) $4 − 9$; (d) $2 − 10$.

Use $a \pm M = a(\bmod M)$:

(a) $9 + 13 = 22 = 22 − 15 = 7$ (c) $4 − 9 = −5 = −5 + 15 = 10$

(b) $7 + 11 = 18 = 18 − 15 = 3$ (d) $2 − 10 = −8 = −8 + 15 = 7$

3.18. Simplify: (a) $\dfrac{n!}{(n-1)!}$; (b) $\dfrac{(n+2)!}{n!}$.

(a) $\dfrac{n!}{(n-1)!} = \dfrac{n(n-1)(n-2)\cdots 3 \cdot 2 \cdot 1}{(n-1)(n-2)\cdots 3 \cdot 2 \cdot 1} = n$ or, simply, $\dfrac{n!}{(n-1)!} = \dfrac{n(n-1)!}{(n-1)!} = n$

(b) $\dfrac{(n+2)!}{n!} = \dfrac{(n+2)(n+1)n!}{n!} = (n+2)(n+1) = n^2 + 3n + 2$

3.19. Evaluate: (a) $\log_2 8$; (b) $\log_2 64$; (c) $\log_{10} 100$; (d) $\log_{10} 0.001$.

(a) $\log_2 8 = 3$ since $2^3 = 8$ (c) $\log_{10} 100 = 2$ since $10^2 = 100$

(b) $\log_2 64 = 6$ since $2^6 = 64$ (d) $\log_{10} 0.001 = −3$ since $10^{-3} = 0.001$

RECURSIVE FUNCTIONS

3.20. Let a and b be positive integers, and suppose Q is defined recursively as follows:

$$Q(a, b) = \begin{cases} 0 & \text{if } a < b \\ Q(a - b, b) + 1 & \text{if } b \leq a \end{cases}$$

(a) Find: (i) $Q(2, 5)$; (ii) $Q(12, 5)$.

(b) What does this function Q do? Find $Q(5861, 7)$.

(a) (i) $Q(2, 5) = 0$ since $2 < 5$.

(ii) $Q(12, 5) = Q(7, 5) + 1$
$= [Q(2, 5) + 1] + 1 = Q(2, 5) + 2$
$= 0 + 2 = 2$

(b) Each time b is subtracted from a, the value of Q is increased by 1. Hence $Q(a, b)$ finds the quotient when a is divided by b. Thus $Q(5861, 7) = 837$.

3.21. Use the definition of the Ackermann function to find $A(1, 3)$.

Figure 3-10 shows the 15 steps that are used to evaluate $A(1, 3)$.

The forward indention indicates that we are postponing an evaluation and are recalling the definition, and the backward indention indicates that we are backtracking. Observe that (a) of the definition is used in Steps 5, 8, 11 and 14; (b) in Step 4; and (c) in Steps 1, 2, and 3. In the other steps we are backtracking with substitutions.

(1)	$A(1,3) = A(0, A(1,2))$	(9)	$A(1,1) = 3$
(2)	$A(1,2) = A(0, A(1,1))$	(10)	$A(1,2) = A(0,3)$
(3)	$A(1,1) = A(0, A(1,0))$	(11)	$A(0,3) = 3+1 = 4$
(4)	$A(1,0) = A(0,1)$	(12)	$A(1,2) = 4$
(5)	$A(0,1) = 1+1 = 2$	(13)	$A(1,3) = A(0,4)$
(6)	$A(1,0) = 2$	(14)	$A(0,4) = 4+1 = 5$
(7)	$A(1,1) = A(0,2)$	(15)	$A(1,3) = 5$
(8)	$A(0,2) = 2+1 = 3$		

Fig. 3-10

MISCELLANEOUS PROBLEMS

3.22. Find the domain D of each of the following real-valued functions of a real variable:

(a) $f(x) = \frac{1}{x-2}$ (c) $f(x) = \sqrt{25 - x^2}$

(b) $f(x) = x^2 - 3x - 4$ (d) x^2 where $0 \le x \le 2$

When a real-valued function of a real variable is given by a formula $f(x)$, then the domain D consists of the largest subset of **R** for which $f(x)$ has meaning and is real, unless otherwise specified.

(a) f is not defined for $x - 2 = 0$, i.e., for $x = 2$; hence $D = \mathbf{R} \backslash \{2\}$.

(b) f is defined for every real number; hence $D = \mathbf{R}$.

(c) f is not defined when $25 - x^2$ is negative; hence $D = [-5, 5] = \{x \mid -5 \le x \le 5\}$.

(d) Here, the domain of f is explicitly given as $D = \{x \mid 0 \le x \le 2\}$.

3.23. For any $n \in \mathbf{N}$, let $D_n = (0, 1/n)$, the open interval from 0 to $1/n$. Find:

(a) $D_3 \cup D_4$;   (b) $D_3 \cap D_{20}$;   (c) $D_s \cup D_t$;   (d) $D_s \cap D_t$.

(a) Since $(0, 1/3)$ is a superset of $(0, 1/7)$, $D_3 \cup D_4 = D_3$.

(b) Since $(0, 1/20)$ is a subset of $(0, 1/3)$, $D_3 \cap D_{20} = D_{20}$.

(c) Let $m = \min(s, t)$, that is, the smaller of the two numbers s and t; then D_m is equal to D_s or D_t contains the other as a subset. Hence $D_s \cap D_t = D_m$.

(d) Let $M = \max(s, t)$, that is, the larger of the two numbers s and t; then $D_s \cap D_t = D_m$.

3.24. Suppose $P(n) = a_0 + a_1 n + a_2 n^2 + \cdots + a_m n^m$ has degree m. Prove $P(n) = O(n^m)$.

Let $b_0 = |a_0|$, $b_1 = |a_1|, \ldots, b_m = |a_m|$. Then for $n \ge 1$,

$$p(n) \le b_0 + b_1 n + b_2 n^2 + \cdots + b_m n^m = \left(\frac{b_0}{n^m} + \frac{b_1}{n^{m-1}} + \cdots + b_m \right) n^m$$
$$\le (b_0 + b_1 + \cdots + b_m) n^m = M n^m$$

where $M = |a_0| + |a_1| + \cdots + |a_m|$. Hence $P(n) = O(n^m)$.

For example, $5x^3 + 3x = O(x^3)$ and $x^5 - 4000000 x^2 = O(x^5)$.

3.25. Prove Theorem 3.4 (Cantor): $|A| < |\text{Power}(A)|$ (where Power(A) is the power set of A).

The function $g: A \to \text{Power}(A)$ defined by $g(a) = \{a\}$ is clearly one-to-one; hence $|A| \le |\text{Power}(A)|$.

If we show that $|A| \ne |\text{Power}(A)|$, then the theorem will follow. Suppose the contrary, that is, suppose $|A| = |\text{Power}(A)|$ and that $f: A \to \text{Power}(A)$ is a function which is both one-to-one and onto. Let $a \in A$ be called a "bad" element if $a \notin f(a)$, and let B be the set of bad elements. In other words,

$$B = \{x : x \in A, \ x \notin f(x)\}$$

Now B is a subset of A. Since $f: A \to \text{Power}(A)$ is onto, there exists $b \in A$ such that $f(b) = B$. Is b a "bad" element or a "good" element? If $b \in B$ then, by definition of B, $b \notin f(b) = B$, which is impossible. Likewise, if $b \notin B$ then $b \in f(b) = B$, which is also impossible. Thus the original assumption that $|A| = |\text{Power}(A)|$ has led to a contradiction. Hence the assumption is false, and so the theorem is true.

3.26. Prove the following equivalent formulation of the Schroeder–Bernstein Theorem 3.5:

Suppose $X \supseteq Y \supseteq X_1$ and $X \simeq X_1$. Then $X \simeq Y$.

Since $X \simeq X_1$ there exists a one-to-one correspondence (bijection) $f: X \to X_1$ Since $X \supseteq Y$, the restriction of f to Y, which we also denote by f, is also one-to-one. Let $f(Y) = Y_1$. Then Y and Y_1 are equipotent,

$$X \supseteq Y \supseteq X_1 \supseteq Y_1$$

and $f: Y \to Y_1$ is bijective. But now $Y \supseteq X_1 \supseteq Y_1$ and $Y \simeq Y_1$. For similar reasons, X_1 and $f(X_1) = X_2$ are equipotent,

$$X \supseteq Y \supseteq X_1 \supseteq Y_1 \supseteq X_2$$

and $f: X_1 \to X_2$ is bijective. Accordingly, there exist equipotent sets $X, X_1, X_2, \ldots$ and equipotent sets $Y, Y_1, Y_2, \ldots$ such that

$$X \supseteq Y \supseteq X_1 \supseteq Y_1 \supseteq X_2 \supseteq Y_2 \supseteq X_3 \supseteq Y_3 \supseteq \cdots$$

and $f: X_k \to X_{k+1}$ and $f: Y_k \to Y_{k+1}$ are bijective.

Let

$$B = X \cap Y \cap X_1 \cap Y_1 \cap X_2 \cap Y_2 \cap \cdots$$

Then

$$X = (X \backslash Y) \cup (Y \backslash X_1) \cup (X_1 \backslash Y_1) \cup \cdots \cup B$$
$$Y = (Y \backslash X_1) \cup (X_1 \backslash Y_1) \cup (Y_1 \backslash X_2) \cup \cdots \cup B$$

Furthermore, $X \backslash Y, X_1 \backslash Y_1, X_2 \backslash Y_2, \ldots$ are equipotent. In fact, the function

$$f: (X_k \backslash Y_k) \to (X_{k+1} \backslash Y_{k+1})$$

is one-to-one and onto.

Consider the function $g: X \to Y$ defined by the diagram in Fig. 3-11. That is,

$$g(x) = \begin{cases} f(x) & \text{if } x \in X_k \backslash Y_k \text{ or } x \in X \backslash Y \\ x & \text{if } x \in Y_k \backslash X_k \text{ or } x \in B \end{cases}$$

Then g is one-to-one and onto. Therefore $X \simeq Y$.

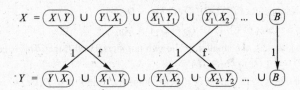

Fig. 3-11

Supplementary Problems

FUNCTIONS

3.27. Let $W = \{a, b, c, d\}$. Decide whether each set of ordered pairs is a function from W into W.

    (a) $\{(b, a), \quad (c, d), \quad (d, a), \quad (c, d) \quad (a, d)\}$ (c) $\{(a, b), \quad (b, b), \quad (c, d), \quad (d, b)\}$

    (b) $\{(d, d), \quad (c, a), \quad (a, b), \quad (d, b)\}$ (d) $\{(a, a), \quad (b, a), \quad (a, b), \quad (c, d)\}$

3.28. Let $V = \{1, 2, 3, 4\}$. For the following functions $f: V \to V$ and $g: V \to V$, find:

(a) $f \circ g$; (b) $g \circ f$; (c) $f \circ f$:

$$f = \{(1, 3), \ (2, 1), \ (3, 4), \ (4, 3)\} \quad \text{and} \quad g = \{(1, 2), \ (2, 3), \ (3, 1), \ (4, 1)\}$$

3.29. Find the composition function $h \circ g \circ f$ for the functions in Fig. 3-9.

ONE-TO-ONE, ONTO, AND INVERTIBLE FUNCTIONS

3.30. Determine if each function is one-to-one.

(a) To each person on the earth assign the number which corresponds to his age.

(b) To each country in the world assign the latitude and longitude of its capital.

(c) To each book written by only one author assign the author.

(d) To each country in the world which has a prime minister assign its prime minister.

3.31. Let functions f, g, h from $V = \{1, 2, 3, 4\}$ into V be defined by: $f(n) = 6 - n$, $g(n) = 3$, $h = \{(1, 2), (2, 3), (3, 4), (4, 1)\}$. Decide which functions are:

(a) one-to-one;   (b) onto;   (c) both;   (d) neither.

3.32. Let functions f, g, h from $\mathbf{N}$ into $\mathbf{N}$ be defined by $f(n) = n + 2$, (b) $g(n) = 2^n$; $h(n) =$ number of positive divisors of n. Decide which functions are:

(a) one-to-one;   (b) onto;   (c) both;   (d) neither;   (e) Find $h'(2) = \{x \mid h(x) = 2\}$.

3.33. Decide which of the following functions are: (a) one-to-one; (b) onto; (c) both; (d) neither.

(1) $f: \mathbf{Z}^2 \to \mathbf{Z}$ where $f(n, m) = n - m$;   (3) $h: \mathbf{Z} \times (\mathbf{Z} \backslash 0) \to \mathbf{Q}$ where $h(n, m) = n/m$;

(2) $g: \mathbf{Z}^2 \to \mathbf{Z}^2$ where $g(n, m) = (m, n)$;   (4) $k: \mathbf{Z} \to \mathbf{Z}^2$ where $k(n) = (n, n)$.

3.34. Let $f: \mathbf{R} \to \mathbf{R}$ be defined by $f(x) = 3x - 7$. Find a formula for the inverse function $f^{-1}: \mathbf{R} \to \mathbf{R}$.

3.35. Consider permutations $\sigma = \begin{pmatrix} 1 & 2 & 3 & 4 & 5 & 6 \\ 2 & 5 & 6 & 1 & 3 & 4 \end{pmatrix}$ and $\tau = \begin{pmatrix} 1 & 2 & 3 & 4 & 5 & 6 \\ 6 & 4 & 3 & 1 & 2 & 5 \end{pmatrix}$ in S_6.

Find: (a) $\tau \circ \sigma$;   (b) $\sigma \circ \tau$;   (c) σ^2;   (d) σ^{-1};   (e) τ^{-1}

PROPERTIES OF FUNCTIONS

3.36. Prove: Suppose $f: A \to B$ and $g: B \to A$ satisfy $g \circ f = 1_A$. Then f is one-to-one and g is onto.

3.37. Prove Theorem 3.1: A function $f: A \to B$ is invertible if and only if f is both one-to-one and onto.

3.38. Prove: Suppose $f: A \to B$ is invertible with inverse function $f^{-1}: B \to A$. Then $f^{-1} \circ f = 1_A$ and $f \circ f^{-1} = 1_B$.

3.39. Suppose $f: A \to B$ is one-to-one and $g: A \to B$ is onto. Let x be a subset of A.

(a) Show $f1_x$, the restriction of f to x, is one-to-one.

(b) Show $g1_x$, need not be onto.

3.40. For each $n \in \mathbf{N}$, consider the open interval $A_n = (0, 1/n) = \{x \mid 0 < x < 1/n\}$. Find:

(a) $A_2 \cup A_8$;   (c) $\cup (A_i \mid i \in J)$;   (e) $\cup (A_i \mid i \in K)$;

(b) $A_3 \cap A_7$;   (d) $\cap (A_i \mid i \in J)$;   (f) $\cap (A_i \mid i \in K)$.

where J is a finite subset of $\mathbf{N}$ and $\mathbf{K}$ is an infinite subset of $\mathbf{N}$.

3.41. For each $n \in N$, let $D_n = \{n, 2n, 3n, \ldots\} = \{$multiples of $n\}$.

(a) Find: (i) $D_2 \cap D_7$; (ii) $D_6 \cap D_8$; (iii) $D_3 \cap D_{12}$; (iv) $D_3 \cup D_{12}$.

(b) Prove that $\cap (D_i \mid i \in K) = \emptyset$ where K is an infinite subset of $\mathbf{N}$.

3.42. Consider an indexed class of sets $\{A_i \mid i \in I\}$, a set $\mathbf{B}$ and an index i_0 in I.

Prove: (a) $B \cap (\cup_i A_i) = \cup_i (B \cap A_i)$; (b) $\cap (A_i \mid i \in I) \subseteq A_{i_0} \subseteq \cup (A_i \mid i \in I)$.

CARDINAL NUMBERS

3.43. Find the cardinal number of each set: (a) $\{x \mid x$ is a letter in "BASEBALL"$\}$;
(b) Power set of $A = \{a, b, c, d, e\}$; (c) $\{x \mid x^2 = 9, 2x = 8\}$.

3.44. Find the cardinal number of:

(a) all functions from $A = \{a, b, c, d\}$ into $B = \{1, 2, 3, 4, 5\}$;

(b) all functions from P into Q where $|P| = r$ and $|Q| = s$;

(c) all relations on $A = \{a, b, c, d\}$;

(d) all relations on P where $|P| = r$.

3.45. Prove:

(a) Every infinite set A contains a denumerable subset D.

(b) Each subset of a denumerable set is finite or denumerable.

(c) If A and B are denumerable, then $A \times B$ is denumerable.

(e) The set **Q** of rational numbers is denumerable.

3.46. Prove: (a) $|A \times B| = |B \times A|$; (b) If $A \subseteq B$ then $|A| \leq |B|$; (c) If $|A| = |B|$ then $|P(A)| = |P(B)|$.

SPECIAL FUNCTIONS

3.47. Find: (a)$\lfloor 13.2 \rfloor$, $\lfloor -0.17 \rfloor$, $\lfloor 34 \rfloor$; (b)$\lceil 13.2 \rceil$, $\lceil -0.17 \rceil$, $\lceil 34 \rceil$.

3.48. Find:

(a) 29 (mod 6); (c) 5 (mod 12); (e) −555 (mod 11).

(b) 200 (mod 20); (d) −347 (mod 6);

3.49. Find: (a) $3! + 4!$; (b) $3!(3! + 2!)$; (c) $6!/5!$; (d) $30!/28!$.

3.50. Evaluate: (a) $\log_2 16$; (b) $\log_3 27$; (c) $\log_{10} 0.01$.

MISCELLANEOUS PROBLEMS

3.51. Let n be an integer. Find $L(25)$ and describe what the function L does where L is defined by:

$$L(n) = \begin{cases} 0 & \text{if } n = 1 \\ L(\lfloor n/2 \rfloor) + 1 & \text{if } n > 1 \end{cases}$$

3.52. Let a and b be integers. Find $Q(2, 7)$, $Q(5, 3)$, and $Q(15, 2)$, where $Q(a, b)$ is defined by:

$$Q(a, b) = \begin{cases} 5 & \text{if } a < b \\ Q(a - b, b + 2) + a & \text{if } a \geq b \end{cases}$$

3.53. Prove: The set P of all polynomials $p(x) = a_0 + a_1 x + \cdots + a_m^{x^m}$ with integral coefficients (that is, where $a_0, a_1, \ldots, a_m$ are integers) is denumerable.

Answers to Supplementary Problems

3.27. (a) Yes; (b) No; (c) Yes; (d) No.

3.28. (a) $\{(1, 1), (2, 4), (3, 3), (4, 3)\}$;
(b) $\{(1, 1), (2, 2), (3, 1), (4, 1)\}$;
(c) $\{(1, 4), (2, 3), (3, 3), (4, 4)\}$.

3.29. $\{(a, 4), (b, 6), (c, 4)\}$

3.30. (a) No, (b) yes, (c) no, (d) yes.

3.31. (a) f, h; (b) f, h; (c) f, h; (d) g.

3.32. (a) f, g; (b) h; (c) none; (d) none; (e) $\{$all prime numbers$\}$.

3.33. (a) g, k; (b) f, g, h; (c) g; (d) none.

3.34. $f^{-1}(x) = (x + 7)/3$

3.35. (a) 425631; (b) 416253; (c) 534261; (d) 415623; (e) 453261.

3.40. (a) A_2; (b) A_7; (c) A_r where r is the smallest integer in J; (d) A_s where s is the largest integer in J; (e) A_r where r is the smallest integer in K; (f) $\varnothing$.

3.41. (i) D_{14}; (ii) D_{24}; (iii) D_{12} (iv) D_3.

3.43. (a) 5; (b) $2^5 = 32$; (c) 0.

3.44. (a) $5^4 = 625$; (b) s^r; (c) $2^{16} = 65\ 536$; (d) 2.

3.47. (a) 13, −1, 34; (b) 14, 0, 34.

3.48. (a) 5; (b) 0; (c) 2; (d) $6 - 5 = 1$; (e) $11 - 5 = 6$.

3.49. (a) 30; (b) 48; (c) 6; (d) 870.

3.50. (a) 4; (b) 3; (c) −2.

3.51. $L(25) = 4$. Each time n is divided by 2, the value of L is increased by 1. Hence L is the greatest integer such that $2^L < N$. Thus $L(n) = \lfloor \log_2 n \rfloor$.

3.52. $Q(2, 7) = 5$, $Q(5, 3) = 10$, $Q(15, 2) = 42$.

3.53. Hint: Let P_k denote the set of polynomials $p(x)$ such that $m \leq k$ and each $|a_i| \leq k$. P_k is finite and $P = \cup_k P_k$.

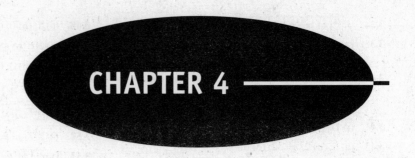

CHAPTER 4

Logic and Propositional Calculus

4.1 INTRODUCTION

Many algorithms and proofs use logical expressions such as:

"IF p THEN q"   or   "If p_1 AND p_2, THEN q_1 OR q_2"

Therefore it is necessary to know the cases in which these expressions are TRUE or FALSE, that is, to know the "truth value" of such expressions. We discuss these issues in this chapter.

We also investigate the truth value of quantified statements, which are statements which use the logical quantifiers "for every" and "there exist."

4.2 PROPOSITIONS AND COMPOUND STATEMENTS

A *proposition* (or *statement*) is a declarative statement which is true or false, but not both. Consider, for example, the following six sentences:

(i) Ice floats in water.   (iii) $2 + 2 = 4$ (v) Where are you going?

(ii) China is in Europe.   (iv) $2 + 2 = 5$ (vi) Do your homework.

The first four are propositions, the last two are not. Also, (i) and (iii) are true, but (ii) and (iv) are false.

Compound Propositions

Many propositions are *composite*, that is, composed of *subpropositions* and various connectives discussed subsequently. Such composite propositions are called *compound propositions*. A proposition is said to be *primitive* if it cannot be broken down into simpler propositions, that is, if it is not composite.

For example, the above propositions (i) through (iv) are primitive propositions. On the other hand, the following two propositions are composite:

"Roses are red and violets are blue." and "John is smart or he studies every night."

70

The fundamental property of a compound proposition is that its truth value is completely determined by the truth values of its subpropositions together with the way in which they are connected to form the compound propositions. The next section studies some of these connectives.

4.3 BASIC LOGICAL OPERATIONS

This section discusses the three basic logical operations of conjunction, disjunction, and negation which correspond, respectively, to the English words "and," "or," and "not."

Conjunction, $p \wedge q$

Any two propositions can be combined by the word "and" to form a compound proposition called the *conjunction* of the original propositions. Symbolically,

$$p \wedge q$$

read "p and q," denotes the conjunction of p and q. Since $p \wedge q$ is a proposition it has a truth value, and this truth value depends only on the truth values of p and q. Specifically:

Definition 4.1: If p and q are true, then $p \wedge q$ is true; otherwise $p \wedge q$ is false.

The truth value of $p \wedge q$ may be defined equivalently by the table in Fig. 4-1(a). Here, the first line is a short way of saying that if p is true and q is true, then $p \wedge q$ is true. The second line says that if p is true and q is false, then $p \wedge q$ is false. And so on. Observe that there are four lines corresponding to the four possible combinations of T and F for the two subpropositions p and q. Note that $p \wedge q$ is true only when both p and q are true.

p	q	$p \wedge q$
T	T	T
T	F	F
F	T	F
F	F	F

(a) "p and q"

p	q	$p \vee q$
T	T	T
T	F	T
F	T	T
F	F	F

(b) "p or q"

p	$\neg p$
T	F
F	T

(c) "not p"

Fig. 4-1

EXAMPLE 4.1 Consider the following four statements:

(i) Ice floats in water and $2 + 2 = 4$.   (iii) China is in Europe and $2 + 2 = 4$.

(ii) Ice floats in water and $2 + 2 = 5$.   (iv) China is in Europe and $2 + 2 = 5$.

Only the first statement is true. Each of the others is false since at least one of its substatements is false.

Disjunction, $p \vee q$

Any two propositions can be combined by the word "or" to form a compound proposition called the *disjunction* of the original propositions. Symbolically,

$$p \vee q$$

read "p or q," denotes the disjunction of p and q. The truth value of $p \vee q$ depends only on the truth values of p and q as follows.

Definition 4.2: If p and q are false, then $p \vee q$ is false; otherwise $p \vee q$ is true.

The truth value of $p \vee q$ may be defined equivalently by the table in Fig. 4-1(b). Observe that $p \vee q$ is false only in the fourth case when both p and q are false.

EXAMPLE 4.2 Consider the following four statements:

(i) Ice floats in water or $2 + 2 = 4$.   (iii) China is in Europe or $2 + 2 = 4$.

(ii) Ice floats in water or $2 + 2 = 5$.   (iv) China is in Europe or $2 + 2 = 5$.

Only the last statement (iv) is false. Each of the others is true since at least one of its sub-statements is true.

Remark: The English word "or" is commonly used in two distinct ways. Sometimes it is used in the sense of "p or q or both," i.e., at least one of the two alternatives occurs, as above, and sometimes it is used in the sense of "p or q but not both," i.e., exactly one of the two alternatives occurs. For example, the sentence "He will go to Harvard or to Yale" uses "or" in the latter sense, called the *exclusive disjunction*. Unless otherwise stated, "or" shall be used in the former sense. This discussion points out the precision we gain from our symbolic language: $p \vee q$ is defined by its truth table and *always* means "p and/or q."

Negation, $\neg p$

Given any proposition p, another proposition, called the *negation* of p, can be formed by writing "It is not true that ..." or "It is false that ..." before p or, if possible, by inserting in p the word "not." Symbolically, the negation of p, read "not p," is denoted by

$$\neg p$$

The truth value of $\neg p$ depends on the truth value of p as follows:

Definition 4.3: If p is true, then $\neg p$ is false; and if p is false, then $\neg p$ is true.

The truth value of $\neg p$ may be defined equivalently by the table in Fig. 4-1(c). Thus the truth value of the negation of p is always the opposite of the truth value of p.

EXAMPLE 4.3 Consider the following six statements:

(a_1) Ice floats in water.   (a_2) It is false that ice floats in water.   (a_3) Ice does not float in water.

(b_1) $2 + 2 = 5$ (b_2) It is false that $2 + 2 = 5$.   (b_3) $2 + 2 \neq 5$

Then (a_2) and (a_3) are each the negation of (a_1); and (b_2) and (b_3) are each the negation of (b_1). Since (a_1) is true, (a_2) and (a_3) are false; and since (b_1) is false, (b_2) and (b_3) are true.

Remark: The logical notation for the connectives "and," "or," and "not" is not completely standardized. For example, some texts use:

$$p \,\&\, q, \; p \cdot q \text{ or } pq \quad \text{for} \quad p \wedge q$$
$$p + q \qquad\qquad \text{for} \quad p \vee q$$
$$p', \; \bar{p} \text{ or } \sim p \quad \text{for} \quad \neg p$$

4.4 PROPOSITIONS AND TRUTH TABLES

Let $P(p, q, \ldots)$ denote an expression constructed from logical variables $p, q, \ldots$, which take on the value TRUE (T) or FALSE (F), and the logical connectives $\wedge$, $\vee$, and $\neg$ (and others discussed subsequently). Such an expression $P(p, q, \ldots)$ will be called a *proposition*.

The main property of a proposition $P(p, q, \ldots)$ is that its truth value depends exclusively upon the truth values of its variables, that is, the truth value of a proposition is known once the truth value of each of its variables is known. A simple concise way to show this relationship is through a *truth table*. We describe a way to obtain such a truth table below.

Consider, for example, the proposition $\neg(p \wedge \neg q)$. Figure 4-2(a) indicates how the truth table of $\neg(p \wedge \neg q)$ is constructed. Observe that the first columns of the table are for the variables $p, q, \ldots$ and that there are enough rows in the table, to allow for all possible combinations of T and F for these *variables*. (For 2 variables, as above, 4 rows are necessary; for 3 variables, 8 rows are necessary; and, in general, for n variables, 2^n rows are required.) There is then a column for each "elementary" stage of the construction of the proposition, the truth value at each step being determined from the previous stages by the definitions of the connectives $\wedge$, $\vee$, $\neg$. Finally we obtain the truth value of the proposition, which appears in the last column.

The actual truth table of the proposition $\neg(p \wedge \neg q)$ is shown in Fig. 4-2(b). It consists precisely of the columns in Fig. 4-2(a) which appear under the variables and under the proposition; the other columns were merely used in the construction of the truth table.

p	q	$\neg q$	$p \wedge \neg q$	$\neg(p \wedge \neg q)$
T	T	F	F ·	T
T	F	T	T ⸱	F
F	T	F	F	T
F	F	T	F	T

p	q	$\neg(p \wedge \neg q)$
T	T	T
T	F	F
F	T	T
F	F	T

(a)   (b)

Fig. 4-2

Remark: In order to avoid an excessive number of parentheses, we sometimes adopt an order of precedence for the logical connectives. Specifically,

$$\neg \text{ has precedence over } \wedge \text{ which has precedence over } \vee$$

For example, $\neg p \wedge q$ means $(\neg p) \wedge q$ and not $\neg(p \wedge q)$.

Alternate Method for Constructing a Truth Table

Another way to construct the truth table for $\neg(p \wedge \neg q)$ follows:

(a) First we construct the truth table shown in Fig. 4-3. That is, first we list all the variables and the combinations of their truth values. Also there is a final row labeled "step." Next the proposition is written on the top row to the right of its variables with sufficient space so there is a column under each variable and under each logical operation in the proposition. Lastly (Step 1), the truth values of the variables are entered in the table under the variables in the proposition.

(b) Now additional truth values are entered into the truth table column by column under each logical operation as shown in Fig. 4-4. We also indicate the step in which each column of truth values is entered in the table.

The truth table of the proposition then consists of the original columns under the variables and the last step, that is, the last column is entered into the table.

p	q	$\neg$	$(p$	$\wedge$	$\neg$	$q)$
T	T		T			T
T	F		T			F
F	T		F			T
F	F		F			F
Step						

Fig. 4-3

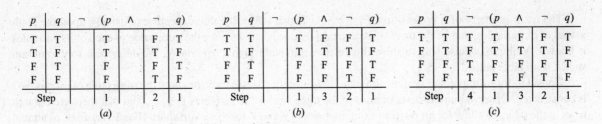

Fig. 4-4

4.5   TAUTOLOGIES AND CONTRADICTIONS

Some propositions $P(p, q, \ldots)$ contain only T in the last column of their truth tables or, in other words, they are true for any truth values of their variables. Such propositions are called *tautologies*. Analogously, a proposition $P(p, q, \ldots)$ is called a *contradiction* if it contains only F in the last column of its truth table or, in other words, if it is false for any truth values of its variables. For example, the proposition "*p* or not *p*," that is, $p \vee \neg p$, is a tautology, and the proposition "*p* and not *p*," that is, $p \wedge \neg p$, is a contradiction. This is verified by looking at their truth tables in Fig. 4-5. (The truth tables have only two rows since each proposition has only the one variable *p*.)

p	$\neg p$	$p \vee \neg p$
T	F	T
F	T	T

p	$\neg p$	$p \wedge \neg p$
T	F	F
F	T	F

       (*a*) $p \vee \neg p$ (*b*) $p \wedge \neg p$

Fig. 4-5

Note that the negation of a tautology is a contradiction since it is always false, and the negation of a contradiction is a tautology since it is always true.

Now let $P(p, q, \ldots)$ be a tautology, and let $P_1(p, q, \ldots)$, $P_2(p, q, \ldots), \ldots$ be any propositions. Since $P(p, q, \ldots)$ does not depend upon the particular truth values of its variables $p, q, \ldots$, we can substitute P_1 for p, P_2 for $q, \ldots$ in the tautology $P(p, q, \ldots)$ and still have a tautology. In other words:

Theorem 4.1 (Principle of Substitution):   If $P(p, q, \ldots)$ is a tautology, then $P(P_1, P_2, \ldots)$ is a tautology for any propositions $P_1, P_2, \ldots$.

4.6   LOGICAL EQUIVALENCE

Two propositions $P(p, q, \ldots)$ and $Q(p, q, \ldots)$ are said to be *logically equivalent*, or simply *equivalent* or *equal*, denoted by

$$P(p, q, \ldots) \equiv Q(p, q, \ldots)$$

if they have identical truth tables. Consider, for example, the truth tables of $\neg(p \wedge q)$ and $\neg p \vee \neg q$ appearing in Fig. 4-6. Observe that both truth tables are the same, that is, both propositions are false in the first case and true in the other three cases. Accordingly, we can write

$$\neg(p \wedge q) \equiv \neg p \vee \neg q$$

In other words, the propositions are logically equivalent.

Remark: Let *p* be "Roses are red" and *q* be "Violets are blue." Let *S* be the statement:

"It is not true that roses are red and violets are blue."

Then *S* can be written in the form $\neg(p \wedge q)$. However, as noted above, $\neg(p \wedge q) \equiv \neg p \vee \neg q$. Accordingly, *S* has the same meaning as the statement:

"Roses are not red, or violets are not blue."

p	q	$p \wedge q$	$\neg(p \wedge q)$
T	T	T	F
T	F	F	T
F	T	F	T
F	F	F	T

(a) $\neg(p \wedge q)$

p	q	$\neg p$	$\neg q$	$\neg p \vee \neg q$
T	T	F	F	F
T	F	F	T	T
F	T	T	F	T
F	F	T	T	T

(b) $\neg p \vee \neg q$

Fig. 4-6

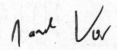

4.7 ALGEBRA OF PROPOSITIONS

Propositions satisfy various laws which are listed in Table 4-1. (In this table, T and F are restricted to the truth values "True" and "False," respectively.) We state this result formally.

Theorem 4.2: Propositions satisfy the laws of Table 4-1.

(Observe the similarity between this Table 4-1 and Table 1-1 on sets.)

Table 4-1   Laws of the algebra of propositions

Idempotent laws:	(1a) $p \vee p \equiv p$	(1b) $p \wedge p \equiv p$
Associative laws:	(2a) $(p \vee q) \vee r \equiv p \vee (q \vee r)$	(2b) $(p \wedge q) \wedge r \equiv p \wedge (q \wedge r)$
Commutative laws:	(3a) $p \vee q \equiv q \vee p$	(3b) $p \wedge q \equiv q \wedge p$
Distributive laws:	(4a) $p \vee (q \wedge r) \equiv (p \vee q) \wedge (p \vee r)$	(4b) $p \wedge (q \vee r) \equiv (p \wedge q) \vee (p \wedge r)$
Identity laws:	(5a) $p \vee F \equiv p$	(5b) $p \wedge T \equiv p$
	(6a) $p \vee T \equiv T$	(6b) $p \wedge F \equiv F$
Involution law:	(7) $\neg\neg p \equiv p$	
Complement laws:	(8a) $p \vee \neg p \equiv T$	(8b) $p \wedge \neg p \equiv F$
	(9a) $\neg T \equiv F$	(9b) $\neg F \equiv T$
DeMorgan's laws:	(10a) $\neg(p \vee q) \equiv \neg p \wedge \neg q$	(10b) $\neg(p \wedge q) \equiv \neg p \vee \neg q$

4.8 CONDITIONAL AND BICONDITIONAL STATEMENTS

Many statements, particularly in mathematics, are of the form "If p then q." Such statements are called *conditional* statements and are denoted by

$$p \to q$$

The conditional $p \to q$ is frequently read "p implies q" or "p only if q."

Another common statement is of the form "p if and only if q." Such statements are called *biconditional* statements and are denoted by

$$p \leftrightarrow q$$

The truth values of $p \to q$ and $p \leftrightarrow q$ are defined by the tables in Fig. 4-7(a) and (b). Observe that:

(a) The conditional $p \to q$ is false only when the first part p is true and the second part q is false. Accordingly, when p is false, the conditional $p \to q$ is true regardless of the truth value of q.

(b) The biconditional $p \leftrightarrow q$ is true whenever p and q have the same truth values and false otherwise.

The truth table of $\neg p \wedge q$ appears in Fig. 4-7(c). Note that the truth table of $\neg p \vee q$ and $p \to q$ are identical, that is, they are both false only in the second case. Accordingly, $p \to q$ is logically equivalent to $\neg p \vee q$; that is,

$$p \to q \equiv \neg p \vee q$$

In other words, the conditional statement "If p then q" is logically equivalent to the statement "Not p or q" which only involves the connectives $\vee$ and $\neg$ and thus was already a part of our language. We may regard $p \to q$ as an abbreviation for an oft-recurring statement.

p	q	$p \to q$
T	T	T
T	F	F
F	T	T
F	F	T

(a) $p \to q$

p	q	$p \leftrightarrow q$
T	T	T
T	F	F
F	T	F
F	F	T

(b) $p \leftrightarrow q$

p	q	$\neg p$	$\neg p \vee q$
T	T	F	T
T	F	F	F
F	T	T	T
F	F	T	T

(c) $\neg p \vee q$

Fig. 4-7

4.9 ARGUMENTS

An *argument* is an assertion that a given set of propositions $P_1, P_2, \ldots, P_n$, called *premises*, yields (has a consequence) another proposition Q, called the *conclusion*. Such an argument is denoted by

$$P_1, P_2, \ldots, P_n \vdash Q$$

The notion of a "logical argument" or "valid argument" is formalized as follows:

Definition 4.4: An argument $P_1, P_2, \ldots, P_n \vdash Q$ is said to be *valid* if Q is true whenever all the premises $P_1, P_2, \ldots, P_n$ are true.

An argument which is not valid is called *fallacy*.

EXAMPLE 4.4

(a) The following argument is valid:

$$p, \ p \to q \vdash q \quad \textbf{(Law of Detachment)}$$

The proof of this rule follows from the truth table in Fig. 4-7(a). Specifically, p and $p \to q$ are true simultaneously only in Case (row) 1, and in this case q is true.

(b) The following argument is a fallacy:

$$p \to q, \ q \vdash p$$

For $p \to q$ and q are both true in Case (row) 3 in the truth table in Fig. 4-7(a), but in this case p is false.

Now the propositions $P_1, P_2, \ldots, P_n$ are true simultaneously if and only if the proposition $P_1 \wedge P_2 \wedge \ldots P_n$ is true. Thus the argument $P_1, P_2, \ldots, P_n \vdash Q$ is valid if and only if Q is true whenever $P_1 \wedge P_2 \wedge \ldots \wedge P_n$ is true or, equivalently, if the proposition $(P_1 \wedge P_2 \wedge \ldots \wedge P_n) \to Q$ is a tautology. We state this result formally.

Theorem 4.3: The argument $P_1, P_2, \ldots, P_n \vdash Q$ is valid if and only if the proposition $(P_1 \wedge P_2 \ldots \wedge P_n) \to Q$ is a tautology.

We apply this theorem in the next example.

EXAMPLE 4.5 A fundamental principle of logical reasoning states:

"If p implies q and q implies r, then p implies r"

$$p \to q, \ q \to r \vdash p \to r$$

p	q	r	[(p	$\rightarrow$	q)	$\wedge$	(q	$\rightarrow$	r)]	$\rightarrow$	(p	$\rightarrow$	r)
T	T	T	T	T	T	T	T	T	T	T	T	T	T
T	T	F	T	T	T	F	T	F	F	T	T	F	F
T	F	T	T	F	F	F	F	T	T	T	T	T	T
T	F	F	T	F	F	F	F	T	F	T	T	F	F
F	T	T	F	T	T	T	T	T	T	T	F	T	T
F	T	F	F	T	T	F	T	F	F	T	F	F	F
F	F	T	F	T	F	T	F	T	T	T	F	T	T
F	F	F	F	T	F	T	F	T	F	T	F	T	F
Step			1	2	1	3	1	2	1	4	1	2	1

Fig. 4-8

That is, the following argument is valid:

$$p \rightarrow q, \; q \rightarrow r \vdash p \rightarrow r \quad \textbf{(Law of Syllogism)}$$

This fact is verified by the truth table in Fig. 4-8 which shows that the following proposition is a tautology:

$$[(p \rightarrow q) \wedge (q \rightarrow r)] \rightarrow (p \rightarrow r)$$

Equivalently, the argument is valid since the premises $p \rightarrow q$ and $q \rightarrow r$ are true simultaneously only in Cases (rows) 1, 5, 7, and 8, and in these cases the conclusion $p \rightarrow r$ is also true. (Observe that the truth table required $2^3 = 8$ lines since there are three variables p, q, and r.)

We now apply the above theory to arguments involving specific statements. We emphasize that the validity of an argument does not depend upon the truth values nor the content of the statements appearing in the argument, but upon the particular form of the argument. This is illustrated in the following example.

EXAMPLE 4.6 Consider the following argument:

S_1 : If a man is a bachelor, he is unhappy.

S_2 : If a man is unhappy, he dies young.

_____

S : Bachelors die young

Here the statement S below the line denotes the conclusion of the argument, and the statements S_1 and S_2 above the line denote the premises. We claim that the argument S_1, $S_2 \vdash S$ is valid. For the argument is of the form

$$p \rightarrow q, \; q \rightarrow r \vdash p \rightarrow r$$

where p is "He is a bachelor," q is "He is unhappy" and r is "He dies young;" and by Example 4.5 this argument (Law of Syllogism) is valid.

4.10 PROPOSITIONAL FUNCTIONS, QUANTIFIERS

Let A be a given set. A _propositional function_ (or an _open sentence_ or _condition_) defined on A is an expression

$$p(x)$$

which has the property that $p(a)$ is true or false for each $a \in A$. That is, $p(x)$ becomes a statement (with a truth value) whenever any element $a \in A$ is substituted for the variable x. The set A is called the _domain_ of $p(x)$, and the set T_p of all elements of A for which $p(a)$ is true is called the _truth set_ of $p(x)$. In other words,

$$T_p = \{x \mid x \in A, \; p(x) \text{ is true}\} \quad \text{or} \quad T_p = \{x \mid p(x)\}$$

Frequently, when A is some set of numbers, the condition $p(x)$ has the form of an equation or inequality involving the variable x.

EXAMPLE 4.7 Find the truth set for each propositional function $p(x)$ defined on the set **N** of positive integers.

(a) Let $p(x)$ be "$x + 2 > 7$." Its truth set is $\{6, 7, 8, \ldots\}$ consisting of all integers greater than 5.

(b) Let $p(x)$ be "$x + 5 < 3$." Its truth set is the empty set $\varnothing$. That is, $p(x)$ is not true for any integer in **N**.

(c) Let $p(x)$ be "$x + 5 > 1$." Its truth set is **N**. That is, $p(x)$ is true for every element in **N**.

Remark: The above example shows that if $p(x)$ is a propositional function defined on a set A then $p(x)$ could be true for all $x \in A$, for some $x \in A$, or for no $x \in A$. The next two subsections discuss quantifiers related to such propositional functions.

Universal Quantifier

Let $p(x)$ be a propositional function defined on a set A. Consider the expression

$$(\forall x \in A)p(x) \quad \text{or} \quad \forall x \, p(x) \tag{4.1}$$

which reads "For every x in A, $p(x)$ is a true statement" or, simply, "For all x, $p(x)$." The symbol

$$\forall$$

which reads "for all" or "for every" is called the *universal quantifier*. The statement *(4.1)* is equivalent to the statement

$$T_p = \{x \mid x \in A, \; p(x)\} = A \tag{4.2}$$

that is, that the truth set of $p(x)$ is the entire set A.

The expression $p(x)$ by itself is an open sentence or condition and therefore has no truth value. However, $\forall x \, p(x)$, that is $p(x)$ preceded by the quantifier $\forall$, does have a truth value which follows from the equivalence of *(4.1)* and *(4.2)*. Specifically:

Q_1: If $\{x \mid x \in A, \; p(x)\} = A$ then $\forall x \, p(x)$ is true; otherwise, $\forall x \, p(x)$ is false.

EXAMPLE 4.8

(a) The proposition $(\forall n \in \mathbf{N})(n + 4 > 3)$ is true since $\{n \mid n + 4 > 3\} = \{1, 2, 3, \ldots\} = \mathbf{N}$.

(b) The proposition $(\forall n \in \mathbf{N})(n + 2 > 8)$ is false since $\{n \mid n + 2 > 8\} = \{7, 8, \ldots\} \neq \mathbf{N}$.

(c) The symbol $\forall$ can be used to define the intersection of an indexed collection $\{A_i \mid i \in I\}$ of sets A_i as follows:

$$\cap(A_i \mid i \in I) = \{x \mid \forall_i \in I, \; x \in A_i\}$$

Existential Quantifier

Let $p(x)$ be a propositional function defined on a set A. Consider the expression

$$(\exists x \in A)p(x) \quad \text{or} \quad \exists x, \; p(x) \tag{4.3}$$

which reads "There exists an x in A such that $p(x)$ is a true statement" or, simply, "For some x, $p(x)$." The symbol

$$\exists$$

which reads "there exists" or "for some" or "for at least one" is called the *existential quantifier*. Statement (4.3) is equivalent to the statement

$$T_p = \{x \mid x \in A, \ p(x)\} \neq \varnothing \tag{4.4}$$

i.e., that the truth set of $p(x)$ is not empty. Accordingly, $\exists x \ p(x)$, that is, $p(x)$ preceded by the quantifier $\exists$, does have a truth value. Specifically:

Q_2: If $\{x \mid p(x)\} \neq \varnothing$ then $\exists x \ p(x)$ is true; otherwise, $\exists x \ p(x)$ is false.

EXAMPLE 4.9

(a) The proposition $(\exists n \in N)(n + 4 < 7)$ is true since $\{n \mid n + 4 < 7\} = \{1, 2\} \neq \varnothing$.

(b) The proposition $(\exists n \in N)(n + 6 < 4)$ is false since $\{n \mid n + 6 < 4\} = \varnothing$.

(c) The symbol $\exists$ can be used to define the union of an indexed collection $\{A_i \mid i \in I\}$ of sets A_i as follows:

$$\cup(A_i \mid i \in I) = \{x \mid \exists i \in I, \ x \mid \in A_i\}$$

4.11 NEGATION OF QUANTIFIED STATEMENTS

Consider the statement: "All math majors are male." Its negation reads:

> "It is not the case that all math majors are male" or, equivalently, "There exists at least one math major who is a female (not male)"

Symbolically, using M to denote the set of math majors, the above can be written as

$$\neg(\forall x \in M)(x \text{ is male}) \equiv (\exists x \in M)(x \text{ is not male})$$

or, when $p(x)$ denotes "x is male,"

$$\neg(\forall x \in M)p(x) \equiv (\exists x \in M)\neg p(x) \quad \text{or} \quad \neg\forall x p(x) \equiv \exists x \neg p(x)$$

The above is true for any proposition $p(x)$. That is:

Theorem 4.4 (DeMorgan): $\neg(\forall x \in A)p(x) \equiv (\exists x \in A)\neg p(x)$.

In other words, the following two statements are equivalent:
(1) It is not true that, for all $a \in A$, $p(a)$ is true. (2) There exists an $a \in A$ such that $p(a)$ is false.

There is an analogous theorem for the negation of a proposition which contains the existential quantifier.

Theorem 4.5 (DeMorgan): $\neg(\exists x \in A)p(x) \equiv (\forall x \in A)\neg p(x)$.

That is, the following two statements are equivalent:
(1) It is not true that for some $a \in A$, $p(a)$ is true. (2) For all $a \in A$, $p(a)$ is false.

EXAMPLE 4.10

(a) The following statements are negatives of each other:

"For all positive integers n we have $n + 2 > 8$"
"There exists a positive integer n such that $n + 2 \not> 8$"

(b) The following statements are also negatives of each other:

"There exists a (living) person who is 150 years old"
"Every living person is not 150 years old"

Remark: The expression $\neg p(x)$ has the obvious meaning:

"The statement $\neg p(a)$ is true when $p(a)$ is false, and vice versa"

Previously, $\neg$ was used as an operation on statements; here $\neg$ is used as an operation on propositional functions. Similarly, $p(x) \wedge q(x)$, read "$p(x)$ and $q(x)$," is defined by:

"The statement $p(a) \wedge q(a)$ is true when $p(a)$ and $q(a)$ are true"

Similarly, $p(x) \vee q(x)$, read "$p(x)$ or $q(x)$," is defined by:

"The statement $p(a) \vee q(a)$ is true when $p(a)$ or $q(a)$ is true"

Thus in terms of truth sets:

   (i) $\neg p(x)$ is the complement of $p(x)$.

  (ii) $p(x) \wedge q(x)$ is the intersection of $p(x)$ and $q(x)$.

 (iii) $p(x) \vee q(x)$ is the union of $p(x)$ and $q(x)$.

One can also show that the laws for propositions also hold for propositional functions. For example, we have DeMorgan's laws:

$$\neg(p(x) \wedge q(x)) \equiv \neg p(x) \vee \neg q(x) \quad \text{and} \quad \neg(p(x) \vee q(x)) \equiv \neg p(x) \wedge \neg q(x)$$

Counterexample

Theorem 4.6 tells us that to show that a statement $\forall x,\, p(x)$ is false, it is equivalent to show that $\exists x \neg p(x)$ is true or, in other words, that there is an element x_0 with the property that $p(x_0)$ is false. Such an element x_0 is called a *counterexample* to the statement $\forall x,\, p(x)$.

EXAMPLE 4.11

(a) Consider the statement $\forall x \in \mathbf{R},\, |x| \neq 0$. The statement is false since 0 is a counterexample, that is, $|0| \neq 0$ is not true.

(b) Consider the statement $\forall x \in \mathbf{R},\, x^2 \geq x$. The statement is not true since, for example, $\frac{1}{2}$ is a counterexample. Specifically, $(\frac{1}{2})^2 \geq \frac{1}{2}$ is not true, that is, $(\frac{1}{2})^2 < \frac{1}{2}$.

(c) Consider the statement $\forall x \in \mathbf{N},\, x^2 \geq x$. This statement is true where $\mathbf{N}$ is the set of positive integers. In other words, there does not exist a positive integer n for which $n^2 < n$.

Propositional Functions with more than One Variable

A propositional function (of n variables) defined over a product set $A = A_1 \times \cdots \times A_n$ is an expression

$$p(x_1, x_2, \ldots, x_n)$$

which has the property that $p(a_1, a_2, \ldots, a_n)$ is true or false for any n-tuple $(a_1, \ldots a_n)$ in A. For example,

$$x + 2y + 3z < 18$$

is a propositional function on $\mathbf{N}^3 = \mathbf{N} \times \mathbf{N} \times \mathbf{N}$. Such a propositional function has no truth value. However, we do have the following:

Basic Principle: A propositional function preceded by a quantifier for each variable, for example,

$$\forall x \exists y, \, p(x, y) \quad \text{or} \quad \exists x \, \forall y \, \exists z, \, p(x, y, z)$$

denotes a statement and has a truth value.

EXAMPLE 4.12 Let $B = \{1, 2, 3, \ldots, 9\}$ and let $p(x, y)$ denote "$x + y = 10$." Then $p(x, y)$ is a propositional function on $A = B^2 = B \times B$.

(a) The following is a statement since there is a quantifier for each variable:

$$\forall x \exists y, \, p(x, y), \quad \text{that is,} \quad \text{"For every } x, \text{ there exists a } y \text{ such that } x + y = 10\text{"}$$

This statement is true. For example, if $x = 1$, let $y = 9$; if $x = 2$, let $y = 8$, and so on.

(b) The following is also a statement:

$$\exists y \forall x, \, p(x, y), \quad \text{that is,} \quad \text{"There exists a } y \text{ such that, for every } x, \text{ we have } x + y = 10\text{"}$$

No such y exists; hence this statement is false.

Note that the only difference between (a) and (b) is the order of the quantifiers. Thus a different ordering of the quantifiers may yield a different statement. We note that, when translating such quantified statements into English, the expression "such that" frequently follows "there exists."

Negating Quantified Statements with more than One Variable

Quantified statements with more than one variable may be negated by successively applying Theorems 4.5 and 4.6. Thus each $\forall$ is changed to $\exists$ and each $\exists$ is changed to $\forall$ as the negation symbol $\neg$ passes through the statement from left to right. For example,

$$\neg[\forall x \exists y \exists z, \, p(x, y, z)] \equiv \exists x \neg[\exists y \exists z, \, p(x, y, z)] \equiv \neg \exists z \forall y [\exists z, \, p(x, y, z)]$$
$$\equiv \exists x \forall y \forall z, \, \neg p(x, y, z)$$

Naturally, we do not put in all the steps when negating such quantified statements.

EXAMPLE 4.13

(a) Consider the quantified statement:

"Every student has at least one course where the lecturer is a teaching assistant."

Its negation is the statement:

"There is a student such that in every course the lecturer is not a teaching assistant."

(b) The formal definition that L is the limit of a sequence $a_1, a_2, \ldots$ follows:

$$\forall \in > 0, \exists n_0 \in \mathbf{N}, \ \forall n > n_0 \text{ we have } |a_n - L| < \in$$

Thus L is not the limit of the sequence $a_1, a_2, \ldots$ when:

$$\exists \in > 0, \forall n_0 \in \mathbf{N}, \ \exists n > n_0 \text{ such that } |a_n - L| \geq \in$$

Solved Problems

PROPOSITIONS AND TRUTH TABLES

4.1. Let p be "It is cold" and let q be "It is raining". Give a simple verbal sentence which describes each of the following statements: (a) $\neg p$; (b) $p \wedge q$; (c) $p \vee q$; (d) $q \vee \neg p$.

In each case, translate $\wedge$, $\vee$, and $\sim$ to read "and," "or," and "It is false that" or "not," respectively, and then simplify the English sentence.

(a) It is not cold.   (c) It is cold or it is raining.

(b) It is cold and raining.   (d) It is raining or it is not cold.

4.2. Find the truth table of $\neg p \wedge q$.

Construct the truth table of $\neg p \wedge q$ as in Fig. 4-9(a).

p	q	$\neg p$	$\neg p \wedge q$
T	T	F	F
T	F	F	F
F	T	T	T
F	F	T	F

(a) $\neg p \wedge q$

p	q	$p \wedge q$	$\neg(p \wedge q)$	$p \vee \neg(p \wedge q)$
T	T	T	F	T
T	F	F	T	T
F	T	F	T	T
F	F	F	T	T

(b) $p \vee \neg(p \wedge q)$

Fig. 4-9

4.3. Verify that the proposition $p \vee \neg(p \wedge q)$ is a tautology.

Construct the truth table of $p \vee \neg(p \wedge q)$ as shown in Fig. 4-9(b). Since the truth value of $p \vee \neg(p \wedge q)$ is T for all values of p and q, the proposition is a tautology.

4.4. Show that the propositions $\neg(p \wedge q)$ and $\neg p \vee \neg q$ are logically equivalent.

Construct the truth tables for $\neg(p \wedge q)$ and $\neg p \vee \neg q$ as in Fig. 4-10. Since the truth tables are the same (both propositions are false in the first case and true in the other three cases), the propositions $\neg(p \wedge q)$ and $\neg p \vee \neg q$ are logically equivalent and we can write

$$\neg(p \wedge q) \equiv \neg p \vee \neg q.$$

p	q	$p \wedge q$	$\neg(p \wedge q)$
T	T	T	F
T	F	F	T
F	T	F	T
F	F	F	T

(a) $\neg(p \wedge q)$

p	q	$\neg p$	$\neg q$	$\neg p \vee \neg q$
T	T	F	F	F
T	F	F	T	T
F	T	T	F	T
F	F	T	T	T

(b) $\neg p \vee \neg q$

Fig. 4-10

4.5. Use the laws in Table 4-1 to show that $\neg(p \wedge q) \vee (\neg p \wedge q) \equiv \neg p$.

	Statement	Reason
(1)	$\neg(p \vee q) \vee (\neg p \wedge q) \equiv (\neg p \wedge \neg q) \vee (\neg p \wedge q)$	DeMorgan's law
(2)	$\equiv \neg p \wedge (\neg q \vee q)$	Distributive law
(3)	$\equiv \neg p \wedge T$	Complement law
(4)	$\equiv \neg p$	Identity law

CONDITIONAL STATEMENTS

4.6. Rewrite the following statements without using the conditional:

(a) If it is cold, he wears a hat.

(b) If productivity increases, then wages rise.

Recall that "If p then q" is equivalent to "Not p or q;" that is, $p \rightarrow q \equiv \neg p \vee q$. Hence,

(a) It is not cold or he wears a hat.

(b) Productivity does not increase or wages rise.

4.7. Consider the conditional proposition $p \rightarrow q$. The simple propositions $q \rightarrow p$, $\neg p \rightarrow \neg q$ and $\neg q \rightarrow \neg p$ are called, respectively, the *converse*, *inverse*, and *contrapositive* of the conditional $p \rightarrow q$. Which if any of these propositions are logically equivalent to $p \rightarrow q$?

Construct their truth tables as in Fig. 4-11. Only the contrapositive $\neg q \rightarrow \neg p$ is logically equivalent to the original conditional proposition $p \rightarrow q$.

p	q	$\neg p$	$\neg q$	Conditional $p \rightarrow q$	Converse $q \rightarrow p$	Inverse $\neg p \rightarrow \neg q$	Contrapositive $\neg q \rightarrow \neg p$
T	T	F	F	T	T	T	T
T	F	F	T	F	T	T	F
F	T	T	F	T	F	F	T
F	F	T	T	T	T	T	T

Fig. 4-11

4.8. Determine the contrapositive of each statement:

(a) If Erik is a poet, then he is poor.

(b) Only if Marc studies will he pass the test.

(a) The contrapositive of $p \rightarrow q$ is $\neg q \rightarrow \neg p$. Hence the contrapositive follows:

If Erik is not poor, then he is not a poet.

(b) The statement is equivalent to: "If Marc passes the test, then he studied." Thus its contrapositive is:

If Marc does not study, then he will not pass the test.

4.9. Write the negation of each statement as simply as possible:

(a) She works and she will not earn money.

(b) He swims if and only if the water is warm.

(c) If it snows, then they do not drive the car.

(a) Note that $\neg(p \rightarrow q) \equiv p \wedge \neg q$; hence the negation of the statement is:

She works or she will not earn money.

(b) Note that $\neg(p \leftrightarrow q) \equiv p \leftrightarrow \neg q \equiv \neg p \leftrightarrow q$; hence the negation of the statement is either of the following:

> He swims if and only if the water is not warm.
> He does not swim if and only if the water is warm.

(c) Note that $\neg(p \rightarrow \neg q) \equiv p \wedge \neg\neg q \equiv p \wedge q$. Hence the negation of the statement is:

> It snows and they drive the car.

ARGUMENTS

4.10. Show that the following argument is a fallacy: $p \rightarrow q, \neg p \vdash \neg q$.

Construct the truth table for $[(p \rightarrow q) \wedge \neg p] \rightarrow \neg q$ as in Fig. 4-12. Since the proposition $[(p \rightarrow q) \wedge \neg p] \rightarrow \neg q$ is not a tautology, the argument is a fallacy. Equivalently, the argument is a fallacy since in the third line of the truth table $p \rightarrow q$ and $\neg p$ are true but $\neg q$ is false.

p	q	$p \rightarrow q$	$\neg p$	$(p \rightarrow q) \wedge \neg p$	$\neg q$	$[(p \rightarrow q) \wedge \neg p] \rightarrow \neg p$
T	T	T	F	F	F	T
T	F	F	F	F	T	T
F	T	T	T	T	F	F
F	F	T	T	T	T	T

Fig. 4-12

4.11. Determine the validity of the following argument: $p \rightarrow q, \neg p \vdash \neg p$.

Construct the truth table for $[(p \rightarrow q) \wedge \neg q] \rightarrow \neg p$ as in Fig. 4-13. Since the proposition $[(p \rightarrow q) \wedge \neg q] \rightarrow \neg p$ is a tautology, the argument is valid.

p	q	$[(p$	$\rightarrow$	$q)$	$\wedge$	$\neg$	$q]$	$\rightarrow$	$\neg$	p
T	T	T	T	T	F	F	T	T	F	T
T	F	T	F	F	F	T	F	T	F	T
F	T	F	T	T	F	F	T	T	T	F
F	F	F	T	F	T	T	F	T	T	F
Step		1	2	1	3	2	1	4	2	1

Fig. 4-13

4.12. Prove the following argument is valid: $p \rightarrow \neg q, r \rightarrow q, r \vdash \neg p$.

Construct the truth table of the premises and conclusions as in Fig. 4-14(a). Now, $p \rightarrow \neg q, r \rightarrow q$, and r are true simultaneously only in the fifth row of the table, where $\neg p$ is also true. Hence the argument is valid.

	p	q	r	$p \rightarrow \neg q$	$r \rightarrow q$	$\neg q$
1	T	T	T	F	T	F
2	T	T	F	F	T	F
3	T	F	T	T	F	F
4	T	F	F	T	T	F
5	F	T	T	T	T	T
6	F	T	F	T	T	T
7	F	F	T	T	F	T
8	F	F	F	T	T	T

p	q	$\neg q$	$p \rightarrow \neg q$	$\neg p$
T	T	F	F	F
T	F	T	T	F
F	T	F	T	T
F	F	T	T	T

(a)　　　　　　　　　　　　　　　　　　(b)

Fig. 4-14

4.13. Determine the validity of the following argument:

> If 7 is less than 4, then 7 is not a prime number.
> 7 is not less than 4.
> _____
> 7 is a prime number.

First translate the argument into symbolic form. Let p be "7 is less than 4" and q be "7 is a prime number." Then the argument is of the form

$$p \rightarrow \neg q, \ \neg q \vdash q$$

Now, we construct a truth table as shown in Fig. 4-14(b). The above argument is shown to be a fallacy since, in the fourth line of the truth table, the premises $p \rightarrow \neg q$ and $\neg p$ are true, but the conclusion q is false.

Remark: The fact that the conclusion of the argument happens to be a true statement is irrelevant to the fact that the argument presented is a fallacy.

4.14. Test the validity of the following argument:

> If two sides of a triangle are equal, then the opposite angles are equal.
> Two sides of a triangle are not equal.
> _____
> The opposite angles are not equal.

First translate the argument into the symbolic form $p \rightarrow q, \ \neg p \vdash \neg q$, where p is "Two sides of a triangle are equal" and q is "The opposite angles are equal." By Problem 4.10, this argument is a fallacy.

Remark: Although the conclusion *does* follow from the second premise and axioms of Euclidean geometry, the above argument does not constitute such a proof since the argument is a fallacy.

QUANTIFIERS AND PROPOSITIONAL FUNCTIONS

4.15. Let $A = \{1, 2, 3, 4, 5\}$. Determine the truth value of each of the following statements:

(a) $(\exists x \in A)(x + 3 = 10)$ (c) $(\exists x \in A)(x + 3 < 5)$

(b) $(\forall x \in A)(x + 3 < 10)$ (d) $(\forall x \in A)(x + 3 \leq 7)$

(a) False. For no number in A is a solution to $x + 3 = 10$.

(b) True. For every number in A satisfies $x + 3 < 10$.

(c) True. For if $x_0 = 1$, then $x_0 + 3 < 5$, i.e., 1 is a solution.

(d) False. For if $x_0 = 5$, then $x_0 + 3$ is not less than or equal 7. In other words, 5 is not a solution to the given condition.

4.16. Determine the truth value of each of the following statements where $\mathbf{U} = \{1, 2, 3\}$ is the universal set:
(a) $\exists x \forall y, \ x^2 < y + 1$;   (b) $\forall x \exists y, \ x^2 + y^2 < 12$;   (c) $\forall x \forall y, \ x^2 + y^2 < 12$.

(a) True. For if $x = 1$, then 1, 2, and 3 are all solutions to $1 < y + 1$.

(b) True. For each x_0, let $y = 1$; then $x_0^2 + 1 < 12$ is a true statement.

(c) False. For if $x_0 = 2$ and $y_0 = 3$, then $x_0^2 + y_0^2 < 12$ is not a true statement.

4.17. Negate each of the following statements:

(a) $\exists x \ \forall y, \ p(x, y)$;   (b) $\forall x \ \forall y, \ p(x, y)$;   (c) $\exists y \ \exists x \ \forall z, \ p(x, y, z)$.

Use $\neg \forall x \ p(x) \equiv \exists x \neg p(x)$ and $\neg \exists x \ p(x) \equiv \forall x \neg p(x)$:

(a) $\neg(\exists x \forall y, \ p(x, y)) \equiv \forall x \exists y \neg p(x, y)$

(b) $\neg(\forall x \forall y, \ p(x, y)) \equiv \exists x \exists y \neg p(x, y)$

(c) $\neg(\exists y \ \exists x \ \forall z, \ p(x, y, z)) \equiv \forall y \ \forall x \ \exists z \neg p(x, y, z)$

4.18. Let $p(x)$ denote the sentence "$x + 2 > 5$." State whether or not $p(x)$ is a propositional function on each of the following sets: (*a*) **N**, the set of positive integers; (*b*) $M = \{-1, -2, -3, \ldots\}$; (*c*) **C**, the set of complex numbers.

(*a*) Yes.

(*b*) Although $p(x)$ is false for every element in M, $p(x)$ is still a propositional function on M.

(*c*) No. Note that $2i + 2 > 5$ does not have any meaning. In other words, inequalities are not defined for complex numbers.

4.19. Negate each of the following statements: (*a*) All students live in the dormitories. (*b*) All mathematics majors are males. (*c*) Some students are 25 years old or older.

Use Theorem 4.4 to negate the quantifiers.

(*a*) At least one student does not live in the dormitories. (Some students do not live in the dormitories.)

(*b*) At least one mathematics major is female. (Some mathematics majors are female.)

(*c*) None of the students is 25 years old or older. (All the students are under 25.)

Supplementary Problems

PROPOSITIONS AND TRUTH TABLES

4.20. Let p denote "He is rich" and let q denote "He is happy." Write each statement in symbolic form using p and q. Note that "He is poor" and "He is unhappy" are equivalent to $\neg p$ and $\neg q$, respectively.

(a) If he is rich, then he is unhappy.   (c) It is necessary to be poor in order to be happy.

(b) He is neither rich nor happy.   (d) To be poor is to be unhappy.

4.21. Find the truth tables for. (a) $p \vee \neg q$; (b) $\neg p \wedge \neg q$.

4.22. Verify that the proposition $(p \wedge q) \wedge \neg(p \vee q)$ is a contradiction.

ARGUMENTS

4.23. Test the validity of each argument:

(a)   If it rains, Erik will be sick.   (b)   If it rains, Erik will be sick.
      It did not rain.   Erik was not sick.
      ─────────────────────   ─────────────────────
      Erik was not sick.   It did not rain.

4.24. Test the validity of the following argument:

If I study, then I will not fail mathematics.
If I do not play basketball, then I will study.
But I failed mathematics.
─────────────────────────────────

Therefore I must have played basketball.

QUANTIFIERS

4.25. Let $A = \{1, 2, \ldots, 9, 10\}$. Consider each of the following sentences. If it is a statement, then determine its truth value. If it is a propositional function, determine its truth set.

(a) $(\forall x \in A)(\exists y \in A)(x + y < 14)$ (c) $(\forall x \in A)(\forall y \in A)(x + y < 14)$

(b) $(\forall y \in A)(x + y < 14)$ (d) $(\exists y \in A)(x + y < 14)$

4.26. Negate each of the following statements:

(a) If the teacher is absent, then some students do not complete their homework.

(b) All the students completed their homework and the teacher is present.

(c) Some of the students did not complete their homework or the teacher is absent.

4.27. Negate each statement in Problem 4.15.

4.28. Find a counterexample for each statement were $U = \{3, 5, 7, 9\}$ is the universal set:

(a) $\forall x, x + 3 \geq 7$, (b) $\forall x, x$ is odd, (c) $\forall x, x$ is prime, (d) $\forall x, |x| = x$

Answers to Supplementary Problems

4.20. (a) $p \to \neg q$; (b) $\neg p \land \neg q$; (c) $q \to \neg p$; (d) $\neg p \to \neg q$.

4.21. (a) T, T, F, T; (b) F, F, F, T.

4.22. Construct its truth table. It is a contradiction since its truth table is false for all values of p and q.

4.23. First translate the arguments into symbolic form: p for "It rains," and q for "Erik is sick:"

(a) $p \to q, \neg p \vdash \neg q$ (b) $p \to q, \neg q \vdash \neg p$

By Problem 4.10, (a) is a fallacy. By Problem 4.11, (b) is valid.

4.24. Let p be "I study," q be "I failed mathematics," and r be "I play basketball." The argument has the form:

$$p \to \neg q, \neg r \to p, q \vdash r$$

Construct the truth tables as in Fig. 4-15, where the premises $p \to \neg q$, $\neg r \to p$, and q are true simultaneously only in the fifth line of the table, and in that case the conclusion r is also true. Hence the argument is valid.

p	q	r	$\neg q$	$p \to \neg q$	$\neg r$	$\neg r \to p$
T	T	T	F	F	F	T
T	T	F	F	F	T	T
T	F	T	T	T	F	T
T	F	F	T	T	T	T
F	T	T	F	T	F	T
F	T	F	F	T	T	F
F	F	T	T	T	F	T
F	F	F	T	T	T	F

Fig. 4-15

4.25. (a) The open sentence in two variables is preceded by two quantifiers; hence it is a statement. Moreover, the statement is true.

(b) The open sentence is preceded by one quantifier; hence it is a propositional function of the other variable. Note that for every $y \in A$, $x_0 + y < 14$ if and only if $x_0 = 1, 2,$ or 3. Hence the truth set is $\{1, 2, 3\}$.

(c) It is a statement and it is false: if $x_0 = 8$ and $y_0 = 9$, then $x_0 + y_0 < 14$ is not true.

(d) It is an open sentence in x. The truth set is A itself.

4.26. (a) The teacher is absent and all the students completed their homework.

(b) Some of the students did not complete their homework or the teacher is absent.

(c) All the students completed their homework and the teacher is present.

4.27. (a) $(\forall x \in A)(x + 3 \neq 10)$ (c) $(\forall x \in A)(x + 3 \geq 5)$

(b) $(\exists x \in A)(x + 3 \geq 10)$ (d) $(\exists x \in A)(x + 3 > 7)$

4.28. (a) Here 3 is a counterexample.

(b) The statement is true; hence no counterexample exists.

(c) Here 9 is the only counterexample.

(d) The statement is true; hence there is no counterexample.

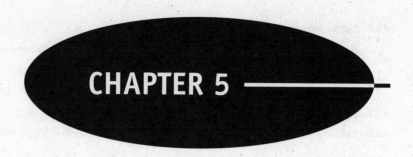

CHAPTER 5

Techniques of Counting

5.1 INTRODUCTION

This chapter develops some techniques for determining, without direct enumeration, the number of possible outcomes of a particular event or the number of elements in a set. Such sophisticated counting is sometimes called *combinatorial analysis*. It includes the study of permutations and combinations.

5.2 BASIC COUNTING PRINCIPLES

There are two basic counting principles used throughout this chapter. The first one involves addition and the second one multiplication.

Sum Rule Principle:

Suppose some event E can occur in m ways and a second event F can occur in n ways, and suppose both events cannot occur simultaneously. Then E or F can occur in $m + n$ ways.

Product Rule Principle:

Suppose there is an event E which can occur in m ways and, independent of this event, there is a second event F which can occur in n ways. Then combinations of E and F can occur in mn ways.

The above principles can be extended to three or more events. That is, suppose an event E_1 can occur in n_1 ways, a second event E_2 can occur in n_2 ways, and, following E_2; a third event E_3 can occur in n_3 ways, and so on. Then:

Sum Rule: If no two events can occur at the same time, then one of the events can occur in:

$$n_1 + n_2 + n_3 + \cdots \text{ ways.}$$

Product Rule: If the events occur one after the other, then all the events can occur in the order indicated in:

$$n_1 \cdot n_2 \cdot n_3 \cdot \ldots \text{ ways.}$$

EXAMPLE 5.1 Suppose a college has 3 different history courses, 4 different literature courses, and 2 different sociology courses.

(a) The number m of ways a student can choose one of each kind of courses is:

$$m = 3(4)(2) = 24$$

one of each kind = multiply

(b) The number n of ways a student can choose just one of the courses is:

$$n = 3 + 4 + 2 = 9$$

• just one = add

There is a set theoretical interpretation of the above two principles. Specifically, suppose $n(A)$ denotes the number of elements in a set A. Then:

(1) **Sum Rule Principle:** Suppose A and B are disjoint sets. Then

$$n(A \cup B) = n(A) + n(B)$$

(2) **Product Rule Principle:** Let $A \times B$ be the Cartesian product of sets A and B. Then

$$n(A \times B) = n(A) \cdot n(B)$$

5.3   MATHEMATICAL FUNCTIONS

We discuss two important mathematical functions frequently used in combinatorics.

Factorial Function

The product of the positive integers from 1 to n inclusive is denoted by $n!$, read "n factorial." Namely:

$$n! = 1 \cdot 2 \cdot 3 \cdot \ldots \cdot (n-2)(n-1)n = n(n-1)(n-2) \cdot \ldots \cdot 3 \cdot 2 \cdot 1$$

Accordingly, $1! = 1$ and $n! = n(n - l)!$. It is also convenient to define $0! = 1$.

EXAMPLE 5.2

(a) $3! = 3 \cdot 2 \cdot 1 = 6, \quad 4! = 4 \cdot 3 \cdot 2 \cdot 1 = 24, \quad 5 = 5 \cdot 4! = 5(24) = 120.$

(b) $\dfrac{12 \cdot 11 \cdot 10}{3 \cdot 2 \cdot 1} = \dfrac{12 \cdot 11 \cdot 10 \cdot 9!}{3 \cdot 2 \cdot 1 \cdot 9!} = \dfrac{12!}{3! \, 9!}$ and, more generally,

$$\frac{n(n - 1) \cdots (n - r + 1)}{r(r - 1) \cdots 3 \cdot 2 \cdot 1} = \frac{n(n - 1) \cdots (n - r + 1)(n - r)!}{r(r - 1) \cdots 3 \cdot 2 \cdot 1 \cdot (n - r)!} = \frac{n!}{r!(n - r)!}$$

(c) For large n, one uses Stirling's approximation (where $e = 2.7128\ldots$):

$$n! = \sqrt{2\pi n} \, n^n e^{-n}$$

Binomial Coefficients

The symbol $\binom{n}{r}$, read "nCr" or "n Choose r," where r and n are positive integers with $r \leq n$, is defined as follows:

$$\binom{n}{r} = \frac{n(n-1)\cdots(n-r+1)}{r(r-1)\dots3\cdot2\cdot1} \quad \text{or equivalently} \quad \binom{n}{r} = \frac{n!}{r!(n-r)!}$$

Note that $n - (n - r) = r$. This yields the following important relation.

Lemma 5.1: $\binom{n}{n-r} = \binom{n}{r}$ or equivalently, $\binom{n}{a} = \binom{n}{b}$ where $a + b = n$.

Motivated by that fact that we defined $0! = 1$, we define:

$$\binom{n}{0} = \frac{n!}{0!n!} = 1 \quad \text{and} \quad \binom{0}{0} = \frac{0!}{0!\,0!} = 1$$

EXAMPLE 5.3

(a) $\binom{8}{2} = \frac{8\cdot7}{2\cdot1} = 28; \quad \binom{9}{4} = \frac{9\cdot8\cdot7\cdot6}{4\cdot3\cdot2\cdot1} = 126; \quad \binom{12}{5} = \frac{12\cdot11\cdot10\cdot9\cdot8}{5\cdot4\cdot3\cdot2\cdot1} = 792.$

Note that $\binom{n}{r}$ has exactly r factors in both the numerator and the denominator.

(b) Suppose we want to compute $\binom{10}{7}$. There will be 7 factors in both the numerator and the denominator. However, $10 - 7 = 3$. Thus, we use Lemma 5.1 to compute:

$$\binom{10}{7} = \binom{10}{3} = \frac{10\cdot9\cdot8}{3\cdot2\cdot1} = 120$$

Binomial Coefficients and Pascal's Triangle

The numbers $\binom{n}{r}$ are called *binomial coefficients*, since they appear as the coefficients in the expansion of $(a + b)^n$. Specifically:

Theorem (Binomial Theorem) 5.2: $\quad (a+b)^n = \sum_{k=0}^{n} \binom{n}{r} a^{n-k}b^k$

The coefficients of the successive powers of $a + b$ can be arranged in a triangular array of numbers, called Pascal's triangle, as pictured in Fig. 5-1. The numbers in Pascal's triangle have the following interesting properties:

(i)   The first and last number in each row is 1.

(ii) Every other number can be obtained by adding the two numbers appearing above it. For example:

$$10 = 4 + 6, \quad 15 = 5 + 10, \quad 20 = 10 + 10.$$

Since these numbers are binomial coefficients, we state the above property formally.

$$
\begin{aligned}
(a+b)^0 &= 1 \\
(a+b)^1 &= a + b \\
(a+b)^2 &= a^2 + 2ab + b^2 \\
(a+b)^3 &= a^3 + 3a^2b + 3ab^2 + b^3 \\
(a+b)^4 &= a^4 + 4a^3b + 6a^2b^2 + 4ab^3 + b^4 \\
(a+b)^5 &= a^5 + 5a^4b + 10a^3b^2 + 10a^2b^3 + 5ab^4 + b^5 \\
(a+b)^6 &= a^6 + 6a^5b + 15a^4b^2 + 20a^3b^3 + 15a^2b^4 + 6ab^5 + b^6
\end{aligned}
$$

```
                    1
                  1   1
                1   2   1
              1   3   3   1
            1   4   6   4   1
          1   5  (10) 10   5   1
        1   6 (15)(20) 15   6   1
```

Fig. 5-1 Pascal's triangle

Theorem 5.3: $\dbinom{n+1}{r} = \dbinom{n}{r-1} + \dbinom{n}{r}$.

5.4 PERMUTATIONS

Any arrangement of a set of n objects in a given order is called a *permutation* of the object (taken all at a time). Any arrangement of any $r \le n$ of these objects in a given order is called an "*r*-permutation" or "a permutation of the n objects taken r at a time." Consider, for example, the set of letters A, B, C, D. Then:

  (i) $BDCA, DCBA$, and $ACDB$ are permutations of the four letters (taken all at a time).
  (ii) BAD, ACB, DBC are permutations of the four letters taken three at a time.
  (iii) AD, BC, CA are permutations of the four letters taken two at a time.

We usually are interested in the number of such permutations without listing them. The number of permutations of n objects taken r at a time will be denoted by

$$P(n, r) \quad \text{(other texts may use } {}_nP_r, \ P_{n,r}, \text{ or } (n)_r).$$

The following theorem applies.

Theorem 5.4: $P(n, r) = n(n-1)(n-2)\cdots(n-r+1) = \dfrac{n!}{(n-r)!}$

We emphasize that there are r factors in $n(n-1)(n-2)\cdots(n-r+1)$.

EXAMPLE 5.4 Find the number m of permutations of six objects, say, A, B, C, D, E, F, taken three at a time. In other words, find the number of "three-letter words" using only the given six letters without repetition.

Let us represent the general three-letter word by the following three positions:

$$\underline{\quad}, \ \underline{\quad}, \ \underline{\quad}$$

The first letter can be chosen in 6 ways; following this the second letter can be chosen in 5 ways; and, finally, the third letter can be chosen in 4 ways. Write each number in its appropriate position as follows:

$$\underline{6}, \ \underline{5}, \ \underline{4}$$

By the Product Rule there are $m = 6 \cdot 5 \cdot 4 = 120$ possible three-letter words without repetition from the six letters. Namely, there are 120 permutations of 6 objects taken 3 at a time. This agrees with the formula in Theorem 5.4:

$$P(6, 3) = 6 \cdot 5 \cdot 4 = 120$$

In fact, Theorem 5.4 is proven in the same way as we did for this particular case.

Consider now the special case of $P(n, r)$ when $r = n$. We get the following result.

Corollary 5.5: There are $n!$ permutations of n objects (taken all at a time).

For example, there are $3! = 6$ permutations of the three letters A, B, C. These are:

$$ABC, \quad ACB, \quad BAC, \quad BCA, \quad CAB, \quad CBA.$$

Permutations with Repetitions

Frequently we want to know the number of permutations of a multiset, that is, a set of objects some of which are alike. We will let

$$P(n; n_1, n_2, \ldots, n_r)$$

denote the number of permutations of n objects of which n_1 are alike, n_2 are alike, ..., n_r are alike. The general formula follows:

Theorem 5.6: $P(n; n_1, n_2, \ldots, n_r) = \dfrac{n!}{n_1! n_2! \ldots n_r!}$

We indicate the proof of the above theorem by a particular example. Suppose we want to form all possible five-letter "words" using the letters from the word "*BABBY*." Now there are $5! = 120$ permutations of the objects B_1, A, B_2, B_3, Y, where the three B's are distinguished. Observe that the following six permutations

$$B_1 B_2 B_3 AY, \quad B_2 B_1 B_3 AY, \quad B_3 B_1 B_2 AY, \quad B_1 B_3 B_2 AY, \quad B_2 B_3 B_1 AY, \quad B_3 B_2 B_1 AY$$

produce the same word when the subscripts are removed. The 6 comes from the fact that there are $3! = 3 \cdot 2 \cdot 1 = 6$ different ways of placing the three B's in the first three positions in the permutation. This is true for each set of three positions in which the B's can appear. Accordingly, the number of different five-letter words that can be formed using the letters from the word "*BABBY*" is:

$$P(5; 3) = \frac{5!}{3!} = 20$$

EXAMPLE 5.5 Find the number m of seven-letter words that can be formed using the letters of the word "*BENZENE*."

We seek the number of permutations of 7 objects of which 3 are alike (the three E's), and 2 are alike (the two N's). By Theorem 5.6,

$$m = P(7; 3, 2) = \frac{7!}{3!2!} = \frac{7 \cdot 6 \cdot 5 \cdot 4 \cdot 3 \cdot 2 \cdot 1}{3 \cdot 2 \cdot 1 \cdot 2 \cdot 1} = 420$$

Ordered Samples

Many problems are concerned with choosing an element from a set S, say, with n elements. When we choose one element after another, say, r times, we call the choice an *ordered sample* of size r. We consider two cases.

(1) Sampling with replacement

Here the element is replaced in the set S before the next element is chosen. Thus, each time there are n ways to choose an element (repetitions are allowed). The Product rule tells us that the number of such samples is:

$$n \cdot n \cdot n \cdots n \cdot n (r \text{ factors}) = n^r$$

(2) Sampling without replacement

Here the element is not replaced in the set S before the next element is chosen. Thus, there is no repetition in the ordered sample. Such a sample is simply an r-permutation. Thus the number of such samples is:

$$P(n, r) = n(n - 1)(n - 2) \cdots (n - r + 1) = \frac{n!}{(n - r)!}$$

w/ replacement = n^r

w/o replacement = $P(n,r)$

EXAMPLE 5.6 Three cards are chosen one after the other from a 52-card deck. Find the number m of ways this can be done: (a) with replacement; (b) without replacement.

(a) Each card can be chosen in 52 ways. Thus $m = 52(52)(52) = 140\,608$.

(b) Here there is no replacement. Thus the first card can be chosen in 52 ways, the second in 51 ways, and the third in 50 ways. Therefore:

$$m = P(52, 3) = 52(51)(50) = 132\,600$$

$C(n,r) = \dfrac{n!}{r!\,(n-r)!}$

5.5   COMBINATIONS

Let S be a set with n elements. A *combination* of these n elements taken r at a time is any selection of r of the elements where order does not count. Such a selection is called an *r-combination*; it is simply a subset of S with r elements. The number of such combinations will be denoted by

$$C(n, r) \qquad \text{(other texts may use } {}_nC_r, \ C_{n,r}, \text{ or } C_r^n \text{)}.$$

Before we give the general formula for $C(n, \ r)$, we consider a special case.

EXAMPLE 5.7 Find the number of combinations of 4 objects, A, B, C, D, taken 3 at a time.

Each combination of three objects determines $3! = 6$ permutations of the objects as follows:

$4\,3\,2 = 12$
24

ABC:	$ABC,$	$ACB,$	$BAC,$	$BCA,$	$CAB,$	CBA
ABD:	$ABD,$	$ADB,$	$BAD,$	$BDA,$	$DAB,$	DBA
ACD:	$ACD,$	$ADC,$	$CAD,$	$CDA,$	$DAC,$	DCA
BCD:	$BDC,$	$BDC,$	$CBD,$	$CDB,$	$DBC,$	DCB

Thus the number of combinations multiplied by $3!$ gives us the number of permutations; that is,

$$C(4, 3) \cdot 3! = P(4, 3) \quad \text{or} \quad C(4, 3) = \frac{P(4, 3)}{3!}$$

But $P(4, \ 3) = 4 \cdot 3 \cdot 2 = 24$ and $3! = 6$; hence $C(4, 3) = 4$ as noted above.

As indicated above, any combination of n objects taken r at a time determines $r!$ permutations of the objects in the combination; that is,

$$P(n, r) = r! \, C(n, r)$$

Accordingly, we obtain the following formula for $C(n, \ r)$ which we formally state as a theorem.

Theorem 5.7: $C(n, r) = \dfrac{P(n, r)}{r!} = \dfrac{n!}{r!(n-r)!}$

Recall that the binomial coefficient $\dbinom{n}{r}$ was defined to be $\dfrac{n!}{r!(n-r)!}$; hence

$$\boxed{C(r, n) = \dbinom{n}{r}}$$

We shall use $C(n, \ r)$ and $\dbinom{n}{r}$ interchangeably.

EXAMPLE 5.8 A farmer buys 3 cows, 2 pigs, and 4 hens from a man who has 6 cows, 5 pigs, and 8 hens. Find the number m of choices that the farmer has.

The farmer can choose the cows in $C(6, \ 3)$ ways, the pigs in $C(5, 2)$ ways, and the hens in $C(8, \ 4)$ ways. Thus the number m of choices follows:

$$m = \dbinom{6}{3}\dbinom{5}{2}\dbinom{8}{4} = \frac{6 \cdot 5 \cdot 4}{3 \cdot 2 \cdot 1} \cdot \frac{5 \cdot 4}{2 \cdot 1} \cdot \frac{8 \cdot 7 \cdot 6 \cdot 5}{4 \cdot 3 \cdot 2 \cdot 1} = 20 \cdot 10 \cdot 70 = 14\,000$$

5.6 THE PIGEONHOLE PRINCIPLE

Many results in combinational theory come from the following almost obvious statement.

Pigeonhole Principle: If n pigeonholes are occupied by $n + 1$ or more pigeons, then at least one pigeonhole is occupied by more than one pigeon.

This principle can be applied to many problems where we want to show that a given situation can occur.

EXAMPLE 5.9

(a) Suppose a department contains 13 professors, then two of the professors (pigeons) were born in the same month (pigeonholes).

(b) Find the minimum number of elements that one needs to take from the set $S = \{1, 2, 3, \ldots, 9\}$ to be sure that two of the numbers add up to 10.
    Here the pigeonholes are the five sets $\{1, 9\}, \{2, 8\}, \{3, 7\}, \{4, 6\}, \{5\}$. Thus any choice of six elements (pigeons) of S will guarantee that two of the numbers add up to ten.

The Pigeonhole Principle is generalized as follows.

Generalized Pigeonhole Principle: If n pigeonholes are occupied by $kn + 1$ or more pigeons, where k is a positive integer, then at least one pigeonhole is occupied by $k + 1$ or more pigeons.

EXAMPLE 5.10 Find the minimum number of students in a class to be sure that three of them are born in the same month.

Here the $n = 12$ months are the pigeonholes, and $k + 1 = 3$ so $k = 2$. Hence among any $kn + 1 = 25$ students (pigeons), three of them are born in the same month.

5.7   THE INCLUSION–EXCLUSION PRINCIPLE

Let A and B be any finite sets. Recall Theorem 1.9 which tells us:

$$n(A \cup B) = n(A) + n(B) - n(A \cap B)$$

In other words, to find the number $n(A \cup B)$ of elements in the union of A and B, we add $n(A)$ and $n(B)$ and then we subtract $n(A \cap B)$; that is, we "include" $n(A)$ and $n(B)$, and we "exclude" $n(A \cap B)$. This follows from the fact that, when we add $n(A)$ and $n(B)$, we have counted the elements of $(A \cap B)$ twice.

The above principle holds for any number of sets. We first state it for three sets.

Theorem 5.8: For any finite sets A, B, C we have

$$n(A \cup B \cup C) = n(A) + n(B) + n(C) - n(A \cap B) - n(A \cap C) - n(B \cap C) + n(A \cap B \cap C)$$

That is, we "include" $n(A), n(B), n(C)$, we "exclude" $n(A \cap B), n(A \cap C), n(B \cap C)$, and finally "include" $n(A \cap B \cap C)$.

EXAMPLE 5.11 Find the number of mathematics students at a college taking at least one of the languages French, German, and Russian, given the following data:

   65 study French,   20 study French and German,
   45 study German,   25 study French and Russian,   8 study all three languages.
   42 study Russian,   15 study German and Russian,

We want to find $n(F \cup G \cup R)$ where F, G, and R denote the sets of students studying French, German, and Russian, respectively.

By the Inclusion–Exclusion Principle,

$$n(F \cup G \cup R) = n(F) + n(G) + n(R) - n(F \cap G) - n(F \cap R) - n(G \cap R) + n(F \cap G \cap R)$$
$$= 65 + 45 + 42 - 20 - 25 - 15 + 8 = 100$$

Namely, 100 students study at least one of the three languages.

Now, suppose we have any finite number of finite sets, say, $A_1, A_2, \ldots, A_m$. Let s_k be the sum of the cardinalities

$$n(A_{i_1} \cap A_{i_2} \cap \cdots \cap A_{i_K})$$

of all possible k-tuple intersections of the given m sets. Then we have the following general Inclusion–Exclusion Principle.

Theorem 5.9: $n(A_1 \cup A_2 \cup \cdots \cup A_m) = s_1 - s_2 + s_3 - \cdots + (-1)^{m-1} s_m.$

5.8   TREE DIAGRAMS

A *tree diagram* is a device used to enumerate all the possible outcomes of a sequence of events where each event can occur in a finite number of ways. The construction of tree diagrams is illustrated in the following example.

EXAMPLE 5.12

(a) Find the product set $A \times B \times C$, where $A = \{1, 2\}$, $B = \{a, b, c\}$, $C = \{x, y\}$.

   The tree diagram for $A \times B \times C$ appears in Fig. 5-2(a). Here the tree is constructed from left to right, and the number of branches at each point corresponds to the possible outcomes of the next event. Each endpoint (leaf) of the tree is labeled by the corresponding element of $A \times B \times C$. As noted previously, $A \times B \times C$ has $n = 2(3)(2) = 12$ elements.

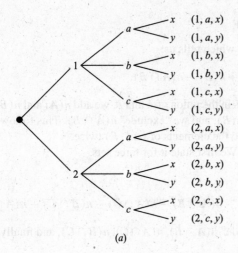
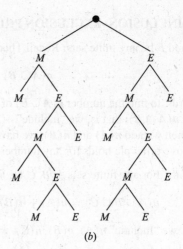

$$(a) \qquad\qquad\qquad\qquad (b)$$

Fig. 5-2

(b) Mark and Erik are to play a tennis tournament. The first person to win two games in a row or who wins a total of three games wins the tournament. Find the number of ways the tournament can occur.

The tree diagram showing the possible outcomes of the tournament appears in Fig. 5-2(b). Here the tree is constructed from top-down rather than from left-right. (That is, the "root" is on the top of the tree.) Note that there are 10 endpoints, and the endpoints correspond to the following 10 ways the tournament can occur:

$$MM, \quad MEMM, \quad MEMEM, \quad MEMEE, \quad MEE, \quad EMM, \quad EMEMM, \quad EMEME, \quad EMEE, \quad EE$$

The path from the beginning (top) of the tree to the endpoint describes who won which game in the tournament.

Solved Problems

FACTORIAL NOTATION AND BINOMIAL COEFFICIENTS

5.1. Compute: (a) 4!, 5!;   (b) 6!, 7!, 8!, 9!;   (c) 50!.

(a) $4! = 4 \cdot 3 \cdot 2 \cdot 1 = 24$, $5! = 5 \cdot 4 \cdot 3 \cdot 2 \cdot 1 = 5(24) = 120$.
(b) Now use $(n + 1)! = (n + 1)n!$:

$$6! = 5(5!) = 6(120) = 720, \qquad 8! = 8(7!) = 8(5040) = 40\,320,$$
$$7! = 7(6!) = 7(720) = 5\,040, \quad 9! = 9(8!) = 9(40\,320) = 362\,880.$$

(c) Since n is very large, we use Sterling's approximation: $n! = \sqrt{2\pi n}\, n^\pi e^{-n}$ (where $e \approx 2.718$). Thus:

$$50! \approx N = \sqrt{100\pi}\, 50^{50} e^{-50}$$

Evaluating N using a calculator, we get $N = 3.04 \times 10^{64}$ (which has 65 digits).

5.2. Compute: (a) $\dfrac{13!}{11!}$;   (b) $\dfrac{7!}{10!}$.

(a) $\dfrac{13!}{11!} = \dfrac{13 \cdot 12 \cdot 11 \cdot 10 \cdot 9 \cdot 8 \cdot 7 \cdot 6 \cdot 5 \cdot 4 \cdot 3 \cdot 2 \cdot 1}{11 \cdot 10 \cdot 9 \cdot 8 \cdot 7 \cdot 6 \cdot 5 \cdot 4 \cdot 3 \cdot 2 \cdot 1} = 13 \cdot 12 = 156.$

Alternatively, this could be solved as follows:

$$\frac{13!}{11!} = \frac{13 \cdot 12 \cdot 11!}{11!} = 13 \cdot 12 = 156.$$

(b) $\dfrac{7!}{10!} = \dfrac{7!}{10 \cdot 9 \cdot 8 \cdot 7!} = \dfrac{1}{10 \cdot 9 \cdot 8} = \dfrac{1}{720}$.

5.3. Simplify: (a) $\dfrac{n!}{(n-1)!}$; (b) $\dfrac{(n+2)!}{n!}$.

(a) $\dfrac{n!}{(n-1)!} = \dfrac{n(n-1)(n-2)\cdots 3 \cdot 2 \cdot 1}{(n-1)(n-2)\cdots 3 \cdot 2 \cdot 1} = n$; alternatively, $\dfrac{n!}{(n-1)!} = \dfrac{n(n-1)!}{(n-1)!} = n$.

(b) $\dfrac{(n+2)!}{n!} = \dfrac{(n+2)(n+1)n!}{n!} = (n+2)(n+1) = n^2 + 3n + 2$.

5.4. Compute: (a) $\dbinom{16}{3}$; (b) $\dbinom{12}{4}$; (c) $\dbinom{8}{5}$.

Recall that there are as many factors in the numerator as in the denominator.

(a) $\dbinom{16}{3} = \dfrac{16 \cdot 15 \cdot 14}{3 \cdot 2 \cdot 1} = 560$; (b) $\dbinom{12}{4} = \dfrac{12 \cdot 11 \cdot 10 \cdot 9}{4 \cdot 3 \cdot 2 \cdot 1} = 495$;

(c) Since $8 - 5 = 3$, we have $\dbinom{8}{5} = \dbinom{8}{3} = \dfrac{8 \cdot 7.6}{3 \cdot 2 \cdot 1} = 56$.

5.5. Prove: $\dbinom{17}{6} = \dbinom{16}{5} + \dbinom{16}{6}$.

Now $\dbinom{16}{5} + \dbinom{16}{6} = \dfrac{16!}{5!11!} + \dfrac{16!}{6!10!}$. Multiply the first fraction by $\frac{6}{6}$ and the second by $\frac{11}{11}$ to obtain the same denominator in both fractions; and then add:

$$\dbinom{16}{5} + \dbinom{16}{6} = \dfrac{6 \cdot 16!}{6 \cdot 5! \cdot 11!} + \dfrac{11 \cdot 16!}{6! \cdot 11 \cdot 10!} = \dfrac{6 \cdot 16!}{6! \cdot 11!} + \dfrac{11 \cdot 16!}{6! \cdot 11!}$$

$$= \dfrac{6 \cdot 16! + 11 \cdot 16!}{6! \cdot 11!} = \dfrac{(6+11) \cdot 16!}{6! \cdot 11!} = \dfrac{17 \cdot 16!}{6! \cdot 11!} = \dfrac{17!}{6! \cdot 11!} = \dbinom{17}{6}$$

5.6. Prove Theorem 5.3: $\dbinom{n+1}{r} = \dbinom{n}{r-1} + \dbinom{n}{r}$.

(The technique in this proof is similar to that of the preceding problem.)

Now $\dbinom{n}{r-1} + \dbinom{n}{r} = \dfrac{n!}{(r-1)! \cdot (n-r+1)!} + \dfrac{n!}{r! \cdot (n-r)!}$.

To obtain the same denominator in both fractions, multiply the first fraction by $\frac{r}{r}$ and the second fraction by $\dfrac{n-r+1}{n-r+1}$. Hence

$$\dbinom{n}{r-1} + \dbinom{n}{r} = \dfrac{r \cdot n!}{r \cdot (r-1)! \cdot (n-r+1)!} + \dfrac{(n-r+1) \cdot n!}{r! \cdot (n-r+1) \cdot (n-r)!}$$

$$= \dfrac{r \cdot n!}{r!(n-r+1)!} + \dfrac{(n-r+1) \cdot n!}{r!(n-r+1)!}$$

$$= \dfrac{r \cdot n! + (n-r+1) \cdot n!}{r!(n-r+1)!} = \dfrac{[r + (n-r+1)] \cdot n!}{r!(n-r+1)!}$$

$$= \dfrac{(n+1)n!}{r!(n-r+1)!} = \dfrac{(n+1)!}{r!(n-r+1)!} = \dbinom{n+1}{r}$$

COUNTING PRINCIPLES

5.7. Suppose a bookcase shelf has 5 History texts, 3 Sociology texts, 6 Anthropology texts, and 4 Psychology texts. Find the number n of ways a student can choose:

(a) one of the texts; (b) one of each type of text.

(a) Here the Sum Rule applies; hence, $n = 5 + 3 + 6 + 4 = 18$.

(b) Here the Product Rule applies; hence, $n = 5 \cdot 3 \cdot 6 \cdot 4 = 360$.

5.8. A history class contains 8 male students and 6 female students. Find the number n of ways that the class can elect: (*a*) 1 class representative; (*b*) 2 class representatives, 1 male and 1 female; (*c*) 1 president and 1 vice president.

(*a*) Here the Sum Rule is used; hence, $n = 8 + 6 = 14$.

(*b*) Here the Product Rule is used; hence, $n = 8 \cdot 6 = 48$.

(*c*) There are 14 ways to elect the president, and then 13 ways to elect the vice president. Thus $n = 14 \cdot 13 = 182$.

5.9. There are four bus lines between A and B, and three bus lines between B and C. Find the number m of ways that a man can travel by bus: (*a*) from A to C by way of B; (*b*) roundtrip from A to C by way of B; (*c*) roundtrip from A to C by way of B but without using a bus line more than once.

(*a*) There are 4 ways to go from A to B and 3 ways from B to C; hence $n = 4 \cdot 3 = 12$.

(*b*) There are 12 ways to go from A to C by way of B, and 12 ways to return. Thus $n = 12 \cdot 12 = 144$.

(*c*) The man will travel from A to B to C to B to A. Enter these letters with connecting arrows as follows:

$$A \to B \to C \to B \to A$$

The man can travel four ways from A to B and three ways from B to C, but he can only travel two ways from C to B and three ways from B to A since he does not want to use a bus line more than once. Enter these numbers above the corresponding arrows as follows:

$$A \overset{4}{\to} B \overset{3}{\to} C \overset{2}{\to} B \overset{3}{\to} A$$

Thus, by the Product Rule, $n = 4 \cdot 3 \cdot 2 \cdot 3 = 72$.

PERMUTATIONS

5.10. State the essential difference between permutations and combinations, with examples.

Order counts with permutations, such as words, sitting in a row, and electing a president, vice president, and treasurer. Order does not count with combinations, such as committees and teams (without counting positions). The product rule is usually used with permutations, since the choice for each of the ordered positions may be viewed as a sequence of events.

5.11. Find: (*a*) $P(7, 3)$; (*b*) $P(14, 2)$.

Recall $P(n, r)$ has r factors beginning with n.

(*a*) $P(7, 3) = 7 \cdot 6 \cdot 5 = 210$;   (*b*) $P(14, 2) = 14 \cdot 13 = 182$.

5.12. Find the number m of ways that 7 people can arrange themselves:

(*a*) In a row of chairs; (*b*) Around a circular table.

(*a*) Here $m = P(7, 7) = 7!$ ways.

(*b*) One person can sit at any place at the table. The other 6 people can arrange themselves in 6! ways around the table; that is $m = 6!$.

This is an example of a *circular permutation*. In general, n objects can be arranged in a circle in $(n - 1)!$ ways.

5.13. Find the number n of distinct permutations that can be formed from all the letters of each word:

(*a*) *THOSE*; (*b*) *UNUSUAL*; (*c*) *SOCIOLOGICAL*.

This problem concerns permutations with repetitions.

(*a*) $n = 5! = 120$, since there are 5 letters and no repetitions.

(*b*) $n = \dfrac{7!}{3!} = 840$, since there are 7 letters of which 3 are U and no other letter is repeated.

(c) $n = \dfrac{12!}{3!2!2!2!}$, since there are 12 letters of which 3 are O, 2 are C, 2 are I, and 2 are L. (We leave the answer using factorials, since the number is very large.)

5.14. A class contains 8 students. Find the number n of samples of size 3:

(a) With replacement; (b) Without replacement.

(a) Each student in the ordered sample can be chosen in 8 ways; hence, there are

$$n = 8 \cdot 8 \cdot 8 = 8^3 = 512 \text{ samples of size 3 with replacement.}$$

(b) The first student in the sample can be chosen in 8 ways, the second in 7 ways, and the last in 6 ways. Thus, there are $n = 8 \cdot 7 \cdot 6 = 336$ samples of size 3 without replacement.

5.15. Find n if $P(n, 2) = 72$.

$$P(n, 2) = n(n - 1) = n^2 - n. \quad \text{Thus, we get}$$

$$n^2 - n = 72 \quad \text{or} \quad n^2 - n - 72 = 0 \quad \text{or} \quad (n - 9)(n + 8) = 0$$

Since n must be positive, the only answer is $n = 9$.

COMBINATIONS

5.16. A class contains 10 students with 6 men and 4 women. Find the number n of ways to:

(a) Select a 4-member committee from the students.

(b) Select a 4-member committee with 2 men and 2 women.

(c) Elect a president, vice president, and treasurer.

(a) This concerns combinations, not permutations, since order does not count in a committee. There are "10 choose 4" such committees. That is:

$$n = C(10, 4) = \binom{10}{4} = \frac{10 \cdot 9 \cdot 8 \cdot 7}{4 \cdot 3 \cdot 2 \cdot 1} = 210$$

(b) The 2 men can be chosen from the 6 men in $C(6, 2)$ ways, and the 2 women can be chosen from the 4 women in $C(4, 2)$ ways. Thus, by the Product Rule:

$$n = \binom{6}{2}\binom{4}{2} = \frac{6 \cdot 5}{2 \cdot 1} \cdot \frac{4 \cdot 3}{2 \cdot 1} = 15(6) = 90$$

(c) This concerns permutations, not combinations, since order does count. Thus,

$$n = P(6, 3) = 6 \cdot 5 \cdot 4 = 120$$

5.17. A box contains 8 blue socks and 6 red socks. Find the number of ways two socks can be drawn from the box if:

(a) They can be any color. (b) They must be the same color.

(a) There are "14 choose 2" ways to select 2 of the 14 socks. Thus:

$$n = C(14, 2) = \binom{14}{2} = \frac{14 \cdot 13}{2 \cdot 1} = 91$$

(b) There are $C(8, 2) = 28$ ways to choose 2 of the 8 blue socks, and $C(6, 2) = 15$ ways to choose 2 of the 4 red socks. By the Sum Rule, $n = 28 + 15 = 43$.

5.18. Find the number m of committees of 5 with a given chairperson that can be selected from 12 people.

The chairperson can be chosen in 12 ways and, following this, the other 4 on the committee can be chosen from the 11 remaining in $C(11, 4)$ ways. Thus $m = 12 \cdot C(11, 4) = 12 \cdot 330 = 3960$.

PIGEONHOLE PRINCIPLE

5.19. Find the minimum number n of integers to be selected from $S = \{1, 2, \ldots, 9\}$ so that: (*a*) The sum of two of the n integers is even. (*b*) The difference of two of the n integers is 5.

    (*a*) The sum of two even integers or of two odd integers is even. Consider the subsets $\{1, 3, 5, 7, 9\}$ and $\{2, 4, 6, 8\}$ of S as pigeonholes. Hence $n = 3$.

    (*b*) Consider the five subsets $\{1, 6\}, \{2, 7\}, \{3, 8\}, \{4, 9\}, \{5\}$ of S as pigeonholes. Then $n = 6$ will guarantee that two integers will belong to one of the subsets and their difference will be 5.

5.20. Find the minimum number of students needed to guarantee that five of them belong to the same class (Freshman, Sophomore, Junior, Senior).

    Here the $n = 4$ classes are the pigeonholes and $k + 1 = 5$ so $k = 4$. Thus among any $kn + 1 = 17$ students (pigeons), five of them belong to the same class.

5.21. Let L be a list (not necessarily in alphabetical order) of the 26 letters in the English alphabet (which consists of 5 vowels, A, E, I, O, U, and 21 consonants).

    (*a*) Show that L has a sublist consisting of four or more consecutive consonants.

    (*b*) Assuming L begins with a vowel, say A, show that L has a sublist consisting of five or more consecutive consonants.

    (*a*) The five letters partition L into $n = 6$ sublists (pigeonholes) of consecutive consonants. Here $k + 1 = 4$ and so $k = 3$. Hence $nk + 1 = 6(3) + 1 = 19 < 21$. Hence some sublist has at least four consecutive consonants.

    (*b*) Since L begins with a vowel, the remainder of the vowels partition L into $n = 5$ sublists. Here $k + 1 = 5$ and so $k = 4$. Hence $kn + 1 = 21$. Thus some sublist has at least five consecutive consonants.

INCLUSION–EXCLUSION PRINCIPLE

5.22. There are 22 female students and 18 male students in a classroom. Find the total number t of students.

    The sets of male and female students are disjoint; hence $t = 22 + 18 = 40$.

5.23. Suppose among 32 people who save paper or bottles (or both) for recycling, there are 30 who save paper and 14 who save bottles. Find the number m of people who:

(*a*) save both; (*b*) save only paper; (*c*) save only bottles.

Let P and B denote the sets of people saving paper and bottles, respectively. Then:

    (*a*) $m = n(P \cap B) = n(P) + n(B) - n(P \cup B) = 30 + 14 - 32 = 12$

    (*b*) $m = n(P \backslash B) = n(P) - n(P \cap B) = 30 - 12 = 18$

    (*c*) $m = n(B \backslash P) = n(B) - n(P \cap B) = 14 - 12 = 2$

5.24. Let A, B, C, D denote, respectively, art, biology, chemistry, and drama courses. Find the number N of students in a dormitory given the data:

12 take A,	5 take A and B,	4 take B and D,	2 take B, C, D,
20 take B,	7 take A and C,	3 take C and D,	3 take A, C, D,
20 take C,	4 take A and D,	3 take A, B, C,	2 take all four,
8 take D,	16 take B and C,	2 take A, B, D,	71 take none.

Let T be the number of students who take at least one course. By the Inclusion–Exclusion Principle Theorem 5.9, $T = s_1 - s_2 + s_3 - s_4$ where:

$$s_1 = 12 + 20 + 20 + 8 = 60, \quad s_2 = 5 + 7 + 4 + 16 + 4 + 3 = 39,$$
$$s_3 = 3 + 2 + 2 + 3 = 10, \quad s_4 = 2.$$

Thus $T = 29$, and $N = 71 + T = 100$.

TREE DIAGRAMS

5.25. Teams A and B play in a tournament. The first team to win three games wins the tournament. Find the number n of possible ways the tournament can occur.

Construct the appropriate tree diagram in Fig. 5-3(a). The tournament can occur in 20 ways:

AAA, $AABA$, $AABBA$, $AABBB$, $ABAA$, $ABABA$, $ABABB$, $ABBAA$, $ABBAB$, $ABBB$;
BBB, $BBAB$, $BBAAB$, $BBAAA$, $BABB$, $BABAB$, $BABAA$, $BAABB$, $BAABA$, $BAAA$

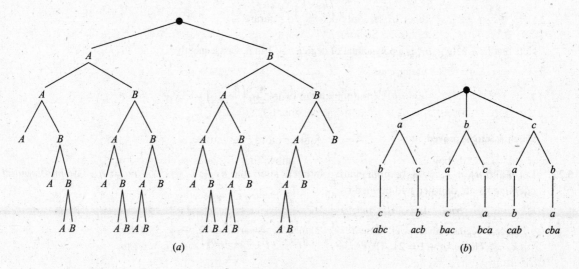

(a) $\qquad\qquad\qquad\qquad\qquad\qquad\qquad\qquad$ (b)

Fig. 5-3

5.26. Construct the tree diagram that gives the permutations of $\{a, b, c\}$.

The tree diagram appears in Fig. 5-3(b). There are six permutations, and they are listed on the bottom of the diagram.

MISCELLANEOUS PROBLEMS

5.27. There are 12 students in a class. Find the number n of ways that the 12 students can take 3 tests if 4 students are to take each test.

There are $C(12, 4) = 495$ ways to choose 4 of the 12 students to take the first test. Following this, there are $C(8, 4) = 70$ ways to choose 4 of the remaining 8 students to take the second test. The remaining students take the third test. Thus:

$$n = 70(495) = 34\,650$$

5.28. Prove Theorem (Binomial Theorem) 5.2: $(a + b)^n = \sum_{r=0}^{n} \binom{n}{r} a^{n-r} b^r$.

The theorem is true for $n = 1$, since

$$\sum_{r=0}^{1} \binom{1}{r} a^{1-r} b^r = \binom{1}{0} a^1 b^0 + \binom{1}{1} a^0 b^1 = a + b = (a + b)^1$$

We assume the theorem is true for $(a + b)^n$ and prove it is true for $(a + b)^n + 1$.

$$(a + b)^{n+1} = (a + b)(a + b)^n$$

$$= (a + b)[a^n + \tbinom{n}{1} a^{n-1} b + \cdots + \tbinom{n}{r-1} a^{n-r+1} b^{r-1} + \tbinom{n}{r} a^{n-r} b^r + \cdots + \tbinom{n}{1} a b^{n-1} + b^n]$$

Now the term in the product which contains b^r is obtained from

$$b[\binom{n}{r-1} a^{n-r+1} b^{r-1}] + a[\binom{n}{r} a^{n-r} b^r] = \binom{n}{r-1} a^{n-r+1} b^r + \binom{n}{r} a^{n-r+1} b^r$$
$$= [\binom{n}{r-1} + \binom{n}{r}] a^{n-r+1} b^r$$

But, by Theorem 5.3, $\binom{n}{r-1} = \binom{n}{r} = \binom{n+1}{r}$. Thus, the term containing b^r is:

$$\binom{n+1}{r} a^{n-r+1} b^r$$

Note that $(a+b)(a+b)^n$ is a polynomial of degree $n+1$ in b. Consequently:

$$(a+b)^{n+1} = (a+b)(a+b)^n = \sum_{r=0}^{n+1} \binom{n+1}{r} a^{n-r+1} b^r$$

which was to be proved.

5.29. Let n and $n_1, n_2, \ldots, n_r$ be nonnegative integers such that $n_1 + n_2 + \cdots + n_r = n$. The *multinomial coefficients* are denoted and defined by:

$$\binom{n}{n_1, n_2, \ldots, n_r} = \frac{n!}{n_1! n_2! \ldots n_r!}$$

Compute the following multinomial coefficients:

$$(a) \quad \binom{6}{3,\ 2,\ 1}; \quad (b) \quad \binom{8}{4,\ 2,\ 2,\ 0}; \quad (c) \quad \binom{10}{5,\ 3,\ 2,\ 2}.$$

(a) $\binom{6}{3,\ 2,\ 1} = \dfrac{6!}{3!\,2!\,1!} = \dfrac{6 \cdot 5 \cdot 4 \cdot 3 \cdot 2 \cdot 1}{3 \cdot 2 \cdot 1 \cdot 2 \cdot 1 \cdot 1} = 60$

(b) $\binom{8}{4,\ 2,\ 2,\ 0} = \dfrac{8!}{4!\,2!\,2!\,0!} = \dfrac{8 \cdot 7 \cdot 6 \cdot 5 \cdot 4 \cdot 3 \cdot 2 \cdot 1}{4 \cdot 3 \cdot 2 \cdot 1 \cdot 2 \cdot 1 \cdot 2 \cdot 1 \cdot 1} = 420$

(c) $\binom{10}{5,\ 3,\ 2,\ 2}$ has no meaning, since $5 + 3 + 2 + 2 \neq 10$.

5.30. A student must take five classes from three areas of study. Numerous classes are offered in each discipline, but the student cannot take more than two classes in any given area.

(a) Using the pigeonhole principle, show that the student will take at least two classes in one area.

(b) Using the Inclusion–Exclusion Principle, show that the student will have to take at least one class in each area.

(a) The three areas are the pigeonholes and the student must take five classes (pigeons). Hence, the student must take at least two classes in one area.

(b) Let each of the three areas of study represent three disjoint sets, A, B, and C. Since the sets are disjoint, $m(A \cup B \cup C) = 5 = n(A) + n(B) + n(C)$. Since the student can take at most two classes in any area of study, the sum of classes in any two sets, say A and B, must be less than or equal to four. Hence, $5 - [n(A) + n(B)] = n(C) \geq 1$. Thus, the student must take at least one class in any area.

Supplementary Problems

FACTORIAL NOTATION, BINOMIAL COEFFICIENTS

5.31. Find: (a) 10!, 11!, 12!; (b) 60!. (Hint: Use Sterling's approximation to $n!$.)

5.32. Evaluate: (a) 16!/14!; (b) 14!/11!; (c) 8!/10!; (d) 10!/13!.

5.33. Simplify: (a) $\dfrac{(n+1)!}{n!}$; (b) $\dfrac{n!}{(n-2)!}$; (c) $\dfrac{(n-1)!}{(n+2)!}$; (d) $\dfrac{(n-r+1)!}{(n-r-1)!}$.

5.34. Find: (a) $\binom{5}{2}$; (b) $\binom{7}{3}$; (c) $\binom{14}{2}$; (d) $\binom{6}{4}$; (e) $\binom{20}{17}$; (f) $\binom{18}{15}$.

5.35. Show that: (a) $\binom{n}{0} + \binom{n}{n} + \binom{n}{2} + \binom{n}{3} + \cdots + \binom{n}{n} = 2^n$

(b) $\binom{n}{0} - \binom{n}{1} + \binom{n}{2} - \binom{n}{3} + \cdots + \binom{n}{n} = 0$

5.36. Given the following eighth row of Pascal's triangle, find: (a) the ninth row; (b) the tenth row.

$$1 \quad 8 \quad 28 \quad 56 \quad 70 \quad 56 \quad 28 \quad 8 \quad 1$$

5.37. Evaluate the following multinomial coefficients (defined in Problem 5.29):

(a) $\begin{pmatrix} 6 \\ 2,\ 3,\ 1 \end{pmatrix}$; (b) $\begin{pmatrix} 7 \\ 3,\ 2,\ 2,\ 0 \end{pmatrix}$; (c) $\begin{pmatrix} 9 \\ 3,\ 5,\ 1 \end{pmatrix}$; (d) $\begin{pmatrix} 8 \\ 4,\ 3,\ 2 \end{pmatrix}$.

COUNTING PRINCIPLES

5.38. A store sells clothes for men. It has 3 kinds of jackets, 7 kinds of shirts, and 5 kinds of pants. Find the number of ways a person can buy: (a) one of the items; (b) one of each of the three kinds of clothes.

5.39. A class has 10 male students and 8 female students. Find the number of ways the class can elect: (a) a class representative; (b) 2 class representatives, one male and one female; (c) a class president and vicepresident.

5.40. Suppose a code consists of five characters, two letters followed by three digits. Find the number of: (a) codes; (b) codes with distinct letter; (c) codes with the same letters.

PERMUTATIONS

5.41. Find the number of automobile license plates where: (a) Each plate contains 2 different letters followed by 3 different digits. (b) The first digit cannot be 0.

5.42. Find the number m of ways a judge can award first, second, and third places in a contest with 18 contestants.

5.43. Find the number of ways 5 large books, 4 medium-size books, and 3 small books can be placed on a shelf where: (a) there are no restrictions; (b) all books of the same size are together.

5.44. A debating team consists of 3 boys and 3 girls. Find the number of ways they can sit in a row where: (a) there are no restrictions; (b) the boys and girls are each to sit together; (c) just the girls are to sit together.

5.45. Find the number of ways 5 people can sit in a row where: (a) there are no restrictions; (b) two of the people insist on sitting next to each other.

5.46. Repeat Problem 5.45 if they sit around a circular table.

5.47. Consider all positive integers with three different digits. (Note that zero cannot be the first digit.) Find the number of them which are: (a) greater than 700; (b) odd; (c) divisible by 5.

5.48. Suppose repetitions are not permitted. (a) Find the number of three-digit numbers that can be formed from the six digits 2, 3, 5, 6, 7, and 9. (b) How many of them are less than 400? (c) How many of them are even?

5.49. Find the number m of ways in which 6 people can ride a toboggan if one of 3 of them must drive.

5.50. Find n if: (a) $P(n, 4) = 42P(n, 2)$; (b) $2P(n, 2) + 50 = P(2n, 2)$.

PERMUTATIONS WITH REPETITIONS, ORDERED SAMPLES

5.51. Find the number of permutations that can be formed from all the letters of each word: (a) QUEUE; (b) COMMITTEE; (c) PROPOSITION; (d) BASEBALL.

5.52. Suppose we are given 4 identical red flags, 2 identical blue flags, and 3 identical green flags. Find the number m of different signals that can be formed by hanging the 9 flags in a vertical line.

5.53. A box contains 12 lightbulbs. Find the number n of ordered samples of size 3:
(a) with replacement; (b) without replacement.

5.54. A class contains 10 students. Find the number n of ordered samples of size 4:
(a) with replacement; (b) without replacement.

COMBINATIONS

5.55. A restaurant has 6 different desserts. Find the number of ways a customer can choose:
(a) 1 dessert; (b) 2 of the desserts; (c) 3 of the desserts.

5.56. A class contains 9 men and 3 women. Find the number of ways a teacher can select a committee of 4 from the class where there is:
(a) no restrictions;   (b) 2 men and 2 women;   (c) exactly one woman;   (d) at least one woman.

5.57. A woman has 11 close friends. Find the number of ways she can invite 5 of them to dinner where:
(a) There are no restrictions.
(b) Two of the friends are married to each other and will not attend separately.
(c) Two of the friends are not speaking with each other and will not attend together.

5.58. A class contains 8 men and 6 women and there is one married couple in the class. Find the number m of ways a teacher can select a committee of 4 from the class where the husband or wife but not both can be on the committee.

5.59. A box has 6 blue socks and 4 white socks. Find the number of ways two socks can be drawn from the box where:
(a) There are no restrictions. (b) They are different colors. (c) They are the same color.

5.60. A women student is to answer 10 out of 13 questions. Find the number of her choices where she must answer:
(a) the first two questions;   (c) exactly 3 out of the first 5 questions;
(b) the first or second question but not both;   (d) at least 3 of the first 5 questions.

INCLUSION–EXCLUSION PRINCIPLE

5.61. Suppose 32 students are in an art class A and 24 students are in a biology class B, and suppose 10 students are in both classes. Find the number of students who are:
(a) in class A or in class B; (b) only in class A; (c) only in class B.

5.62. A survey of 80 car owners shows that 24 own a foreign-made car and 60 own a domestic-made car. Find the number of them who own:
(a) both a foreign made car and a domestic made car;
(b) only a foreign made car;
(c) only a domestic made car.

5.63. Consider all integers from 1 up to and including 100. Find the number of them that are:
(a) odd or the square of an integer; (b) even or the cube of an integer.

5.64. In a class of 30 students, 10 got A on the first test, 9 got A on a second test, and 15 did not get an A on either test. Find: the number of students who got:

(a) A on both tests;

(b) A on the first test but not the second;

(c) A on the second test but not the first.

5.65. Consider all integers from 1 up to and including 300. Find the number of them that are divisible by:
(a) at least one of 3, 5, 7;   (c) by 5, but by neither 3 nor 7;
(b) 3 and 5 but not by 7;   (d) by none of the numbers 3, 5, 7.

5.66. In a certain school, French (F), Spanish (S), and German (G) are the only foreign languages taught. Among 80 students:

    (i)   20 study F,   25 study S,   15 study G.

    (ii) 8 study F and S,   6 study S and G,   5 study F and G.

    (iii) 2 study all three languages.

Find the number of the 80 students who are studying:

(a) none of the languages;   (c) only one language;   (e) exactly two of the languages.

(b) only French;   (d) only Spanish and German;

5.67. Find the number m of elements in the union of sets A, B, C, D with the following 4 conditions:

    (i) A, B, C, D have 50, 60, 70, 80 elements, respectively.

    (ii) Each pair of sets has 20 elements in common.

    (iii) Each three of the sets has 10 elements in common.

    (iv) All four of the sets have 5 elements in common.

PIGEONHOLE PRINCIPLE

5.68. Find the minimum number of students needed to guarantee that 4 of them were born: (a) on the same day of the week; (b) in the same month.

5.69. Find the minimum number of students needed to guarantee that 3 of them:

(a) have last names which begin with the same first letter;

(b) were born on the same day of a month (with 31 days).

5.70. Consider a tournament with n players where each player plays against every other player. Suppose each player wins at least once. Show that at least 2 of the players have the same number of wins.

5.71. Suppose 5 points are chosen at random in the interior of an equilateral triangle T where each side has length two inches. Show that the distance between two of the points must be less than one inch.

5.72. Consider any set $X = \{x_1, x_2, \ldots, x_7\}$ of seven distinct integers. Show that there exist $x, y \in X$ such that $x + y$ or $x - y$ is divisible by 10.

MISCELLANEOUS PROBLEMS

5.73. Find the number m of ways 10 students can be divided into three teams where one team has 4 students and the other teams have 3 students.

5.74. Assuming a cell can be empty, find the number n of ways that a set with 3 elements can be partitioned into:
(a) 3 ordered cells; (b) 3 unordered cells.

5.75. Assuming a cell can be empty, find the number n of ways that a set with 4 elements can be partitioned into:
(a) 3 ordered cells; (b) 3 unordered cells.

5.76. The English alphabet has 26 letters of which 5 are vowels. Consider only 5-letter "words" consisting of 3 different consonants and 2 different vowels. Find the number of such words which:

(a) have no restrictions;   (c) contain the letters B and C;

(b) contain the letter B;   (d) begin with B and contain the letter C.

5.77. Teams A and B play in the World Series of baseball, where the team that first wins four games wins the series. Suppose A wins the first game, and that the team that wins the second game also wins the fourth game.

(a) Find and list the number n of ways the series can occur.

(b) Find the number of ways that B wins the series.

(c) Find the number of ways the series lasts seven games.

5.78. Find the number of ways a coin can be tossed:

(a) 6 times so that there is exactly 3 heads and no two heads occur in a row.

(b) $2n$ times so that there is exactly n heads and no two heads occur in a row.

5.79. Find the number of ways 3 elements a, b, c, can be assigned to 3 cells, so exactly 1 cell is empty.

5.80. Find the number of ways n distinct elements can be assigned to n cells so exactly 1 cell is empty.

Answers to Supplementary Problems

5.31. (a) 3 628 800; 39 916 800; 479 001 600;
(b) $\log(60!) = 81.92$, so $60! = 6.59 \times 10^{81}$.

5.32. (a) 240; (b) 2 184; (c) 1/90; (d) 1/1716.

5.33. (a) $n + 1$; (b) $n(n - 1)$; (c) $1/[n(n + 1)(n + 2)]$;
(d) $(n - r)(n - r + 1)$.

5.34. (a) 10; (b) 35; (c) 91; (d) 15; (e) 1140; (f) 816.

5.35. Hints: (a) Expand $(1 + 1)^n$; (b) Expand $(1 - 1)^n$.

5.36. (a) 1, 9, 36, 84, 126, 126, 84, 36, 9, 1;
(b) 1, 10, 45, 120, 210, 252, 210, 120, 45, 10, 1.

5.37. (a) 60; (b) 210; (c) 504; (d) not defined.

5.38. (a) 15; (b) 105.

5.39. (a) 18; (b) 80; (c) 306.

5.40. (a) $26^2 \cdot 10^3$; (b) $26 \cdot 25 \cdot 10^3$; (b); (c) $26 \cdot 10^3$.

5.41. (a) $26 \cdot 25 \cdot 10 \cdot 9 \cdot 8 = 468\,000$; (b) $26 \cdot 25 \cdot 9 \cdot 9 \cdot 8 = 421\,200$.

5.42. $m = 18 \cdot 17 \cdot 16 = 4896$.

5.43. (a) 12!; (b) $3!5!4!3! = 103\,680$.

5.44. (a) $6! = 720$; (b) $2 \cdot 3! \cdot 3! = 72$; (c) $4 \cdot 3! \cdot 3! = 144$.

5.45. (a) 120; (b) 48.

5.46. (a) 24; (b) 12.

5.47. (a) $3 \cdot 9 \cdot 8$; (b) $9 \cdot 8 \cdot 5$; (c) $9 \cdot 8 \cdot 7/2$; (d) $9 \cdot 8 \cdot 7/5$.

5.48. (a) $P(6, 3) = 120$; (b) $2 \cdot 5 \cdot 4 = 40$; (c) $2 \cdot 5 \cdot 4 = 40$.

5.49. $m = 360$.

5.50. (a) 9; (b) 5.

5.51. (a) 30; (b) $9!/[2!2!2!] = 45\,360$; (c) $11!/[2!3!2!] = 1\,663\,200$; (d) $8!/[2!2!2!] = 5040$.

5.52. $m = 9!/[4!2!3!] = 1260$.

5.53. (a) $12^3 = 1\,728$; (b) $P(12, 3) = 1320$.

5.54. (a) $10^4 = 10\,000$; (b) $P(10, 4) = 5040$.

5.55. (a) 6; (b) 15; (c) 20.

5.56. (a) $C(12, 4)$; (b) $C(9, 2) \cdot C(3, 2) = 108$;
(c) $C(9, 3) \cdot 3 = 252$; (d) $9 + 108 + 252 = 369$ or $C(12, 4) - C(9, 4) = 369$.

5.57. (a) $C(11, 5) = 462$; (b) $126 + 84 = 210$;
(c) $C(9, 5) + 2C(9, 4) = 378$.

5.58. $m = C(12, 4) + 2C(12, 3) = 935$.

5.59. (a) $C(10, 2) = 45$; (b) $6 \cdot 4 = 24$; (c) $C(6, 2) + C(4, 2) = 21$ or $45 - 24 = 21$.

5.60. (a) 165; (b) 110; (c) 80; (d) 276.

5.61. (a) 46; (b) 22; (c) 14.

5.62. (a) 4; (b) 20; (c) 56.

5.63. (a) 55; (b) 52.

5.64. (a) 4; (b) 6; (c) 5.

5.65. (a) $100 + 60 + 42 - 20 - 14 - 8 + 2 = 162$;
(b) $20 - 2 = 18$; (c) $60 - 20 - 8 + 2 = 34$;
(d) $300 - 162 = 138$.

5.66. (a) 37; (b) 9; (c) 28; (d) 4; (e) 13.

5.67. $m = 175$

5.68. (a) 22; (b) 37.

5.69. (a) 53; (b) 63.

5.70. Each player will win anywhere from 1 up to $n - 1$ games (pigeonholes). There are n players (pigeons).

5.71. Draw three lines between the midpoints of the sides of T. This partitions T into 4 equilateral triangles (pigeonholes) where each side has length 1. Two of the 5 points (pigeons) must lie in one of the triangles.

5.72. Let r_i be the remainder when x_i is divisible by 10. Consider the six pigeonholes: $H_1 = \{x_i \mid r_i = 0\}$, $H_2 = \{x_i \mid r_i = 5\}$, $H_3 = \{x_i \mid r_i = 1 \text{ or } 9\}$, $H_4 = \{x_i \mid r_i = 2 \text{ or } 8\}$, $H_5 = \{x_i \mid r_i = 3 \text{ or } 7\}$, $H_6 = \{x_i \mid r_i = 4 \text{ or } 6\}$. Then some x and y belong to some H_k.

5.73. $m = C(10, 4) \cdot C(6, 3) = 420$

5.74. (a) $n = 3^3 = 27$ (Each element can be placed in any of the three cells.) (b) The number of elements in three cells can be distributed as follows: [3, 0, 0], [2,1,0], or [1,1,1]. Thus $n = 1 + 3 + 1 = 5$.

5.75. (a) $n = 3^4 = 81$ (Each element can be placed in any of the three cells.) (b) The number of elements in three cells can be distributed as follows: [4, 0, 0], [3, 1, 0], [2, 2, 0], or [2, 1, 1]. Thus $n = 1 + 4 + 3 + 6 = 14$.

5.76. (a) $C(21, 3) \cdot C(5, 2) \cdot 5!$; (b) $C(20, 2) \cdot C(5, 2) \cdot 5!$; (c) $19 \cdot C(5, 2) \cdot 5!$; (d) $19 \cdot C(5, 2) \cdot 4!$.

5.77. Draw tree diagram T as in Fig. 5-4. Note T begins at A, the winner of the first game, and there is only one choice in the fourth game, the winner of the second game.

(a) $n = 15$ as listed below; (b) 6; (c) 8:
AAAA, AABAA, AABABA, AABABBA,
AABABBB, ABABAA, ABABABA, ABABABB,
ABABBAA, ABABBAB, ABABBB, ABBBAAA,
ABBBAAB, ABBBAB, ABBBB.

5.78. (a) 4, *HTHTHT, HTTHTH, HTHTTH, THTHTH*;
(b) $n + 1$.

5.79. 18.

5.80. $n! C(n, 2)$.

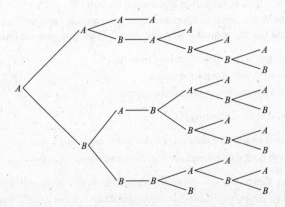

Fig. 5-4

CHAPTER 6

Advanced Counting Techniques, Recursion

6.1 INTRODUCTION

Here we consider more sophisticated counting techniques and problems. This includes problems involving combinations with repetition, ordered and unordered partitions, and the Inclusion–Exclusion Principle and the Pigeonhole Principle.

We also discuss recursion in this chapter.

6.2 COMBINATIONS WITH REPETITIONS

Consider the following problem. A bakery makes only $M = 4$ kinds of cookies: apple (a), banana (b), carrot (c), dates (d). Find the number of ways a person can buy $r = 8$ of the cookies.

Observe that order does not count. This is an example of combinations with repetitions. In particular, each combination can be listed with the a's first, then the b's, then the c's, and finally the d's. Four such combinations follow:

$$r_1 = aa, \ bb, \ cc, \ dd; \quad r_2 = aaa, \ c, \ ddd; \quad r_3 = bbbb, \ c, \ ddd; \quad r_4 = aaaaa, \ ddd.$$

Counting the number m of such combinations may not be easy.

Suppose we want to code the above combinations using only two symbols, say 0 and 1. This can be done by letting 0 denote a cookie, and letting 1 denote a change from one kind of cookie to another. Then each combination will require $r = 8$ zeros, one for each cookie, and $M - 1 = 3$ ones, where the first one denotes the change from a to b, the second one from b to c, and the third one from c to d. Thus the above four combinations will be coded as follows:

$$r_1 = 00100100100, \quad r_2 = 00001101000, \quad r_3 = 10000101000, \quad r_4 = 00000111000.$$

Counting the number m of these "codewords" is easy. Each codeword contains $R + M - 1 = 11$ digits where $r = 8$ are 0's and hence $M - 1 = 3$ are 1's. Accordingly,

$$M = C(11, 8) = C(11, 3) = \frac{11 \cdot 10 \cdot 9}{3 \cdot 2 \cdot 1} = 165$$

A similar argument gives us the following theorem.

Theorem 6.1: Suppose there are M kinds of objects. Then the number of combinations of r such objects is
$$C(r + M - 1, r) = C(r + M - 1, M - 1).$$

EXAMPLE 6.1 Find the number m of nonnegative integer solutions of $x + y + z = 18$.

We can view each solution, say $x = 3$, $y = 7$, $z = 8$, as a combination of $r = 18$ objects consisting of 3 a's, 7 b's, and 8 c's, where there are $M = 3$ kinds of objects, a's, b's, and c's. By Theorem 6.1,

$$m = C(r + M - 1, M - 1) = C(20, 2) = 190.$$

6.3 ORDERED AND UNORDERED PARTITIONS

Suppose a set has 7 elements. We want to find the number m of ordered partitions of S into three cells, say $[A_1, A_2, A_3]$, so they contain 2, 3, and 2 elements, respectively.

Since S has 7 elements, there are $C(7, 2)$ ways of choosing the first two elements for A_1. Following this, there are $C(5, 3)$ ways of choosing the 3 elements for A_2. Lastly, there are $C(2, 2)$ ways of choosing the 2 elements for A_3 (or, the last 2 elements form the cell A_3). Thus:

$$m = C(7, 2)C(5, 3)C(2, 2) = \binom{7}{2}\binom{5}{3}\binom{2}{2} = \frac{7 \cdot 6}{2 \cdot 1} \cdot \frac{5 \cdot 4 \cdot 3}{3 \cdot 2 \cdot 1} \cdot \frac{2 \cdot 1}{2 \cdot 1} = 210$$

Observe that

$$m = \binom{7}{2}\binom{5}{3}\binom{2}{2} = \frac{7!}{2!5!} \cdot \frac{5!}{3!2!} \cdot \frac{2!}{2!0!} = \frac{7!}{2!3!2!}$$

since each numerator after the first is cancelled by a term in the denominator of the previous factor.

The above discussion can be shown to be true in general. Namely:

Theorem 6.2: The number m of ordered partitions of a set S with n elements into r cells $[A_1, A_2, \ldots, A_r]$ where, for each i, $n(A_i) = n_i$, follows:

$$m = \frac{n!}{n_1! n_2! \ldots n_r!}$$

Unordered Partitions

Frequently, we want to partition a set S into cells $[A_1, A_2, \ldots, A_r]$ where the cells are now unordered. The number m of such unordered partitions is obtained from the number m' of ordered partitions by dividing m' by each $k!$ where k of the cells have the same number of elements.

EXAMPLE 6.2 Find the number m of ways to partition 10 students into four teams $[A_1, A_2, A_3, A_4]$ so that two teams contain 3 students and two teams contain 2 students.

By Theorem 6.2, there are $m' = 10!/(3!3!2!2!) = 25\,200$ such ordered partitions.

Since the teams form an unordered partition, we divide m' by 2! because of the two cells with 3 elements each and 2! because of the two cells with 2 elements each.

Thus $m = 25\,200/(2!2!) = 6300$.

6.4 INCLUSION–EXCLUSION PRINCIPLE REVISITED

Let $A_1, A_2, \ldots, A_r$ be subsets of a universal set U. Suppose we let s_k denote the sum of the cardinalities of all possible k-tuple intersections of the sets, that is, the sum of all of the cardinalities

$$n(A_{i_1} \cap A_{i_2} \cap \cdots \cap A_{i_k})$$

For example,

$$s_1 = \sum_i n(A_i), \quad s_2 = \sum_{i<j} n(A_i \cap A_j), \quad s_3 = \sum_{i_1<i_2<i_3} n(A_{i_1} \cap A_{i_2} \cap A_{i_3})$$

The Inclusion–Exclusion Principle, which appears in Section 5.7, gave a formula for the number of elements in the union of the sets. Specifically, (Theorem 5.9) we have

$$n(A_1 \cup A_2 \cup \cdots \cup A_r) = s_1 - s_2 + s_3 - \cdots + (-1)^{r-1} s_r$$

On the other hand, using DeMorgan's law,

$$n(A_1^C \cap A_2^C \cap \ldots \cap A_r^C) = n([A_1 \cup A_2 \cup \ldots \cup A_r]^C) = |\mathbf{U}| - n(A_1 \cup A_2 \cup \ldots \cup A_r)$$

Accordingly, we obtain an alternate form for Theorem 5.9:

Theorem (Inclusion–Exclusion Principle) 6.3: Let $A_1, A_2, \ldots, A_r$ be subsets of a universal set $\mathbf{U}$. Then the number m of elements which do not appear in any of the subsets $A_1, A_2, \ldots, A_r$ of $\mathbf{U}$ is:

$$m = n(A_1^C \cap A_2^C \cap \ldots \cap A_r^C) = |\mathbf{U}| - s_1 + s_2 - s_3 + \cdots + (-1)^r s_r$$

EXAMPLE 6.3 Let $\mathbf{U}$ be the set of positive integers not exceeding 1000. Then $|U| = 1000$. Find $|S|$ where S is the set of such integers which are not divisible by 3, 5, or 7.

Let A be the subset of integers which are divisible by 3, B which are divisible by 5, and C which are divisible by 7. Then $S = A^C \cap B^C \cap C^C$ since each element of S is not divisible by 3, 5 or 7. By integer division,

$$|A| = 1000/3 = 333, \quad |B| = 1000/5 = 200, \quad |C| = 1000/7 = 142,$$
$$|A \cap B| = 1000/15 = 66, \quad |A \cap C| = 1000/21 = 47, \quad |B \cap C| = 1000/35 = 28,$$
$$|A \cap B \cap C| = 1000/105 = 9$$

Thus, by the Inclusion–Exclusion Principle Theorem 6.3,

$$|S| = 1000 - (333 + 200 + 142) + (66 + 47 + 28) - 9 = 1000 - 675 + 141 - 9 = 457$$

Number of Onto Functions

Let A and B be sets such that $|A| = 6$ and $|B| = 4$. We want to find the number of surjective (onto) functions from A onto B.

Let b_1, b_2, b_3, b_4 be the four elements in B. Let $\mathbf{U}$ be the set of all functions from A into B. Furthermore, let F_1 be the set of functions which do not send any element of A into b_1 that is, b_1 is not in the range of any function in F_1. Similarly, let F_2, F_3, and F_4 be the sets of functions which do not send any element of A into b_2, b_3, and b_4, respectively.

We are looking for the number of functions in $S = F_1^C \cap F_2^C \cap F_3^C \cap F_4^C$, that is, those functions which do send at least one element of A into b_1, at least one element of A into b_2, and so on. We will use the Inclusion–Exclusion Principle Theorem 6.3 as follows.

  (i) For each function in $\mathbf{U}$, there are 4 choices for each of the 6 elements in A; hence $|U| = 4^6 = 4096$
  (ii) There are $C(4, 1) = 4$ functions F_i. In each case, there are 3 choices for each of the 6 elements in A, hence $|F_i| = 3^6 = 729$.
  (iii) There are $C(4, 2) = 6$ pairs $F_i \cap F_j$. In each case, there are 2 choices for each of the 6 elements in A, hence $|F_i \cap F_j| = 2^6 = 64$.
  (iv) There are $C(4, 3) = 4$ triplets $F_i \cap F_j \cap F_k$. In each case, there is only one choice for each of the 6 elements in A. Hence $|F_i \cap F_j \cap F_k| = 1^6 = 1$.

(v) $F_1 \cap F_2 \cap F_3 \cap F_4$ has no element, that is, is empty. Hence $|F_1 \cap F_2 \cap F_3 \cap F_4| = 0$. By the Inclusion–Exclusion Principle Theorem 6.3,

$$|S| = |F_1^C \cap F_2^C \cap F_3^C \cap F_4^C| = 4^6 - C(4,1)3^6 + C(4,2)2^6 - C(4,3)1^7$$
$$= 4096 - 2916 + 384 - 1 = 795$$

The above result is true in general. Namely:

Theorem 6.4: Suppose $|A| = m$ and $|B| = n$ where $m \geq n$. Then the number N of surjective (onto) functions from A onto B is:

$$N = n^m - C(n,1)(n-1)^m + C(n,2)(n-2)^m - \cdots + (-1)^{n-1}C(n,n-1)1^m$$

Derangements

A *derangement* is a permutation of objects where each object is not in its original position. For example, 453162 is not a derangement of 123456 since 3 is in its correct position, but 264531 is a derangement of 123456. (Alternately, a permutation $\sigma: X \to X$ is a derangement if $\sigma(i) \neq i$ for every $i \in X = \{1, 2, \ldots, n\}$.)

Let D_n denote the number of derangements of n objects. For example, 231 and 312 are the only derangements of 123. Hence $D_3 = 2$. The following theorem, proved in Problem 6.6, applies.

Theorem 6.5: $D_n = n![1 - \frac{1}{1!} + \frac{1}{2!} - \frac{1}{3!} + \cdots + (-1)^n \frac{1}{n!}]$

The probability (Chapter 7) that a derangement of n objects occurs equals D_n divided by $n!$, the number of permutations of the n objects. Thus Theorem 6.5 yields:

Corollary 6.6: Let p be the probability of a derangement of n objects. Then

$$p = 1 - \frac{1}{1!} + \frac{1}{2!} - \frac{1}{3!} + \cdots + (-1)^n \frac{1}{n!}$$

EXAMPLE 6.4 (Hat Check Problem)   Suppose $n = 5$ people check in their hats at a restaurant and they are given back their hats at random. Find the probability p that no person receives his/her own hat.

This is an example of a derangement with $n = 5$. By Corollary 6.6,

$$p = 1 - 1 + 1/2 - 1/6 + 1/24 - 1/120 = 44/120 = 11/30 \approx 0.367$$

Note that the signs alternate and the terms get very, very small in Corollary 6.6. Figure 6-1 gives the values of p for the first few values of n. Note that, for $n > 4$, p is very close to the following value (where $e = 2.718$):

$$e^{-1} = 1 - \frac{1}{1!} + \frac{1}{2!} - \frac{1}{3!} + \cdots + (-1)^n \frac{1}{n!} + \cdots \approx 0.368$$

n	1	2	3	4	5	6	7
$p = D_n/n!$	0.0000	0.5000	0.3333	0.3750	0.3667	0.3681	0.3679

Fig. 6-1

6.5 PIGEONHOLE PRINCIPLE REVISITED

The Pigeonhole Principle (with its generalization) is stated with simple examples in Section 5.6. Here we give examples of more sophisticated applications of this principle.

EXAMPLE 6.5 Consider six people, where any two of them are either friends or strangers. Show that there are three of them which are either mutual friends or mutual strangers.

Let A be one of the people. Let X consist of those which are friends of A, and Y consist of those which are strangers of A. By the Pigeonhole Principle, either X or Y has at least three people. Suppose X has three people. If two of them are friends, then the two with A are three mutual friends. If not, then X has three mutual strangers. Alternately, suppose Y has three people. If two of them are strangers, then the two with A are three mutual strangers. If not, then X has three mutual friends.

EXAMPLE 6.6 Consider five *lattice* points $(x_1, y_1), \ldots, (x_5, y_5)$ in the plane, that is, points with integer coordinates. Show that the midpoint of one pair of the points is also a lattice point.

The midpoint of points $P(a, b)$ and $Q(c, d)$ is $([a + c]/2, [b + d]/2)$. Note that $(r + s)/2$ is an integer if r and s are integers with the same *parity*, that is, both are odd or both are even. There are four pairs of parities: (odd, odd), (odd, even), (even, odd), and (even, even). There are five points. By the Pigeonhole Principle, two of the points have the same pair of parities. The midpoint of these two points has integer coordinates.

An important application of the Pigeonhole Principle follows.

Theorem 6.7: Every sequence of distinct $n^2 + 1$ real numbers contains a subsequence of length $n + 1$ which is strictly increasing or strictly decreasing.

For example, consider the following sequence of $10 = 3^2 + 1$ numbers (where $n = 3$): 2, 1, 8, 6, 7, 5, 9, 4, 12, 3. There are many subsequences of length $n + 1 = 4$ which are strictly increasing or strictly decreasing; for example,

$$2, 6, 9, 12; \quad 1, 5, 9, 12; \quad 8, 6, 5, 4; \quad 7, 5, 4, 3.$$

On the other hand, the following sequence of $9 = 3^2$ numbers has no subsequence of length $n + 1 = 4$ which is strictly increasing or strictly decreasing:

$$3, \quad 2, \quad 1, \quad 6, \quad 5, \quad 4, \quad 9, \quad 8, \quad 7.$$

The proof of Theorem 6.7 appears in Problem 6.10.

6.6 RECURRENCE RELATIONS

Previously, we discussed recursively defined functions such as

(a) Factorial function,   (b) Fibonacci sequence,   (c) Ackermann function.

Here we discuss certain kinds of recursively defined sequences $\{a_n\}$ and their solution. We note that a *sequence* is simply a function whose domain is

$$\mathbf{N} = \{1, 2, 3, \ldots\} \quad \text{or} \quad \mathbf{N}_0 = \mathbf{N} \cup \{0\} = \{0, 1, 2, 3, \ldots\}$$

We begin with some examples.

EXAMPLE 6.7 Consider the following sequence which begins with the number 3 and for which each of the following terms is found by multiplying the previous term by 2:

$$3, \quad 6, \quad 12, \quad 24, \quad 48, \quad \ldots$$

It can be defined recursively by:

$$a_0 = 3, \quad a_k = 2a_{k-1} \text{ for } k \geq 1 \quad \text{or} \quad a_0 = 3, \quad a_{k+1} = 2a_k \quad \text{for } k \geq 0$$

The second definition may be obtained from the first by setting $k = k + 1$. Clearly, the formula $a_n = 3(2^n)$ gives us the nth term of the sequence without calculating any previous term.

The following remarks about the above example are in order.

(1) The equation $a_k = 2a_{k-1}$ or, equivalently, $a_{k+1} = 2a_k$, where one term of the sequence is defined in terms of previous terms of the sequence, is called a *recurrence relation*.
(2) The equation $a_0 = 3$, which gives a specific value to one of the terms, is called an *initial condition*.
(3) The function $a_n = 3(2^n)$, which gives a formula for a_n as a function of n, not of previous terms, is called a *solution* of the recurrence relation.
(4) There may be many sequences which satisfy a given recurrence relation. For example, each of the following is a solution of the recurrence relation $a_k = 2a_{k-1}$.

$$1, 2, 4, 8, 16, \ldots \quad \text{and} \quad 7, 14, 28, 56, 112, \ldots$$

All such solutions form the so-called *general solution* of the recurrence relation.
(5) On the other hand, there may be only a unique solution to a recurrence relation which also satisfies given initial conditions. For example, the initial condition $a_0 = 3$ uniquely yields the solution $3, 6, 12, 24, \ldots$ of the recurrence relation $a_k = 2a_{k-1}$.

This chapter shows how to solve certain recurrence relations. First we give two important sequences the reader may have previously studied.

EXAMPLE 6.8

(a) *Arithmetic Progression*

An arithmetic progression is a sequence of the form

$$a, a + d, a + 2d, a + 3d, \ldots$$

That is, the sequence begins with the number a and each successive term is obtained from the previous term by adding d (the common difference between any two terms). For example:

  (i) $a = 5, d = 3$: $5, 8, 9, 11, \ldots$
  (ii) $a = 2, d = 5$: $2, 7, 12, 17, \ldots$
  (iii) $a = 1, d = 0$: $1, 1, 1, 1, 1, \ldots$

We note that the general arithmetic progression may be defined recursively by:

$$a_1 = a \quad \text{and} \quad a_{k+1} = a_k + d \quad \text{for } k \geq 1$$

where the solution is $a_n = a + (n - 1)d$.

(b) *Geometric Progression*

A geometric progression is a sequence of the form

$$a, ar, ar^2, ar^3, \ldots$$

That is, the sequence begins with the number a and each successive term is obtained from the previous term by multiplying by r (the common ratio between any two terms) for example:

  (i) $a = 1, r = 3$: $1, 3, 9, 27, 81, \ldots$
  (ii) $a = 5, r = 2$: $5, 10, 20, 40, \ldots$
  (iii) $a = 1, r = \frac{1}{2}$: $1, \frac{1}{2}, \frac{1}{4}, \frac{1}{8}, \ldots$

We note that the general geometric progression may be defined recursively by:

$$a_1 = a \quad \text{and} \quad a_{k+1} = ra_k \quad \text{for } k \geq 1$$

where the solution is $a_{n+1} = ar^n$.

6.7   LINEAR RECURRENCE RELATIONS WITH CONSTANT COEFFICIENTS

A *recurrence relation of order k* is a function of the form

$$a_n = \Phi(a_{n-1}, a_{n-2}, \dots, a_{n-k}, n)$$

that is, where the nth term a_n of a sequence is a function of the preceding k terms $a_{n-1}, a_{n-2}, \dots, a_{n-k}$ (and possibly n). In particular, a *linear kth-order recurrence relation with constant coefficients* is a recurrence relation of the form

$$a_n = C_1 a_{n-1} + C_2 a_{n-2} + \cdots + C_k a_{n-k} + f(n)$$

where $C_1, C_2, \dots, C_k$ are constants with $C_k \neq 0$, and $f(n)$ is a function of n. The meanings of the names linear and constant coefficients follow:

   Linear: There are no powers or products of the a_j's.
   Constant coefficients: The $C_1 C_2, \dots, C_k$ are constants (do not depend on n).

If $f(n) = 0$, then the relation is also said to be *homogeneous*.

Clearly, we can uniquely solve for a_n if we know the values of $a_{n-1}, a_{n-2}, \dots, a_{n-k}$. Accordingly, by mathematical induction, there is a unique sequence satisfying the recurrence relation if we are given *initial values* for the first k elements of the sequence.

EXAMPLE 6.9   Consider each of the following recurrence relations.

(a) $a_n = 5a_{n-1} - 4a_{n-2} + n^2$

This is a second-order recurrence relation with constant coefficients. It is nonhomogeneous because of the n^2. Suppose we are given the initial conditions $a_1 = 1$, $a_2 = 2$. Then we can find sequentially the next few elements of the sequence:

$$a_3 = 5(2) - 4(1) + 3^2 = 15, \quad a_4 = 5(15) - 4(2) + 4^2 = 83$$

(b) $a_n = 2a_{n-1}a_{n-2} + n^2$

The product $a_{n-1}a_{n-2}$ means the recurrence relation is not linear. Given initial conditions $a_1 = 1$, $a_2 = 2$, we can still find the next few elements of the sequence:

$$a_3 = 2(2)(1) + 3^2 = 13, \quad a_4 = 2(13)(2) + 4^2 = 68$$

(c) $a_n = na_{n-1} + 3a_{n-2}$

This is a homogeneous linear second-order recurrence relation but it does not have constant coefficients because the coefficient of a_{n-1} is n, not a constant. Given initial conditions $a_1 = 1$, $a_2 = 2$, the next few elements of the sequence follow:

$$a_3 = 3(2) + 3(1) = 9, \quad a_4 = 4(9) + 3(2) = 42$$

(d) $a_n = 2a_{n-1} + 5a_{n-2} - 6a_{n-3}$

This is a homogeneous linear third-order recurrence relation with constant coefficients. Thus we need three, not two, initial conditions to yield a unique solution of the recurrence relation. Suppose we are given the initial conditions $a_1 = 1$, $a_2 = 2$, $a_3 = 1$. Then, the next few elements of the sequence follow:

$$a_4 = 2(1) + 5(2) - 6(1) = 6, \quad a_5 = 2(2) + 5(1) - 6(6) = -37$$
$$a_6 = 2(1) + 5(6) - 6(-37) = 254$$

This chapter will investigate the solutions of homogeneous linear recurrence relations with constant coefficients. The theory of nonhomogeneous recurrence relations and recurrence relations without constant coefficients lies beyond the scope of this text.

For computational convenience, most of our sequences will begin with a rather than a. The theory is not affected at all.

6.8 SOLVING SECOND-ORDER HOMOGENEOUS LINEAR RECURRENCE RELATIONS

Consider a homogeneous second-order recurrence relation with constant coefficients which has the form

$$a_n = sa_{n-1} + ta_{n-2} \quad \text{or} \quad a_n - sa_{n-1} - ta_{n-2} = 0$$

where s and t are constants with $t \neq 0$. We associate the following quadratic polynomial with the above recurrence relation:

$$\Delta(x) = x^2 - sx - t$$

This polynomial $\Delta(x)$ is called the *characteristic polynomial* of the recurrence relation, and the roots of $\Delta(x)$ are called its *characteristic roots*.

Theorem 6.8: Suppose the characteristic polynomial $\Delta(x) = x^2 - sx - t$ of the recurrence relation

$$a_n = sa_{n-1} + ta_{n-2}$$

has distinct roots r_1 and r_2. Then the general solution of the recurrence relation follows, where c_1 and c_2 are arbitrary constants:

$$a_n = c_1 r_1^n + c_2 r_2^n$$

We emphasize that the constants c_1 and c_2 may be uniquely computed using initial conditions. We note that the theorem is true even when the roots are not real. Such cases lie beyond the scope of this text.

EXAMPLE 6.10 Consider the following homogeneous recurrence relation:

$$a_n = 2a_{n-1} + 3a_{n-2}$$

The general solution is obtained by first finding its characteristic polynomial $\Delta(x)$ and its roots r_1 and r_2:

$$\Delta(x) = x^2 - 2x - 3 = (x - 3)(x + 1); \quad \text{roots } r_1 = 3, r_2 = -1$$

Since the roots are distinct, we can use Theorem 6.8 to obtain the general solution:

$$a_n = c_1 3^n + c_2(-1)^n$$

Thus any values for c_1 and c_2 will give a solution to the recurrence relation.

Suppose we are also given the initial conditions $a_0 = 1$, $a_1 = 2$. Using the recurrence relation we can compute the next few terms of the sequence:

$$1, \quad 2, \quad 8, \quad 28, \quad 100, \quad 356, \quad 1268, \quad 3516, \quad \ldots$$

The unique solution is obtained by finding c_1 and c_2 using the initial conditions. Specifically:

For $n = 0$ and $a_0 = 1$, we get: $c_1 3^0 + c_2(-1)^0 = 1$ or $c_1 + c_2 = 1$

For $n = 1$ and $a_1 = 2$, we get: $c_1 3^1 + c_2(-1)^1 = 2$ or $3c_1 - c_2 = 2$

Solving the system of the two equations in the unknowns c_1 and c_2 yields:

$$c_1 = \frac{3}{4} \quad \text{and} \quad c_2 = \frac{1}{4}$$

Thus the following is the unique solution of the given recurrence relation with the given initial conditions $a_0 = 1$, $a = 2$:

$$a_n = \frac{3}{4} 3^n + \frac{1}{4}(-1)^n = \frac{3^{n+1} + (-1)^n}{4}$$

EXAMPLE 6.11 Consider the celebrated Fibonacci sequence:

$$a_n = a_{n-1} + a_{n-2}, \quad \text{with} \quad a_0 = 0, a_1 = 1$$

The first 10 terms of the sequence follow:

$$0, 1, 1, 2, 3, 5, 8, 13, 21, 34, \ldots$$

Sometimes the Fibonacci sequence is defined using the initial conditions $a_0 = 1$, $a_1 = 1$ or the initial conditions $a_1 = 1$, $a_2 = 2$. We use $a_0 = 0$, $a_1 = 1$ for computational convenience. (All three initial conditions yield the same sequence after the pair of terms 1, 2.)

Observe that the Fibonacci sequence is a homogeneous linear second-order recurrence relation, so it can be solved using Theorem 6.8. Its characteristic polynomial follows:

$$\Delta(x) = x^2 - x - 1$$

Using the quadratic formula, we obtain the roots:

$$r_1 = \frac{1 + \sqrt{5}}{2}, \quad r_2 = \frac{1 - \sqrt{5}}{2}$$

By Theorem 6.8, we obtain the general solution:

$$a_n = c_1 \left(\frac{1 + \sqrt{5}}{2} \right)^n + c_2 \left(\frac{1 - \sqrt{5}}{2} \right)^n$$

The initial conditions yield the following system of two linear equations in c_1 and c_2

For $n = 0$ and $a_0 = 0$, we get: $\quad 0 = c_1 + c_2$

For $n = 1$ and $a_1 = 1$, we get: $\quad 1 = c_1 \left(\frac{1 + \sqrt{5}}{2} \right) + c_2 \left(\frac{1 - \sqrt{5}}{2} \right)$

The solution of the system follows:

$$c_1 = \frac{1}{\sqrt{5}}, c_2 = -\frac{1}{\sqrt{5}}$$

Accordingly, the following is the solution of the Fibonacci recurrence relation:

$$a_n = \frac{1}{\sqrt{5}} \left(\frac{1 + \sqrt{5}}{2} \right)^n - \frac{1}{\sqrt{5}} \left(\frac{1 - \sqrt{5}}{2} \right)^n$$

One can show that the absolute value of the above second term for a_n is always less than 1/2. Thus a_n is also the closest integer to the number

$$\frac{1}{\sqrt{5}} \left(\frac{1 + \sqrt{5}}{2} \right)^n \approx (0.4472)(1.6180)^n$$

Solution when Roots of the Characteristic Polynomial are Equal

Suppose the roots of the characteristic polynomial are not distinct. Then we have the following result.

Theorem 6.9: Suppose the characteristic polynomial $\Delta(x) = x^2 - sx - t$ of the recurrence relation

$$a_n = sa_{n-1} + ta_{n-2}$$

has only one root r_0. Then the general solution of the recurrence relation follows, where c_1 and c_2 are arbitrary constants:

$$a_n = c_1 r_0^n + c_2 n r_0^n$$

The constants c_1 and c_2 may be uniquely computed using initial conditions.

EXAMPLE 6.12 Consider the following homogeneous recurrence relation:

$$a_n = 6a_{n-1} - 9a_{n-2}$$

The characteristic polynomial $\Delta(x)$ follows:

$$\Delta(x) = x^2 - 6x + 9 = (x - 3)^2$$

Thus $\Delta(x)$ has only the one root $r_0 = 3$. Now we use Theorem 6.9 to obtain the following general solution of the recurrence relation:

$$a_n = c_1 3^n + c_2 n 3^n$$

Thus any values for c_1 and c_2 will give a solution to the recurrence relation.

Suppose we are also given the initial conditions $a_1 = 3$, $a_2 = 27$. Using the recurrence relation we can compute the next few terms of the sequence:

$$3, \quad 27, \quad 135, \quad 567, \quad 2187, \quad 8109, \quad \ldots$$

The unique solution is obtained by finding c_1 and c_2 using the initial conditions. Specifically:

For $n = 1$ and $a_1 = 3$,   we get: $c_1 3^1 + c_2 (1)(3)^1 = 3$ or $3c_1 + 3c_2 = 3$
For $n = 2$ and $a_2 = 27$,   we get: $c_1 3^2 + c_2 (2)(3)^2 = 27$ or $9c_1 + 18c_2 = 27$

Solving the system of the two equations in the unknowns c_1 and c_2 yields:

$$c_1 = -1 \quad \text{and} \quad c_2 = 2$$

Thus the following is the unique solution of the recurrence relation with the given initial conditions:

$$a_n = -3^n + 2n 3^n = 3^n (2n - 1)$$

6.9   SOLVING GENERAL HOMOGENEOUS LINEAR RECURRENCE RELATIONS

Consider now a general linear homogeneous kth-order recurrence relation with constant coefficients which has the form

$$a_n = C_1 a_{n-1} + C_2 a_{n-2} + C_3 a_{n-3} + \cdots + C_k a_{n-k} = \sum_{i=1}^{k} C_i a_{n-1} \tag{6.1}$$

where $C_1, C_2, \ldots, C_k$ are constants with $C_k \neq 0$. The *characteristic polynomial* $\Delta(x)$ of the recurrence relation (*6.1*) follows:

$$\Delta(x) = x^k - C_1 x^{k-1} - C_2 x^{k-2} - C_3 x^{k-3} - \cdots - C_k = x^k - \sum_{i=1}^{k} C_1 x^{k-1}$$

The roots of $\Delta(x)$ are called the *characteristic roots* of the recurrence relation.

The following remarks are in order.

Remark 1: If $p(n)$ and $q(n)$ are solutions of (*6.1*), then any linear combination

$$c_1 p(n) + c_2 q(n)$$

of $p(n)$ and $q(n)$ is also a solution. (This is not true if the recurrence relation is nonhomogeneous.)

Remark 2: If r is a root of multiplicity m of the characteristic polynomial $\Delta(x)$ of (6.1), then each of the following

$$r^n, nr^n, n^2 r^n, \ldots, n^{n-1} r^n$$

is a solution of (6.1). Thus any linear combination

$$c_1 r^n + c_2 nr^n + c_3 n^2 r^n + \cdots + c_m n^{m-1} r^n = (c_1 + c_2 n + c_3 n^2 + \cdots + c_m n^{m-1}) r^n$$

is also a solution.

EXAMPLE 6.13 Consider the following third-order homogeneous recurrence relation:

$$a_n = 11 a_{n-1} - 39 a_{n-2} + 45 a_{n-3}$$

The characteristic polynomial $\Delta(x)$ of the recurrence relation follows:

$$\Delta(x) = x^3 - 11 x^2 + 39 x - 45 = (x - 3)^2 (x - 5)$$

Thus $\Delta(x)$ has two roots, $r_1 = 3$ of multiplicity 2 and $r_2 = 5$ of multiplicity 1. Thus, by the above remarks, the following is the general solution of the recurrence relation:

$$a_n = c_1 (3^n) + c_2 n (3^n) + c_3 (5^n) = (c_1 + c_2 n)(3^n) + c_3 (5^n)$$

Thus any values for c_1, c_2, c_3 will give a solution to the recurrence relation.

Suppose we are also given the initial conditions $a_0 = 5, a_1 = 11, a_2 = 25$. Using the recurrence relation we can compute the next few terms of the sequence:

$$5, \quad 11, \quad 25, \quad 71, \quad 301, \quad 1667, \quad \ldots$$

The unique solution is obtained by finding c_1, c_2, c_3 using the initial conditions. Specifically:

$$
\begin{aligned}
&\text{For } n = 0 \text{ and } a_0 = 5, &&\text{we get:} &&c_1 + c_3 = 5 \\
&\text{For } n = 1 \text{ and } a_1 = 11, &&\text{we get:} &&3 c_1 + 3 c_2 + 5 c_3 = 11 \\
&\text{For } n = 2 \text{ and } a_2 = 25, &&\text{we get:} &&9 c_1 + 18 c_2 + 25 c_3 = 25
\end{aligned}
$$

Solving the system of the three equations in the unknowns c_1, c_2, c_3 yields:

$$c_1 = 4, \quad c_2 = -2, \quad c_3 = 1$$

Thus the following is the unique solution of the recurrence relation with the given initial conditions:

$$a_n = (4 - 2n)(3^n) + 5^n$$

Remark: Finding the roots of the characteristic polynomial $\Delta(x)$ is an important step in solving recurrence relations. Generally speaking, this may be difficult when the degree of $\Delta(x)$ is greater than 2. (Example B.16 indicates one way to find the roots of some polynomials of degree 3 or more.)

Solved Problems

ADVANCED COUNTING TECHNIQUES, INCLUSION–EXCLUSION

6.1. A bagel shop sells $M = 5$ kinds of bagels. Find the number m of ways a customer can buy: (a) 8 bagels; (b) a dozen bagels.

Use $m = C(r + M - 1, r) = C(r + M - 1, M - 1)$, that is, Theorem 6.1, since this problem concerns combinations with repetitions.

(a) Here $r = 8$, so $m = C(8 + 4, 4) = C(12, 4) = 494$.

(b) Here $r = 12$, so $m = C(12 + 4, 4) = C(16, 4) = 1820$.

6.2. Find the number m of nonnegative solutions to $x + y + z = 18$ with the conditions that $x \geq 3$, $y \geq 2$, $z \geq 1$.

Let $x' = x - 3$, $y' = y - 2$ and $z' = z - 1$. Then m is also the number of nonnegative solutions to $x' + y' + z' = 12$. As in Example 6.1, this second problem concerns combinations with repetitions with $M = 3$ and $r = 12$. Thus

$$m = C(12 + 2, 2) = C(14, 2) = 91.$$

6.3. Let E be the equation $x + y + z = 18$. Find the number m of nonnegative solutions to E with the conditions that $x < 7$, $y < 8$, $z < 9$.

Let S be the set of all nonnegative solutions of E. Let A be the set of solutions for which $x \geq 7$, let B be the set of solutions for which $y \geq 8$, and let C be the set of solutions for which $z \geq 9$. Then

$$m = |A^C \cap B^C \cap C^C|$$

As in Problem 6.2, we obtain:

$$\begin{aligned}
|A| &= C(11 + 2, 2) = 78, & |A \cap B| &= C(3 + 2, 2) = 10 \\
|B| &= C(10 + 2, 2) = 66, & |A \cap C| &= C(2 + 2, 2) = 6 \\
|C| &= C(9 + 2, 2) = 55, & |B \cap C| &= C(1 + 2, 2) = 3
\end{aligned}$$

Also, $|S| = C(18 + 2, 2) = 190$ and $|A \cap B \cap C| = 0$. By the Inclusion–Exclusion Principle,

$$m = 190 - (78 + 66 + 55) + (10 + 6 + 3) - 0 = 10$$

6.4. There are 9 students in a class. Find the number m of ways: (a) the 9 students can take 3 different tests if 3 students are to take each test; (b) the 9 students can be partitioned into 3 teams A, B, C so that each team contains 3 students,

(a) Method 1: We seek the number m of partitions of the 9 students into cells containing 3 students. By Theorem 6.2, $m = 9!/(3!3!3!) = 5040$.

Method 2: There are $C(9, 3)$ to choose three students to take the first test; then there are $C(6, 3)$ ways to choose 3 students to take the second test; and the remaining students take the third test. Thus $m = C(9, 3)C(6, 3) = 5040$.

(b) Each partition $\{A, B, C\}$ of the students can be arranged in $3! = 6$ ways as an ordered partition. By (a), there are 5040 such ordered partitions. Hence $m = 5040/6 = 840$.

6.5. Find the number N of ways a company can assign 7 projects to 4 people so that each person gets at least one project.

We want to find the number N of onto functions from a set with $m = 7$ elements onto a set with $n = 4$ elements. We use Theorem 6.4:

$$\begin{aligned}
N &= 4^2 - C(4, 1)(3^7) + C(4, 2)(2^7) - C(4, 3)(1^7) \\
&= 4^7 - 4(3^7) + 6(2^7) - 4(1^7) = 16\,384 - 8748 + 768 - 4 = 8400
\end{aligned}$$

6.6 Prove Theorem 6.5: $D_n = n![1 - \frac{1}{1!} + \frac{1}{2!} - \frac{1}{3!} + \cdots + (-1)^n \frac{1}{n!}]$

Recall (Section 3.3) that S_n denotes the set of permutations on $X = \{1, 2, \ldots, n\}$ and $|S_n| = n!$. For $i = 1, \ldots, n$, let F_i denote all permutations in S_n which "fix i," that is, $F_i = \{\sigma \in S_n \mid \sigma(i) = i\}$. Then, for distinct subscripts,

$$|F_r| = (n-1)!, \quad |F_i \cap F_i| = (n-2)!, \ldots |F_{i_1} \cap F_{i_2} \cap \ldots \cap F_{i_r}| = (n-r)!$$

Let Y denote the set of all derangements in S_n. Then

$$D_n = |Y| = |F_1^C \cap F_2^C \cap \cdots \cap F_n^C|$$

By the Inclusion–Exclusion principle,

$$D_n = |S_n| - s_1 + s_2 - s_3 + \cdots + (-1)^n s_n$$

where

$$s_r = \sum_{i_1 < i_2 < \ldots < i_n} |F_{i_1} \cap F_{i_2} \cap \ldots \cap F_{i_r}| = C(n, r)(n-r)! = \frac{n!}{r!}$$

Setting $|S_n| = n!$ and $s_r = n!/r!$ in the formula for D_n gives us our theorem.

PIGEONHOLE PRINCIPLE

6.7. Suppose five points are chosen from the interior of a square S where each side has length two inches. Show that the distance between two of the points must be less than $\sqrt{2}$ inches.

Draw two lines between the opposite sides of S which partitions S into four subsquares each whose sides have length one inch. By the Pigeonhole Principle, two of the points lie in one of the subsquares. The diagonal of each subsquare is $\sqrt{2}$ inches, so the distance between the two points is less than $\sqrt{2}$ inches.

6.8. Let p and q be positive integers. A number r is said to satisfy the (p, q)-Ramsey property if a set of r people must have a subset of p mutual friends or a subset of q mutual strangers. The Ramsey number $R(p, q)$ is the smallest such integer r. Show that R(3, 3) = 6.

By Example 6.5, $R(3, 3) \geq 6$. We show that $R(3, 3) > 5$. Consider five people who are sitting around a circular table, and suppose each person is only friends with the people sitting next to him/her. No three people can be strangers since two of the three people must be sitting next to each other. Also no three people can be mutual friends since they cannot be sitting next to each other. Thus $R(3, 3) > 5$. Accordingly $R(3, 3) = 6$.

6.9. Suppose a team X plays 18 games in a two-week 14-day period, and plays at least one game a day. Show that there is a period of days in which exactly 9 games were played.

Let $S = \{s_1, s_2, \ldots, s_{14}\}$ where s_i is the number of games X played from the first day to the ith day. Then $s_{14} = 18$, and all the s_i are distinct. Let $T = \{t_1, t_2, \ldots, t_{14}\}$ where $t_i = s_i + 9$. Then $t_{14} = 18 + 9 = 27$, and the t_i are distinct. Together S and T have $14 + 14 = 28$ numbers, which lie between 1 and 27. By the Pigeonhole Principle, two of the numbers must be equal. However the entries in S and the entries in T are distinct. Thus there is $s_j \in S$ and $t_n \in T$ such that $s_j = t_k = s_k + 9$. Therefore,

$$9 = s_j - s_n = \text{number of games played in days } k+1, k+2, \ldots, j-1, j$$

6.10. Prove Theorem 6.7: Every sequence of distinct $n^2 + 1$ real numbers contains a subsequence of length $n + 1$ which is strictly increasing or strictly decreasing.

Let $a_1, a_2, \ldots, a_{n^2+i}$ be a sequence of $n^2 + 1$ distinct real numbers. To each a_t we associate the pair (i_t, d_t) where: (1) i_t is the longest increasing subsequence beginning at a_t and (2) d_t is the longest decreasing subsequence beginning at a_t. Thus there are $n^2 + 1$ such ordered pairs, one for each number in the sequence.

Now suppose that no subsequence is longer than n. Then i_t and d_t cannot exceed n. Thus there are at most n^2 distinct pairs (i_t, d_t). By the Pigeonhole Principle, two of the $n^2 + 1$ pairs are equal, that is, there are two distinct points a_r and a_s such that $(i_r, d_r) = (i_s, d_s)$. WLOG, we can assume $r < s$. Then a_r occurs before a_s in the sequence (See Fig. 6-2(a)). Then a_r followed by the increasing subsequence of i_s numbers beginning at a_s gives a subsequence of length $i_s + 1 = i_r + 1$ beginning a_r (See Fig. 6-2(b)). This contradicts the definition of i_r. Similarly, suppose $a_r > a_s$. Then a_n followed by the decreasing subsequence of d_s numbers beginning at a_s gives a subsequence of length $d_r + 1 = d_s + 1$ beginning at a_r which contradicts the definition of d_r (See Fig. 6-2(c)). In each case we get a contradiction. Thus the assumption that no subsequence exceeds n is not true, and the theorem is proved.

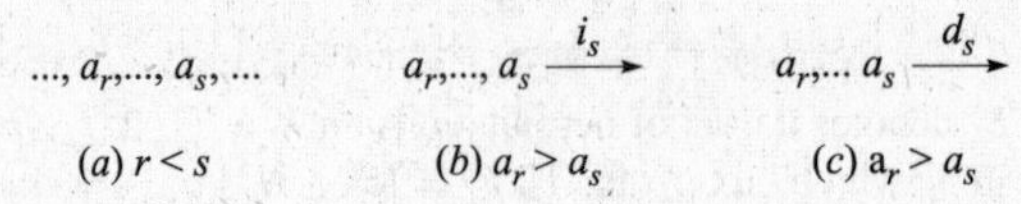

$$..., a_r,..., a_s, ... \qquad a_r,..., a_s \xrightarrow{\ i_s\ } \qquad a_r,... a_s \xrightarrow{\ d_s\ }$$

$$(a)\ r < s \qquad\qquad (b)\ a_r > a_s \qquad\qquad (c)\ a_r > a_s$$

Fig. 6-2

RECURSION

6.11. Consider the second-order homogeneous recurrence relation $a_n = a_{n-1} + 2a_{n-2}$ with initial conditions $a_0 = 2, a_1 = 7$,

(a) Find the next three terms of the sequence.

(b) Find the general solution.

(c) Find the unique solution with the given initial conditions.

(a) Each term is the sum of the preceding term plus twice its second preceding term. Thus:

$$a_2 = 7 + 2(2) = 11, \quad a_3 = 11 + 2(7) = 25, \quad a_4 = 25 + 2(11) = 46$$

(b) First we find the characteristic polynomial $\Delta(t)$ and its roots:

$$\Delta(x) = x^2 - x - 2 = (x - 2)(x + 1); \quad \text{roots } r_1 = 2, r_2 = -1$$

Since the roots are distinct, we use Theorem 6.8 to obtain the general solution:

$$a_n = c_1(2^n) + c_2(-1)^n$$

(c) The unique solution is obtained by finding c_1 and c_2 using the initial conditions:

For $n = 0, a_0 = 2$, we get: $\quad c_1(2^0) + c_2(-1)^0 = 2 \quad$ or $\quad c_1 + c_2 = 2$
For $n = 1, a_1 = 7$, we get: $\quad c_1(2^1) + c_2(-1)^1 = 7 \quad$ or $\quad 2c_1 - c_2 = 7$

Solving the two equations for c_1 and c_2 yields $c_1 = 3$ and $c_2 = 1$. The unique solution follows:

$$a_n = 3(2^n) - (-1)^n$$

6.12. Consider the third-order homogeneous recurrence relation $a_n = 6a_{n-1} - 12a_{n-2} + 8a_{n-3}$

(a) Find the general solution.

(b) Find the solution with initial conditions $a_0 = 3, a_1 = 4, a_2 = 12$.

(a) First we find the characteristic polynomial

$$\Delta(x) = x^3 - 6x^2 + 12x - 8 = (x - 2)^3$$

Then $\Delta(x)$ has only one root $r_0 = 2$ which has multiplicity 3. Thus the general solution of the recurrence relation follows:

$$a_n = c_1(2^n) + c_2 n(2^n) + c_3 n^2(2^n) = (c_1 + c_2 n + c_3 n^2)(2^n)$$

(b) We find the values for c_1, c_2, and c_3 as follows:

For $n = 0, a_0 = 3 \quad$ we get: $\quad c_1 = 3$
For $n = 1, a_1 = 4 \quad$ we get: $\quad 2c_1 + 2c_2 + 2c_3 = 4$
For $n = 2, a_2 = 12 \quad$ we get: $\quad 4c_1 + 8c_2 + 16c_3 = 12$

Solving the system of three equations in c_1, c_2, c_3 yields the solution

$$c_1 = 3, \quad c_2 = -2, \quad c_3 = 1$$

Thus the unique solution of the recurrence relation follows:

$$a_n = (3 - 2n + n^2)(2^n)$$

Supplementary Problems

ADVANCED COUNTING TECHNIQUES, INCLUSION–EXCLUSION

6.13. A store sells $M = 4$ kinds of cookies. Find the number of ways a customer can buy:
(a) 10 cookies;   (b) 15 cookies.

6.14. Find the number m of nonnegative solutions to $x + y + z = 20$ with the conditions that $x \geq 5$, $y \geq 3$, and $z \geq 1$.

6.15. Let E be the equation $x + y + z = 20$. Find the number m of nonnegative solutions to E with the conditions that $x < 8$, $y < 9$, $z < 10$.

6.16. Find the number m of positive integers not exceeding 1000 which are not divisible by 3, 7, or 11.

6.17. Find the number of ways that 14 people can be partitioned into 6 committees, such that 2 committees contain 3 people and the other committees contain 2 people.

6.18. Assume that a cell can be empty. Find the number m of ways that a set:
(a) With 3 people can be partitioned into: (i) three ordered cells; (ii) three unordered cells.
(b) With 4 people can be partitioned into: (i) three ordered cells; (ii) three unordered cells.

6.19. Find the number N of surjective (onto) functions from a set A to a set B where:

(a) $|A| = 8, |B| = 3$;   (b) $|A| = 6, |B| = 4$;   (c) $|A| = 5, |B| = 5$;   (d) $|A| = 5, |B| = 7$.

6.20. Find the number of derangements of $X = \{1, 2, 3, \ldots, 2m\}$ such that the first m elements of each derangement are:
(a) the first m elements of X; (b) the last m elements of X.

PIGEONHOLE PRINCIPLE

6.21. Find the minimum number of students that can be admitted to a college so that there are at least 15 students from each of the 50 states.

6.22. Consider nine lattice points in space. Show that the midpoint of two of the points is also a lattice point.

6.23. Find an increasing subsequence of maximum length and a decreasing subsequence of maximum length in the sequence: 14, 2, 8, 3, 25, 15, 10, 20, 9, 4.

6.24. Consider a line of 50 people with distinct heights. Show there is a subline of 8 people which is either increasing or decreasing.

6.25. Give an example of a sequence of 25 distinct integers which does not have a subsequence of 6 integers which is either increasing or decreasing.

6.26. Suppose a team X plays 19 games in a two-week period of 14 days, and plays at least one game per day. Show there is a period of consecutive days that X played exactly 8 games.

6.27. Suppose 10 points are chosen at random in the interior of an equilateral triangle T where each side has length three inches. Show that the distance between two of the points must be less than one inch.

6.28. Let $X = \{x_i\}$ be a set of n positive integers. Show that the sum of the integers of a subset of X is divisible by n.

6.29. Consider a group of 10 people (where each pair are either friends or strangers). Show that there is either a subgroup of 4 mutual friends or a subgroup of 3 mutual strangers.

6.30. For the Ramsey numbers $R(p, q)$ show that: (a) $R(p, q) = R(q, p)$; (b) $R(p, 1) = 1$; (c) $R(p, 2) = p$.

RECURSION

6.31. For each recurrence relation and initial conditions, find: (i) general solution; (ii) unique solution with the given initial conditions:
(a) $a_n = 3a_{n-1} + 10a_{n-2}; a_0 = 5, a_1 = 11$ (d) $a_n = 5a_{n-1} - 6a_{n-2}; a_0 = 2, a_1 = 8$
(b) $a_n = 4a_{n-1} + 21a_{n-2}; a_0 = 9, a_1 = 13$ (e) $a_n = 3a_{n-1} - a_{n-2}; a_0 = 0, a_1 = 1$
(c) $a_n = 3a_{n-1} - 2a_{n-2}; a_0 = 5, a_1 = 8$ (f) $a_n = 5a_{n-1} - 3a_{n-2}; a_0 = 0, a_1 = 1$

6.32. Repeat Problem 6.31 for the following recurrence relations and initial conditions:

(a) $a_n = 6a_{n-1}; a_0 = 5$ (c) $a_n = 4a_{n-1} - 4a_{n-2}; a_0 = 1, a_1 = 8$
(b) $a_n = 7a_{n-1}; a_0 = 5$ (d) $a_n = 10a_{n-1} - 25a_{n-2}; a_0 = 2, a_1 = 15$

6.33. Find the unique solution to each recurrence relation with the given initial conditions:

(a) $a_n = 10a_{n-1} - 32a_{n-2} + 32a_{n-3}$ with $a_0 = 5$, $a_1 = 18$, $a_2 = 76$

(b) $a_n = 9a_{n-1} - 27a_{n-2} + 27a_{n-3}$ with $a_0 = 5$, $a_1 = 24$, $a_2 = 117$

6.34. Consider the following second-order recurrence relation and its characteristic polynomial $\Delta(x)$:

$$a_n = sa_{n-1} + ta_{n-2} \quad \text{and} \quad \Delta(x) = x^2 - sx - t \qquad (*)$$

(a) Suppose $p(n)$ and $q(n)$ are solutions of $(*)$. Show that, for any constants c_1 and c_2, $c_1 p(n) + c_2 q(n)$ is also a solution of $(*)$.

(b) Suppose r is a root of $\Delta(x)$. Show that $a_n = r^n$ is a solution to $(*)$.

(c) Suppose r is a double root of $\Delta(x)$. Show that: (i) $s = 2r$ and $t = -r^2$; (ii) $a_n = nr^n$ is also a root of $(*)$.

6.35. Repeat Problem 6.34(a) and (b) for any linear kth-order homogeneous recurrence relation with constant coefficients and its characteristic polynomial $\Delta(x)$ which have the form:

$$a_n = C_1 a_{n-1} + C_2 a_{n-2} + \cdots + C_k a_{n-k} = \sum_{i=1}^{k} C_1 a_{n-1} \quad \text{and} \quad \Delta(x) = x^k - \sum_{i=1}^{k} C_1 x^{k-i}$$

Answers to Supplementary Problems

6.13. (a) 286; (b) 646.

6.14. 78.

6.15. 15.

6.16. 520.

6.17. $(14!)/[(3!3!2!2!2!)(2!4!)] = 3\,153\,150$.

6.18. (a) (i) $3^3 = 27$; (ii) They may be distributed as: [3, 0, 0], [2, 1, 0], or [1, 1, 1]. Hence $m = 1 + 3 + 1 = 5$.
(b) (i) $3^4 = 81$; (ii) They may be distributed as: [4, 0, 0], [3, 1, 0], [2, 2, 0] or [2, 1, 1]. Hence $m = 1 + 4 + 3 + 6 = 14$.

6.19. (a) 5796; (b) 1560; (c) $5! = 120$; (d) 0.

6.20. (a) $(D_m)^2$; (b) $(m!)^2$.

6.21. 701.

6.22. There are eight triplets of parities: (odd, odd, odd), (odd, odd, even), ... Thus 2 of the 9 points have the same triplet of parities.

6.23. 2, 3, 10, 20; 25, 15, 10, 8, 4.

6.24. Use Theorem 6.7 with $n = 9$.

6.25. 5, 4, 3, 2, 1, 10, 9, 8, 7, 6, ..., 25, 24, 23, 22, 21.

6.26. (Hint: See Problem 6.9.)

6.27. (Hint: Partition T into 9 equilateral triangles where each side has length one inch.)

6.28. Let $s_i = x_1 + \cdots + x_i$. The result is true if n divides some s_i. Otherwise, let r' be the remainder when s_i is divided by n. Two of the rs must be equal. Say $r_p = r_q$ where $p < q$. Then n divides $s_q - s_p = x_{p+1} + \cdots + x_q$.

6.31. (a) $a_n = c_1(5^n) + c_2(-2)^n$; $c_1 = 3, c_2 = 2$
(b) $a_n = c_1(7^n) + c_2(-3)^n$; $c_1 = 4, c_2 = 5$
(c) $a_n = c_1 + c_2(2^n)$; $c_1 = 2, c_2 = 3$
(d) $a_n = c_1(2^n) + c_2(3^n)$; $c_1 = -2, c_2 = 4$
(e) $a_n = c_1[(3+t)/2]^n + c_2[(3-t)/2]^n$; $c_1 = 1/t$, $c_2 = -1/t$ where $t = \sqrt{5}$
(f) $a_n = c_1[(5+s)/2]^n + c_2[(5-s)/2]^n$; $c_1 = 1/s$, $c_2 = -1/s$ where $= \sqrt{13}$

6.32. (a) $a_n = c_1(6^n)$, $c_1 = 5$
(b) $a_n = c_1(7^n)$, $c_1 = 5$
(c) $a_n = c_1(2^n) + c_2 n(2^n)$, $c_1 = 1, c_2 = 3$
(d) $a_n = c_1(5^n) + c_2 n(5^n)$, $c_1 = 2, c_2 = 1$

6.33. (a) $a_n = 2(4^n) + n(4^n) + 3(2^n)$; (b) $a_n = 5(3^n) + 2n(3^n) + n^2(3^n) = (5 + 2n + n^2)3^n$.

6.34. (b) r is a root of $\Delta(x)$ so $r^2 - sr - t = 0$ or $r^2 = sr + t$. Let $a_n = r^n$. Then $sa_{n-1} + ta_{n-2} = sr^{n-1} + tr^{n-2} = (sr + t)r^{n-2} = r^2(r^{n-2}) = r^n = a_n$
(c) (i) r is a double root of $\Delta(x)$; hence $\Delta(x) = (x - r)^2 = x^2 - 2rx + r^2 = x^2 - sx - t$. Thus $s = 2r$ and $t = -r^2$. (ii) Let $a_n \; nr_n$. Then $sa_{n-1} + ta_{n-2} = nr^n = an$.

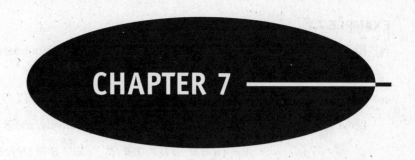

CHAPTER 7

Probability

7.1 INTRODUCTION

Probability theory is a mathematical modeling of the phenomenon of chance or randomness. If a coin is tossed in a random manner, it can land heads or tails, but we do not know which of these will occur in a single toss. However, suppose we let s be the number of times heads appears when the coin is tossed n times. As n increases, the ratio $f = s/n$, called the *relative frequency* of the outcome, becomes more stable. If the coin is perfectly balanced, then we expect that the coin will land heads approximately 50% of the time or, in other words, the relative frequency will approach $\frac{1}{2}$. Alternatively, assuming the coin is perfectly balanced, we can arrive at the value $\frac{1}{2}$ deductively. That is, any side of the coin is as likely to occur as the other; hence the chance of getting a head is 1 in 2 which means the probability of getting heads is $\frac{1}{2}$. Although the specific outcome on any one toss is unknown, the behavior over the long run is determined. This stable long-run behavior of random phenomena forms the basis of probability theory.

A probabilistic mathematical model of random phenomena is defined by assigning "probabilities" to all the possible outcomes of an experiment. The reliability of our mathematical model for a given experiment depends upon the closeness of the assigned probabilities to the actual limiting relative frequencies. This then gives rise to problems of testing and reliability, which form the subject matter of statistics and which lie beyond the scope of this text.

7.2 SAMPLE SPACE AND EVENTS

The set S of all possible outcomes of a given experiment is called the *sample space*. A particular outcome, i.e., an element in S, is called a *sample point*. An *event* A is a set of outcomes or, in other words, a subset of the sample space S. In particular, the set $\{a\}$ consisting of a single sample point $a \in S$ is called an *elementary event*. Furthermore, the empty set $\varnothing$ and S itself are subsets of S and so $\varnothing$ and S are also events; $\varnothing$ is sometimes called the *impossible event* or the *null event*.

Since an event is a set, we can combine events to form new events using the various set operations:

   (i) $A \cup B$ is the event that occurs iff A occurs *or* B occurs (or both).
  (ii) $A \cap B$ is the event that occurs iff A occurs *and* B occurs.
 (iii) A^c, the complement of A, also written $\overline{A}$, is the event that occurs iff A does *not* occur.

Two events A and B are called *mutually exclusive* if they are disjoint, that is, if $A \cap B = \varnothing$. In other words, A and B are mutually exclusive iff they cannot occur simultaneously. Three or more events are mutually exclusive if every two of them are mutually exclusive.

EXAMPLE 7.1

(a) **Experiment**: Toss a coin three times and observe the sequence of heads (H) and tails (T) that appears.

The sample space consists of the following eight elements:

$$S = \{HHH, HHT, HTH, HTT, THH, THT, TTH, TTT\}$$

Let A be the event that two or more heads appear consecutively, and B that all the tosses are the same:

$$A = \{HHH, HHT, THH\} \quad \text{and} \quad B = \{HHH, TTT\}$$

Then $A \cap B = \{HHH\}$ is the elementary event that only heads appear. The event that five heads appears is the empty set $\varnothing$.

(b) **Experiment**: Toss a (six-sided) die, pictured in Fig. 7-1(a), and observe the number (of dots) that appear on top.

The sample space S consists of the six possible numbers, that is, $S = \{1, 2, 3, 4, 5, 6\}$. Let A be the event that an even number appears, B that an odd number appears, and C that a prime number appears. That is, let

$$A = \{2, 4, 6\}, \quad B = \{1, 3, 5\}, \quad C = \{2, 3, 5\}$$

Then

$A \cup C = \{2, 3, 4, 5, 6\}$ is the event that an even or a prime number occurs.

$B \cap C = \{3, 5\}$ is the event that an odd prime number occurs.

$C^c = \{1, 4, 6\}$ is the event that a prime number does not occur.

Note that A and B are mutually exclusive: $A \cap B = \varnothing$. In other words, an even number and an odd number cannot occur simultaneously.

(c) **Experiment**: Toss a coin until a head appears, and count the number of times the coin is tossed.

The sample space S of this experiment is $S = \{1, 2, 3, \ldots\}$. Since every positive integer is an element of S, the sample space is infinite.

Remark: The sample space S in Example 7.1(c), as noted, is not finite. The theory concerning such sample spaces lies beyond the scope of this text. Thus, unless otherwise stated, all our sample spaces S shall be finite.

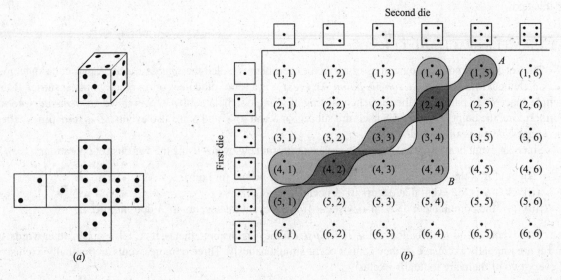

(a)

(b)

Fig. 7-1

EXAMPLE 7.2 (Pair of dice) Toss a pair of dice and record the two numbers on the top.

There are six possible numbers, 1, 2, ..., 6, on each die. Thus S consists of the pairs of numbers from 1 to 6, and hence $n(S) = 36$. Figure 7-1(b) shows these 36 pairs of numbers arranged in an array where the rows are labeled by the first die and the columns by the second die.

Let A be the event that the sum of the two numbers is 6, and let B be the event that the largest of the two numbers is 4. That is, let

$$A = \{(1, 5), (2, 4), (3, 3), (4, 2), (5, 1)\}, \quad B = \{(1, 4), (2, 4), (3, 4), (4, 4), (4, 3), (4, 2), (4, 1)\}$$

Then the event "A and B" consists of those pairs of integers whose sum is 6 and whose largest number is 4 or, in other words, the intersection of A and B. Thus

$$A \cap B = \{(2, 4), (4, 2)\}$$

Similarly, "A or B," the sum is 6 or the largest is 4, shaded in Fig. 7-1(b), is the union $A \cup B$.

EXAMPLE 7.3 (Deck of cards) A card is drawn from an ordinary deck of 52 cards which is pictured in Fig. 7-2(a).

The sample space S consists of the four *suits*, clubs (C), diamonds (D), hearts (H), and spades (S), where each suit contains 13 cards which are numbered 2 to 10, and jack (J), queen (Q), king (K), and ace (A). The hearts (H) and diamonds (D) are red cards, and the spades (S) and clubs (C) are black cards. Figure 7-2(b) pictures 52 points which represent the deck S of cards in the obvious way. Let E be the event of a *picture card, or face card*, that is, a Jack (J), Queen (Q), or King (K), and let F be the event of a heart. Then $E \cap F = \{JH, QH, KH\}$, as shaded in Fig. 7-2($b$).

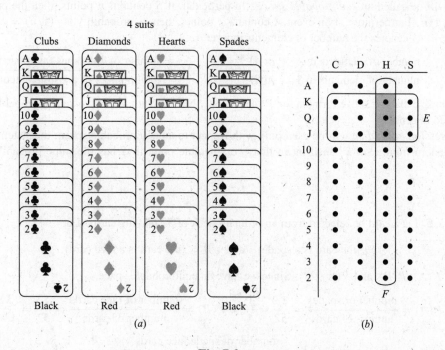

Fig. 7-2

7.3 FINITE PROBABILITY SPACES

The following definition applies.

Definition 7.1: Let S be a finite sample space, say $S = \{a_1, a_2, \ldots, a_n\}$. A *finite probability space*, or *probability model*, is obtained by assigning to each point a_i in S a real number p_i, called the *probability* of a_i satisfying the following properties:

(i) Each p_i is nonnegative, that is, $p_i \geq 0$.

(ii) The sum of the p_i is 1, that is, is $p_1 + p_2 + \cdots + p_n = 1$.

The *probability* of an event A written $P(A)$, is then defined to be the sum of the probabilities of the points in A.

The singleton set $\{a_i\}$ is called an *elementary* event and, for notational convenience, we write $P(a_i)$ for $P(\{a_i\})$.

EXAMPLE 7.4 (Experiment) Suppose three coins are tossed, and the number of heads is recorded. (Compare with the above Example 7.1(a).)

The sample space is $S = \{0, 1, 2, 3\}$. The following assignments on the elements of S define a probability space:

$$P(0) = \tfrac{1}{8}, \quad P(1) = \tfrac{3}{8}, \quad P(2) = \tfrac{3}{8}, \quad P(3) = \tfrac{1}{8}$$

That is, each probability is nonnegative, and the sum of the probabilities is 1. Let A be the event that at least one head appears, and let B be the event that all heads or all tails appear; that is, let $A = \{1, 2, 3\}$ and $B = \{0, 3\}$. Then, by definition,

$$P(A) = P(1) + P(2) + P(3) = \tfrac{3}{8} + \tfrac{3}{8} + \tfrac{1}{8} = \tfrac{7}{8} \quad \text{and} \quad P(B) = P(0) + P(3) = \tfrac{1}{8} + \tfrac{1}{8} = \tfrac{1}{4}$$

Equiprobable Spaces

Frequently the physical characteristics of an experiment suggest that the various outcomes of the sample space be assigned equal probabilities. Such a finite probability space S, where each sample point has the same probability, will be called an *equiprobable space*. In particular, if S contains n points, then the probability of each point is $1/n$. Furthermore, if an event A contains r points, then its probability is $r(1/n) = r/n$. In other words, where $n(A)$ denotes the number of elements in a set A,

$$P(A) = \frac{\text{number of elements in } A}{\text{number of elements in } S} = \frac{n(A)}{n(S)} \quad \text{or} \quad P(A) = \frac{\text{number of outcomes favorable to } A}{\text{total number of possible outcomes}}$$

We emphasize that the above formula for $P(A)$ can only be used with respect to an equiprobable space, and cannot be used in general.

The expression *at random* will be used only with respect to an equiprobable space; the statement "choose a point at random from a set S" shall mean that every sample point in S has the same probability of being chosen.

EXAMPLE 7.5 Let a card be selected from an ordinary deck of 52 playing cards. Let

$$A = \{\text{the card is a spade}\} \quad \text{and} \quad B = \{\text{the card is a face card}\}.$$

We compute $P(A)$, $P(B)$, and $P(A \cap B)$. Since we have an equiprobable space,

$$P(A) = \frac{\text{number of spades}}{\text{number of cards}} = \frac{13}{52} = \frac{1}{4}, \quad P(B) = \frac{\text{number of face cards}}{\text{number of cards}} = \frac{12}{52} = \frac{3}{13}$$

$$P(A \cap B) = \frac{\text{number of spade face cards}}{\text{number of cards}} = \frac{3}{52}$$

Theorems on Finite Probability Spaces

The following theorem follows directly from the fact that the probability of an event is the sum of the probabilities of its points.

Theorem 7.1: The probability function P defined on the class of all events in a finite probability space has the following properties:

   $[\mathbf{P_1}]$ For every event A, $0 \leq P(A) \leq 1$.

   $[\mathbf{P_2}]$ $P(S) = 1$.

   $[\mathbf{P_3}]$ If events A and B are mutually exclusive, then $P(A \cup B) = P(A) + P(B)$.

The next theorem formalizes our intuition that if p is the probability that an event E occurs, then $1 - p$ is the probability that E does not occur. (That is, if we hit a target $p = 1/3$ of the times, then we miss the target $1 - p = 2/3$ of the times.)

Theorem 7.2: Let A be any event. Then $P(A^c) = 1 - P(A)$.

The following theorem (proved in Problem 7.13) follows directly from Theorem 7.1.

Theorem 7.3: Consider the empty set $\varnothing$ and any events A and B. Then:

   (i) $P(\varnothing) = 0$.

   (ii) $P(A \backslash B) = P(A) - P(A \cap B)$.

   (iii) If $A \subseteq B$, then $P(A) \leq P(B)$.

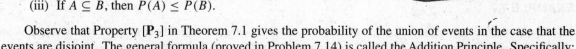

Observe that Property $[\mathbf{P_3}]$ in Theorem 7.1 gives the probability of the union of events in the case that the events are disjoint. The general formula (proved in Problem 7.14) is called the Addition Principle. Specifically:

Theorem 7.4 (Addition Principle): For any events A and B,

$$P(A \cup B) = P(A) + P(B) - P(A \cap B)$$

EXAMPLE 7.6 Suppose a student is selected at random from 100 students where 30 are taking mathematics, 20 are taking chemistry, and 10 are taking mathematics and chemistry. Find the probability p that the student is taking mathematics or chemistry.

Let $M = \{$students taking mathematics$\}$ and $C = \{$students taking chemistry$\}$. Since the space is equiprobable,

$$P(M) = \frac{30}{100} = \frac{3}{10}, \quad P(C) = \frac{20}{100} = \frac{1}{5}, \quad P(M \text{ and } C) = P(M \cap C) = \frac{10}{100} = \frac{1}{10}$$

Thus, by the Addition Principle (Theorem 7.4),

$$p = P(M \text{ or } C) = P(M \cup C) = P(M) + P(C) - P(M \cap C) = \frac{3}{10} + \frac{1}{5} - \frac{1}{10} = \frac{2}{5}$$

7.4 CONDITIONAL PROBABILITY

Suppose E is an event in a sample space S with $P(E) > 0$. The probability that an event A occurs once E has occurred or, specifically, the *conditional probability of A given E*, written $P(A|E)$, is defined as follows:

$$P(A|E) = \frac{P(A \cap E)}{P(E)}$$

As pictured in the Venn diagram in Fig. 7-3, $P(A|E)$ measures, in a certain sense, the relative probability of A with respect to the reduced space E.

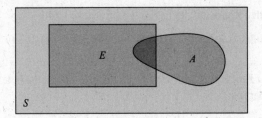

Fig. 7-3

Suppose S is an equiprobable space, and $n(A)$ denotes the number of elements in A. Then:

$$P(A \cap E) = \frac{n(A \cap E)}{n(S)}, \quad P(E) = \frac{n(E)}{n(S)}, \quad \text{and so} \quad P(A|E) = \frac{P(A \cap E)}{P(E)} = \frac{n(A \cap E)}{n(E)}$$

We state this result formally.

Theorem 7.5: Suppose S is an equiprobable space and A and E are events. Then

$$P(A|E) = \frac{\text{number of elements in } A \cap E}{\text{number of elements in } E} = \frac{n(A \cap E)}{n(E)}$$

EXAMPLE 7.7

(a) A pair of fair dice is tossed. The sample space S consists of the 36 ordered pairs (a, b), where a and b can be any of the integers from 1 to 6. (See Example 7.2.) Thus the probability of any point is $\frac{1}{36}$. Find the probability that one of the dice is 2 if the sum is 6. That is, find $P(A|E)$ where:

$$E = \{\text{sum is } 6\} \quad \text{and} \quad A = \{2 \text{ appears on at least one die}\}$$

Now E consists of 5 elements and $A \cap E$ consists of two elements; namely

$$E = \{(1, 5), (2, 4), (3, 3), (4, 2), (5, 1)\} \quad \text{and} \quad A \cap E = \{(2, 4), (4, 2)\}$$

By Theorem 7.5, $P(A|E) = 2/5$.

On the other hand A itself consists of 11 elements, that is,

$$A = \{(2, 1), (2, 2), (2, 3), (2, 4), (2, 5), (2, 6), (1, 2), (3, 2), (4, 2), (5, 2), (6, 2)\}$$

Since S consists of 36 elements, $P(A) = 11/36$.

(b) A couple has two children; the sample space is $S = \{bb, bg, gb, gg\}$ with probability $\frac{1}{4}$ for each point. Find the probability p that both children are boys if it is known that: (i) at least one of the children is a boy; (ii) the older child is a boy.

  (i) Here the reduced space consists of three elements, $\{bb, bg, gb\}$; hence $p = \frac{1}{3}$.

  (ii) Here the reduced space consists of only two elements $\{bb, bg\}$; hence $p = \frac{1}{2}$.

Multiplication Theorem for Conditional Probability

Suppose A and B are events in a sample space S with $P(A) > 0$. By definition of conditional probability,

$$P(B|A) = \frac{P(A \cap B)}{P(A)}$$

Multiplying both sides by $P(A)$ gives us the following useful result:

Theorem 7.6 (Multiplication Theorem for Conditional Probability):

$$P(A \cap B) = P(A)P(B|A)$$

The multiplication theorem gives us a formula for the probability that events A and B both occur. It can easily be extended to three or more events $A_1, A_2, \ldots A_m$; that is,

$$P(A_1 \cap A_2 \cap \cdots A_m) = P(A_1) \cdot P(A_2|A_1) \cdots P(A_m|A_1 \cap A_2 \cap \cdots \cap A_{m-1})$$

EXAMPLE 7.8 A lot contains 12 items of which 4 are defective. Three items are drawn at random from the lot one after the other. Find the probability p that all three are nondefective.

The probability that the first item is nondefective is $\frac{8}{12}$ since 8 of 12 items are nondefective. If the first item is nondefective, then the probability that the next item is nondefective is $\frac{7}{11}$ since only 7 of the remaining 11 items are nondefective. If the first 2 items are nondefective, then the probability that the last item is nondefective is $\frac{6}{10}$ since only 6 of the remaining 10 items are now nondefective. Thus by the multiplication theorem,

$$p = \frac{8}{12} \cdot \frac{7}{11} \cdot \frac{6}{10} = \frac{14}{55} \approx 0.25$$

7.5 INDEPENDENT EVENTS

Events A and B in a probability space S are said to be *independent* if the occurrence of one of them does not influence the occurrence of the other. More specifically, B is independent of A if $P(B)$ is the same as $P(B|A)$. Now substituting $P(B)$ for $P(B|A)$ in the Multiplication Theorem $P(A \cap B) = P(A)P(B|A)$ yields

$$P(A \cap B) = P(A)P(B).$$

We formally use the above equation as our definition of independence.

Definition 7.2: Events A and B are *independent* if $P(A \cap B) = P(A)P(B)$; otherwise they are *dependent*.

We emphasize that independence is a symmetric relation. In particular, the equation

$$P(A \cap B) = P(A)P(B) \quad \text{implies both} \quad P(B|A) = P(B) \quad \text{and} \quad P(A|B) = P(A)$$

EXAMPLE 7.9 A fair coin is tossed three times yielding the equiprobable space

$$S = \{HHH, HHT, HTH, HTT, THH, THT, TTH, TTT\}$$

Consider the events:

$$A = \{\text{first toss is heads}\} = \{HHH, HHT, HTH, HTT\}$$
$$B = \{\text{second toss is heads}\} = \{HHH, HHT, THH, THT\}$$
$$C = \{\text{exactly two heads in a row}\} = \{HHT, THH\}$$

Clearly A and B are independent events; this fact is verified below. On the other hand, the relationship between A and C and between B and C is not obvious. We claim that A and C are independent, but that B and C are dependent. We have:

$$P(A) = \frac{4}{8} = \frac{1}{2}, \quad P(B) = \frac{4}{8} = \frac{1}{2}, \quad P(C) = \frac{2}{8} = \frac{1}{4}$$

Also,

$$P(A \cap B) = P(\{HHH, HHT\}) = \tfrac{1}{4}, \quad P(A \cap C) = P(\{HHT\}) = \tfrac{1}{8}, \quad P(B \cap C) = P(\{HHT, THH\}) = \tfrac{1}{4}$$

Accordingly,

$$P(A)P(B) = \tfrac{1}{2} \cdot \tfrac{1}{2} = \tfrac{1}{4} = P(A \cap B), \quad \text{and so } A \text{ and } B \text{ are independent}$$

$$P(A)P(C) = \tfrac{1}{2} \cdot \tfrac{1}{4} = \tfrac{1}{8} = P(A \cap C), \quad \text{and so } A \text{ and } C \text{ are independent}$$

$$P(B)P(C) = \tfrac{1}{2} \cdot \tfrac{1}{4} = \tfrac{1}{8} \neq P(B \cap C), \quad \text{and so } B \text{ and } C \text{ are dependent}$$

Frequently, we will postulate that two events are independent, or the experiment itself will imply that two events are independent.

EXAMPLE 7.10 The probability that A hits a target is $\tfrac{1}{4}$, and the probability that B hits the target is $\tfrac{2}{5}$. Both shoot at the target. Find the probability that at least one of them hits the target, i.e., that A or B (or both) hit the target.

We are given that $P(A) = \tfrac{1}{2}$ and $P(B) = \tfrac{2}{5}$, and we seek $P(A \cup B)$. Furthermore, the probability that A or B hits the target is not influenced by what the other does; that is, the event that A hits the target is independent of the event that B hits the target, that is, $P(A \cap B) = P(A)P(B)$. Thus

$$P(A \cup B) = P(A) + P(B) - P(A \cap B) = P(A) + P(B) - P(A)P(B) = \tfrac{1}{4} + \tfrac{2}{5} - \left(\tfrac{1}{4}\right)\left(\tfrac{2}{5}\right) = \tfrac{11}{20}$$

7.6 INDEPENDENT REPEATED TRIALS, BINOMIAL DISTRIBUTION

We have previously discussed probability spaces which were associated with an experiment repeated a finite number of times, as the tossing of a coin three times. This concept of repetition is formalized as follows:

Definition 7.3: Let S be a finite probability space. By the space of n *independent repeated trials*, we mean the probability space S_n consisting of ordered n-tuples of elements of S, with the probability of an n-tuple defined to be the product of the probabilities of its components:

$$P((s_1, s_2, \ldots, s_n)) = P(s_1)P(s_2) \ldots P(s_n)$$

EXAMPLE 7.11 Whenever three horses a, b, and c race together, their respective probabilities of winning are $\tfrac{1}{2}$, $\tfrac{1}{3}$, and $\tfrac{1}{6}$. In other words, $S = \{a, b, c\}$ with $P(a) = \tfrac{1}{2}$, $P(b) = \tfrac{1}{3}$, and $P(c) = \tfrac{1}{6}$. If the horses race twice, then the sample space of the two repeated trials is

$$S_2 = \{aa, ab, ac, ba, bb, bc, ca, cb, cc\}$$

For notational convenience, we have written ac for the ordered pair (a, c). The probability of each point in S_2 is

$$P(aa) = P(a)P(a) = \tfrac{1}{2}\left(\tfrac{1}{2}\right) = \tfrac{1}{4}, \quad P(ba) = \tfrac{1}{6}, \quad P(ca) = \tfrac{1}{12}$$

$$P(ab) = P(a)P(b) = \tfrac{1}{2}\left(\tfrac{1}{3}\right) = \tfrac{1}{6}, \quad P(bb) = \tfrac{1}{9}, \quad P(cb) = \tfrac{1}{18}$$

$$P(ac) = P(a)P(c) = \tfrac{1}{2}\left(\tfrac{1}{6}\right) = \tfrac{1}{12}, \quad P(bc) = \tfrac{1}{18}, \quad P(cc) = \tfrac{1}{36}$$

Thus the probability of c winning the first race and a winning the second race is $P(ca) = \tfrac{1}{12}$.

Repeated Trials with Two Outcomes, Bernoulli Trials, Binomial Experiment

Now consider an experiment with only two outcomes. Independent repeated trials of such an experiment are called Bernoulli trials, named after the Swiss mathematician Jacob Bernoulli (1654–1705). The term independent trials means that the outcome of any trial does not depend on the previous outcomes (such as tossing a coin). We will call one of the outcomes *success* and the other outcome *failure*.

Let p denote the probability of success in a Bernoulli trial, and so $q = 1 - p$ is the probability of failure. A *binomial experiment* consists of a fixed number of Bernoulli trials. A binomial experiment with n trials and probability p of success will be denoted by

$$B(n, p)$$

Frequently, we are interested in the number of successes in a binomial experiment and not in the order in which they occur. The next theorem (proved in Problem 7.27) applies. We note that the theorem uses the following binomial coefficient which is discussed in detail in Chapter 5:

$$\binom{n}{k} = \frac{n(n-1)(n-2)\ldots(n-k+1)}{k(k-1)(k-2)\ldots 3\cdot 2\cdot 1} = \frac{n!}{k!(n-k)!}$$

Theorem 7.7: The probability of exactly k successes in a binomial experiment $B(n, p)$ is given by

$$P(k) = P(k \text{ successes}) = \binom{n}{k} p^k q^{n-k}$$

The probability of one or more successes is $1 - q^n$.

EXAMPLE 7.12 A fair coin is tossed 6 times; call heads a success. This is a binomial experiment with $n = 6$ and $p = q = \frac{1}{2}$.

(a) The probability that exactly two heads occurs (i.e., $k = 2$) is

$$P(2) = \binom{6}{2}\left(\frac{1}{2}\right)^2\left(\frac{1}{2}\right)^4 = \frac{15}{64} \approx 0.23$$

(b) The probability of getting at least four heads (i.e., $k = 4, 5$ or 6) is

$$P(4) + P(5) + P(6) = \binom{6}{4}\left(\frac{1}{2}\right)^4\left(\frac{1}{2}\right)^2 + \binom{6}{4}\left(\frac{1}{2}\right)^4\left(\frac{1}{2}\right)^2 + \binom{6}{6}\left(\frac{1}{2}\right)^6$$

$$= \frac{15}{64} + \frac{6}{64} + \frac{1}{64} = \frac{11}{32} \approx 0.34$$

(c) The probability of getting no heads (i.e., all failures) is $q^6 = \left(\frac{1}{2}\right)^6 = \frac{1}{64}$, so the probability of one or more heads is $1 - q^n = 1 - \frac{1}{64} = \frac{63}{64} \approx 0.94$.

Remark: The function $P(k)$ for $k = 0, 1, 2, \ldots, n$, for a binomial experiment $B(n, p)$, is called the *binomial distribution* since it corresponds to the successive terms of the binomial expansion:

$$(q + p)^n = q^n + \binom{n}{1}q^{n-1}p + \binom{n}{2}q^{n-2}p^2 + \cdots + p^n$$

The use of the term *distribution* will be explained later in the chapter.

7.7   RANDOM VARIABLES

Let S be a sample of an experiment. As noted previously, the outcome of the experiment, or the points in S, need not be numbers. For example, in tossing a coin the outcomes are H (heads) or T (tails), and in tossing a pair of dice the outcomes are pairs of integers. However, we frequently wish to assign a specific number to each outcome of the experiment. For example, in coin tossing, it may be convenient to assign 1 to H and 0 to T; or, in the tossing of a pair of dice, we may want to assign the sum of the two integers to the outcome. Such an assignment of numerical values is called a *random variable*. More generally, we have the following definition.

Definition 7.4:   A *random variable* X is a rule that assigns a numerical value to each outcome in a sample space S.

We shall let R_X denote the set of numbers assigned by a random variable X, and we shall refer to R_X as the *range space*.

Remark: In more formal terminology, X is a function from S to the real numbers $\mathbf{R}$, and R_X is the range of X. Also, for some infinite sample spaces S, not all functions from S to $\mathbf{R}$ are considered to be random variables. However, the sample spaces here are finite, and every real-valued function defined on a finite sample space is a random variable.

EXAMPLE 7.13   A pair of fair dice is tossed. (See Example 7.2.) The sample space S consists of the 36 ordered pairs (a, b) where a and b can be any of the integers from 1 to 6.
Let X assign to each point in S the sum of the numbers; then X is a random variable with range space

$$R_X = \{2, 3, 4, 5, 6, 7, 8, 9, 10, 11, 12\}$$

Let Y assign to each point the maximum of the two numbers; then Y is a random variable with range space

$$R_Y = \{1, 2, 3, 4, 5, 6\}$$

Sums and Products of Random Variables, Notation

Suppose X and Y are random variables on the same sample space S. Then $X + Y$, kX and XY are functions on S defined as follows (where $s \in S$):

$$(X + Y)(s) = X(s) + Y(s), \quad (kX)(s) = kX(s), \quad (XY)(s) = X(s)Y(s)$$

More generally, for any polynomial or exponential function $h(x, y, \ldots, z)$, we define $h(X, Y, \ldots, Z)$ to be the function on S defined by

$$[h(X, Y, \ldots, Z)](s) = h[X(s), Y(s), \ldots, Z(s)]$$

It can be shown that these are also random variables. (This is trivial in the case that every subset of S is an event.)
The short notation $P(X = a)$ and $P(a \le X \le b)$ will be used, respectively, for the probability that "X maps into a" and "X maps into the interval $[a, b]$." That is, for $s \in S$:

$$P(X = a) \equiv P(\{s \mid X(s) = a\}) \quad \text{and} \quad P(a \le X \le b) \equiv P(\{s \mid a \le X(s) \le b\})$$

Analogous meanings are given to $P(X \le a)$, $P(X = a, Y = b)$, $P(a \le X \le b, c \le Y \le d)$, and so on.

Probability Distribution of a Random Variable

Let X be a random variable on a finite sample space S with range space $R_x = \{x_1, x_2, \ldots, x_t\}$. Then X induces a function f which assigns probabilities p_k to the points x_k in R_x as follows:

$$f(x_k) = p_k = P(X = x_k) = \text{sum of probabilities of points in } S \text{ whose image is } x_k.$$

The set of ordered pairs $(x_1, f(x_1)), (x_2, f(x_2)), \ldots, (x_t, f(x_t))$ is called the *distribution* of the random variable X; it is usually given by a table as in Fig. 7-4. This function f has the following two properties:

$$\text{(i) } f(x_k) \geq 0 \quad \text{and} \quad \text{(ii) } \sum_k f(x_k) = 1$$

Thus R_X with the above assignments of probabilities is a probability space. (Sometimes we will use the pair notation $[x_k, p_k]$ to denote the distribution of X instead of the functional notation $[x, f(x)]$).

Outcome x	x_1	x_2	x_3	...	x_t
Probability f(x)	$f(x_1)$	$f(x_2)$	$f(x_3)$	...	$f(x_t)$

Fig. 7-4   Distribution f of a random variable X

In the case that S is an equiprobable space, we can easily obtain the distribution of a random variable from the following result.

Theorem 7.8:   Let S be an equiprobable space, and let f be the distribution of a random variable X on S with the range space $R_X = \{x_1, x_2, \ldots, x_t\}$. Then

$$p_i = f(x_i) = \frac{\text{number of points in } S \text{ whose image is } x_i}{\text{number of points in } S}$$

EXAMPLE 7.14   Let X be the random variable in Example 7.13 which assigns the sum to the toss of a pair of dice. Note $n(S) = 36$, and $R_x = \{2, 3, \ldots, 12\}$. Using Theorem 7.8, we obtain the distribution f of X as follows:

$f(2) = 1/36$, since there is one outcome $(1, 1)$ whose sum is 2.
$f(3) = 2/36$, since there are two outcomes, $(1, 2)$ and $(2,1)$, whose sum is 3.
$f(4) = 3/36$, since there are three outcomes, $(1, 3), (2, 2)$ and $(3, 1)$, whose sum is 4.

Similarly, $f(5) = 4/36$, $f(6) = 5/36, \ldots, f(12) = 1/36$. Thus the distribution of X follows:

x	2	3	4	5	6	7	8	9	10	11	12
$f(x)$	1/36	2/36	3/36	4/36	5/36	6/36	5/36	4/36	3/36	2/36	1/36

Expectation of a Random Variable

Let X be a random variable on a probability space $S = \{s_1, s_2, \ldots, s_m\}$ Then the *mean* or *expectation* of X is denoted and defined by:

$$\mu = E(X) = X(s_1)P(s_1) + X(s_2)P(sa_2) + \cdots + X(s_m)P(s_m) = \sum X(s_k)P(s_k)$$

In particular, if X is given by the distribution f in Fig. 7-4, then the expectation of X is:

$$\mu = E(X) = x_1 f(x_1) + x_2 f(x_2) + \cdots + x_t f(x_t) = \sum x_k f(x_k)$$

Alternately, when the notation $[x_k, p_k]$ is used instead of $[x_k, f(x_k)]$,

$$\mu = E(X) = x_1 p_1 + x_2 p_2 + \cdots + x_t p_t = \sum x_i p_i$$

(For notational convenience, we have omitted the limits in the summation symbol Σ.)

EXAMPLE 7.15

(a) Suppose a fair coin is tossed six times. The number of heads which can occur with their respective probabilities follows:

x_i	0	1	2	3	4	5	6
p_1	1/64	6/64	15/64	20/64	15/64	6/64	1/64

Then the mean or expectation (or expected number of heads) is:

$$\mu = E(X) = 0\left(\tfrac{1}{64}\right) + 1\left(\tfrac{6}{64}\right) + 2\left(\tfrac{15}{64}\right) + 3\left(\tfrac{20}{64}\right) + 4\left(\tfrac{15}{64}\right) + 5\left(\tfrac{6}{64}\right) + 6\left(\tfrac{1}{64}\right) = 3$$

(This agrees with our intuition that we expect that half of the tosses to be heads.)

(b) Three horses a, b, and c are in a race; suppose their respective probabilities of winning are $\tfrac{1}{2}$, $\tfrac{1}{3}$, and $\tfrac{1}{6}$. Let X denote the payoff function for the winning horse, and suppose X pays \$2, \$6, or \$9 according as a, b, or c wins the race. The expected payoff for the race is

$$E(X) = X(a)P(a) + X(b)P(b) + X(c)P(c)$$
$$= 2\left(\tfrac{1}{2}\right) + 6\left(\tfrac{1}{3}\right) + 9\left(\tfrac{1}{6}\right) = 4.5$$

Variance and Standard Deviation of a Random Variable

Let X be a random variable with mean μ and distribution f as in Fig. 7-4. Then the *variance* of X, denoted by $Var(X)$, is defined by:

$$Var(X) = (x_1 - \mu)^2 f(x_1) + (x_2 - \mu)^2 f(x_2) + \cdots + (x_t - \mu)^2 f(x_t) = \Sigma(x_k - \mu)^2 f(x_k) = E((X - \mu)^2)$$

Alternately, when the notation $[x_k, p_k]$ is used instead of $[x_k, f(x_k)]$,

$$Var(X) = (x_1 - \mu)^2 p_1 + (x_2 - \mu)^2 p_2 + \cdots + (x_t - \mu)^2 p_t = \Sigma(x_k - \mu)^2 p_k = E((X - \mu)^2)$$

The *standard deviation* of X, denoted by σ_x or simply σ, is the nonnegative square root of $Var(X)$:

$$\sigma_x = \sqrt{Var(X)}$$

Accordingly, $Var(X) = \sigma_x^2$. Both $Var(X)$ and σ_x^2 or simply σ^2 are used to denote the variance of X.

The following formulas are usually more convenient for computing $Var(X)$ than the above:

$$Var(X) = x_1^2 f(x_1) + x_2^2 f(x_2) + \cdots + x_t^2 f(x_t) - \mu^2 = \left[\sum x_k^2 f(x_k)\right] - \mu^2 = E(X^2) - \mu^2$$

or

$$Var(X) = x_1^2 p_1 + x_2^2 p_2 + \cdots + x_t^2 p_t - \mu^2 = \left[\sum x_k^2 p_k\right] - \mu^2 = E(X^2) - \mu^2$$

EXAMPLE 7.16 Let X denote the number of times heads occurs when a fair coin is tossed six times. The distribution of X appears in Example 7.15(a), where its mean $\mu = 3$ is computed. The variance of X is computed as follows:

$$Var(X) = (0 - 3)^2 \tfrac{1}{64} + (1 - 3)^2 \tfrac{6}{64} + (2 - 3)^2 \tfrac{15}{64} + \cdots + (6 - 3) \tfrac{1}{64} = 1.5$$

Alternatively:

$$Var(X) = 0^2 \tfrac{1}{64} + 1^2 \tfrac{6}{64} + 2^2 \tfrac{15}{64} + 3^2 \tfrac{20}{64} + 4^2 \tfrac{15}{64} + 5^2 \tfrac{6}{64} + 6^2 \tfrac{1}{64} - 3^2 = 1.5$$

Thus the standard deviation is $\sigma = \sqrt{1.5} \approx 1.225$ (heads).

Binomial Distribution

Consider a binomial experiment $B(n, p)$. That is, $B(n, p)$ consists of n independent repeated trials with two outcomes, success or failure, and p is the probability of success (and $q = (1 - p)$ is the probability of failure). The number X of k successes is a random variable with distribution appearing in Fig. 7-5.

Number of successes k	0	1	2	...	n
Probability P(k)	q^n	$\binom{n}{1}q^{n-1}p$	$\binom{n}{2}q^{n-2}p^2$	...	p^n

Fig. 7-5

The following theorem applies.

Theorem 7.9: Consider the binomial distribution $B(n, p)$. Then:

       (i) Expected value $E(X) = \mu = np$.

       (ii) Variance $Var(X) = \sigma^2 = npq$.

       (iii) Standard deviation $\sigma = \sqrt{npq}$.

EXAMPLE 7.17

(a) The probability that a man hits a target is $p = 1/5$. He fires 100 times. Find the expected number μ of times he will hit the target and the standard deviation σ.

Here $p = \frac{1}{5}$ and so $q = \frac{4}{5}$. Hence

$$\mu = np = 100 \cdot \frac{1}{5} = 20 \quad \text{and} \quad \sigma = \sqrt{npq} = \sqrt{100 \cdot \frac{1}{5} \cdot \frac{4}{5}} = 4$$

(b) Find the expected number $E(X)$ of correct answers obtained by guessing in a five-question true–false test.

Here $p = \frac{1}{2}$. Hence $E(X) = np = 5 \cdot \frac{1}{2} = 2.5$.

7.8   CHEBYSHEV'S INEQUALITY, LAW OF LARGE NUMBERS

The standard deviation σ of a random variable X measures the weighted spread of the values of X about the mean μ. Thus, for smaller σ, we would expect that X will be closer to μ. A more precise statement of this expectation is given by the following inequality, named after the Russian mathematician P. L. Chebyshev (1821–1894).

Theorem 7.10 (Chebyshev's Inequality): Let X be a random variable with mean μ and standard deviation σ. Then for any positive number k, the probability that a value of X lies in the interval $[\mu - k\sigma, \mu + k\sigma]$ is at least $1 - 1/k^2$. That is,

$$P(\mu - k\sigma \leq X \leq \mu + k\sigma) \geq 1 - \frac{1}{k^2}$$

EXAMPLE 7.18 Suppose X is a random variable with mean $\mu = 75$ and standard deviation $\sigma = 5$. What conclusion about X can be drawn from Chebyshev's inequality for $k = 2$ and $k = 3$?

Setting $k = 2$, we obtain:

$$\mu - k\sigma = 75 - 2(5) = 65 \quad \text{and} \quad \mu + k\sigma = 75 + 2(5) = 85$$

Thus we can conclude that the probability that a value of X lies between 65 and 85 is at least $1 - (1/2)^2 = 3/4$; that is:

$$P(65 \leq X \leq 85) \geq 3/4$$

Similarly, by letting $k = 3$, we can conclude that the probability that a value of X lies between 60 and 90 is at least $1 - (1/3)^2 = 8/9$.

Sample Mean and Law of Large Numbers

Consider a finite number of random variable $X, Y, \ldots, Z$ on a sample space S. They are said to be *independent* if, for any values $x_i, y_j, \ldots, z_k$,

$$P(X = x_i, Y = y_j, \ldots, Z = z_k) \equiv P(X = x_i)P(Y = y_j) \ldots P(Z = z_k)$$

In particular, X and Y are independent if

$$P(X = x_i, Y = y_j) \equiv P(X = x_i)P(Y = y_j)$$

Now let X be a random variable with mean μ. We can consider the numerical outcome of each of n independent trials to be a random variable with the same distribution as X. The random variable corresponding to the ith outcome will be denoted by $X_i (i = 1, 2, \ldots, n)$. (We note that the X_i are independent with the same distribution as X.) The average value of all n outcomes is also a random variable which is denoted by $\overline{X_n}$ and called the *sample mean*. That is:

$$\overline{X_n} = \frac{X_1 + X_2 + \cdots + X_n}{n}$$

The law of large numbers says that as n increases the value of the sample mean $\overline{X_n}$ approaches the mean value μ. Namely:

Theorem 7.11 (Law of Large Numbers): For any positive number α, no matter how small, the probability that the sample mean $\overline{X_n}$ has a value in the interval $[\mu - \alpha, \mu + \alpha]$ approaches 1 as n approaches infinity. That is:

$$P([\mu - \alpha \leq X \leq \mu + \alpha]) \to 1 \quad \text{as} \quad n \to \infty.$$

EXAMPLE 7.19 Suppose a die is tossed 5 times with outcomes:

$$x_1 = 3, \quad x_2 = 4, \quad x_3 = 6, \quad x_4 = 1, \quad x_5 = 4$$

Then the corresponding value $\overline{x}$ of the sample mean $\overline{X_5}$ follows:

$$\overline{x} = \frac{3 + 4 + 6 + 1 + 4}{5} = 3.6$$

For a fair die, the mean $\mu = 3.5$. The law of large numbers tells us that, as n gets larger, there is a greater likelihood that $\overline{X_n}$ will get closer to 3.5.

Solved Problems

SAMPLE SPACES AND EVENTS

7.1. Let a coin and a die be tossed; and let the sample space S consists of the 12 elements:

$$S = \{H1, H2, H3, H4, H5, H6, T1, T2, T3, T4, T5, T6\}$$

$A = \{H2, H4, H6\}$

(a) Express explicitly the following events:

    $A = \{$heads and an even number$\}$, $B = \{$prime number$\}$, $C = \{$tails and an odd number$\}$

(b) Express explicitly the events: (i) A or B occurs; (ii) B and C occur: (iii) only B occurs.

(c) Which pair of the events A, B, and C are mutually exclusive?

(a) The elements of A are those elements of S consisting of an H and an even number:

$$A = \{H2, H4, H6\}$$

The elements of B are those points in S whose second component is a prime number (2, 3, or 5):

$$B = \{H2, H3, H5, T2, T3, T5\}$$

The elements of C are those points in S consisting of a T and an odd number; $C = \{T1, T3, T5\}$.

(b) (i) $A \cup B = \{H2, H4, H6, H3, H5, T2, T3, T5\}$
    (ii) $B \cap C = \{T3, T5\}$
    (iii) $B \cap A^c \cap C^c = \{H3, H5, T2\}$

(c) A and C are mutually exclusive since $A \cap C = \varnothing$.

7.2. A pair of dice is tossed. (See Example 7.2.) Find the number of elements in each event:

    (a) $A = \{$two numbers are equal$\}$ (c) $C = \{$5 appears on the first die$\}$
    (b) $B = \{$sum is 10 or more$\}$ (d) $D = \{$5 appears on at least one die$\}$

Use Fig. 7-1(b) to help count the number of elements in the event.

(a) $A = \{(1, 1), (2, 2), \ldots, (6, 6)\}$, so $n(A) = 6$.

(b) $B = \{(6, 4), (5, 5), (4, 6), (6, 5), (5, 6), (6, 6)\}$, so $n(B) = 6$.

(c) $C = \{(5, 1), (5, 2), \ldots, (5, 6)\}$, so $n(C) = 6$.

(d) There are six pairs with 5 as the first element, and six pairs with 5 as the second element. However, $(5, 5)$ appears in both places. Hence

$$n(D) = 6 + 6 - 1 = 11$$

Alternately, count the pairs in Fig. 7-1(b) which are in D to get $n(D) = 11$.

FINITE EQUIPROBABLE SPACES

7.3 Determine the probability p of each event:

(a) An even number appears in the toss of a fair die;

(b) One or more heads appear in the toss of three fair coins;

(c) A red marble appears in a random drawing of one marble from a box containing four white, three red, and five blue marbles.

    Each sample space S is an equiprobable space. Thus, for each event E, use:

$$P(E) = \frac{\text{number of elements in } E}{\text{number of elements in } S} = \frac{n(E)}{n(S)}$$

(a) The event can occur in three ways (2, 4 or 6) out of 6 cases; hence $p = \frac{3}{6} = \frac{1}{2}$.

(b) There are 8 cases:

$$HHH, HHT, HTH, HTT, THH, THT, TTH, TTT$$

Only the last case is not favorable; hence $p = 7/8$.

(c) There are $4 + 3 + 5 = 12$ marbles of which three are red; hence $p = \frac{3}{12} = \frac{1}{4}$.

7.4. A single card is drawn from an ordinary deck of 52 cards. (See Fig. 7-2.) Find the probability p that the card is a:

(a) face card (jack, queen or king);　　(c) face card and a heart;

(b) heart;　　　　　　　　　　　　　　(d) face card or a heart.

　　Here $n(S) = 52$.

(a) There are $4(3) = 12$ face cards; hence $p = \dfrac{12}{52} = \dfrac{3}{13}$.

(b) There are 13 hearts; hence $p = \dfrac{13}{52} = \dfrac{1}{4}$.

(c) There are three face cards which are hearts; hence $p = \dfrac{3}{52}$.

(d) Letting $F = \{$face cards$\}$ and $H = \{$hearts$\}$, we have

$$n(F \cup H) = n(F) + n(H) - n(F \cap H) = 12 + 13 - 3 = 22$$

　　Hence $p = \frac{22}{52} = \frac{11}{26}$.

7.5. Two cards are drawn at random from an ordinary deck of 52 cards. Find the probability p that: (a) both are spades; (b) one is a spade and one is a heart.

　　There are $\binom{52}{2} = 1326$ ways to draw 2 cards from 52 cards.

(a) There are $\binom{13}{2} = 78$ ways to draw 2 spades from 13 spades; hence

$$p = \frac{\text{number of ways 2 spades can be drawn}}{\text{number of ways 2 cards can be drawn}} = \frac{78}{1326} = \frac{3}{51}$$

(b) There are 13 spades and 13 hearts, so there are $13 \cdot 13 = 169$ ways to draw a spade and a heart. Thus $p = \frac{169}{1326} = \frac{13}{102}$.

7.6. Consider the sample space in Problem 7.1. Assume the coin and die are fair; hence S is an equiprobable space. Find:

(a) $P(A)$, $P(B)$, $P(C)$

(b) $P(A \cup B)$, $P(B \cap C)$, $P(B \cap A^C \cap C^C)$

　　Since S is an equiprobable space, use $P(E) = n(E)/n(S)$. Here $n(S) = 12$. So we need only count the number of elements in the given set.

(a) $P(A) = \frac{3}{12}$, $P(B) = \frac{6}{12}$, $P(C) = \frac{3}{12}$

(b) $P(A \cup B) = \frac{8}{12}$, $P(B \cap C) = \frac{2}{12}$, $P(B \cap A^C \cap C^C) = \frac{3}{12}$

7.7. A box contains two white socks and two blue socks. Two socks are drawn at random. Find the probability p they are a match (the same color).

　　There are $\binom{4}{2} = 6$ ways to draw two of the socks. Only two pairs will yield a match. Thus $p = \frac{2}{6} = \frac{1}{3}$.

7.8. Five horses are in a race. Audrey picks two of the horses at random, and bets on them. Find the probability p that Audrey picked the winner.

　　There are $\binom{5}{2} = 10$ ways to pick two of the horses. Four of the pairs will contain the winner. Thus $p = \frac{4}{10} = \frac{2}{5}$.

FINITE PROBABILITY SPACES

7.9. A sample space S consists of four elements; that is, $S = \{a_1, a_2, a_3, a_4\}$. Under which of the following functions does S become a probability space?

(a) $P(a_1) = \frac{1}{2}$ $P(a_2) = \frac{1}{3}$ $P(a_3) = \frac{1}{4}$ $P(a_4) = \frac{1}{5}$

(b) $P(a_1) = \frac{1}{2}$ $P(a_2) = \frac{1}{4}$ $P(a_3) = -\frac{1}{4}$ $P(a_4) = \frac{1}{2}$

(c) $P(a_1) = \frac{1}{2}$ $P(a_2) = \frac{1}{4}$ $P(a_3) = \frac{1}{8}$ $P(a_4) = \frac{1}{8}$

(d) $P(a_1) = \frac{1}{2}$ $P(a_2) = \frac{1}{4}$ $P(a_3) = \frac{1}{4}$ $P(a_4) = 0$

(a)   Since the sum of the values on the sample points is greater than one, the function does not define S as a probability space.

(b)   Since $P(a_3)$ is negative, the function does not define S as a probability space.

(c)   Since each value is nonnegative and the sum of the values is one, the function does define S as a probability space.

(d)   The values are nonnegative and add up to one; hence the function does define S as a probability space.

7.10. A coin is weighted so that heads is twice as likely to appear as tails. Find $P(T)$ and $P(H)$.

Let $P(T) = p$; then $P(H) = 2p$. Now set the sum of the probabilities equal to one, that is, set $p + 2p = 1$. Then $p = \frac{1}{3}$. Thus $P(H) = \frac{1}{3}$ and $P(T) = \frac{2}{3}$.

7.11. Suppose A and B are events with $P(A) = 0.6$, $P(B) = 0.3$, and $P(A \cap B) = 0.2$. Find the probability that:

(a) A does not occur;   (c) A or B occurs;

(b) B does not occur;   (d) Neither A nor B occurs.

(a) $P(\text{not } A) = P(A^C) = 1 - P(A) = 0.4.$

(b) $P(\text{not } B) = P(B^C) = 1 - P(B) = 0.7.$

(c)   By the Addition Principle,

$$P(A \text{ or } B) = P(A \cup B) = P(A) + P(B) - P(A \cap B)$$
$$= 0.6 + 0.3 - 0.2 = 0.7$$

(d)   Recall (DeMorgan's Law) that neither A nor B is the complement of $A \cup B$. Thus:

$$P(\text{neither } A \text{ nor } B) = P((A \cup B)^C) = 1 - P(A \cup B) = 1 - 0.7 = 0.3$$

7.12. Prove Theorem 7.2: $P(A^c) = 1 - P(A)$.

$S = A \cup A^c$ where A and A^c are disjoint. Our result follows from the following:

$$1 = P(S) = P(A \cup A^c) = P(A) + P(A^c)$$

7.13. Prove Theorem 7.3: (i) $P(\varnothing) = 0$; (ii) $P(A \backslash B) = P(A) - P(A \cap B)$; (iii) If $A \subseteq B$, then $P(A) \le P(B)$.

(i) $\varnothing = S^c$ and $P(S) = 1$. Thus $P(\varnothing) = 1 - 1 = 0$.

(ii)   As indicated by Fig. 7-6(a), $A = (A \backslash B) \cup (A \cap B)$ where $A \backslash B$ and $A \cap B$ are disjoint. Hence

$$P(A) = P(A \backslash B) + P(A \cap B)$$

From which our result follows.

(iii)   If $A \subseteq B$, then, as indicated by Fig. 7-6(b), $B = A \cup (B \backslash A)$ where A and $B \backslash A$ are disjoint. Hence

$$P(B) = P(A) + P(B \backslash A)$$

Since $P(B \backslash A) \ge 0$, we have $P(A) \le P(B)$.

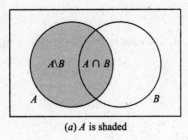
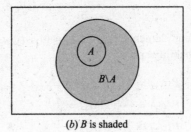
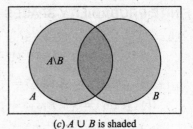

(a) A is shaded   (b) B is shaded   (c) A ∪ B is shaded

Fig. 7-6

7.14. Prove Theorem 7.4 (Addition Principle): For any events A and B,

$$P(A \cup B) = P(A) + P(B) - P(A \cap B)$$

As indicated by Fig. 7-6(c), $A \cup B = (A \backslash B) \cup B$ where $A \backslash B$ and B are disjoint sets. Thus, using Theorem 7.3(ii),

$$P(A \cup B) = P(A \backslash B) + P(B) = P(A) - P(A \cap B) + P(B)$$
$$= P(A) + P(B) - P(A \cap B)$$

CONDITIONAL PROBABILITY

7.15. A pair of fair dice is thrown. (See Fig. 7-1(b).) Find the probability that the sum is 10 or greater if: (a) 5 appears on the first die; (b) 5 appears on at least one die.

(a) If a 5 appears on the first die, then the reduced sample space is

$$A = \{(5, 1), (5, 2), (5, 3), (5, 4), (5, 5), (5, 6)\}$$

The sum is 10 or greater on two of the six outcomes: (5, 5), (5, 6). Hence $p = \frac{2}{6} = \frac{1}{3}$.

(b) If a 5 appears on at least one of the dice, then the reduced sample space has eleven elements.

$$B = \{(5, 1), (5, 2), (5, 3), (5, 4), (5, 5), (5, 6), (1, 5), (2, 5), (3, 5), (4, 5), (6, 5)\}$$

The sum is 10 or greater on three of the eleven outcomes: (5, 5), (5, 6), (6, 5). Hence $p = \frac{3}{11}$.

7.16. In a certain college town, 25% of the students failed mathematics (M), 15% failed chemistry (C), and 10% failed both mathematics and chemistry. A student is selected at random.

(a) If he failed chemistry, find the probability that he also failed mathematics.

(b) If he failed mathematics, find the probability that he also failed chemistry.

(c) Find the probability that he failed mathematics or chemistry.

(d) Find the probability that he failed neither mathematics nor chemistry.

(a) The probability that a student failed mathematics, given that he failed chemistry, is

$$P(M|C) = \frac{P(M \cap C)}{P(C)} = \frac{0.10}{0.15} = \frac{2}{3}$$

(b) The probability that a student failed chemistry, given that he failed mathematics is

$$P(C|M) = \frac{P(C \cap M)}{P(M)} = \frac{0.10}{0.25} = \frac{2}{5}$$

(c) By the Addition Principle (Theorem 7.4),

$$P(M \cup C) = P(M) + P(C) - P(M \cap C) = 0.25 + 0.15 - 0.10 = 0.30$$

(d) Students who failed neither mathematics nor chemistry form the complement of the set $M \cup C$, that is, they form the set $(M \cup C)^C$. Hence

$$P((M \cup C)^C) = 1 - P(M \cup C) = 1 - 0.30 = 0.70$$

7.17. A pair of fair dice is thrown. Given that the two numbers appearing are different, find the probability p that: (a) the sum is 6; (b) an one appears; (c) the sum is 4 or less.

There are 36 ways the pair of dice can be thrown, and six of them, $(1, 1), (2, 2), \ldots, (6, 6)$, have the same numbers. Thus the reduced sample space will consist of $36 - 6 = 30$ elements.

(a) The sum 6 can appear in four ways: $(1, 5), (2, 4), (4, 2), (5, 1)$. (We cannot include $(3, 3)$ since the numbers are the same.) Hence $p = \frac{4}{30} = \frac{2}{15}$.

(b) An one can appear in 10 ways: $(1, 2), (1, 3), \ldots, (1, 6)$ and $(2, 1), (3, 1), \ldots, (6, 1)$. Therefore $p = \frac{10}{30} = \frac{1}{3}$.

(c) The sum of 4 or less can occur in four ways: $(3, 1), (1, 3), (2, 1), (1, 2)$. Thus $p = \frac{4}{30} = \frac{2}{15}$.

7.18. A class has 12 boys and 4 girls. Suppose three students are selected at random from the class. Find the probability p that they are all boys.

The probability that the first student selected is a boy is $12/16$ since there are 12 boys out of 16 students. If the first student is a boy, then the probability that the second is a boy is $11/15$ since there are 11 boys left out of 15 students. Finally, if the first two students selected were boys, then the probability that the third student is a boy is $10/14$ since there are 10 boys left out of 14 students. Thus, by the multiplication theorem, the probability that all three are boys is

$$p = \frac{12}{16} \cdot \frac{11}{15} \cdot \frac{10}{14} = \frac{11}{28}$$

Another Method

There are $C(16, 3) = 560$ ways to select three students out of the 16 students, and $C(12, 3) = 220$ ways to select three boys out of 12 boys; hence

$$p = \frac{220}{560} = \frac{11}{28}$$

Another Method

If the students are selected one after the other, then there are $16 \cdot 15 \cdot 14$ ways to select three students, and $12 \cdot 11 \cdot 10$ ways to select three boys; hence

$$p = \frac{2 \cdot 11 \cdot 10}{16 \cdot 15 \cdot 14} = \frac{11}{28}$$

INDEPENDENCE

7.19. The probability that A hits a target is $\frac{1}{3}$ and the probability that B hits a target is $\frac{1}{5}$. They both fire at the target. Find the probability that:

(a) A does not hit the target;   (c) one of them hits the target;

(b) both hit the target;   (d) neither hits the target.

We are given $P(A) = \frac{1}{3}$ and $P(B) = \frac{1}{5}$ (and we assume the events are independent).

(a) $P(\text{not } A) = P(A^C) = 1 - P(A) = 1 - \frac{1}{3} = \frac{2}{3}$.

(b) Since the events are independent,

$$P(A \text{ and } B) = P(A \cap B) = P(A) \cdot P(B) = \frac{1}{3} \cdot \frac{1}{5} = \frac{1}{15}$$

(c) By the Addition Principle (Theorem 7.4),

$$P(A \text{ or } B) = P(A \cup B) = P(A) + P(B) - P(A \cap B) = \frac{1}{3} + \frac{1}{5} - \frac{1}{15} = \frac{7}{15}$$

(d) We have

$$P(\text{neither } A \text{ nor } B) = P((A \cup B)^C) = 1 - P(A \cup B) = 1 - \tfrac{7}{15} = \tfrac{8}{15}$$

7.20. Consider the following events for a family with children:

$$A = \{\text{children of both sexes}\}, \quad B = \{\text{at most one boy}\}.$$

(a) Show that A and B are independent events if a family has three children.

(b) Show that A and B are dependent events if a family has only two children.

(a) We have the equiprobable space $S = \{bbb, bbg, bgb, bgg, gbb, gbg, ggb, ggg\}$. Here

$$A = \{bbg, bgb, bgg, gbb, gbg, ggb\} \quad \text{and so} \quad P(A) = \tfrac{6}{8} = \tfrac{3}{4}$$
$$B = \{bgg, gbg, ggb, ggg\} \quad \text{and so} \quad P(B) = \tfrac{4}{8} = \tfrac{1}{2}$$
$$A \cap B = \{bgg, gbg, ggb\} \quad \text{and so} \quad P(A \cap B) = \tfrac{3}{8}$$

Since $P(A)P(B) = \tfrac{3}{4} \cdot \tfrac{1}{2} = \tfrac{3}{8} = P(A \cap B)$ A and B are independent.

(b) We have the equiprobable space $S = \{bb, bg, gb, gg\}$. Here

$$A = \{bg, gb\} \quad \text{and so} \quad P(A) = \tfrac{1}{2}$$
$$B = \{bg, gb, gg\} \quad \text{and so} \quad P(B) = \tfrac{3}{4}$$
$$A \cap B = \{bg, gb\} \quad \text{and so} \quad P(A \cap B) = \tfrac{1}{2}$$

Since $P(A)P(B) \neq P(A \cap B)$, A and B are dependent.

7.21. Box A contains five red marbles and three blue marbles, and box B contains three red and two blue. A marble is drawn at random from each box.

(a) Find the probability p that both marbles are red.

(b) Find the probability p that one is red and one is blue.

(a) The probability of choosing a red marble from A is $\tfrac{5}{8}$ and from B is $\tfrac{3}{5}$. Since the events are independent, $P = \tfrac{5}{8} \cdot \tfrac{3}{5} = \tfrac{3}{8}$.

(b) The probability p_1 of choosing a red marble from A and a blue marble from B is $\tfrac{5}{8} \cdot \tfrac{2}{5} = \tfrac{1}{4}$. The probability p_2 of choosing a blue marble from A and a red marble from B is $\tfrac{3}{8} \cdot \tfrac{3}{5} = \tfrac{9}{40}$. Hence $p = p_1 + p_2 = \tfrac{1}{4} + \tfrac{9}{40} = \tfrac{19}{40}$.

7.22. Prove: If A and B are independent events, then A^c and B^c are independent events.

Let $P(A) = x$ and $P(B) = y$. Then $P(A^c) = 1 - x$ and $P(B^c) = 1 - y$. Since A and B are independent. $P(A \cap B) = P(A)P(B) = xy$. Furthermore,

$$P(A \cup B) = P(A) + P(B) - P(A \cap B) = x + y - xy$$

By DeMorgan's law, $(A \cup B)^c = A^c \cap B^c$; hence

$$P(A^c \cap B^c) = P((A \cup B)^c) = 1 - P(A \cup B) = 1 - x - y + xy$$

On the other hand,

$$P(A^c)P(B^c) = (1 - x)(1 - y) = 1 - x - y + xy$$

Thus $P(A^c \cap B^c) = P(A^c)P(B^c)$, and so A^c and B^c are independent.

In similar fashion, we can show that A and B^c, as well as A^c and B, are independent.

REPEATED TRIALS, BINOMIAL DISTRIBUTION

7.23. Suppose that, whenever horses a, b, c, d race together, their respective probabilities of winning are 0.2, 0.5, 0.1, 0.2. That is, $S = \{a, b, c, d\}$ where $P(a) = 0.2$, $P(b) = 0.5$, $P(c) = 0.1$, $P(d) = 0.2$. They race three times.

   (a) Describe and find the number of elements in the product probability space S_3.

   (b) Find the probability that the same horse wins all three races.

   (c) Find the probability that a, b, c each win one race.

   For notational convenience, we write xyz for (x, y, z).

   (a) By definition, $S_3 = S \times S \times S = \{xyz \mid x, y, z \in S\}$ and $P(xyz) = P(x)P(y)P(z)$.
   Thus, in particular, S_3 contains $4^3 = 64$ elements.

   (b) We seek the probability of the event $A = \{aaa, bbb, ccc, ddd\}$. By definition,

$$P(aaa) = (0.2)^3 = 0.008, \quad P(ccc) = (0.1)^3 = 0.001$$
$$P(bbb) = (0.5)^3 = 0.125, \quad P(ddd) = (0.2)^3 = 0.008$$

   Thus $P(A) = 0.0008 + 0.125 + 0.001 + 0.008 = 0.142$.

   (c) We seek the probability of the event $B = \{abc, acb, bac, bca, cab, cba\}$. Every element in B has the same probability, the product $(0.2)(0.5)(0.1) = 0.01$. Thus $P(B) = 6(0.01) = 0.06$.

7.24. The probability that John hits a target is $p = \frac{1}{4}$. He fires $n = 6$ times. Find the probability that he hits the target: (a) exactly two times; (b) more than four times; (c) at least once.

This is a binomial experiment with $n = 6$, $p = \frac{1}{4}$, and $q = 1 - p = \frac{3}{4}$; that is, $B(6, \frac{1}{4})$. Accordingly, we use Theorem 7.7.

   (a) $P(2) = \binom{6}{2} \left(\frac{1}{4}\right)^2 \left(\frac{3}{4}\right)^4 = 15(3^4)/(4^6) = \frac{1215}{4096} \approx 0.297$.

   (b) $P(5) + P(6) = \binom{6}{5} \left(\frac{1}{4}\right)^5 \left(\frac{3}{4}\right)^1 + \left(\frac{1}{4}\right)^6 = \frac{18}{4^6} + \frac{1}{4^6} = \frac{19}{4^6} = \frac{19}{4096} \approx 0.0046$.

   (c) $P(0) = \left(\frac{3}{4}\right)^6 = \frac{729}{4096}$, so $P(X > 0) = 1 - \frac{729}{4096} = \frac{3367}{4096} \approx 0.82$.

7.25. A family has six children. Find the probability p that there are: (a) three boys and three girls; (b) fewer boys than girls. Assume that the probability of any particular child being a boy is $\frac{1}{2}$.

Here $n = 6$ and $p = q = \frac{1}{2}$.

   (a) $p = P(3 \text{ boys}) = \binom{6}{3} \left(\frac{1}{2}\right)^3 \left(\frac{1}{2}\right)^2 = \frac{20}{64} = \frac{5}{16}$.

   (b) There are fewer boys than girls if there are zero, one, or two boys. Hence

$$p = P(0 \text{ boys}) + P(1 \text{ boy}) + P(2 \text{ boys}) = \left(\frac{1}{2}\right)^6 + \binom{6}{1} \left(\frac{1}{2}\right)^5 + \binom{6}{2} \left(\frac{1}{2}\right)^2 \left(\frac{1}{2}\right)^4 = \frac{11}{32} = 0.34$$

7.26. A man fires at a target $n = 6$ times and hits it $k = 2$ times, (a) List the different ways that this can happen, (b) How many ways are there?

   (a) List all sequences with two S's (successes) and four F's (failures):

   SSFFFF, SFSFFF, SFFSFF, SFFFSF, SFFFFS, FSSFFF, FSFSFF, FSFFSF,
   FSFFFS, FFSSFF, FFSFSF, FFSFFS, FFFSSF, FFFSFS, FFFFSS.

   (b) There are 15 different ways as indicated by the list. Observe that this is equal to $\binom{6}{2}$ since we are distributing $k = 2$ letters S among the $n = 6$ positions in the sequence.

7.27. Prove Theorem 7.7: The probability of exactly k successes in a binomial experiment $B(n, p)$ is given by

$$P(k) = p(k \text{ successes}) = \binom{n}{k} p^k q^{n-k}$$

The probability of one or more successes is $1 - q^n$.

The sample space of the n repeated trials consists of all n-tuples (i.e., n-element sequences) whose components are either S (success) or F (failure). Let A be the event of exactly k successes. Then A consists of all n-tuples of which k components are S and $n - k$ components are F. The number of such n-tuples in the event A is equal to the number of ways that k letters S can be distributed among the n components of an n-tuple; hence A consists of $C(n, k) = \binom{n}{k}$ sample points. The probability of each point in A is $p^k q^{n-k}$; hence

$$P(A) = \binom{n}{k} p^k q^{n-k}$$

In particular, the probability of no successes is

$$P(0) = \binom{n}{0} p^0 q^n = q^n$$

Thus the probability of one or more successes is $1 - q^n$.

RANDOM VARIABLES, EXPECTATION

7.28. A player tosses two fair coins. He wins \$2 if two heads occur, and \$1 if one head occurs. On the other hand, he loses \$3 if no heads occur. Find the expected value E of the game. Is the game fair? (The game is fair, favorable, or unfavorable to the player according as $E = 0$, $E > 0$ or $E < 0$.)

The sample space $S = \{HH, HT, TH, TT\}$, and each sample point has probability $1/4$. For the player's gain, we have

$$X(HH) = \$2, \quad X(HT) = X(TH) = \$1, \quad X(TT) = -\$3$$

Hence the distribution of X follows:

x_i	2	1	-3
p_i	1/4	2/4	1/4

Thus $E = E(X) = 2(1/4) + 1(2/4) - 3(1/4) = \$0.25$. Since $E(X) > 0$, the game is favorable to the player.

7.29. You have won a contest. Your prize is to select one of three envelopes and keep what is in it. Each of two of the envelopes contains a check for \$30, but the third envelope contains a check for \$3000. Find the expectation E of your winnings (as a probability distribution).

Let X denote your winnings. Then $X = 30$ or 3000, and $P(30) = \frac{2}{3}$ and $P(3000) = \frac{1}{3}$. Hence

$$E = E(X) = 30 \cdot \frac{2}{3} + 3000 \cdot \frac{1}{3} = 20 + 1000 = 1020$$

7.30. A random sample with replacement of size $n = 2$ is drawn from the set $\{1, 2, 3\}$, yielding the following 9-element equiprobable sample space:

$$S = \{(1, 1), (1, 2), (1, 3), (2, 1), (2, 2), (2, 3), (3, 1), (3, 2), (3, 3)\}$$

(a) Let X denote the sum of the two numbers. Find the distribution f of X, and find the expected value $E(X)$.

(b) Let Y denote the minimum of the two numbers. Find the distribution g of X, and find the expected value $E(Y)$.

(a) The random variable X assumes the values 2, 3, 4, 5, 6. We compute the distribution f of X:

(i) One point $(1, 1)$ has sum 2; hence $f(2) = \frac{1}{9}$.

(ii) Two points, (1, 2), (2, 1), have sum 3; hence $f(3) = \frac{2}{9}$.

(iii) Three points, (1, 3), (2, 2), (1, 3), have sum 4; hence $f(4) = \frac{3}{9}$.

(iv) Two points, (2, 3), (3, 2), have sum 5; hence $f(5) = \frac{2}{9}$.

(v) One point (3, 3) has sum 6; hence $f(6) = \frac{1}{9}$.

Thus the distribution f of X is:

x	2	3	4	5	6
$f(x)$	1/9	2/9	3/9	2/9	1/9

The expected value $E(X)$ of X is obtained by multiplying each value of x by its probability $f(x)$ and taking the sum. Hence

$$E(X) = 2\left(\frac{1}{9}\right) + 3\left(\frac{2}{9}\right) + 4\left(\frac{3}{9}\right) + 5\left(\frac{2}{9}\right) + 6\left(\frac{1}{9}\right) = 4$$

(b) The random variable Y only assumes the values 1, 2, 3. We compute the distribution g of Y:

(i) Five points, (1, 1), (1, 2), (1, 3), (2, 1), (3, 1), have minimum 1; hence $g(1) = \frac{5}{9}$.

(ii) Three points, (2, 2), (2, 3), (3, 2), have minimum 2; hence $g(2) = \frac{3}{9}$.

(iii) One point (3, 3) has minimum 3; hence $g(3) = \frac{1}{9}$.

Thus the following is the distribution g of Y:

y	1	2	3
$g(y)$	5/9	3/9	1/9

The expected value $E(Y)$ of Y is:

$$E(Y) = 1\left(\frac{5}{9}\right) + 2\left(\frac{3}{9}\right) + 3\left(\frac{1}{9}\right) = \frac{12}{9} \approx 1.33$$

7.31. A linear array EMPLOYEE has n elements. Suppose NAME appears randomly in the array, and there is a linear search to find the location K of NAME, that is, to find K such that EMPLOYEE$[K]$ = NAME. Let $f(n)$ denote the number of comparisons in the linear search.

(a) Find the expected value of $f(n)$.

(b) Find the maximum value (worst case) of $f(n)$.

(a) Let X denote the number of comparisons. Since NAME can appear in any position in the array with the same probability of $1/n$, we have $X = 1, 2, 3, \ldots, n$, each with probability $1/n$. Hence

$$f(n) = E(X) = 1 \cdot \frac{1}{n} + 2 \cdot \frac{1}{n} + 3 \cdot \frac{1}{n} + \cdots + n \cdot \frac{1}{n}$$
$$= (1 + 2 + \cdots + n) \cdot \frac{1}{n} = \frac{n(n+1)}{2} \cdot \frac{1}{n} = \frac{n+1}{2}$$

(b) If NAME appears at the end of the array, then $f(n) = n$.

MEAN, VARIANCE, STANDARD DEVIATION

7.32. Find the mean $\mu = E(X)$, variance $\sigma^2 = Var(X)$, and standard deviation $\sigma = \sigma_x$ of each distribution:

(a)
x_i	2	3	11
p_i	1/3	1/2	1/6

(b)
x_i	1	3	4	5
p_i	0.4	0.1	0.2	0.3

Use the formulas:

$$\mu = E(X) = x_1 p_1 + x_2 p_2 + \cdots + x_m p_m = \Sigma x_i p_i, \quad \sigma^2 = Var(X) = E(X^2) - \mu^2$$

$$E(X^2) = x_1^2 p_1 + x_2^2 p_2 + \cdots + x_m^2 p_m = \Sigma x_i^2 p_i, \qquad \sigma = \sigma_x = \sqrt{Var(X)}$$

(a) $\quad \mu = \Sigma x_i p_i = 2\left(\frac{1}{3}\right) + 3\left(\frac{1}{2}\right) + 11\left(\frac{1}{6}\right) = 4$

$\quad E(X^2) = \Sigma x_i^2 p_i = 2^2\left(\frac{1}{3}\right) + 3^2\left(\frac{1}{2}\right) + 11^2\left(\frac{1}{6}\right) = 26$

$\quad \sigma^2 = Var(X) = E(X^2) - \mu^2 = 26 - 4^2 = 10$

$\quad \sigma = \sqrt{Var(X)} = \sqrt{10} = 3.2$

(b) $\quad \mu = \Sigma x_i p_i = 1(0.4) + 3(0.1) + 4(0.2) + 5(0.3) = 3$

$\quad E(X^2) = \Sigma x_i^2 p_i = 1(0.4) + 9(0.1) + 16(0.2) + 25(0.3) = 12$

$\quad \sigma^2 = Var(X) = E(X^2) - \mu^2 = 12 - 9 = 3$

$\quad \sigma = \sqrt{Var(X)} = \sqrt{3} = 1.7$

7.33. A fair die is tossed yielding the equiprobable sample space $S = \{1, 2, 3, 4, 5, 6\}$ where $n(S) = 6$ and each point has probability $1/6$.

(a) Let X be the random variable which denotes twice the number that occurs. Find the distribution f of X and its expectation $E(X)$.

(b) Let Y be the random variable which assigns 1 or 3 according as an odd or even number occurs. Find the distribution g of Y and its expectation $E(Y)$.

(a) Here the range space $R_X = \{2, 4, 6, 8, 10, 12\}$ since

$$X(1) = 2, \quad X(2) = 4, \quad X(3) = 6, \quad X(4) = 8, \quad X(5) = 10, \quad X(6) = 12$$

Also, each number occurs with probability $1/6$. Thus the distribution f of X follows:

x	2	4	6	8	10	12
$f(x)$	1/6	1/6	1/6	1/6	1/6	1/6

Hence

$$E(X) = \sum xf(x) = \frac{2}{6} + \frac{4}{6} + \frac{6}{6} + \frac{8}{6} + \frac{10}{6} + \frac{12}{6} = 7$$

(b) Here the range space $R_Y = \{1, 3\}$ since

$$Y(1) = 1, \quad Y(2) = 3, \quad Y(3) = 1, \quad Y(4) = 3, \quad Y(5) = 1, \quad Y(6) = 3$$

We compute the distribution g of Y using the fact that $n(S) = 6$:

(i) Three points 1, 3, 5 are odd and have image 1; hence $g(1) = 3/6$.

(ii) Three points 2, 4, 6 are even and have image 3; hence $g(3) = 3/6$.

Thus the distribution g of Y is:

y	1	3
$g(y)$	3/6	3/6

Hence

$$E(Y) = \sum yg(y) = \frac{3}{6} + \frac{9}{6} = 2$$

7.34. Let $Z = X + Y$ where X and Y are the random variables in Problem 7.33. Find the distribution h of Z, and find $E(Z)$. Verify that $E(X + Y) = E(X) + E(Y)$.

The sample space is still $S = \{1, 2, 3, 4, 5, 6\}$ and each point still has probability $1/6$. We obtain using $Z(s) = (X + Y)(s) = X(s) + Y(s)$

$$Z(1) = X(1) + Y(1) = 2 + 1 = 3; \quad Z(4) = X(4) + Y(4) = 8 + 3 = 11,$$

$$Z(2) = X(2) + Y(2) = 4 + 3 = 7; \quad Z(5) = X(5) + Y(5) = 10 + 1 = 11,$$

$$Z(3) = X(3) + Y(3) = 6 + 1 = 7; \quad Z(6) = X(6) + Y(6) = 12 + 3 = 15.$$

Thus the range space $R_z = \{3, 7, 11, 15\}$. We compute the distribution h of Z using the fact that $n(S) = 6$:

(i) One point has image 3, so $h(3) = 1/6$;   (iii) Two points have image 11, so $h(11) = 2/6$;

(ii) Two points have image 7, so $h(7) = 2/6$;   (iv) One point has image 15, so $h(15) = 1/6$.

Thus the distribution h of Z follows:

z	3	7	11	15
$h(z)$	1/6	2/6	2/6	1/6

Hence

$$E(Z) = \sum zh(z) = \frac{3}{6} + \frac{14}{6} + \frac{22}{6} + \frac{15}{6} = 9$$

Accordingly,

$$E(X + Y) = E(Z) = 9 = 7 + 2 = E(X) + E(Y).$$

BINOMIAL DISTRIBUTION

7.35. The probability that a man hits a target is $p = 0.1$. He fires $n = 100$ times. Find the expected number μ of times he will hit the target, and the standard deviation σ.

This is a binomial experiment $B(n, p)$ where $n = 100$, $p = 0.1$, and $q = 1 - p = 0.9$. Accordingly, we apply Theorem 7.9 to obtain

$$\mu = np = 100(0.1) = 10 \quad \text{and} \quad \sigma = \sqrt{npq} = \sqrt{100(0.1)(0.9)} = 3$$

7.36. A student takes an 18-question multiple-choice exam, with four choices per question. Suppose one of the choices is obviously incorrect, and the student makes an "educated" guess of the remaining choices. Find the expected number $E(X)$ of correct answers, and the standard deviation σ.

This is a biomioal experiment $B(n, p)$ where $n = 18$, $p = \frac{1}{3}$, and $q = 1 - p = \frac{2}{3}$. Hence

$$E(X) = np = 18 \cdot \frac{1}{3} = 6 \quad \text{and} \quad \sigma = \sqrt{npq} = \sqrt{18 \cdot \frac{1}{3} \cdot \frac{2}{3}} = 2$$

7.37. The expectation function $E(X)$ on the space of random variables on a sample space S can be proved to be *linear*, that is,

$$E(X_1 + X_2 + \cdots + X_n) = E(X_1) + E(X_2) + \cdots + E(X_n)$$

Use this property to prove $\mu = np$ for a binomial experiment $B(n, p)$.

On the sample space of n Bernoulli trials, let X_i (for $i = 1, 2, \ldots, n$) be the random variable which has the value 1 or 0 according as the ith trial is a success or a failure. Then each X_i has the distribution

x	0	1
$p(x)$	q	p

Thus $E(X_i) = 0(q) + 1(p) = p$. The total number of successes in n trials is

$$X = X_1 + X_2 + \cdots + X_n$$

Using the linearity property of E, we have

$$\begin{aligned} E(X) &= E(X_1 + X_2 + \cdots + X_n) \\ &= E(X_1) + E(X_2) + \cdots + E(X_n) \\ &= p + p + \cdots + p = np \end{aligned}$$

MISCELLANEOUS PROBLEMS

7.38. Suppose X is a random variable with mean $\mu = 75$ and standard deviation $\sigma = 5$.

Estimate the probability that X lies between $75 - 20 = 55$ and $75 + 20 = 95$.

Recall Chebyshev's Inequality states

$$P(\mu - k\sigma \leq X \leq \mu + k\sigma) \geq 1 - \frac{1}{k^2}$$

Here $k\sigma = 20$. Since $\sigma = 5$, we get $k = 4$. Then, by Chebyshev's Inequality,

$$P(55 \leq X \geq 95) = 1 - \frac{1}{4^2} = \frac{15}{16} \approx 0.94$$

7.39. Let X be a random variable with mean $\mu = 40$ and standard deviation $\sigma = 2$. Use Chebyshev's Inequality to find a b for which $P(40 - b \leq X \leq 40 + b) \geq 0.95$.

First solve $1 - 1/k^2 = 0.95$ for k as follows:

$$0.05 = \frac{1}{k^2} \quad \text{or} \quad k^2 = \frac{1}{0.05} = 20 \quad \text{or} \quad k = \sqrt{20} = 2\sqrt{5}$$

Then, by Chebyshev's Inequality, $b = k\sigma = 10\sqrt{5} \approx 23.4$. Hence $[P(16.6 \leq X \leq 63.60) \geq 0.95]$

7.40. Let X be a random variable with distribution f. The rth *moment* M_r of X is defined by

$$M_r = E(X^r) = \sum x_i^r f(x_i)$$

Find the first four moments of X if X has the distribution:

x	-2	1	3
$f(x)$	$1/2$	$1/4$	$1/4$

Note M_1 is the mean of X, and M_2 is used in computing the standard deviation of X.

Use the formula for M_r to obtain:

$$M_1 = \sum x_i f(x_i) = -2\left(\tfrac{1}{2}\right) + 1\left(\tfrac{1}{4}\right) + 3\left(\tfrac{1}{4}\right) = 0$$
$$M_2 = \sum x_i^2 f(x_i) = 4\left(\tfrac{1}{2}\right) + 1\left(\tfrac{1}{4}\right) + 9\left(\tfrac{1}{4}\right) = 4.5$$
$$M_3 = \sum x_i^3 f(x_i) = -8\left(\tfrac{1}{2}\right) + 1\left(\tfrac{1}{4}\right) + 27\left(\tfrac{1}{4}\right) = 3$$
$$M_4 = \sum x_i^4 f(x_i) = 16\left(\tfrac{1}{2}\right) + 1\left(\tfrac{1}{4}\right) + 81\left(\tfrac{1}{4}\right) = 28.5$$

7.41. Prove Theorem 7.10 (Chebyshev's Inequality): For $k > 0$,

$$P(\mu - k\sigma \leq X \leq \mu + k\sigma) \geq 1 - \frac{1}{k^2}$$

By definition

$$\sigma^2 = Var(X) = \sum (x_i - \mu)^2 p_i$$

Delete all terms from the summation for which x_i is in the interval $[\mu - k\sigma, \mu + k\sigma]$; that is, delete all terms for which $|x_i - \mu| \leq k\sigma$. Denote the summation of the remaining terms by $\sum^* (x_i - \mu)^2 pi$. Then

$$\left[\sigma^2 \geq \sum{}^* (x_i - \mu)^2 p_i \geq \sum{}^* k^2 \sigma^2 p_i = k^2 \sigma^2 \sum{}^* p_i = k^2 \sigma^2 P(|X - \mu| > k\sigma) \right]$$
$$= k^2 \sigma^2 [1 - P(|X - \mu| \leq k\sigma)] = k^2 \sigma^2 [1 - P(\mu - k\sigma \leq X \leq \mu + k\sigma)]$$

If $\sigma > 0$, then dividing by $k^2 \sigma^2$ gives

$$\tfrac{1}{k^2} \geq 1 - P(\mu - k\sigma \leq X \leq \mu + k\sigma) \quad \text{or} \quad P(\mu - k\sigma \leq X \leq \mu + k\sigma) \geq 1 - \tfrac{1}{k^2}$$

which proves Chebyshev's Inequality for $\sigma > 0$. If $\sigma = 0$, then $x_i = \mu$ for all $p_i > 0$, and

$$P(\mu - k \cdot 0 \leq X \leq \mu + k \cdot 0) = P(X = \mu) = 1 > 1 - \tfrac{1}{k^2}$$

which completes the proof.

Supplementary Problems

SAMPLE SPACES AND EVENTS

7.42. Let A, B, and C be events. Rewrite each of the following events using set notation:
   (a) A and B but not C occurs;   (c) none of the events occurs;
   (b) A or C, but not B occurs;   (d) at least two of the events occur.

7.43. A penny, a dime, and a die are tossed.

   (a) Describe a suitable sample space S, and find $n(S)$.

   (b) Express explicitly the following events:
   $A = \{$two heads and an even number$\}$
   $B = \{$2 appears$\}$
   $C = \{$exactly one head and an odd number$\}$

   (c) Express explicitly the events: (i) A and B;   (ii) only B;   (iii) B and C.

FINITE EQUIPROBABLE SPACES

7.44. Determine the probability of each event:

   (a) An odd number appears in the toss of a fair die.

   (b) One or more heads appear in the toss of four fair coins.

   (c) One or both numbers exceed 4 in the toss of two fair dice.

7.45. One card is selected at random from 50 cards numbered 1 to 50. Find the probability that the number on the card is:
   (a) greater than 10;   (c) greater than 10 and divisible by 5;
   (b) divisible by 5;   (d) greater than 10 or divisible by 5.

7.46. Of 10 girls in a class, three have blue eyes. Two of the girls are chosen at random. Find the probability that:
   (a) both have blue eyes;   (c) at least one has blue eyes;
   (b) neither has blue eyes;   (d) exactly one has blue eyes.

7.47. Ten students, A, B, ..., are in a class. A committee of three is chosen at random to represent the class. Find the probability that:
   (a) A belongs to the committee;   (c) A and B belong to the committee;
   (b) B belongs to the committee;   (d) A or B belong to the committee.

7.48. Three bolts and three nuts are in a box. Two parts are chosen at random. Find the probability that one is a bolt and one is a nut.

7.49. A box contains two white socks, two blue socks, and two red socks. Two socks are drawn at random. Find the probability they are a match (the same color).

7.50. Of 120 students, 60 are studying French, 50 are studying Spanish, and 20 are studying both French and Spanish. A student is chosen at random. Find the probability that the student is studying: (a) French or Spanish; (b) neither French nor Spanish; (c) only French; (d) exactly one of the two languages.

FINITE PROBABILITY SPACES

7.51. Decide which of the following functions defines a probability space on $S = \{a_1, a_2, a_3\}$:

   (a) $P(a_1) = \frac{1}{4}$, $P(a_2) = \frac{1}{3}$, $P(a_3) = \frac{1}{2}$ (c) $P(a_1) = \frac{1}{6}$, $P(a_2) = \frac{1}{3}$, $P(a_3) = \frac{1}{2}$

   (b) $P(a_1) = \frac{2}{3}$, $P(a_2) = -\frac{1}{3}$, $P(a_3) = \frac{2}{3}$ (d) $P(a_1) = 0$, $P(a_2) = \frac{1}{3}$, $P(a_3) = \frac{2}{3}$

7.52. A coin is weighted so that heads is three times as likely to appear as tails. Find $P(H)$ and $P(T)$.

7.53. Three students A, B, and C are in a swimming race. A and B have the same probability of winning and each is twice as likely to win as C. Find the probability that: (a) B wins; (b) C wins; (c) B or C wins.

7.54. Consider the following probability distribution:

Outcome x	1	2	3	4	5
Probability $P(x)$	0.2	0.4	0.1	0.1	0.2

Consider the events $A = \{$even number$\}$, $B = \{2, 3, 4, 5\}$, $C = \{1, 2\}$. Find:

(a) $P(A)$, $P(B)$, $P(C)$;   (b) $P(A \cap B)$, $P(A \cap C)$, $P(B \cap C)$.

7.55. Suppose A and B are events with $P(A) = 0.7$, $P(B) = 0.5$, and $P(A \cap B) = 0.4$. Find the probability that:

(a) A does not occur;   (c) A but not B occurs;

(b) A or B occurs;   (d) neither A nor B occurs.

CONDITIONAL PROBABILITY, INDEPENDENCE

7.56. A fair die is tossed. Consider events $A = \{2, 4, 6\}$, $B = \{1, 2\}$, $C = \{1, 2, 3, 4\}$. Find:

(a) $P(A$ and $B)$ and $P(A$ or $C)$,   (c) $P(A|C)$ and $P(C|A)$

(b) $P(A|B)$ and $P(B|A)$ (d) $P(B|C)$ and $P(C|B)$

Decide whether the following are independent: (i) A and B; (ii) A and C; (iii) B and C.

7.57. A pair of fair dice is tossed. If the numbers appearing are different, find the probability that: (a) the sum is even; (b) the sum exceeds nine.

7.58. Let A and B be events with $P(A) = 0.6$, $P(B) = 0.3$, and $P(A \cap B) = 0.2$. Find:

(a) $P(A \cup B)$;   (b) $P(A|B)$;   (c) $P(B|A)$.

7.59. Let A and B be events with $P(A) = 1/3$, $P(B) = \frac{1}{4}$, and $P(A \cup B) = \frac{1}{2}$.

(a) Find $P(A|B)$ and $P(B|A)$.   (b) Are A and B independent?

7.60. Let A and B be events with $P(A) = 0.3$, $P(A \cup B) = 0.5$, and $P(B) = p$. Find p if:

(a) A and B are mutually disjoint;   (b) A and B are independent;   (c) A is a subset of B.

7.61. Let A and B be independent events with $P(A) = 0.3$ and $P(B) = 0.4$. Find:

(a) $P(A \cap B)$ and $P(A \cup B)$;   (b)$P(A|B)$ and $P(B|A)$.

7.62. In a country club, 60% of the women play tennis, 40% play golf, and 20% play both tennis and golf. A woman is chosen at random.

(a) Find the probability that she plays neither tennis nor golf.

(b) If she plays tennis, find the probability that she plays golf.

(c) If she plays golf, find the probability that she plays tennis.

7.63. Box A contains six red marbles and two blue marbles, and box B contains two red and four blue. A marble is drawn at random from each box.

(a) Find the probability p that both marbles are red.

(b) Find the probability p that one is red and one is blue.

7.64. The probability that A hits a target is $\frac{1}{4}$ and the probability that B hits a target is $\frac{1}{3}$.

(a) If each fires twice, what is the probability that the target will be hit at least once?

(b) If each fires once and the target is hit only once, what is the probability that A hits the target?

7.65. Three fair coins are tossed. Consider the events:

$A = \{$all heads or all tails$\}$, $B = \{$at least two heads$\}$, $C = \{$at most two heads$\}$.

Of the pairs (A, B), (A, C), and (B, C), which are independent? Which are dependent?

7.66. Find $P(B|A)$ if: (a) A is a subset of B; (b) A and B are mutually exclusive. (Assume $P(A) > 0$.)

REPEATED TRIALS, BINOMIAL DISTRIBUTION

7.67. Whenever horses a, b, and c race together, their respective probabilities of winning are 0.3, 0.5, and 0.2. They race three times.

(a) Find the probability that the same horse wins all three races.

(b) Find the probability that a, b, c each win one race.

7.68. The batting average of a baseball player is 0.300. He comes to bat four times. Find the probability that he will get: (a) exactly two hits; (b) at least one hit.

7.69. The probability that Tom scores on a three-point basketball shot is $p = 0.4$. He shoots $n = 5$ times. Find the probability that he scores: (a) exactly two times; (b) at least once.

7.70. A certain type of missile hits its target with probability $P = \frac{1}{3}$

(a) If three missiles are fired, find the probability that the target is hit at least once.

(b) Find the number of missiles that should be fired so that there is at least a 90% probability of hitting the target.

RANDOM VARIABLES

7.71. A pair of dice is thrown. Let X denote the minimum of the two numbers which occur. Find the distributions and expectation of X.

7.72. A fair coin is tossed four times. Let X denote the longest string of heads. Find the distribution and expectation of X.

7.73. A fair coin is tossed until a head or five tails occurs. Find the expected number E of tosses of the coin.

7.74. A coin is weighted so that $P(H) = \frac{3}{4}$ and $P(T) = \frac{1}{4}$. The coin is tossed three times. Let X denote the number of heads that appear.

(a) Find the distribution f of X.   (b) Find the expectation $E(X)$.

7.75. The probability of team A winning any game is $\frac{1}{2}$. Suppose A plays B in a tournament. The first team to win two games in a row or three games wins the tournament. Find the expected number of games in the tournament.

7.76. A box contains 10 transistors of which two are defective. A transistor is selected from the box and tested until a nondefective one is chosen. Find the expected number of transistors to be chosen.

7.77. A lottery with 500 tickets gives one prize of $100, three prizes of $50 each, and five prizes of $25 each. (a) Find the expected winnings of a ticket. (b) If a ticket costs $1, what is the expected value of the game?

7.78. A player tosses three fair coins. He wins $5 if three heads occur, $3 if two heads occur, and $1 if only one head occurs. On the other hand, he loses $15 if three tails occur. Find the value of the game to the player.

MEAN, VARIANCE, AND STANDARD DEVIATION

7.79. Find the mean μ, variance σ^2, and standard deviation σ of each distribution:

(a)

x	2	3	8
$f(x)$	$\frac{1}{4}$	$\frac{1}{2}$	$\frac{1}{4}$

(b)

y	-1	0	1	2	3
$g(y)$	0.3	0.1	0.1	0.3	0.2

7.80. Find the mean μ, variance σ^2, and standard deviation σ of the following two-point distribution where $p + q = 1$:

x	a	b
$f(x)$	p	q

7.81. Let $W = XY$ where X and Y are the random variables in Problem 7.33. (Recall $W(s) = (XY)(s) = X(s)Y(s)$.) Find: (a) the distribution h of W; (b) find $E(W)$.

Does $E(W) = E(X)E(Y)$?

7.82. Let X be a random variable with the distribution:

x	-1	1	2
$f(x)$	0.2	0.5	0.3

(a) Find the mean, variance, and standard deviation of X.

(b) Find the distribution, mean, variance, and standard deviation of Y where:

(i) $Y = X^4$; (ii) $Y = 3^X$.

BINOMIAL DISTRIBUTION

7.83. The probability that a women hits a target is $p = 1/3$. She fires 50 times. Find the expected number μ of times she will hit the target and the standard deviation σ.

7.84. Team A has probability $p = 0.8$ of winning each time it plays. Let X denote the number of times A will win in $n = 100$ games. Find the mean μ, variance σ^2, and standard deviation σ of X.

7.85. An unprepared student takes a five-question true–false quiz and guesses every answer. Find the probability that the student will pass the quiz if at least four correct answers is the passing grade.

7.86. Let X be a binomially distributed random variable $B(n, p)$ with $E(X) = 2$ and $Var(X) = \frac{4}{3}$. Find n and p.

CHEBYSHEV'S INEQUALITY

7.87. Let X be a random variable with mean μ and standard deviation σ.
Use Chebyshev's Inequality to estimate $P(\mu - 3\sigma \leq X \leq \mu + 3\sigma)$.

7.88. Let Z be the normal random variable with mean $\mu = 0$ and standard deviation $\sigma = 1$.
Use Chebyshev's Inequality to find a value b for which $P(-b \leq Z \leq b) = 0.9$.

7.89. Let X be a random variable with mean $\mu = 0$ and standard deviation $\sigma = 1.5$.
Use Chebyshev's Inequality to estimate $P(-3 \leq X \leq 3)$.

7.90. Let X be a random variable with mean $\mu = 70$.
For what value of σ will Chebyshev's Inequality give $P(65 \leq X \leq 75) \geq 0.95$.

Answers to Supplementary Problems

The notation $[x_1, \ldots, x_n; f(x_1), \ldots, f(x_n)]$ will be used for the distribution $f = \{(x_i, f(X_i)\}$.

7.42. (a) $A \cap B \cap C^C$;   (c) $(A \cup B \cup B)^C = A^C \cap B^C C^C$;
(b) $(A \cup C) \cap B^C$;   (d) $(A \cap B) \cup (A \cap C) \cup (B \cap C)$.

7.43. (a) $n(S) = 24$; $S = \{H, T\} \times \{H, T\} \times \{1, 2, \ldots, 6\}$
(b) $A = \{HH2, HH4, HH6\}$; $B = \{HH2, HT2, TH2, TT2\}$; $C = \{HT1, HT3, HT5, TH1, TH3, TH5\}$
(c) (i) $HH2$; (ii) $HT2, TH2, TT2$; (iii) $\varnothing$.

7.44. (a) 3/6; (b) 15/16; (c) 20/36.

7.45. (a) 40/50; (b) 10/50; (c) 8/50; (d) 42/50.

7.46. (a) 1/15; (b) 7/15; (c) 8/15; (d) 7/15.

7.47. (a) 3/10; (b) 3/10; (c) 1/15; (d) 8/15.

7.48. 3/5.

7.49. 1/5.

7.50. (a) 3/4; (b) 1/4; (c) 1/3; (d) 7/12.

7.51. (c) and (d).

7.52. $P(H) = 3/4$; $P(T) = 1/4$.

7.53. (a) 2/5; (b) 1/5; (c) 3/5.

7.54. (a) 0.6, 0.8, 0.5; (b) 0.5, 0.7, 0.4.

7.55. (a) 0.3; (b) 0.8; (c) 0.3; (d) 0.2.

7.56. (a) 1/6, 5/6; (b) 1/2, 1/3; (c) 1/2, 2/3; (d) 1/2,
(i) Yes; (ii) yes (iii) no.

7.57. (a) 12/30; (b) 4/30.

7.58. (a) 0.7; (b) 2/3; (c) 1/3.

7.59. (a) 1/3, 1/4; (b) yes.

7.60. (a) 0.2; (b) 2/7; (c) 0.5.

7.61. (a) 0.12, 0.58; (b) 3/10, 4/10.

7.62. (a) 20%; (b) 1/3; (c) 1/2.

7.63. (a) 1/4; (b) 7/12.

7.64. (a) 3/4; (b) 1/3.

7.65. Only (A, B) are independent.

7.66. (a) 1, (b) 0.

7.67. (a) 0.16; (b) 0.18.

7.68. (a) $6(0.3)^2(0.7)^2 = 0.2646$; (b) $1 - (0.7)^4 = 0.7599$.

7.69. (a) $10(0.4)^2(0.6)^3$; (b) $1 - (0.6)^5$.

7.70. (a) $1 - (2/3)^5 = 211/243$; (b) Six times.

7.71. $[1, 2, 3, 4, 5, 6; 11/36, 9/36, 7/36, 5/36, 3/36, 1/36]$;
$E(X) = 91/36 \approx 2.5$.

7.72. $[0, 1, 2, 3, 4; 1/16, 7/16, 5/16, 2/16, 1/16]$;
$E(X) = 27/16 \approx 1.7$.

7.73. $E = 1.9$.

7.74. (a) $[0, 1, 2, 3; 1/64, 9/64, 27/64, 27/64]$;
(b) $E(X) = 2.25$.

7.75. $23/8 \approx 2.9$.

7.76. $11/9 \approx 1.2$.

7.77. (a) 0.75; (b) -0.25.

7.78. 0.25.

7.79. (a) $\mu = 4$, $\sigma^2 = 5.5$, $\sigma = 2.3$; (b) $\mu = 1$, $\sigma^2 = 2.4$, $\sigma = 1.5$.

7.80. $\mu = ap + bq$; $\sigma^2 = pq(a - b)^2$; $\sigma = |a - b|\sqrt{pq}$

7.81. (a) [2, 6, 10, 12, 24, 36; 1/6, ..., 1/6];
(b) $E(W) = 15$. No.

7.82. (a) 0.9, 1.09, 1.04; (b) (i) [1, 1, 16; 0.2, 0.5, 0.3], 5.5, 47.25, 6.87; (ii) [1/3, 3, 9; 0.2, 0.5, 0.3], 4.67, 5.21, 3.26.

7.83. $\mu = 50/3 = 16.67$; $\sigma = 10/3 = 3.33$

7.84. $\mu = 80$; $\sigma^2 = 16$; $\sigma = 4$

7.85. 6/32.

7.86. $n = 6$, $p = 1/3$

7.87. $P \geq 1 - 1/8 \approx 8.75$

7.88. $b = \sqrt{10} \approx 3.16$

7.89. $P \geq 0.75$

7.90. $\sigma = 5/\sqrt{20} \approx 1.12$

CHAPTER 8

Graph Theory

8.1 INTRODUCTION, DATA STRUCTURES

Graphs, directed graphs, trees and binary trees appear in many areas of mathematics and computer science. This and the next two chapters will cover these topics. However, in order to understand how these objects may be stored in memory and to understand algorithms on them, we need to know a little about certain data structures. We assume the reader does understand linear and two-dimensional arrays; hence we will only discuss linked lists and pointers, and stacks and queues below.

Linked Lists and Pointers

Linked lists and pointers will be introduced by means of an example. Suppose a brokerage firm maintains a file in which each record contains a customer's name and salesman; say the file contains the following data:

Customer	Adams	Brown	Clark	Drew	Evans	Farmer	Geller	Hiller	Infeld
Salesman	Smith	Ray	Ray	Jones	Smith	Jones	Ray	Smith	Ray

There are two basic operations that one would want to perform on the data:

Operation A: Given the name of a customer, find his salesman.

Operation B: Given the name of a salesman, find the list of his customers.

We discuss a number of ways the data may be stored in the computer, and the ease with which one can perform the operations A and B on the data.

Clearly, the file could be stored in the computer by an array with two rows (or columns) of nine names. Since the customers are listed alphabetically, one could easily perform operation A. However, in order to perform operation B one must search through the entire array.

One can easily store the data in memory using a two-dimensional array where, say, the rows correspond to an alphabetical listing of the customers and the columns correspond to an alphabetical listing of the salesmen, and where there is a 1 in the matrix indicating the salesman of a customer and there are 0's elsewhere. The main drawback of such a representation is that there may be a waste of a lot of memory because many 0's may be in the matrix. For example, if a firm has 1000 customers and 20 salesmen, one would need 20 000 memory locations for the data, but only 1000 of them would be useful.

We discuss below a way of storing the data in memory which uses linked lists and pointers. By a *linked list*, we mean a linear collection of data elements, called *nodes*, where the linear order is given by means of a field of pointers. Figure 8-1 is a schematic diagram of a linked list with six nodes. That is, each node is divided into two parts: the first part contains the information of the element (e.g., NAME, ADDRESS, . . .), and the second part, called the *link field* or *nextpointer field*, contains the address of the next node in the list. This pointer field is indicated by an arrow drawn from one node to the next node in the list. There is also a variable pointer, called START in Fig. 8-1, which gives the address of the first node in the list. Furthermore, the pointer field of the last node contains an invalid address, called a *null pointer*, which indicates the end of the list.

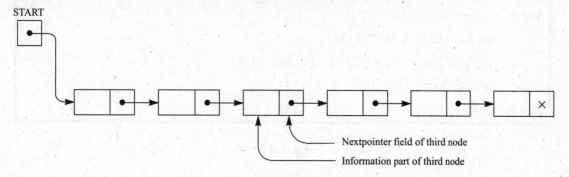

Fig. 8-1   Linked list with 6 nodes

One main way of storing the original data pictured in Fig. 8-2, uses linked lists. Observe that there are separate (sorted alphabetically) arrays for the customers and the salesmen. Also, there is a pointer array SLSM parallel to CUSTOMER which gives the location of the salesman of a customer, hence operation A can be performed very easily and quickly. Furthermore, the list of customers of each salesman is a linked list as discussed above. Specifically, there is a pointer array START parallel to SALESMAN which points to the first customer of a salesman, and there is an array NEXT which points to the location of the next customer in the salesman's list (or contains a 0 to indicate the end of the list). This process is indicated by the arrows in Fig. 8-2 for the salesman Ray.

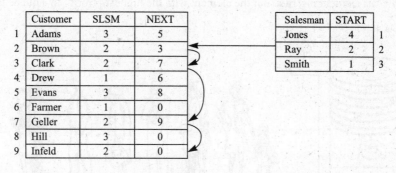

Fig. 8-2

Operation B can now be performed easily and quickly; that is, one does not need to search through the list of all customers in order to obtain the list of customers of a given salesman. Figure 8-3 gives such an algorithm (which is written in pseudocode).

Stacks, Queues, and Priority Queues

There are data structures other than arrays and linked lists which will occur in our graph algorithms. These structures, stacks, queues, and priority queues, are briefly described below.

(a) **Stack**: A *stack*, also called a *last-in first-out* (LIFO) system, is a linear list in which insertions and deletions can take place only at one end, called the "top" of the list. This structure is similar in its operation to a stack of dishes on a spring system, as pictured in Fig. 8-4(*a*). Note that new dishes are inserted only at the top of the stack and dishes can be deleted only from the top of the stack.

Algorithm 8.1   The name of a salesman is read and the list of his customers is printed.

Step 1.   Read XXX.

Step 2.   Find K such that SALESMAN[K] = XXX.   [Use binary search.]

Step 3.   Set PTR := START[K].   [Initializes pointer PTR.]

Step 4.   Repeat while PTR $\neq$ NULL.

(a) Print CUSTOMER[PTR].

(b) Set PTR := NEXT[PTR].   [Update PTR.]

[End of loop.]

Step 5.   Exit.

Fig. 8-3

(b) **Queue**: A *queue*, also called a *first-in first-out* (FIFO) system, is a linear list in which deletions can only take place at one end of the list, the "front" of the list, and insertions can only take place at the other end of the list, the "rear" of the list. The structure operates in much the same way as a line of people waiting at a bus stop, as pictured in Fig. 8-4(*b*). That is, the first person in line is the first person to board the bus, and a new person goes to the end of the line.

(c) **Priovity queue**: Let S be a set of elements where new elements may be periodically inserted, but where the current largest element (element with the "highest priority") is always deleted. Then S is called a *priority queue*. The rules "women and children first" and "age before beauty" are examples of priority queues. Stacks and ordinary queues are special kinds of priority queues. Specifically, the element with the highest priority in a stack is the last element inserted, but the element with the highest priority in a queue is the first element inserted.

(*a*)   **Stack of dishes.**   (*b*)   **Queue waiting for a bus.**

Fig. 8-4

8.2   GRAPHS AND MULTIGRAPHS

A *graph G* consists of two things:

(i)   A set $V = V(G)$ whose elements are called *vertices*, *points*, or *nodes* of G.

(ii)   A set $E = E(G)$ of unordered pairs of distinct vertices called *edges* of G.

We denote such a graph by $G(V, E)$ when we want to emphasize the two parts of G.

Vertices u and v are said to be *adjacent* or *neighbors* if there is an edge $e = \{u, v\}$. In such a case, u and v are called the *endpoints* of e, and e is said to *connect* u and v. Also, the edge e is said to be *incident* on each of its endpoints u and v. Graphs are pictured by diagrams in the plane in a natural way. Specifically, each vertex v in V is represented by a dot (or small circle), and each edge $e = \{v_1, v_2\}$ is represented by a curve which connects its endpoints v_1 and v_2 For example, Fig. 8-5(a) represents the graph $G(V, E)$ where:

(i) V consists of vertices A, B, C, D.

(ii) E consists of edges $e_1 = \{A, B\}, e_2 = \{B, C\}, e_3 = \{C, D\}, e_4 = \{A, C\}, e_5 = \{B, D\}$.

In fact, we will usually denote a graph by drawing its diagram rather than explicitly listing its vertices and edges.

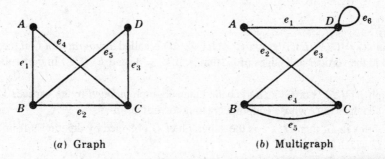

(a) Graph   (b) Multigraph

Fig. 8-5

Multigraphs

Consider the diagram in Fig. 8-5(b). The edges e_4 and e_5 are called *multiple edges* since they connect the same endpoints, and the edge e_6 is called a *loop* since its endpoints are the same vertex. Such a diagram is called a *multigraph*; the formal definition of a graph permits neither multiple edges nor loops. Thus a graph may be defined to be a multigraph without multiple edges or loops.

Remark: Some texts use the term *graph* to include multigraphs and use the term *simple graph* to mean a graph without multiple edges and loops.

Degree of a Vertex

The *degree* of a vertex v in a graph G, written deg (v), is equal to the number of edges in G which contain v, that is, which are incident on v. Since each edge is counted twice in counting the degrees of the vertices of G, we have the following simple but important result.

Theorem 8.1: The sum of the degrees of the vertices of a graph G is equal to twice the number of edges in G.

Consider, for example, the graph in Fig. 8-5(a). We have

$$\deg(A) = 2, \quad \deg(B) = 3, \quad \deg(C) = 3, \quad \deg(D) = 2.$$

The sum of the degrees equals 10 which, as expected, is twice the number of edges. A vertex is said to be *even* or *odd* according as its degree is an even or an odd number. Thus A and D are even vertices whereas B and C are odd vertices.

Theorem 8.1 also holds for multigraphs where a loop is counted twice toward the degree of its endpoint. For example, in Fig. 8-5(b) we have $\deg(D) = 4$ since the edge e_6 is counted twice; hence D is an even vertex.

A vertex of degree zero is called an *isolated* vertex.

Finite Graphs, Trivial Graph

A multigraph is said to be *finite* if it has a finite number of vertices and a finite number of edges. Observe that a graph with a finite number of vertices must automatically have a finite number of edges and so must be finite. The finite graph with one vertex and no edges, i.e., a single point, is called the *trivial graph*. Unless otherwise specified, the multigraphs in this book shall be finite.

8.3   SUBGRAPHS, ISOMORPHIC AND HOMEOMORPHIC GRAPHS

This section will discuss important relationships between graphs.

Subgraphs

Consider a graph $G = G(V, E)$. A graph $H = H(V', E')$ is called a *subgraph* of G if the vertices and edges of H are contained in the vertices and edges of G, that is, if $V' \subseteq V$ and $E' \subseteq E$. In particular:

(i)   A subgraph $H(V', E')$ of $G(V, E)$ is called the subgraph *induced* by its vertices V' if its edge set E' contains all edges in G whose endpoints belong to vertices in H.

(ii) If v is a vertex in G, then $G - v$ is the subgraph of G obtained by deleting v from G and deleting all edges in G which contain v.

(iii) If e is an edge in G, then $G - e$ is the subgraph of G obtained by simply deleting the edge e from G.

Isomorphic Graphs

Graphs $G(V, E)$ and $G(V^*, E^*)$ are said to be *isomorphic* if there exists a one-to-one correspondence $f: V \rightarrow V^*$ such that $\{u, v\}$ is an edge of G if and only if $\{f(u), f(v)\}$ is an edge of G^*. Normally, we do not distinguish between isomorphic graphs (even though their diagrams may "look different"). Figure 8-6 gives ten graphs pictured as letters. We note that A and R are isomorphic graphs. Also, F and T are isomorphic graphs, K and X are isomorphic graphs and M, S, V, and Z are isomorphic graphs.

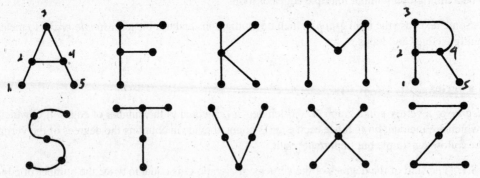

Fig. 8-6

Homeomorphic Graphs

Given any graph G, we can obtain a new graph by dividing an edge of G with additional vertices. Two graphs G and G^* are said to *homeomorphic* if they can be obtained from the same graph or isomorphic graphs by this method. The graphs (*a*) and (*b*) in Fig. 8-7 are not isomorphic, but they are homeomorphic since they can be obtained from the graph (*c*) by adding appropriate vertices.

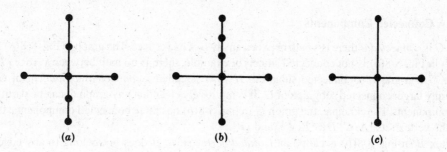

Fig. 8-7

8.4   PATHS, CONNECTIVITY

A *path* in a multigraph G consists of an alternating sequence of vertices and edges of the form

$$v_0, \quad e_1, \quad v_1, \quad e_2, \quad v_2, \quad \ldots, \quad e_{n-1}, \quad v_{n-1}, \quad e_n, \quad v_n$$

where each edge e_i contains the vertices v_{i-1} and v_i (which appear on the sides of e_i in the sequence). The number n of edges is called the *length* of the path. When there is no ambiguity, we denote a path by its sequence of vertices $(v_0, v_1, \ldots, v_n)$. The path is said to be *closed* if $v_0 = v_n$. Otherwise, we say the path is from v_0, to v_n or *between* v_0 and v_n, or *connects* v_0 to v_n.

A *simple path* is a path in which all vertices are distinct. (A path in which all edges are distinct will be called a *trail*.) A *cycle* is a closed path of length 3 or more in which all vertices are distinct except $v_0 = v_n$. A cycle of length k is called a *k-cycle*.

EXAMPLE 8.1   Consider the graph G in Fig. 8-8(a). Consider the following sequences:

$$\alpha = (P_4, P_1, P_2, P_5, P_1, P_2, P_3, P_6), \quad \beta = (P_4, P_1, P_5, P_2, P_6),$$
$$\gamma = (P_4, P_1, P_5, P_2, P_3, P_5, P_6), \quad \delta = (P_4, P_1, P_5, P_3, P_6).$$

The sequence α is a path from P_4 to P_6; but it is not a trail since the edge $\{P_1, P_2\}$ is used twice. The sequence β is not a path since there is no edge $\{P_2, P_6\}$. The sequence γ is a trail since no edge is used twice; but it is not a simple path since the vertex P_5 is used twice. The sequence δ is a simple path from P_4 to P_6; but it is not the shortest path (with respect to length) from P_4 to P_6. The shortest path from P_4 to P_6 is the simple path (P_4, P_5, P_6) which has length 2.

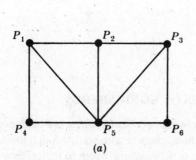

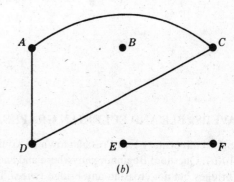

Fig. 8-8

By eliminating unnecessary edges, it is not difficult to see that any path from a vertex u to a vertex v can be replaced by a simple path from u to v. We state this result formally.

Theorem 8.2:   There is a path from a vertex u to a vertex v if and only if there exists a simple path from u to v.

Connectivity, Connected Components

A graph G is *connected* if there is a path between any two of its vertices. The graph in Fig. 8-8(a) is connected, but the graph in Fig. 8-8(b) is not connected since, for example, there is no path between vertices D and E.

Suppose G is a graph. A connected subgraph H of G is called a *connected component* of G if H is not contained in any larger connected subgraph of G. It is intuitively clear that any graph G can be partitioned into its connected components. For example, the graph G in Fig. 8-8(b) has three connected components, the subgraphs induced by the vertex sets $\{A, C, D\}$, $\{E, F\}$, and $\{B\}$.

The vertex B in Fig. 8-8(b) is called an *isolated vertex* since B does not belong to any edge or, in other words, $\deg(B) = 0$. Therefore, as noted, B itself forms a connected component of the graph.

Remark: Formally speaking, assuming any vertex u is connected to itself, the relation "u is connected to v" is an equivalence relation on the vertex set of a graph G and the equivalence classes of the relation form the connected components of G.

Distance and Diameter

Consider a connected graph G. The *distance* between vertices u and v in G, written $d(u, v)$, is the length of the shortest path between u and v. The *diameter* of G, written $\mathrm{diam}(G)$, is the maximum distance between any two points in G. For example, in Fig. 8-9(a), $d(A, F) = 2$ and $\mathrm{diam}(G) = 3$, whereas in Fig. 8-9(b), $d(A, F) = 3$ and $\mathrm{diam}(G) = 4$.

Cutpoints and Bridges

Let G be a connected graph. A vertex v in G is called a *cutpoint* if $G - v$ is disconnected. (Recall that $G - v$ is the graph obtained from G by deleting v and all edges containing v.) An edge e of G is called a *bridge* if $G - e$ is disconnected. (Recall that $G - e$ is the graph obtained from G by simply deleting the edge e). In Fig. 8-9(a), the vertex D is a cutpoint and there are no bridges. In Fig. 8-9(b), the edge $= \{D, F\}$ is a bridge. (Its endpoints D and F are necessarily cutpoints.)

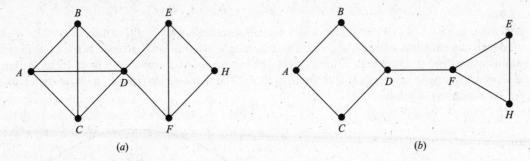

(a) (b)

Fig. 8-9

8.5   TRAVERSABLE AND EULERIAN GRAPHS, BRIDGES OF KÖNIGSBERG

The eighteenth-century East Prussian town of Königsberg included two islands and seven bridges as shown in Fig. 8-10(a). Question: Beginning anywhere and ending anywhere, can a person walk through town crossing all seven bridges but not crossing any bridge twice? The people of Königsberg wrote to the celebrated Swiss mathematician L. Euler about this question. Euler proved in 1736 that such a walk is impossible. He replaced the islands and the two sides of the river by points and the bridges by curves, obtaining Fig. 8-10(b).

Observe that Fig. 8-10(b) is a multigraph. A multigraph is said to be *traversable* if it "can be drawn without any breaks in the curve and without repeating any edges," that is, if there is a path which includes all vertices and uses each edge exactly once. Such a path must be a trail (since no edge is used twice) and will be called a *traversable trail*. Clearly a traversable multigraph must be finite and connected.

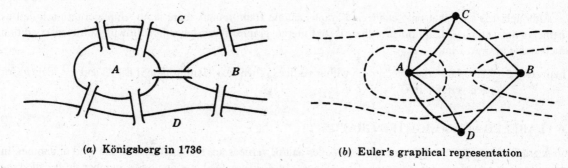

(a) Königsberg in 1736   (b) Euler's graphical representation

Fig. 8-10

We now show how Euler proved that the multigraph in Fig. 8-10(b) is not traversable and hence that the walk in Königsberg is impossible. Recall first that a vertex is even or odd according as its degree is an even or an odd number. Suppose a multigraph is traversable and that a traversable trail does not begin or end at a vertex P. We claim that P is an even vertex. For whenever the traversable trail enters P by an edge, there must always be an edge not previously used by which the trail can leave P. Thus the edges in the trail incident with P must appear in pairs, and so P is an even vertex. Therefore if a vertex Q is odd, the traversable trail must begin or end at Q. Consequently, a multigraph with more than two odd vertices cannot be traversable. Observe that the multigraph corresponding to the Königsberg bridge problem has four odd vertices. Thus one cannot walk through Königsberg so that each bridge is crossed exactly once.

Euler actually proved the converse of the above statement, which is contained in the following theorem and corollary. (The theorem is proved in Problem 8.9.) A graph G is called an *Eulerian* graph if there exists a closed traversable trail, called an *Eulerian* trail.

Theorem 8.3 (Euler): A finite connected graph is Eulerian if and only if each vertex has even degree.

Corollary 8.4: Any finite connected graph with two odd vertices is traversable. A traversable trail may begin at either odd vertex and will end at the other odd vertex.

Hamiltonian Graphs

The above discussion of Eulerian graphs emphasized traveling edges; here we concentrate on visiting vertices. A *Hamiltonian circuit* in a graph G, named after the nineteenth-century Irish mathematician William Hamilton (1803–1865), is a closed path that visits every vertex in G exactly once. (Such a closed path must be a cycle.) If G does admit a Hamiltonian circuit, then G is called a *Hamiltonian graph*. Note that an Eulerian circuit traverses every edge exactly once, but may repeat vertices, while a Hamiltonian circuit visits each vertex exactly once but may repeat edges. Figure 8-11 gives an example of a graph which is Hamiltonian but not Eulerian, and vice versa.

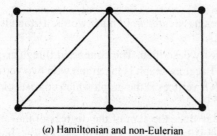

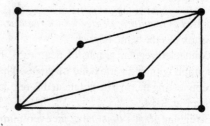

(a) Hamiltonian and non-Eulerian   (b) Eulerian and non-Hamiltonian

Fig. 8-11

Although it is clear that only connected graphs can be Hamiltonian, there is no simple criterion to tell us whether or not a graph is Hamiltonian as there is for Eulerian graphs. We do have the following sufficient condition which is due to G. A. Dirac.

Theorem 8.5: Let G be a connected graph with n vertices. Then G is Hamiltonian if $n \geq 3$ and $n \leq \deg(v)$ for each vertex v in G.

8.6 LABELED AND WEIGHTED GRAPHS

A graph G is called a *labeled graph* if its edges and/or vertices are assigned data of one kind or another. In particular, G is called a *weighted graph* if each edge e of G is assigned a nonnegative number $w(e)$ called *the weight* or *length* of v. Figure 8-12 shows a weighted graph where the weight of each edge is given in the obvious way. The *weight* (or *length*) of a path in such a weighted graph G is defined to be the sum of the weights of the edges in the path. One important problem in graph theory is to find a *shortest path*, that is, a path of minimum weight (length), between any two given vertices. The length of a shortest path between P and Q in Fig. 8-12 is 14; one such path is

$$(P, A_1, A_2, A_5, A_3, A_6, Q)$$

The reader can try to find another shortest path.

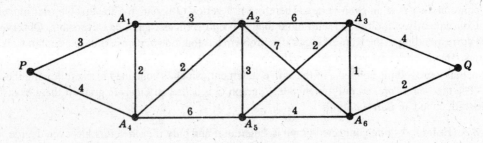

Fig. 8-12

8.7 COMPLETE, REGULAR, AND BIPARTITE GRAPHS

There are many different types of graphs. This section considers three of them: complete, regular, and bipartite graphs.

Complete Graphs

A graph G is said to be *complete* if every vertex in G is connected to every other vertex in G. Thus a complete graph G must be connected. The complete graph with n vertices is denoted by K_n. Figure 8-13 shows the graphs K_1 through K_6.

Regular Graphs

A graph G is *regular of degree k* or *k-regular* if every vertex has degree k. In other words, a graph is regular if every vertex has the same degree.

The connected regular graphs of degrees 0, 1, or 2 are easily described. The connected 0-regular graph is the trivial graph with one vertex and no edges. The connected 1-regular graph is the graph with two vertices and one edge connecting them. The connected 2-regular graph with n vertices is the graph which consists of a single n-cycle. See Fig. 8-14.

The 3-regular graphs must have an even number of vertices since the sum of the degrees of the vertices is an even number (Theorem 8.1). Figure 8-15 shows two connected 3-regular graphs with six vertices. In general, regular graphs can be quite complicated. For example, there are nineteen 3-regular graphs with ten vertices. We note that the complete graph with n vertices K_n is regular of degree $n - 1$.

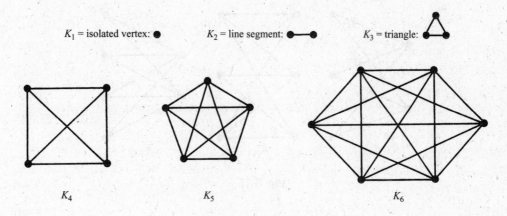

K_1 = isolated vertex: ● K_2 = line segment: ●—● K_3 = triangle: △

K_4 K_5 K_6

Fig. 8-13

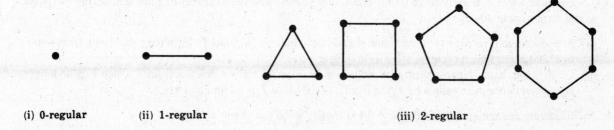

(i) 0-regular   (ii) 1-regular   (iii) 2-regular

Fig. 8-14

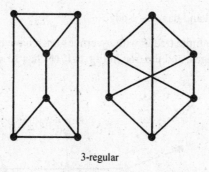

3-regular

Fig. 8-15

Bipartite Graphs

A graph G is said to be *bipartite* if its vertices V can be partitioned into two subsets M and N such that each edge of G connects a vertex of M to a vertex of N. By a complete bipartite graph, we mean that each vertex of M is connected to each vertex of N; this graph is denoted by $K_{m,n}$ where m is the number of vertices in M and n is the number of vertices in N, and, for standardization, we will assume $m \leq n$. Figure 8-16 shows the graphs $K_{2,3}$, $K_{3,3}$, and $K_{2,4}$, Clearly the graph $K_{m,n}$ has mn edges.

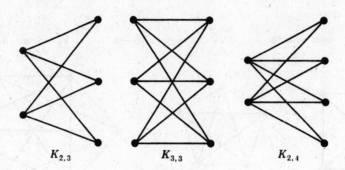

Fig. 8-16

8.8 TREE GRAPHS

A graph T is called a *tree* if T is connected and T has no cycles. Examples of trees are shown in Fig. 8-17. A *forest* G is a graph with no cycles; hence the connected components of a forest G are trees. A graph without cycles is said to be *cycle-free*. The tree consisting of a single vertex with no edges is called the *degenerate tree*.

Consider a tree T. Clearly, there is only one simple path between two vertices of T; otherwise, the two paths would form a cycle. Also:

(a) Suppose there is no edge $\{u, v\}$ in T and we add the edge $e = \{u, v\}$ to T. Then the simple path from u to v in T and e will form a cycle; hence T is no longer a tree.

(b) On the other hand, suppose there is an edge $e = \{u, v\}$ in T, and we delete e from T. Then T is no longer connected (since there cannot be a path from u to v); hence T is no longer a tree.

The following theorem (proved in Problem 8.14) applies when our graphs are finite.

Theorem 8.6: Let G be a graph with $n > 1$ vertices. Then the following are equivalent:

     (i) G is a tree.

     (ii) G is a cycle-free and has $n - 1$ edges.

     (iii) G is connected and has $n - 1$ edges.

This theorem also tells us that a finite tree T with n vertices must have $n - 1$ edges. For example, the tree in Fig. 8-17(a) has 9 vertices and 8 edges, and the tree in Fig. 8-17(b) has 13 vertices and 12 edges.

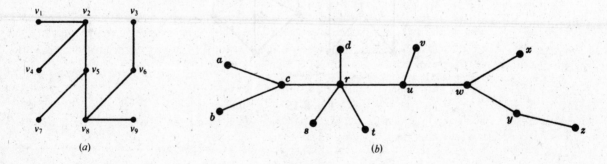

(a)   (b)

Fig. 8-17

Spanning Trees

A subgraph T of a connected graph G is called a *spanning tree* of G if T is a tree and T includes all the vertices of G. Figure 8-18 shows a connected graph G and spanning trees T_1, T_2, and T_3 of G.

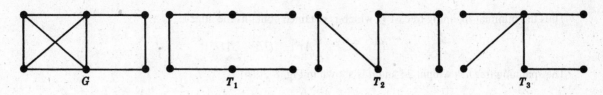

Fig. 8-18

Minimum Spanning Trees

Suppose G is a connected weighted graph. That is, each edge of G is assigned a nonnegative number called the *weight* of the edge. Then any spanning tree T of G is assigned a total weight obtained by adding the weights of the edges in T. A *minimal spanning tree* of G is a spanning tree whose total weight is as small as possible.

Algorithms 8.2 and 8.3, which appear in Fig. 8-19, enable us to find a minimal spanning tree T of a connected weighted graph G where G has n vertices. (In which case T must have $n - 1$ vertices.)

Algorithm 8.2:   The input is a connected weighted graph G with n vertices.

Step 1.   Arrange the edges of G in the order of decreasing weights.

Step 2.   Proceeding sequentially, delete each edge that does not disconnect the graph until $n - 1$ edges remain.

Step 3.   Exit.

Algorithm 8.3 (Kruskal):   The input is a connected weighted graph G with n vertices.

Step 1.   Arrange the edges of G in order of increasing weights.

Step 2.   Starting only with the vertices of G and proceeding sequentially, add each edge which does not result in a cycle until $n - 1$ edges are added.

Step 3.   Exit.

Fig. 8-19

The weight of a minimal spanning tree is unique, but the minimal spanning tree itself is not. Different minimal spanning trees can occur when two or more edges have the same weight. In such a case, the arrangement of the edges in Step 1 of Algorithms 8.2. or 8.3 is not unique and hence may result in different minimal spanning trees as illustrated in the following example.

EXAMPLE 8.2 Find a minimal spanning tree of the weighted graph Q in Fig. 8-20(a). Note that Q has six vertices, so a minimal spanning tree will have five edges.

(a) Here we apply Algorithm 8.2.

First we order the edges by decreasing weights, and then we successively delete edges without disconnecting Q until five edges remain. This yields the following data:

Edges	BC	AF	AC	BE	CE	BF	AE	DF	BD
Weight	8	7	7	7	6	5	4	4	3
Delete	Yes	Yes	Yes	No	No	Yes			

Thus the minimal spanning tree of Q which is obtained contains the edges

$$BE, \quad CE, \quad AE, \quad DF, \quad BD$$

The spanning tree has weight 24 and it is shown in Fig. 8-20(b).

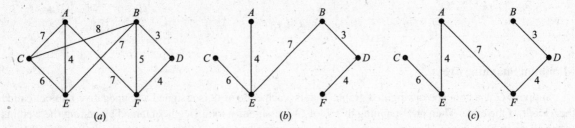

Fig. 8-20

(b) Here we apply Algorithm 8.3.

First we order the edges by increasing weights, and then we successively add edges without forming any cycles until five edges are included. This yields the following data:

Edges	BD	AE	DF	BF	CE	AC	AF	BE	BC
Weight	3	4	4	5	6	7	7	7	8
Add?	Yes	Yes	Yes	No	Yes	No	Yes		

Thus the minimal spanning tree of Q which is obtained contains the edges

$$BD, \quad AE, \quad DF, \quad CE, \quad AF$$

The spanning tree appears in Fig. 8-20(c). Observe that this spanning tree is not the same as the one obtained using Algorithm 8.2 as expected it also has weight 24.

Remark: The above algorithms are easily executed when the graph G is relatively small as in Fig. 8-20(a). Suppose G has dozens of vertices and hundreds of edges which, say, are given by a list of pairs of vertices. Then even deciding whether G is connected is not obvious; it may require some type of depth-first search (DFS) or breadth-first search (BFS) graph algorithm. Later sections and the next chapter will discuss ways of representing graphs G in memory and will discuss various graph algorithms.

8.9   PLANAR GRAPHS

A graph or multigraph which can be drawn in the plane so that its edges do not cross is said to be *planar*. Although the complete graph with four vertices K_4 is usually pictured with crossing edges as in Fig. 8-21(a), it can also be drawn with noncrossing edges as in Fig. 8-21(b); hence K_4 is planar. Tree graphs form an important class of planar graphs. This section introduces our reader to these important graphs.

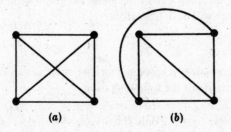

Fig. 8-21

Maps, Regions

A particular planar representation of a finite planar multigraph is called a *map*. We say that the map is *connected* if the underlying multigraph is connected. A given map divides the plane into various regions. For example, the map in Fig. 8-22 with six vertices and nine edges divides the plane into five regions. Observe that four of the regions are bounded, but the fifth region, outside the diagram, is unbounded. Thus there is no loss in generality in counting the number of regions if we assume that our map is contained in some large rectangle rather than in the entire plane.

Observe that the border of each region of a map consists of edges. Sometimes the edges will form a cycle, but sometimes not. For example, in Fig. 8-22 the borders of all the regions are cycles except for r_3. However, if we do move counterclockwise around r_3 starting, say, at the vertex C, then we obtain the closed path

$$(C, D, E, F, E, C)$$

where the edge $\{E, F\}$ occurs twice. By the *degree* of a region r, written $\deg(r)$, we mean the length of the cycle or closed walk which borders r. We note that each edge either borders two regions or is contained in a region and will occur twice in any walk along the border of the region. Thus we have a theorem for regions which is analogous to Theorem 8.1 for vertices.

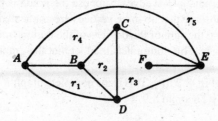

Fig. 8-22

Theorem 8.7: The sum of the degrees of the regions of a map is equal to twice the number of edges.

The degrees of the regions of Fig. 8-22 are:

$$\deg(r_1) = 3, \quad \deg(r_2) = 3, \quad \deg(r_3) = 5, \quad \deg(r_4) = 4, \quad \deg(r_5) = 3$$

The sum of the degrees is 18, which, as expected, is twice the number of edges.

For notational convenience we shall picture the vertices of a map with dots or small circles, or we shall assume that any intersections of lines or curves in the plane are vertices.

Euler's Formula

Euler gave a formula which connects the number V of vertices, the number E of edges, and the number R of regions of any connected map. Specifically:

Theorem 8.8 (Euler): $V - E + R = 2$.

(The proof of Theorem 8.8 appears in Problem 8.18.)
Observe that, in Fig. 8-22, $V = 6$, $E = 9$, and $R = 5$; and, as expected by Euler's formula.

$$V - E + R = 6 - 9 + 5 = 2$$

We emphasize that the underlying graph of a map must be connected in order for Euler's formula to hold.

Let G be a connected planar multigraph with three or more vertices, so G is neither K_1 nor K_2. Let M be a planar representation of G. It is not difficult to see that (1) a region of M can have degree 1 only if its border is a loop, and (2) a region of M can have degree 2 only if its border consists of two multiple edges. Accordingly, if G is a graph, not a multigraph, then every region of M must have degree 3 or more. This comment together with Euler's formula is used to prove the following result on planar graphs.

Theorem 8.9: Let G be a connected planar graph with p vertices and q edges, where $p \geq 3$. Then $q \geq 3p - 6$.

Note that the theorem is not true for K_1 where $p = 1$ and $q = 0$, and is not true for K_2 where $p = 2$ and $q - 1$.

Proof: Let r be the number of regions in a planar representation of G. By Euler's formula, $p - q + r = 2$.

Now the sum of the degrees of the regions equals $2q$ by Theorem 8.7. But each region has degree 3 or more; hence $2q \geq 3r$. Thus $r \geq 2q/3$. Substituting this in Euler's formula gives

$$2 = p - q + r \leq p - q + \frac{2q}{3} \quad \text{or} \quad 2 \leq p - \frac{q}{3}$$

Multiplying the inequality by 3 gives $6 \leq 3p - q$ which gives us our result.   □

Nonplanar Graphs, Kuratowski's Theorem

We give two examples of nonplanar graphs. Consider first the *utility graph*; that is, three houses A_1, A_2, A_3 are to be connected to outlets for water, gas and electricity, B_1, B_2, B_3, as in Fig. 8-23(a). Observe that this is the graph $K_{3,3}$ and it has $p = 6$ vertices and $q = 9$ edges. Suppose the graph is planar. By Euler's formula a planar representation has $r = 5$ regions. Observe that no three vertices are connected to each other; hence the degree of each region must be 4 or more and so the sum of the degrees of the regions must be 20 or more. By Theorem 8.7 the graph must have 10 or more edges. This contradicts the fact that the graph has $q = 9$ edges. Thus the utility graph $K_{3,3}$ is nonplanar.

Consider next the *star graph* in Fig. 8-23(b). This is the complete graph K_5 on $p = 5$ vertices and has $q = 10$ edges. If the graph is planar, then by Theorem 8.9.

$$10 = q \leq 3p - 6 = 15 - 6 = 9$$

which is impossible. Thus K_5 is nonplanar.

For many years mathematicians tried to characterize planar and nonplanar graphs. This problem was finally solved in 1930 by the Polish mathematician K. Kuratowski. The proof of this result, stated below, lies beyond the scope of this text.

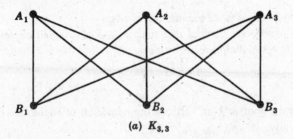
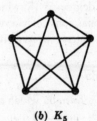

(a) $K_{3,3}$ (b) K_5

Fig. 8-23

Theorem 8.10: **(Kuratowski)** A graph is nonplanar if and only if it contains a subgraph homeomorphic to $K_{3,3}$ or K_5.

8.10 GRAPH COLORINGS

Consider a graph G. *A vertex coloring*, or simply a *coloring* of G is an assignment of colors to the vertices of G such that adjacent vertices have different colors. We say that G is n-colorable if there exists a coloring of G which uses n colors. (Since the word "color" is used as a noun, we will try to avoid its use as a verb by saying, for example, "paint" G rather than "color" G when we are assigning colors to the vertices of G.) The minimum number of colors needed to paint G is called the *chromatic number* of G and is denoted by $\chi(G)$.

Fig. 8-24 gives an algorithm by Welch and Powell for a coloring of a graph G. We emphasize that this algorithm does not always yield a minimal coloring of G.

Algorithm 8.4 (Welch-Powell):   The input is a graph G.

Step 1.   Order the vertices of G according to decreasing degrees.

Step 2.   Assign the first color C_1 to the first vertex and then, in sequential order, assign C_1 to each vertex which is not adjacent to a previous vertex which was assigned C_1.

Step 3.   Repeat Step 2 with a second color C_2 and the subsequence of noncolored vertices.

Step 4.   Repeat Step 3 with a third color C_3, then a fourth color C_4, and so on until all vertices are colored.

Step 5.   Exit.

Fig. 8-24

EXAMPLE 8.3

(a) Consider the graph G in Fig. 8-25. We use the Welch-Powell Algorithm 8.4 to obtain a coloring of G. Ordering the vertices according to decreasing degrees yields the following sequence:

$$A_5, \quad A_3, \quad A_7, \quad A_1, \quad A_2, \quad A_4, \quad A_6, \quad A_8$$

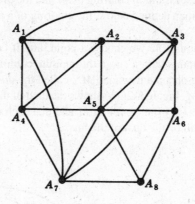

Fig. 8-25

The first color is assigned to vertices A_5 and A_1. The second color is assigned to vertices A_3, A_4, and A_8. The third color is assigned to vertices A_7, A_2, and A_6. All the vertices have been assigned a color, and so G is 3-colorable. Observe that G is not 2-colorable since vertices A_1, A_2, and A_3, which are connected to each other, must be assigned different colors. Accordingly, $\chi(G) = 3$.

(b) Consider the complete graph K_n with n vertices. Since every vertex is adjacent to every other vertex, K_n requires n colors in any coloring. Thus $\chi(K_n) = n$.

There is no simple way to actually determine whether an arbitrary graph is n-colorable. However, the following theorem (proved in Problem 8.19) gives a simple characterization of 2-colorable graphs.

Theorem 8.11: The following are equivalent for a graph G:

        (i) G is 2-colorable.

        (ii) G is bipartite.

        (iii) Every cycle of G has even length.

There is no limit on the number of colors that may be required for a coloring of an arbitrary graph since, for example, the complete graph K_n requires n colors. However, if we restrict ourselves to planar graphs, regardless of the number of vertices, five colors suffice. Specifically, in Problem 8.20 we prove:

Theorem 8.12: Any planar graph is 5-colorable.

Actually, since the 1850s mathematicians have conjectured that planar graphs are 4-colorable since every known planar graph is 4-colorable. Kenneth Appel and Wolfgang Haken finally proved this conjecture to be true in 1976. That is:

Four Color Theorem (Appel and Haken): Any planar graph is 4-colorable.

We discuss this theorem in the next subsection.

Dual Maps and the Four Color Theorem

Consider a map M, say the map M in Fig. 8-26(a). In other words, M is a planar representation of a planar multigraph. Two regions of M are said to be *adjacent* if they have an edge in common. Thus the regions r_2 and r_5 in Fig. 8-26(a) are adjacent, but the regions r_3 and r_5 are not. By a *coloring* of M we mean an assignment of a color to each region of M such that adjacent regions have different colors. A map M is *n-colorable* if there exists a coloring of M which uses n colors. Thus the map M in Fig. 8-26(a) is 3-colorable since the regions can be assigned the following colors:

$$r_1 \text{ red}, \quad r_2 \text{ white}, \quad r_3 \text{ red}, \quad r_4 \text{ white}, \quad r_5 \text{ red}, \quad r_6 \text{ blue}$$

Observe the similarity between this discussion on coloring maps and the previous discussion on coloring graphs. In fact, using the concept of the dual map defined below, the coloring of a map can be shown to be equivalent to the vertex coloring of a planar graph.

Consider a map M. In each region of M we choose a point, and if two regions have an edge in common then we connect the corresponding points with a curve through the common edge. These curves can be drawn so that they are noncrossing. Thus we obtain a new map M^*, called the *dual* of M, such that each vertex of M^* corresponds to exactly one region of M. Figure 8-26(b) shows the dual of the map of Fig. 8-26(a). One can prove that each region of M^* will contain exactly one vertex of M and that each edge of M^* will intersect exactly one edge of M and vice versa. Thus M will be the dual of the map M^*.

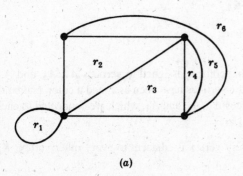

(a)

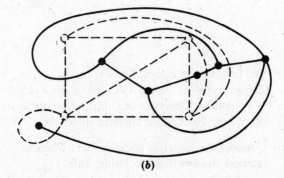

(b)

Fig. 8-26

Observe that any coloring of the regions of a map M will correspond to a coloring of the vertices of the dual map M^*. Thus M is n-colorable if and only if the planar graph of the dual map M^* is vertex n-colorable. Thus the above theorem can be restated as follows:

Four Color Theorem (Appel and Haken): If the regions of any map M are colored so that adjacent regions have different colors, then no more than four colors are required.

The proof of the above theorem uses computers in an essential way. Specifically, Appel and Haken first showed that if the four color theorem was false, then there must be a counterexample among one of approximately 2000 different types of planar graphs. They then showed, using the computer, that none of these types of graphs has such a counterexample. The examination of each different type of graph seems to be beyond the grasp of human beings without the use of a computer. Thus the proof, unlike most proofs in mathematics, is technology dependent; that is, it depended on the development of high-speed computers.

8.11   REPRESENTING GRAPHS IN COMPUTER MEMORY

There are two standard ways of maintaining a graph G in the memory of a computer. One way, called the *sequential representation* of G, is by means of its adjacency matrix A. The other way, called the *linked representation* or *adjacency structure* of G, uses linked lists of neighbors. Matrices are usually used when the graph G is dense, and linked lists are usually used when G is sparse. (A graph G with m vertices and n edges is said to be *dense* when $m = O(n^2)$ and *sparse* when $m = O(n)$ or even $O(n \log n)$.)

Regardless of the way one maintains a graph G in memory, the graph G is normally input into the computer by its formal definition, that is, as a collection of vertices and a collection of pairs of vertices (edges).

Adjacency Matrix

Suppose G is a graph with m vertices, and suppose the vertices have been ordered, say, $v_1, v_2, \ldots, v_m$. Then the *adjacency matrix* $A = [a_{ij}]$ of the graph G is the $m \times m$ matrix defined by

$$a_{ij} = \begin{cases} 1 & \text{if } v_i \text{ is adjacent to } v_j \\ 0 & \text{otherwise} \end{cases}$$

Figure 8-27(b) contains the adjacency matrix of the graph G in Fig. 8-27(a) where the vertices are ordered A, B, C, D, E. Observe that each edge $\{v_i, v_j\}$ of G is represented twice, by $a_{ij} = 1$ and $a_{ji} = 1$. Thus, in particular, the adjacency matrix is symmetric.

The adjacency matrix A of a graph G does depend on the ordering of the vertices of G, that is, a different ordering of the vertices yields a different adjacency matrix. However, any two such adjacency matrices are closely related in that one can be obtained from the other by simply interchanging rows and columns. On the other hand, the adjacency matrix does not depend on the order in which the edges (pairs of vertics) are input into the computer.

There are variations of the above representation. If G is a multigraph, then we usually let a_{ij} denote the number of edges $\{v_i, v_j\}$. Moreover, if G is a weighted graph, then we may let a_{ij} denote the weight of the edge $\{v_i, v_j\}$.

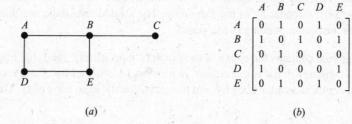

$$
\begin{array}{c@{\quad}c}
& \begin{array}{ccccc} A & B & C & D & E \end{array} \\
\begin{array}{c} A \\ B \\ C \\ D \\ E \end{array} &
\left[\begin{array}{ccccc}
0 & 1 & 0 & 1 & 0 \\
1 & 0 & 1 & 0 & 1 \\
0 & 1 & 0 & 0 & 0 \\
1 & 0 & 0 & 0 & 1 \\
0 & 1 & 0 & 1 & 0
\end{array}\right]
\end{array}
$$

(a)   (b)

Fig. 8-27

Linked Representation of a Graph *G*

Let *G* be a graph with *m* vertices. The representation of *G* in memory by its adjacency matrix *A* has a number of major drawbacks. First of all it may be difficult to insert or delete vertices in *G*. The reason is that the size of *A* may need to be changed and the vertices may need to be reordered, so there may be many, many changes in the matrix *A*. Furthermore, suppose the number of edges is $O(m)$ or even $O(m \log m)$, that is, suppose *G* is sparse. Then the matrix *A* will contain many zeros; hence a great deal of memory space will be wasted. Accordingly, when *G* is sparse, *G* is usually represented in memory by some type of *linked representation*, also called an *adjacency structure*, which is described below by means of an example.

Consider the graph *G* in Fig. 8-28(*a*). Observe that *G* may be equivalently defined by the table in Fig. 8-28(*b*) which shows each vertex in *G* followed by its *adjacency list*, i.e., its list of adjacent vertices (*neighbors*). Here the symbol $\varnothing$ denotes an empty list. This table may also be presented in the compact form

$$G = [A{:}B, D; \quad B{:}A, C, D; \quad C{:}B; \quad D{:}A, B; \quad E{:}\varnothing]$$

where a colon ":" separates a vertex from its list of neighbors, and a semicolon ";" separates the different lists.

Remark: Observe that each edge of a graph *G* is represented twice in an adjacency structure; that is, any edge, say {*A*, *B*}, is represented by *B* in the adjacency list of *A*, and also by *A* in the adjacency list of *B*. The graph *G* in Fig. 8-28(*a*) has four edges, and so there must be 8 vertices in the adjacency lists. On the other hand, each vertex in an adjacency list corresponds to a unique edge in the graph *G*.

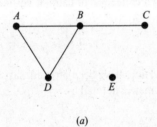

Vertex	Adjacency list
A	B, D
B	A, C, D
C	B
D	A, B
E	$\varnothing$

(*a*)   (*b*)

Fig. 8-28

The *linked representation* of a graph *G*, which maintains *G* in memory by using its adjacency lists, will normally contain two files (or sets of records), one called the Vertex File and the other called the Edge File, as follows.

(a) *Vertex File:* The Vertex File will contain the list of vertices of the graph *G* usually maintained by an array or by a linked list. Each record of the Vertex File will have the form

VERTEX	NEXT-V	PTR	

Here VERTEX will be the name of the vertex, NEXT-V points to the next vertex in the list of vertices in the Vertex File when the vertices are maintained by a linked list, and **PTR** will point to the first element in the adjacency list of the vertex appearing in the Edge File. The shaded area indicates that there may be other information in the record corresponding to the vertex.

(b) *Edge File:* The Edge File contains the edges of the graph *G*. Specifically, the Edge File will contain all the adjacency lists of *G* where each list is maintained in memory by a linked list. Each record of the Edge File will correspond to a vertex in an adjacency list and hence, indirectly, to an edge of *G*. The record will usually have the form

EDGE	ADJ	NEXT	

Here:

(1) EDGE will be the name of the edge (if it has one).

(2) ADJ points to the location of the vertex in the Vertex File.

(3) NEXT points to the location of the next vertex in the adjacency list.

We emphasize that each edge is represented twice in the Edge File, but each record of the file corresponds to a unique edge. The shaded area indicates that there may be other information in the record corresponding to the edge.

Figure 8-29 shows how the graph G in Fig. 8-28(a) may appear in memory. Here the vertices of G are maintained in memory by a linked list using the variable START to point to the first vertex. (Alternatively, one could use a linear array for the list of vertices, and then NEXT-V would not be required.) Note that the field EDGE is not needed here since the edges have no name. Figure 8-29 also shows, with the arrows, the adjacency list $[D, C, A]$ of the vertex B.

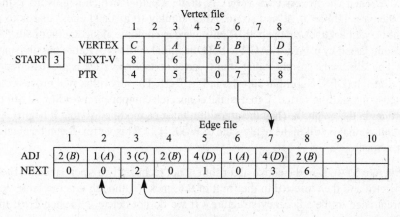

Fig. 8-29

8.12   GRAPH ALGORITHMS

This section discusses two important graph algorithms which systematically examine the vertices and edges of a graph G. One is called a *depth-first search* (DFS) and the other is called a *breadth-first search* (BFS). Other graph algorithms will be discussed in the next chapter in connection with directed graphs. Any particular graph algorithm may depend on the way G is maintained in memory. Here we assume G is maintained in memory by its adjacency structure. Our test graph G with its adjacency structure appears in Fig. 8-30 where we assume the vertices are ordered alphabetically.

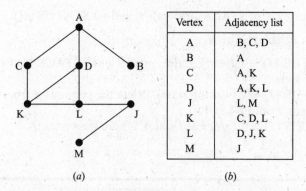

Vertex	Adjacency list
A	B, C, D
B	A
C	A, K
D	A, K, L
J	L, M
K	C, D, L
L	D, J, K
M	J

(a)   (b)

Fig. 8-30

During the execution of our algorithms, each vertex (node) N of G will be in one of three states, called the *status* of N, as follows:

STATUS = 1: (Ready state) The initial state of the vertex N.

STATUS = 2: (Waiting state) The vertex N is on a (waiting) list, waiting to be processed.

STATUS = 3: (Processed state) The vertex N has been processed.

The waiting list for the depth-first seach (DFS) will be a (modified) STACK (which we write horizontally with the top of STACK on the left), whereas the waiting list for the breadth-first search (BFS) will be a QUEUE.

Depth-first Search

The general idea behind a depth-first search beginning at a starting vertex A is as follows. First we process the starting vertex A. Then we process each vertex N along a path P which begins at A; that is, we process a neighbor of A, then a neighbor of A, and so on. After coming to a "dead end," that is to a vertex with no unprocessed neighbor, we backtrack on the path P until we can continue along another path P'. And so on. The backtracking is accomplished by using a STACK to hold the initial vertices of future possible paths. We also need a field STATUS which tells us the current status of any vertex so that no vertex is processed more than once.

The depth-first search (DFS) algorithm appears in Fig. 8-31. The algorithm will process only those vertices which are connected to the starting vertex A, that is, the connected component including A. Suppose one wants to process all the vertices in the graph G. Then the algorithm must be modified so that it begins again with another vertex (which we call B) that is still in the ready state (STATE = 1). This vertex B can be obtained by traversing through the list of vertices.

Remark: The structure STACK in the above algorithm is not technically a stack since, in Step 5(b), we allow a vertex J to be deleted and then inserted in the front of the stack. (Although it is the same vertex J, it usually represents a different edge in the adjacency structure.) If we do not delete J in Step 5(b), then we obtain an alternate form of DFS.

Algorithm 8.5 (Depth-first Search):   This algorithm executes a depth-first search on a graph G beginning with a starting vertex A.

Step 1.   Initialize all vertices to the ready state (STATUS = 1).

Step 2.   Push the starting vertex A onto STACK and change the status of A to the waiting state (STATUS = 2).

Step 3.   Repeat Steps 4 and 5 until STACK is empty.

Step 4.   Pop the top vertex N of STACK. Process N, and set STATUS (N) = 3, the processed state.

Step 5.   Examine each neighbor J of N.

(a)   If STATUS (J) = 1 (ready state), push J onto STACK and reset STATUS (J) = 2 (waiting state).

(b)   If STATUS (J) = 2 (waiting state), delete the previous J from the STACK and push the current J onto STACK.

(c)   If STATUS (J) = 3 (processed state), ignore the vertex J.

[End of Step 3 loop.]

Step 6.   Exit.

Fig. 8-31

EXAMPLE 8.4 Suppose the DFS Algorithm 8.5 in Fig. 8-31 is applied to the graph in Fig. 8-30. The vertices will be processed in the following order:

$$A, \quad D, \quad L, \quad K, \quad C, \quad J, \quad M, \quad B$$

Specifically, Fig. 8-32(a) shows the sequence of vertices being processed and the sequence of waiting lists in STACK. (Note that after vertex A is processed, its neighbors, B, C, and D are added to STACK in the order first B, then C, and finally D; hence D is on the top of the STACK and D is the next vertex to be processed.) Each vertex, excluding A, comes from an adjacency list and hence corresponds to an edge of the graph. These edges form a spanning tree of G which is pictured in Fig. 8-32(b). The numbers indicate the order that the edges are added to the spanning tree, and the dashed lines indicate backtracking.

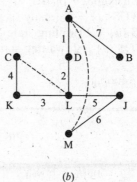

STACK	Vertex
A	A
D, C, B	D
L, K, C, B	L
K, J, K̸, C, B	K
C, J, C̸, B	C
J, B	J
M, B	M
B	B
∅	

(a)   (b)

Fig. 8-32

Breadth-first Search

The general idea behind a breadth-first search beginning at a starting vertex A is as follows. First we process the starting vertex A. Then we process all the neighbors of A. Then we process all the neighbors of neighbors of A. And so on. Naturally we need to keep track of the neighbors of a vertex, and we need to guarantee that no vertex is processed twice. This is accomplished by using a QUEUE to hold vertices that are waiting to be processed, and by a field STATUS which tells us the current status of a vertex.

The breadth-first search (BFS) algorithm appears in Fig. 8-33, Again the algorithm will process only those vertices which are connected to the starting vertex A, that is, the connected component including A. Suppose one wants to process all the vertices in the graph G. Then the algorithm must be modified so that it begins again with another vertex (which we call B) that is still in the ready state (STATUS = 1). This vertex B can be obtained by traversing through the list of vertices.

EXAMPLE 8.5 Suppose the breadth-first search (BFS) Algorithm 8.6 in Fig. 8-33 is applied to the graph in Fig. 8-30. The vertices will be processed in the following order:

$$A, \quad B, \quad C, \quad D, \quad K, \quad L, \quad J, \quad M$$

Specifically, Fig. 8-34(a) shows the sequence of waiting lists in QUEUE and the sequence of vertices being processed (Note that after vertex A is processed, its neighbors, B, C, and D are added to QUEUE in the order first B, then C, and finally D; hence B is on the front of the QUEUE and so B is the next vertex to be processed.) Again, each vertex, excluding A, comes from an adjacency list and hence corresponds to an edge of the graph. These edges form a spanning tree of G which is pictured in Fig. 8-34(b). Again, the numbers indicate the order that the edges are added to the spanning tree. Observe that this spanning tree is different from the one in Fig. 8.32(b) which came from a depth-first search.

Algorithm 8.6 (Breadth-first Search):   This algorithm executes a breadth-first search on a graph
G beginning with a starting vertex A.

Step 1.   Initialize all vertices to the ready state (STATUS $= 1$).

Step 2.   Put the starting vertex A in QUEUE and change the status of A to the waiting state
(STATUS $= 2$).

Step 3.   Repeat Steps 4 and 5 until QUEUE is empty.

Step 4.   Remove the front vertex N of QUEUE. Process N, and set STATUS $(N) = 3$, the
processed state.

Step 5.   Examine each neighbor J of N.
  (a)   If STATUS $(J) = 1$ (ready state), add J to the rear of QUEUE and reset
STATUS $(J) = 2$ (waiting state).
  (b)   If STATUS $(J) = 2$ (waiting state) or STATUS $(J) = 3$ (processed state), ignore the
vertex J.
    [End of Step 3 loop.]
Step 6.   Exit.

Fig. 8-33

QUEUE	Vertex
A	A
D, C, B	B
D, C	C
D	D
L, K	K
L	L
J	J
M	M
∅	

(a)   (b)

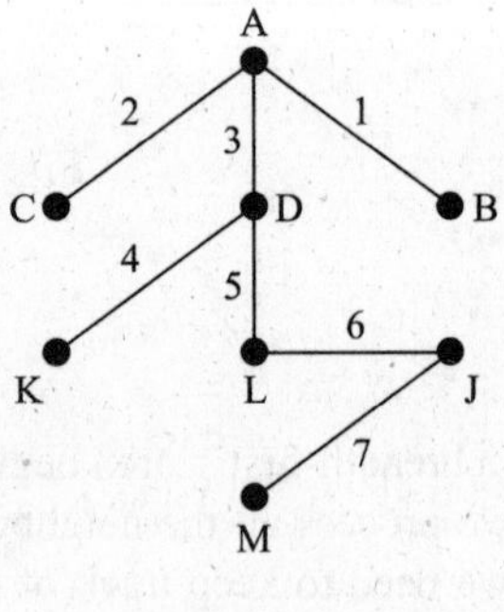

Fig. 8-34

8.13   TRAVELING-SALESMAN PROBLEM

Let G be a complete weighted graph. (We view the vertices of G as cities, and the weighted edges of G as
the distances between the cities.) The "traveling-salesman" problem refers to finding a Hamiltonian circuit for G
of minimum weight.

First we note the following theorem, proved in Problem 8.33:

Theorem 8.13:   The complete graph K_n with $n \geq 3$ vertices has $H = (n-1)!/2$ Hamiltonian circuits (where
we do not distinguish between a circuit and its reverse).

Consider the complete weighted graph G in Fig. 8-35(a). It has four vertices, A, B, C, D. By Theorem 8.13
it has $H = 3!/2 = 3$ Hamiltonian circuits. Assuming the circuits begin at the vertex A, the following are the
three circuits and their weights:

$$|ABCDA| = 3 + 5 + 6 + 7 = 21$$
$$|ACDBA| = 2 + 6 + 9 + 3 = 20$$
$$|ACBDA| = 2 + 5 + 9 + 7 = 23$$

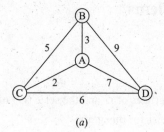

	P	Q	R	S	T
P		18	22	15	20
Q	18		11	12	22
R	22	11		16	10
S	15	12	16		13
T	20	22	10	13	

(a)   (b)

Fig. 8-35

Thus *ACDBA* with weight 20 is the Hamiltonian circuit of minimum weight.

We solved the "traveling-salesman problem" for the weighted complete graph in Fig. 8-35(a) by listing and finding the weights of its three possible Hamiltonian circuits. However, for a graph with many vertices, this may be impractical or even impossible. For example, a complete graph with 15 vertices has over 40 million Hamiltonian circuits. Accordingly, for circuits with many vertices, a strategy of some kind is needed to solve or give an approximate solution to the traveling-salesman problem. We discuss one of the simplest algorithms here.

Nearest-Neighbor Algorithm

The nearest-neighbor algorithm, starting at a given vertex, chooses the edge with the least weight to the next possible vertex, that is, to the "closest" vertex. This strategy is continued at each successive vertex until a Hamiltonian circuit is completed.

EXAMPLE 8.6 Let G be the weighted graph given by the table in Fig. 8-35(b). That is, G has the vertices $P, Q, \ldots, T$, and the distance from P to Q is 18, from P to R is 22, and so on until the distance from T to S is 13. We apply the nearest-neighbor algorithm to G starting at: (a) P, (b) Q.

(a) Starting at P, the first row of the table shows us that the closest vertex to P is S with distance 15. The fourth row shows that the closest vertex to S is Q with distance 12. The closest vertex to Q is R with distance 11. From R, there is no choice but to go to T with distance 10. Finally, from T, there is no choice but to go back to P with distance 20. Accordingly, the nearest-neighbor algorithm beginning at P yields the following weighted Hamiltonian circuit:

$$|PSQRTP| = 15 + 12 + 11 + 10 + 20 = 68$$

(b) Starting at Q, the closest vertex is R with distance 11; from R the closest is T with distance 10; and from T the closest is S with distance 13. From S we must go to P with distance 15; and finally from P we must go back to Q with distance 18. Accordingly, the nearest-neighbor algorithm beginning at Q yields the following weighted Hamiltonian circuit:

$$|QRTSPQ| = 11 + 10 + 13 + 15 + 18 = 67$$

The idea behind the nearest-neighbor algorithm is to minimize the total weight by minimizing the weight at each step. Although this may seem reasonable, Example 8.6 shows that we may not get a Hamiltonian circuit of minimum weight; that is, it cannot be both 68 and 67. Only by checking all $H = (n - 1)!/2 = 12$ Hamiltonian circuits of G will we really know the one with minimum weight. In fact, the nearest-neighbor algorithm beginning at A in Fig. 8-35(a) yields the circuit *ACBDA* which has the maximum weight. However, the nearest-neighbor algorithm usually gives a Hamiltonian circuit which is relatively close to the one with minimum weight.

Solved Problems

GRAPH TERMINOLOGY

8.1. Consider the graph G in Fig. 8-36(a).

(a) Describe G formally, that is, find the set $V(G)$ of vertices of G and the set $E(G)$ of edges of G.

(b) Find the degree of each vertex and verify Theorem 8.1 for this graph.

(a) There are five vertices so $V(G) = \{A, B, C, D, E\}$. There are seven pairs $\{x, y\}$ of vertices where the vertex x is connected with the vertex y, hence

$$E(G) = [\{A, B\}, \{A, C\}, \{A, D\}, \{B, C\}, \{B, E\}, \{C, D\}, \{C, E\}]$$

(b) The degree of a vertex is equal to the number of edges to which it belongs; e.g., $\deg(A) = 3$ since A belongs to the three edges $\{A, B\}$, $\{A, C\}$, $\{A, D\}$. Similarly,

$$\deg(B) = 3, \ \deg(C) = 4, \ \deg(D) = 2, \ \deg(E) = 2$$

The sum of the degrees is $3 + 3 + 4 + 2 + 2 = 14$ which does equal twice the number of edges.

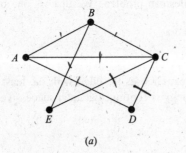
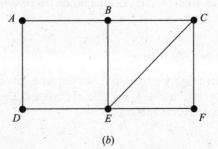

(a)   (b)

Fig. 8-36

8.2. Consider the graph G in Fig. 8-36(b). Find:

(a) all simple paths from A to F;   (d) diam(G), the diameter of G;
(b) all trails from A to F;   (e) all cycles which include vertex A;
(c) $d(A, F)$, the distance from A to F;   (f) all cycles in G.

(a) A simple path from A to F is a path such that no vertex, and hence no edge, is repeated. There are seven such paths, four beginning with the edges $\{A, B\}$ and three beginning with the ege $\{A, D\}$:

$$(A, B, C, F), \quad (A, B, C, E, F), \quad (A, B, E, F), \quad (A, B, E, C, F),$$
$$(A, D, E, F), \quad (A, D, E, B, C, F), \quad (A, D, E, C, F).$$

(b) A trail from A to F is a path such that no edge is repeated. There are nine such trails, the seven simple paths from (a) together with

$$(A, D, E, B, C, E, F) \quad \text{and} \quad (A, D, E, C, B, E, F).$$

(c) There is a path, e.g., (A, B, C, F), from A to F of length 3 and no shorter path from A to F; hence $d(A, F) = 3$.

(d) The distance between any two vertices is not greater than 3, and the distance from A to F is 3; hence diam(G) = 3.

(e) A cycle is a closed path in which no vertex is repeated (except the first and last). There are three cycles which include vertex A:

$$(A, B, E, D, A), \quad (A, B, C, E, D, A), \quad (A, B, C, F, E, D, A).$$

(f) There are six cycles in G; the three in (e) and

$$(B, C, E, B), \quad (C, F, E, C), \quad (B, C, F, E, B).$$

8.3. Consider the multigraphs in Fig. 8-37.

    (*a*) Which of them are connected? If a graph is not connected, find its connected components.

    (*b*) Which are cycle-free (without cycles)?

    (*c*) Which are loop-free (without loops)?

    (*d*) Which are (simple) graphs?

    (*a*) Only (1) and (3) are connected, (2) is disconnected; its connected components are $\{A, D, E\}$ and $\{B, C\}$. (4) is disconnected; its connected components are $\{A, B, E\}$ and $\{C, D\}$.

    (*b*) Only (1) and (4) are cycle-free. (2) has the cycle (A, D, E, A), and (3) has the cycle (A, B, E, A).

    (*c*) Only (4) has a loop which is $\{B, B\}$.

    (*d*) Only (1) and (2) are graphs. Multigraph (3) has multiple edges $\{A, E\}$ and $\{A, E\}$; and (4) has both multiple edges $\{C, D\}$ and $\{C, D\}$ and a loop $\{B, B\}$.

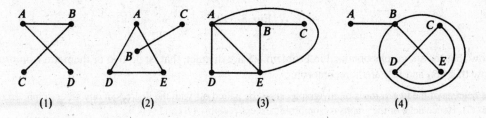

Fig. 8-37

8.4. Let G be the graph in Fig. 8-38(*a*). Find:

    (*a*) all simple paths from A to C;   (*d*) $G - Y$;

    (*b*) all cycles;   (*e*) all cut points;

    (*c*) subgraph H generated by $V' = \{B, C, X, Y\}$;   (*f*) all bridges.

    (*a*) There are two simple paths from A to C: (A, X, Y, C) and (A, X, B, Y, C).

    (*b*) There is only one cycle: (B, X, Y, B).

    (*c*) As pictured in Fig. 8-38(*b*), H consists of the vertices V' and the set E' of all edges whose endpoints belong to V', that is, $E' = [\{B, X\}, \{X, Y\}, \{B, Y\}, \{C, Y\}]$.

    (*d*) Delete vertex Y from G and all edges which contain Y to obtain the graph $G - Y$ in Fig. 8-38(*c*). (Note Y is a cutpoint since $G - Y$ is disconnected.)

    (*e*) Vertices A, X, and Y are cut points.

    (*f*) An edge e is a bridge if $G - e$ is disconnected. Thus there are three bridges: $\{A, Z\}$, $\{A, X\}$, and $\{C, Y\}$.

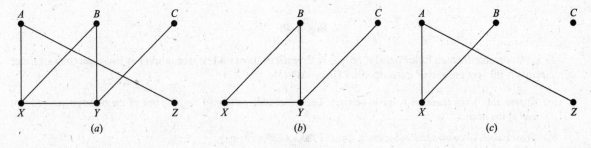

Fig. 8-38

8.5. Consider the graph G in Fig. 8-36(b). Find the subgraphs obtained when each vertex is deleted. Does G have any cut points?

When we delete a vertex from G, we also have to delete all edges which contain the vertex. The six graphs obtained by deleting each of the vertices of G are shown in Fig. 8-39. All six graphs are connected; hence no vertex is a cut point.

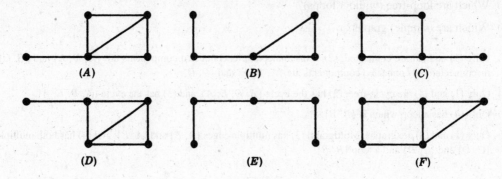

Fig. 8-39

8.6. Show that the six graphs obtained in Problem 8.5 are distinct, that is, no two of them are isomorphic. Also show that (B) and (C) are homeomorphic.

The degrees of the five vertices of any graph cannot be paired off with the degrees of any other graph, except for (B) and (C). Hence none of the graphs is isomorphic except possibly (B) and (C).

However if we delete the vertex of degree 3 in (B) and (C), we obtain distinct subgraphs. Thus (B) and (C) are also nonisomorphic; hence all six graphs are distinct. However, (B) and (C) are homeomorphic since they can be obtained from isomorphic graphs by adding appropriate vertices.

TRAVERSABLE GRAPHS, EULER AND HAMILTONIAN CIRCUITS

8.7. Consider each graph in Fig. 8-40. Which of them are traversable, that is, have Euler paths?

Which are Eulerian, that is, have an Euler circuit? For those that do not, explain why.

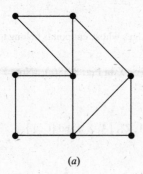 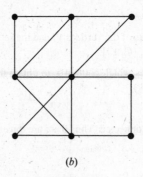 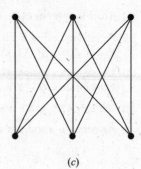

(a)   (b)   (c)

Fig. 8-40

G is traversable (has an Euler path) if only 0 or 2 vertices have odd degree, and G is Eulerian (has an Euler circuit) if all vertices are of even degree (Theorem 8.3).

(a) Traversable, since there are two odd vertices. The traversable path must begin at one of the odd vertices and will end at the other.

(b) Traversable, since all vertices are even. Thus G has an Euler circuit.

(c) Since six vertices have odd degrees, G is not traversable.

8.8. Which of the graphs in Fig. 8-40 have a Hamiltonian circuit? If not, why not?

Graphs (*a*) and (*c*) have Hamiltonian circuits. (The reader should be able to easily find one of them.) However, graph (*b*) has no Hamiltonian circuit. For if α is a Hamiltonian circuit, then α must connect the middle vertex with the lower right vertex, then proceed along the bottom row to the lower right vertex, then vertically to the middle right, but then is forced back to the central vertex before visiting the remaining vertices.

8.9. Prove Theorem 8.3 (Euler): A finite connected graph G is Eulerian if and only if each vertex has even degree.

Suppose G is Eulerian and T is a closed Eulerian trail. For any vertex v of G, the trail T enters and leaves v the same number of times without repeating any edge. Hence v has even degree.

Suppose conversely that each vertex of G has even degree. We construct an Eulerian trail. We begin a trail T_1 at any edge e. We extend T_1 by adding one edge after the other. If T_1 is not closed at any step, say, T_1 begins at u but ends at $v \neq u$, then only an odd number of the edges incident on v appear in T_1; hence we can extend T_1 by another edge incident on v. Thus we can continue to extend T_1 until T_1 returns to its initial vertex u, i.e., until T_1 is closed. If T_1 includes all the edges of G, then T_1 is our Eulerian trail.

Suppose T_1 does not include all edges of G. Consider the graph H obtained by deleting all edges of T_1 from G. H may not be connected, but each vertex of H has even degree since T_1 contains an even number of the edges incident on any vertex. Since G is connected, there is an edge e' of H which has an endpoint u' in T_1. We construct a trail T_2 in H beginning at u' and using e'. Since all vertices in H have even degree, we can continue to extend T_2 in H until T_2 returns to u' as pictured in Fig. 8-41. We can clearly put T_1 and T_2 together to form a larger closed trail in G. We continue this process until all the edges of G are used. We finally obtain an Eulerian trail, and so G is Eulerian.

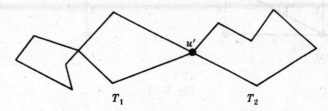

Fig. 8-41

TREES, SPANNING TREES

8.10. Draw all trees with exactly six vertices.

There are six such trees which are exhibited in Fig. 8-42. The first tree has diameter 5, the next two diameter 4, the next two diameter 3, and the last one diameter 2. Any other tree with 6 nodes is isormorphic to one of these trees.

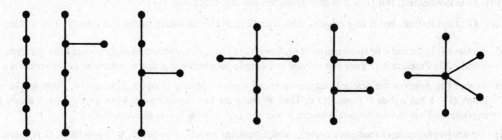

Fig. 8-42

8.11. Find all spanning trees of the graph G shown in Fig. 8-43(*a*).

There are eight such spanning trees as shown in Fig. 8-43(*b*). Each spanning tree must have $4 - 1 = 3$ edges since G has four vertices. Thus each tree can be obtained by deleting two of the five edges of G. This can be done in 10 ways,

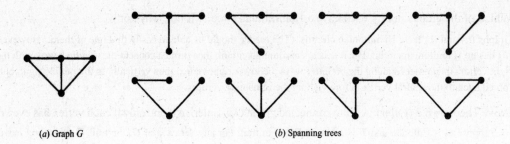

(a) Graph G   (b) Spanning trees

Fig. 8-43

except that two of the ways lead to disconnected graphs. Hence the above eight spanning trees are all the spanning trees of G.

8.12. Find a minimal spanning tree T for the weighted graph G in Fig. 8-44(a).

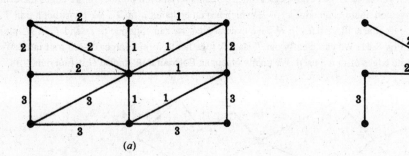

(a)   (b)

Fig. 8-44

Since G has n = 9 vertices, T must have n − 1 = 8 edges. Apply Algorithm 8.2, that is, keep deleting edges with maximum length and without disconnecting the graph until only n − 1 = 8 edges remain. Alternatively, apply Algorithm 8.3, that is, beginning with the nine vertices, keep adding edges with minimum length and without forming any circle until n − 1 = 8 edges are added. Both methods give a minimum spanning tree such as that shown in Fig. 8-44(b).

8.13. Let G be a graph with more than one vertex. Prove the following are equivalent.

   (i) G is a tree.

 (ii) Each pair of vertices is connected by exactly one simple path.

(iii) G is connected; but G − e is disconnected for any edge e of G.

(iv) G is cycle-free, but if any edge is added to G then the resulting graph has exactly one cycle.

  (i) *implies* (ii) Let u and v be two vertices in G. Since G is a tree, G is connected so there is at least one path between u and v. By Problem 8.37 there can only be one simple path between u and v, otherwise G will contain a cycle.

 (ii) *implies* (iii) Suppose we delete an edge e = {u, v} from G. Note e is a path from u to v. Suppose the resulting graph G − e has a path P from u to v. Then P and e are two distinct paths from u to v, which contradicts the hypothesis. Thus there is no path between u and v in G − e, so G − e is disconnected.

(iii) *implies* (iv) Suppose G contains a cycle C which contains an edge e = {u, v}. By hypothesis, G is connected but G' = G − e is disconnected, with u and v belonging to different components of G' (Problem 8.41) This contradicts the fact that u and v are connected by the path P = C − e which lies in G'. Hence G is cycle-free. Now let x and y be vertices of G and let H be the graph obtained by adjoining the edge e = {x, y} to G. Since G is connected, there is a path P from x to y in G; hence C = Pe forms a cycle in H. Suppose H contains another cycle C'. Since G is cycle-free, C' must contain the edge e, say C' = P'e. Then P and P' are two simple paths in G from x to y. (See Fig. 8-45.) By Problem 8.37, G contains a cycle, which contradicts the fact that G is cycle-free. Hence H contains only one cycle.

(iv) *implies* (i) Since adding any edge $e = \{x, y\}$ to G produces a cycle, the vertices x and y must already be connected in G. Hence G is connected and by hypothesis G is cycle-free; that is, G is a tree.

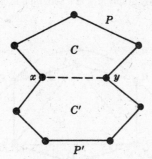

Fig. 8-45

8.14. Prove Theorem 8.6: Let G be a finite graph with $n \geq 1$ vertices. Then the following are equivalent. (i) G is a tree, (ii) G is a cycle-free and has $n - 1$ edges, (iii) G is connected and has $n - 1$ edges.

The proof is by induction on n. The theorem is certainly true for the graph with only one vertex and hence no edges. That is, the theorem holds for $n = 1$. We now assume that $n > 1$ and that the theorem holds for graphs with less than n vertices.

(i) *implies* (ii) Suppose G is a tree. Then G is cycle-free, so we only need to show that G has $n - 1$ edges. By Problem 8.38, G has a vertex of degree 1. Deleting this vertex and its edge, we obtain a tree T which has $n - 1$ vertices. The theorem holds for T, so T has $n - 2$ edges. Hence G has $n - 1$ edges.

(ii) *implies* (iii) Suppose G is cycle-free and has $n - 1$ edges. We only need show that G is connected. Suppose G is disconnected and has k components, $T_1, \ldots, T_k$, which are trees since each is connected and cycle-free. Say T_i has n_i vertices. Note $n_i < n$. Hence the theorem holds for T_i, so T_i has $n_i - 1$ edges. Thus

$$n = n_1 + n_2 + \cdots + n_k$$

and

$$n - 1 = (n_1 - 1) + (n_2 - 1) + \cdots + (n_k - 1) = n_1 + n_2 + \cdots + n_k - k = n - k$$

Hence $k = 1$. But this contradicts the assumption that G is disconnected and has $k > 1$ components. Hence G is connected.

(iii) *implies* (i) Suppose G is connected and has $n - 1$ edges. We only need to show that G is cycle-free. Suppose G has a cycle containing an edge e. Deleting e we obtain the graph $H = G - e$ which is also connected. But H has n vertices and $n - 2$ edges, and this contradicts Problem 8.39. Thus G is cycle-free and hence is a tree.

PLANAR GRAPHS

8.15. Draw a planar representation, if possible, of the graphs (*a*), (*b*), and (*c*) in Fig. 8-46.

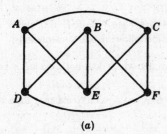
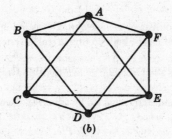
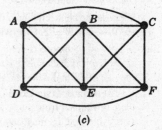

Fig. 8-46

(a) Redrawing the positions of B and E, we get a planar representation of the graph as in Fig. 8-47(a).

(b) This is not the star graph K_5. This has a planar representation as in Fig. 8-47(b).

(c) This graph is non-planar. The utility graph $K_{3,3}$ is a subgraph as shown in Fig. 8-47(c) where we have redrawn the positions of C and F.

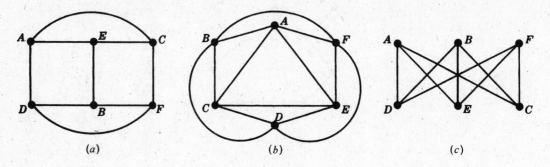

Fig. 8-47

8.16. Count the number V of vertices, the number E of edges, and the number R of regions of each map in Fig. 8-48; and verify Euler's formula. Also find the degree d of the outside region.

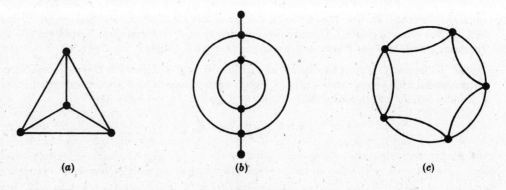

Fig. 8-48

(a) $V = 4$, $E = 6$, $R = 4$. Hence $V - E + R = 4 - 6 + 4 = 2$. Also $d = 3$.

(b) $V = 6$, $E = 9$, $R = 5$; so $V - E + R = 6 - 9 + 5 = 2$. Here $d = 6$ since two edges are counted twice.

(c) $V = 5$, $E = 10$, $R = 7$. Hence $V - E + R = 5 - 10 + 7 = 2$. Here $d = 5$.

8.17. Find the minimum number n of colors required to paint each map in Fig. 8-48.

(a) $n = 4$;   (b) $n = 3$;   (c) $n = 2$.

8.18. Prove Theorem 8.8 (Euler): $V - E + R = 2$.

Suppose the connected map M consists of a single vertex P as in Fig. 8-49(a). Then $V = 1$, $E = 0$, and $R = 1$. Hence $V - E + R = 2$. Otherwise M can be built up from a single vertex by the following two constructions:

(1) Add a new vertex Q_2 and connect it to an existing vertex Q_1 by an edge which does not cross any existing edge as in Fig. 8-49(b).

(2) Connect two existing vertices Q_1 and Q_2 by an edge e which does not cross any existing edge as in Fig. 8-49(c).

Neither operation changes the value of $V - E + R$. Hence M has the same value of $V - E + R$ as the map consisting of a single vertex, that is, $V - E + R = 2$. Thus the theorem is proved.

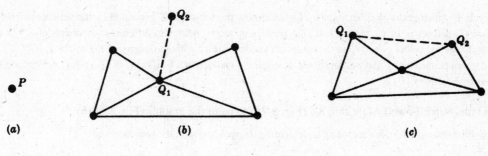

Fig. 8-49

8.19. Prove Theorem 8.11: The following are equivalent for a graph G: (i) G is 2-colorable. (ii) G is bipartite. (iii) Every cycle of G has even length.

   (i) *implies* (ii). Suppose G is 2-colorable. Let M be the set of vertices painted the first color, and let N be the set of vertices painted the second color. Then M and N form a bipartite partition of the vertices of G since neither the vertices of M nor the vertices of N can be adjacent to each other since they are of the same color.

   (ii) *implies* (iii). Suppose G is bipartite and M and N form a bipartite partition of the vertices of G. If a cycle begins at a vertex u of, say, M, then it will go to a vertex of N, and then to a vertex of M, and then to N and so on. Hence when the cycle returns to u it must be of even length. That is, every cycle of G will have even length.

   (iii) *implies* (i). Lastly, suppose every cycle of G has even length. We pick a vertex in each connected component and paint it the first color, say red. We then successively paint all the vertices as follows: If a vertex is painted red, then any vertex adjacent to it will be painted the second color, say blue. If a vertex is painted blue, then any vertex adjacent to it will be painted red. Since every cycle has even length, no adjacent vertices will be painted the same color. Hence G is 2-colorable, and the theorem is proved.

8.20. Prove Theorem 8.12: A planar graph G is 5-colorable.

   The proof is by induction on the number p of vertices of G. If $p \leq 5$, then the theorem obviously holds. Suppose $p > 5$, and the theorem holds for graphs with less than p vertices. By the preceding problem, G has a vertex v such that $\deg(v) \leq 5$. By induction, the subgraph $G - v$ is 5-colorable. Assume one such coloring. If the vertices adjacent to v use less than the five colors, than we simply paint v with one of the remaining colors and obtain a 5-coloring of G. We are still left with the case that v is adjacent to five vertices which are painted different colors. Say the vertices, moving counterclockwise about v, are $v_1, \ldots, v_5$ and are painted respectively by the colors $c_1, \ldots, c_5$. (See Fig. 8-50(a).)

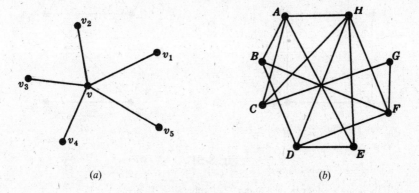

(a)   (b)

Fig. 8-50

Consider now the subgraph H of G generated by the vertices painted c_1 and c_3. Note H includes v_1 and v_3. If v_1 and v_3 belong to different components of H, then we can interchange the colors c_1 and c_3 in the component containing v_1 without destroying the coloring of $G - v$. Then v_1 and v_3 are painted by c_3, c_1 can be chosen to paint v, and we have a 5-coloring of G. On the other hand, suppose v_1 and v_3 are in the same component of H. Then there is a path P from v_1 to v_3 whose vertices are painted either c_1 or c_3. The path P together with the edges $\{v, v_1\}$ and $\{v, v_3\}$ form a cycle C which encloses either v_2 or v_4. Consider now the subgraph K generated by the vertices painted c_3 or c_4. Since C encloses v_2 or v_4, but not both, the vertices v_2 and v_4 belong to different components of K. Thus we can interchange the colors c_2 and c_4 in the component containing v_2 without destroying the coloring of $G - v$. Then v_2 and v_4 are painted by c_4, and we can choose c_2 to paint v and obtain a 5-coloring of G. Thus G is 5-colorable and the theorem is proved.

8.21. Use the Welch-Powell Algorithm 8.4 (Fig. 8-24) to paint the graph in Fig. 8-50(b).

First order the vertices according to decreasing degrees to obtain the sequence

$$H, \quad A, \quad D, \quad F, \quad B, \quad C, \quad E, \quad G$$

Proceeding sequentially, we use the first color to paint the vertices H, B, and then G. (We cannot paint A, D, or F the first color since each is connected to H, and we cannot paint C or E the first color since each is connected to either H or B.) Proceeding sequentially with the unpainted vertices, we use the second color to paint the vertices A and D. The remaining vertices F, C, and E can be painted with the third color. Thus the chromatic number n cannot be greater than 3. However, in any coloring, H, D, and E must be painted different colors since they are connected to each other. Hence $n = 3$.

8.22. Let G be a finite connected planar graph with at least three vertices. Show that G has at least one vertex of degree 5 or less.

Let p be the number of vertices and q the number of edges of G, and suppose $\deg(u) \geq 6$ for each vertex u of G. But $2q$ equals the sum of the degrees of the vertices of G (Theorem 8.1); so $2q \geq 6p$. Therefore

$$q \geq 3p > 3p - 6$$

This contradicts Theorem 8.9. Thus some vertex of G has degree 5 or less.

SEQUENTIAL REPRESENTATION OF GRAPHS

8.23. Find the adjacency matrix $A = [a_{ij}]$ of each graph G in Fig. 8-51.

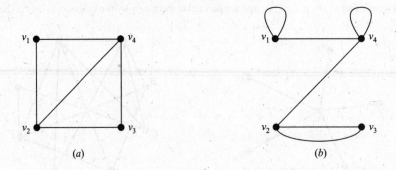

(a)   (b)

Fig. 8-51

Set $a_{ij} = n$ if there are n edges $\{v_i, v_j\}$ and $a_{ij} = 0$ otherwise. Hence:

$$(a)\ A = \begin{bmatrix} 0 & 1 & 0 & 1 \\ 1 & 0 & 1 & 1 \\ 0 & 1 & 0 & 1 \\ 1 & 1 & 1 & 0 \end{bmatrix}; \quad (b)\ A = \begin{bmatrix} 1 & 0 & 0 & 1 \\ 0 & 0 & 2 & 1 \\ 0 & 2 & 0 & 0 \\ 1 & 1 & 0 & 1 \end{bmatrix}$$

(Since (a) has no multiple edges and no loops, the entries in A are either 0 or 1, and are 0 on the diagonal.)

8.24. Draw the graph G corresponding to each adjacency matrix:

$$(a)\ A = \begin{bmatrix} 0 & 1 & 0 & 1 & 0 \\ 1 & 0 & 0 & 1 & 1 \\ 0 & 0 & 0 & 1 & 1 \\ 1 & 1 & 1 & 0 & 1 \\ 0 & 1 & 1 & 1 & 0 \end{bmatrix}; \quad (b)\ A = \begin{bmatrix} 1 & 3 & 0 & 0 \\ 3 & 0 & 1 & 1 \\ 0 & 1 & 2 & 2 \\ 0 & 1 & 2 & 0 \end{bmatrix}$$

    (a) Since A is a 5-square matrix, G has five vertices, say, $v_1, v_2 \ldots, v_5$. Draw an edge from v_i to v_j when $a_{ij} = 1$. The graph appears in Fig. 8-52(a).

    (b) Since A is a 4-square matrix, G has four vertices, say, $v_1, \ldots, v_4$. Draw n edges from v_i to v_j when $a_{ij} = n$. Also, draw n loops at v_i when $a_i = n$. The graph appears in Fig. 8-52(b).

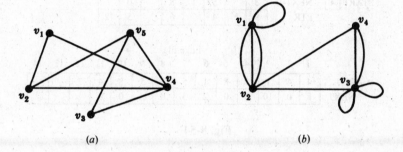

(a)   (b)

Fig. 8-52

8.25. Find the weight matrix $W = [w_{ij}]$ of the weighted graph G in Fig. 8-53(a) where the vertices are stored in the array DATA as follows: DATA: A, B, C, X, Y.

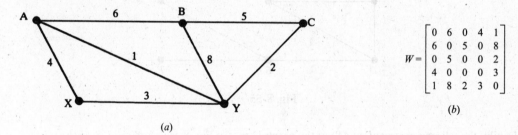

(a) $$W = \begin{bmatrix} 0 & 6 & 0 & 4 & 1 \\ 6 & 0 & 5 & 0 & 8 \\ 0 & 5 & 0 & 0 & 2 \\ 4 & 0 & 0 & 0 & 3 \\ 1 & 8 & 2 & 3 & 0 \end{bmatrix}$$

                                                 (b)

Fig. 8-53

The vertices are numbered according to the way they are stored in the array DATA; so $v_1 = A$, $v_2 = B, \ldots$, $v_5 = Y$. Then set $W_{ij} = w$, where w is the weight of the edge from v_i to v_j. This yields the matrix W in Fig. 8-53(b).

LINKED REPRESENTATION OF GRAPHS

8.26. A graph G with vertices $A, B, \ldots, F$ is stored in memory using a linked representation with a vertex file and an edge file as in Fig. 8-54.

    (a) List the vertices in the order they appear in memory.

    (b) Find the adjacency list adj(v) of each vertex v of G.

    (a) Since START $= 4$, the list begins with the vertex D. The NEXT-V tells us to go to 1(B), then 3(F), then 5(A), then 8(E), and then 7(C); that is,

$$D, \quad B, \quad F, \quad A, \quad E, \quad C$$

(b) Here adj(D) = [5(A), 1(B), 8(E)]. Specifically, PTR[4(D)] = 7 and ADJ[7] = 5(A) tells us that adj(D) begins with A. Then NEXT[7] = 3 and ADJ[3] = 1(B) tells us that B is the next vertex in adj(D). Then NEXT[3] = 10 and ADJ[10] = 8(E) tells us that E in the next vertex in adj(D). However, NEXT[10] = 0 tells us that there are no more neighbors of D. Similarly.

$$\text{adj}(B) = [A, D], \quad \text{adj}(F) = [E], \quad \text{adj}(A) = [B, D], \quad \text{adj}(E) = [C, D, F], \quad \text{adj}(C) = [E]$$

In other words, the following is the adjacency structure of G:

$$G = [A{:}B, D; \quad B{:}A, D; \quad C{:}E; \quad D{:}A, B, E; \quad E{:}C, D, F; \quad F{:}E]$$

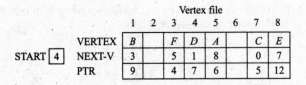

Vertex file

	1	2	3	4	5	6	7	8
VERTEX	B		F	D	A		C	E
NEXT-V	3		5	1	8		0	7
PTR	9		4	7	6		5	12

START [4]

Edge file

	1	2	3	4	5	6	7	8	9	10	11	12	13	14
ADJ	4	4	1	8	8	1	5	3	5	8	4	7		
NEXT	8	0	10	0	0	2	3	0	11	0	0	1		

Fig. 8-54

8.27. Draw the diagram of the graph G whose linked representation appears in Fig. 8-54.

Use the vertex list obtained in Problem 8.26(a) and the adjacency lists obtained in Problem 8.26(b) to draw the graph G in Fig. 8-55.

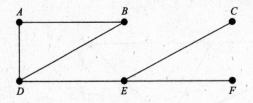

Fig. 8-55

8.28. Exhibit the adjacency structure (AS) of the graph G in: (a) Fig. 8-56(a), (b) Fig. 8-56(b).

The adjacency structure of a graph G consists of the adjacency lists of the vertices where we use a colon ":" to separate a vertex from its adjacency list, and a semicolon ";" to separate the different lists. Thus:

(a) $G = [A{:}B, C, D; \quad B{:}A, C, E; \quad C{:}A, B, D, E; \quad D{:}A, C; \quad E{:}B, C]$

(b) $G = [A{:}B, D; \quad B{:}A, C, E; \quad C{:}B, E, F; \quad D{:}A, E; \quad E{:}B, C, D, F; \quad F{:}C, E]$

GRAPH ALGORITHMS

8.29. Consider the graph G in Fig. 8-56(a) (where the vertices are ordered alphabetically).

(a) Find the adjacency structure of G.

(b) Find the order in which the vertices of G are processed using a DFS (depth-first search) algorithm beginning at vertex A.

(a) List the neighbors of each vertex as follows:

$$G = [A{:}B, C, D; \quad B{:}A, J; \quad C{:}A; \quad D{:}A, K; \quad J{:}B, K, M; \quad K{:}D, J, L; \quad L{:}K, M; \quad M{:}J, L]$$

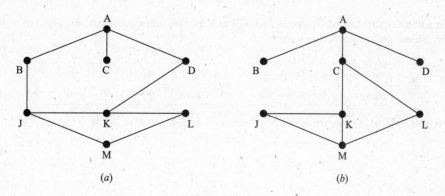

Fig. 8-56

(b) During the DFS algorithm, the first vertex N in STACK is processed and the neighbors of N (which have not been previously processed) are then pushed onto STACK. Initially, the beginning vertex A is pushed onto STACK. The following shows the sequence of waiting lists in STACK and the vertices being processed:

STACK	A	DCB	KCB	$LJCB$	$MJCB$	JCB	CB	B	$\emptyset$
Vertex	A	D	K	L	M	J	C	B	

In other words the vertices are processed in the order: A, D, K, L, M, J, C, B.

8.30. Repeat Problem 8.29 for the graph G in Fig. 8-56(b).

(a) List the neighbors of each vertex as follows:

$$G = [A{:}B, C, D; \quad B{:}A; \quad C{:}A, K, L; \quad D{:}A; \quad J{:}K, M; \quad K{:}C, J, M; \quad L{:}C, M; \quad M{:}J, K, L]$$

(b) The following shows the sequence of waiting lists in STACK and the vertices being processed:

STACK	A	DCB	CB	LKB	MKB	KJB	JB	B	$\emptyset$
Vertex	A	D	C	L	M	K	J	B	

In other words the vertices are processed in the order: A, D, C, L, M, K, J, B.

8.31. Beginning at vertex A and using a BFS (breadth-first search) algorithm, find the order the vertices are processed for the graph G: (a) in Fig. 8-56(a), (b) in Fig. 8-56(b).

(a) The adjacency structure of G appears in Problem 8.29. During the BFS algorithm, the first vertex N in QUEUE is processed and the neighbors of N (which have not appeared previously) are then added onto QUEUE. Initially, the beginning vertex A is assigned to QUEUE. The following shows the sequence of waiting lists in QUEUE and the vertices being processed:

QUEUE	A	DCB	JDC	JD	KJ	MK	LM	L	$\emptyset$
Vertex	A	B	C	D	J	K	M	L	

In other words the vertices are processed in the order: A, B, C, D, J, K, M, L.

(b) The adjacency structure of G appears in Problem 8.30. The following shows the sequence of waiting lists in QUEUE and the vertices being processed:

QUEUE	A	DCB	DC	LKD	LK	MJL	MJ	M	∅
Vertex	A	B	C	D	K	L	J	M	

In other words the vertices are processed in the order: A, B, C, D, K, L, J, M.

TRAVELING–SALESMAN PROBLEM

8.32. Apply the nearest-neighbor algorithm to the complete weighted graph G in Fig. 8-57 beginning at: (a) vertex A; (b) vertex D.

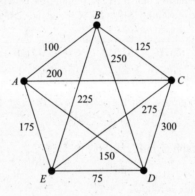

Fig. 8-57

(a) Starting at A, the closest vertex is B with distance 100; from B, the closest is C with distance 125; and from C the closest is E with distance 275. From E we must go to D with distance 75; and finally from D we must go back to A with distance 150. Accordingly, the nearest-neighbor algorithm beginning at A yields the following weighted Hamiltonian circuit:

$$|ABCEDA| = 100 + 125 + 275 + 75 + 150 = 725$$

(b) Starting at D, we must go to E, then A, then B then C and finally back to D. Accordingly, the nearest-neighbor algorithm beginning at D yields the following weighted Hamiltonian circuit:

$$|DEABCD| = 75 + 175 + 100 + 125 + 300 = 775$$

8.33. Prove Theorem 8.13. The complete graph K_n with $n \geq 3$ vertices has $H = (n - 1)!/2$ Hamiltonian circuits.

The counting convention for Hamiltonian circuits enables us to designate any vertex in a circuit as the starting point. From the starting point, we can go to any $n - 1$ vertices, and from there to any one of $n - 2$ vertices, and so on until arriving at the last vertex and then returning to the starting point. By the basic counting principle, there are a total of $(n-1)(n-2)\cdots 2 \cdot 1 = (n-1)!$ circuits that can be formed from a stating point. For $n \geq 3$, any circuit can be paired with one in the opposite direction which determines the same Hamiltonian circuit. Accordingly, there are a total of $H = (n - 1)!/2$ Hamiltonian circuits.

Supplementary Problems

GRAPH TERMINOLOGY

8.34. Consider the graph G in Fig. 8-58. Find:

(a) degree of each vertex (and verify Theorem 8.1);

(b) all simple paths from A to L;

(c) all trails (distinct edges) from B to C;

(d) $d(A, C)$, distance from A to C;

(e) diam(G), the diameter of G.

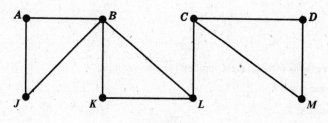

Fig. 8-58

8.35. Consider the graph in Fig. 8-58. Find (if any): (a) all cycles; (b) all cut points; (c) all bridges.

8.36. Consider the graph in Fig. 8-58. Find the subgraph $H = H(V', E')$ of G where V' equals:

(a) $\{B, C, D, J, K\}$ (b) $\{A, C, J, L, M\}$ (c) $\{B, D, J, M\}$ (d) $\{C, K, L, M\}$

Which of them are isomorphic and which homeomorphic?

8.37. Suppose a graph G contains two distinct paths from a vertex u to a vertex v. Show that G has a cycle.

8.38. Suppose G is a finite cycle-free graph with at least one edge. Show that G has at least two vertices of degree 1.

8.39. Show that a connected graph G with n vertices must have at least $n - 1$ edges.

8.40. Find the number of connected graphs with four vertices. (Draw them.)

8.41. Let G be a connected graph. Prove:

(a) If G contains a cycle C which contains an edge e, then $G - e$ is still connected.

(b) If $e = \{u, v\}$ is an edge such that $G - e$ is disconnected, then u and v belong to different components of $G - e$.

8.42. Suppose G has V vertices and E edges. Let M and m denote, respectively, the maximum and minimum of the degrees of the vertices in G. Show that $m \leq 2E/V \leq M$.

8.43. Consider the following two steps on a graph G: (1) Delete an edge. (2) Delete a vertex and all edges containing that vertex. Show that every subgraph H of a finite graph G can be obtained by a sequence consisting of these two steps.

TRAVERSABLE GRAPHS, EULER AND HAMILTONIAN CIRCUITS

8.44. Consider the graphs K_5, $K_{3,3}$ and $K_{2,3}$ in Fig. 8-59. Find an Euler (traversable) path or an Euler circuit of each graph, if it exists. If it does not, why not?

8.45. Consider each graph in Fig. 8-59. Find a Hamiltonian path or a Hamiltonian circuit, if it exists. If it does not, why not?

8.46. Show that K_n has $H = (n - 1)!/2$ Hamiltonian circuits. In particular, find the number of Hamiltonian circuits for the graph K_5 in Fig. 8-59(a).

8.47. Suppose G and G^* are homeomorphic graphs. Show that G is traversable (Eulerian) if and only if G^* is traversable (Eulerian).

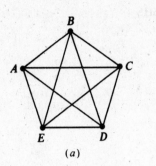

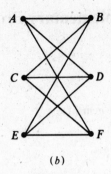

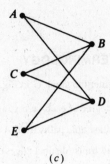

(a)   (b)   (c)

Fig. 8-59

SPECIAL GRAPHS

8.48. Draw two 3-regular graphs with: (a) eight vertices; (b) nine vertices.

8.49. Consider the complete graph K_n.

(a) Find the diameter of K_n.

(b) Find the number m of edges in K_n.

(c) Find the degree of each vertex in K_n.

(d) Find those values of n for which K_n is: (i) traversable; (ii) regular.

8.50. Consider the complete graph $K_{m,n}$.

(a) Find the diameter of $K_{m,n}$.

(b) Find the number E of edges in $K_{m,n}$.

(c) Find those $K_{m,n}$ which are traversable.

(d) Which of the graphs $K_{m,n}$ are isomorphic and which homeomorphic?

8.51. The *n-cube*, denoted by Q_n, is the graph whose vertices are the 2^n bit strings of length n, and where two vertices are adjacent if they differ in only one position. Figure 8-60(a) and (b) show the *n*-cubes Q_2 and Q_3.

(a) Find the diameter of Q_n.

(b) Find the number m of edges in Q_n.

(c) Find the degree of each vertex in Q_n.

(d) Find those values of n for which Q_n is traversable.

(e) Find a Hamiltonian circuit (called a *Gray code*) for (i) Q_3; (ii) Q_4.

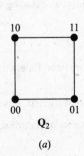

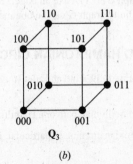

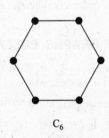

Q_2 Q_3 C_6

(a)   (b)   (c)

Fig. 8-60

8.52. The *n-cycle*, denoted by C_n, is the graph which consists of only a single cycle of length n. Figure 8-60(c) shows the 6-cycle C_6. (a) Find the number of vertices and edges in C_n. (b) Find the diameter of C_n.

8.53. Describe those connected graphs which are both bipartite and regular.

TREES

8.54. Draw all trees with five or fewer vertices.

8.55. Find the number of trees with seven vertices.

8.56. Find the number of spanning trees in Fig. 8-61(a).

8.57. Find the weight of a minimum spanning tree in Fig. 8-61(b)

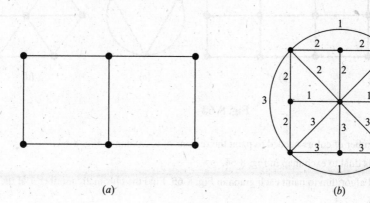

(a)   (b)

Fig. 8-61

8.58. Show that any tree is a bipartite graph.

8.59. Which complete bipartite graphs $K_{m,n}$ are trees.

PLANAR GRAPHS, MAPS, COLORINGS

8.60. Draw a planar representation of each graph G in Fig. 8-62, if possible; otherwise show that it has a subgraph homeomorphic to K_5 or $K_{3,3}$.

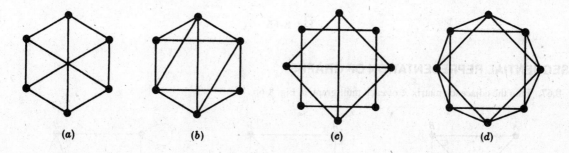

(a)   (b)   (c)   (d)

Fig. 8-62

8.61. Show that the 3-cube Q_3 (Fig. 8-60(b)) is planar.

8.62. For the map in Fig. 8-63, find the degree of each region and verify that the sum of the degrees of the regions is equal to twice the number of edges.

8.63. Count the number V of vertices, the number E of edges, and the number R of regions of each of the maps in Fig. 8-64, and verify Euler's formula.

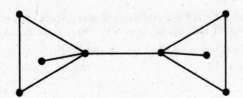

Fig. 8-63

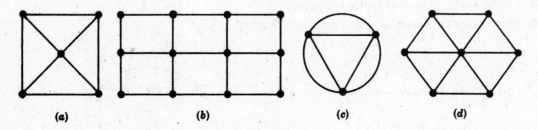

     (a)   (b)   (c)   (d)

Fig. 8-64

8.64. Find the minimum number of colors needed to paint the regions of each map in Fig. 8-64.

8.65. Draw the map which is dual to each map in Fig. 8-64.

8.66. Use the Welch-Powell algorithm to paint each graph in Fig. 8-65. Find the chromatic number n of the graph.

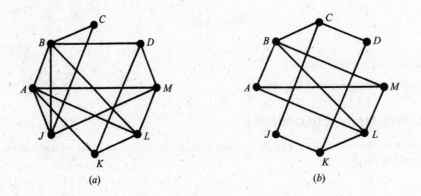

     (a)   (b)

Fig. 8-65

SEQUENTIAL REPRESENTATION OF GRAPHS

8.67. Find the adjacency matrix A of each multigraph in Fig. 8-66.

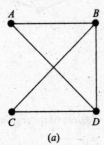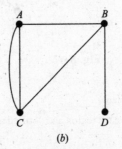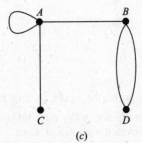

     (a)   (b)   (c)

Fig. 8-66

8.68. Draw the multigraph G corresponding to each of the following adjacency matrices:

$$(a)\ A = \begin{bmatrix} 0 & 2 & 0 & 1 \\ 2 & 1 & 1 & 1 \\ 0 & 1 & 0 & 1 \\ 1 & 1 & 1 & 0 \end{bmatrix}; \quad (b)\ A = \begin{bmatrix} 1 & 1 & 1 & 2 \\ 1 & 0 & 0 & 0 \\ 1 & 0 & 0 & 2 \\ 2 & 0 & 2 & 2 \end{bmatrix}$$

8.69. Suppose a graph G is bipartite. Show that one can order the vertices of G so that its adjacency matrix A has the form:
$$A = \begin{bmatrix} 0 & B \\ C & 0 \end{bmatrix}$$

LINKED REPRESENTATION OF GRAPHS

8.70. Suppose a graph G is stored in memory as in Fig. 8-67.

Vertex file

		1	2	3	4	5	6	7	8
START 7	VERTEX	C		F	E	A		B	D
	NEXT-V	0		5	1	8		3	4
	PTR	2		11	6	12		4	1

Edge file

	1	2	3	4	5	6	7	8	9	10	11	12
ADJ	7	7	4	5		7	1		8	3	1	7
NEXT	0	10	0	7		0	9		3	0	0	0

Fig. 8-67

(a) List the vertices in the order in which they appear in memory.

(b) Find the adjacency structure of G, that is, find the adjacency list adj(v) of each vertex v of G.

8.71. Exhibit the adjacency structure (AS) for each graph G in Fig. 8-59.

8.72. Figure 8-68(a) shows a graph G representing six cities A, B, ..., F connected by seven highways numbered 22, 33, ..., 88. Show how G may be maintained in memory using a linked representation with sorted arrays for the cities and for the numbered highways. (Note that VERTEX is a sorted array and so the field NEXT-V is not needed.)

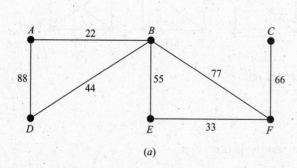

(a)

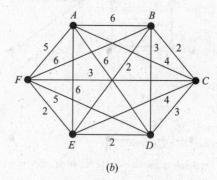

(b)

Fig. 8-68

TRAVELING–SALESMAN PROBLEM

8.73. Apply the nearest-neighbor algorithm to the complete weighted graph G in Fig. 8-68(b) beginning at: (a) vertex A; (b) vertex B.

8.74. Consider the complete weighted graph G in Fig. 8-57 with 5 vertices.

    (a) Beginning at vertex A, list the $H = (n-1)!/2 = 12$ Hamiltonian circuits of G, and find the weight of each of them.

    (b) Find a Hamiltonian circuit of minimal weight.

GRAPH ALGORITHMS

8.75. Consider the graph G in Fig. 8-57 (where the vertices are ordered alphabetically).

    (a) Find the adjacency structure (AS) of G.

    (b) Using the DFS (depth-first search) Algorithm 8.5 on G and beginning at vertex C, find the STACK sequence and the order in which the vertices are processed.

    (c) Repeat (b) beginning at vertex K.

8.76. Using the BFS (breadth-first search) Algorithm 8.6 on the graph G in Fig. 8-57, find the QUEUE sequence and the order the vertices are processed beginning at: (a) vertex C; (b) vertex K.

8.77. Repeat Problem 8.75 for the graph G in Fig. 8-65(a).

8.78. Repeat Problem 8.76 for the graph G in Fig. 8-65(a).

8.79. Repeat Problem 8.75 for the graph G in Fig. 8-65(b).

8.80. Repeat Problem 8.76 for the graph G in Fig. 8-65(b).

Answers to Supplementary Problems

8.34. (a) 2, 4, 3, 2, 2, 2, 3, 2; (b) ABL, $ABKL$, $AJBL$, $AJBKL$; (c) BLC, $BKLC$, $BAJBLC$, $BAJBKLC$; (d) 3; (e) 4.

8.35. (a) $AJBA$, $BKLB$, $CDMC$; (b) B, C, L; (c) only $\{C, L\}$.

8.36. (a) $E' = \{BJ, BK, CD\}$; (b) $E' = \{AJ, CM, LC\}$; (c) $E' = \{BJ, DM\}$; (d) $E' = \{KL, LC, CM\}$, Also, (a) and (b) are isomorphic, and (a), (b), and (c) are homeomorphic.

8.38. *Hint*: Consider a maximal simple path α, and show that its endpoints have degree 1.

8.40. There are five of them, as shown in Fig. 8-69.

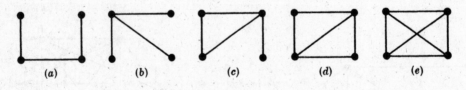

       (a)   (b)   (c)   (d)   (e)

Fig. 8-69

8.42. *Hint*: Use Theorem 8.1.

8.43. First delete all edges in G not in H, then delete all vertrices in G not in H.

8.44. (a) Eulerian since all vertices are even: $ABCDEACEBDA$. (b) None, since four vertices are odd. (c) Euler path beginning at B and ending at D (or vice versa): $BADCBED$.

8.45. (a) $ABCDEA$; (b) $ABCDEFA$; (c) none, since B or D must be visited twice in any closed path including all vertices.

8.46. $(5-1)!/2 = 12$.

8.47. *Hint*: Adding a vertex by dividing an edge does not change the degree of the original vertices and simply adds a vertex of even degree.

8.48. (a) The two 3-regular graphs in Fig. 8-70 are not isomorphic: (b) has a 5-cycle, but (a) does not. (b) There are none. The sum of the degrees of an r-regular graph with s vertices equals rs, and rs must be even.

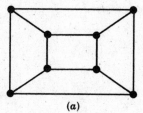 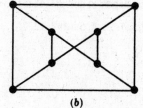

(a)   (b)

Fig. 8-70

8.49. (a) $\text{diam}(K_1) = 0$; all others have diameter 1; (b) $m = C(n, 2) = n(n-1)/2$; (c) $n-1$; (d) (i) $n = 2$ and n odd; (ii) all n.

8.50. (a) $\text{diam}(K_{1,1}) = 1$; all others have diameter 2; (b) $E = mn$; (c) $K_{1,1}$, $K_{1,2}$, and all $K_{m,n}$ where m and n are even; (d) none are isomorphic; only $K_{1,1}$ and $K_{1,2}$ are homeomorphic.

8.51. (a) n; (b) $n2^{n-1}$; (c) n; (d) $n = 1$, even; (e) consider the 4×16 matrix:

$$M = \begin{bmatrix} 0 & 1 & 1 & 0 & 0 & 1 & 1 & 0 & 0 & 1 & 1 & 0 & 0 & 1 & 1 & 0 \\ 0 & 0 & 1 & 1 & 1 & 1 & 0 & 0 & 0 & 0 & 1 & 1 & 1 & 1 & 0 & 0 \\ 0 & 0 & 0 & 0 & 1 & 1 & 1 & 1 & 1 & 1 & 1 & 1 & 0 & 0 & 0 & 0 \\ 0 & 0 & 0 & 0 & 0 & 0 & 0 & 0 & 1 & 1 & 1 & 1 & 1 & 1 & 1 & 1 \end{bmatrix}$$

which shows how $\mathbf{Q}_4$ (the columns of M) is obtained from $\mathbf{Q}_3$. That is, the upper left 3×8 submatrix of M is $\mathbf{Q}_3$, the upper right 3×8 submatrix of M is $\mathbf{Q}_3$ written in reverse, and the last row consists of eight 0s followed by eight 1s.

8.52. (a) n and n; (b) $n/2$ when n is even, $(n+1)/2$ when n is odd.

8.53. $K_{m,m}$ is bipartite and m-regular. Also, beginning with $K_{m,m}$, delete m disjoint edges to obtain a bipartite graph which is $(m-1)$-regular, delete another m disjoint edges to obtain a bipartite graph which is $(m-2)$-regular, and so on. These graphs may be disconnected, but their connected components have the desired properties.

8.54. There are eight such trees, as shown in Fig. 8-71. The graph with one vertex and no edges is called the *trivial tree*.

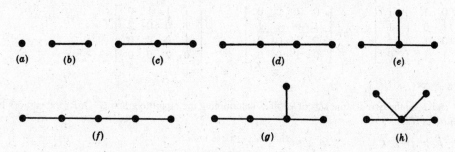

Fig. 8-71

8.55. 10

8.56. 15

8.57. $1 + 1 + 1 + 1 + 1 + 2 + 2 + 3 = 12$.

8.59. $m = 1$.

8.60. Only (a) is nonplanar, and $K_{3,3}$ is a subgraph.

8.61. Figure 8-70(a) is a planar representation of $Q3$.

8.62. The outside region has degree 8, and the other two regions have degree 5.

8.63. (a) 5, 8, 5; (b) 12, 17, 7; (c) 3, 6, 5; (d) 7, 12, 7.

8.64. (a) 3; (b) 3; (c) 2; (d)3.

8.65. See Fig. 8-72.

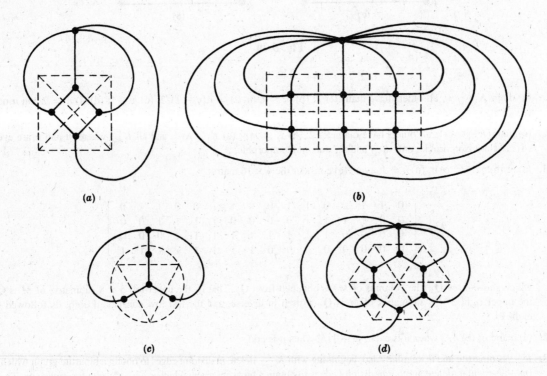

Fig. 8-72

8.66. (a) $n = 3$; (b) $n = 4$.

8.67. (a) $\begin{bmatrix} 0 & 1 & 0 & 1 \\ 1 & 0 & 1 & 1 \\ 0 & 1 & 0 & 1 \\ 1 & 1 & 1 & 0 \end{bmatrix}$; (b) $\begin{bmatrix} 0 & 1 & 2 & 0 \\ 1 & 0 & 1 & 1 \\ 2 & 1 & 0 & 0 \\ 0 & 1 & 0 & 0 \end{bmatrix}$; (c) $\begin{bmatrix} 1 & 1 & 1 & 0 \\ 1 & 0 & 0 & 2 \\ 1 & 0 & 0 & 0 \\ 0 & 2 & 0 & 0 \end{bmatrix}$

8.68. See Fig. 8-73.

8.69. Let M and N be the two disjoint sets of vertices determining the bipartite graph G. Order the vertices in M first and then those in N.

8.70. (a) B, F, A, D, E, C.

  (b) $G = [A{:}B;\ B{:}A, C, D, E;\ C{:}F;\ D{:}B;\ E{:}B;\ F{:}C]$.

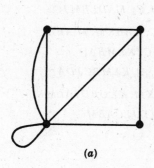
(a)

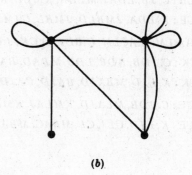
(b)

Fig. 8-73

8.71. (a) Each vertex is adjacent to the other four vertices.

(b) $G = [A:B, D, F; B:A, C, E; C:B, D, F; D:A, C, E; E:B, D, F; F:A, C, E]$.

(c) $G = [A:B, D; B:A, C, E; C:B, D; D:A, C, E; E:B, D]$.

8.72. See Fig. 8-74.

Vertex file

	1	2	3	4	5	6	7	8
VERTEX	A	B	C	D	E	F		
PTR	1	2	9	14	8	12		

Edge file

	1	2	3	4	5	6	7	8	9	10	11	12	13	14	15
NUMBER	22	22	33	33	44	44	55	55	66	66	77	77	88	88	
ADJ	2	1	6	5	4	2	5	2	6	3	6	2	4	1	
NEXT	13	5	0	0	7	0	11	3	0	4	0	10	0	6	

Fig. 8-74

8.73. (a) $|ACBEDFA| = 20$ or $|ACBEFDA| = 21$; (b) $|BCFEDAB| = 21$ or $|BCDEFAB| = 20$

8.74. (a) $|ABCDEA| = 775$, $|ABCEDA| = 725$, $|ABDCEA| = 1100$, $|ABDECA| = 900$, $|ABECDA| = 1050$, $|ABEDCA| = 900$, $|ACBDEA| = 825$, $|ACBEDA| = 775$, $|ACDBEA| = 1150$, $|ACEBDA| = 1100$, $|ADBCEA| = 975$, $|ADCBEA| = 975$

(b) $|ABCEDA| = 725$

8.75. (a) $G = [A:BJ; B:AJKL; C:DLM; D:CM; J:AB; K:BL; L:BCK; M:CD]$

(b) $[STACK: C, MLD, DL, L, KB, B, J, A], CMDLKBJA$

(c) $[STACK: K, LB, CB, MDB, DB, B, JA, A], KLCMDBJA$

8.76. (a) $[QUEUE: C, MLD, ML, L, KB, JAK, JA, J], CDMLBKAJ$

(b) $[QUEUE: K, LB, JAL, CJA, CJ, C, MD, M], KBLAJCDM$

8.77. (a) $G = [A:BMJKL; B:ACDJL; C:BJ; D:BKM; J:ABCM; K:ADL; L:ABKM; M:ADJL]$

(b) $[STACK: C, JB, MBA, LDAB, KBAD, DAB, AB, B], CJMLKDAB$

(c) $[STACK: K, LDA, MBAD, JDAB, CBAD, BAD, AD, D], KLMJCBAD$

8.78. (a) [QUEUE : $C, JB, LDAJ, MLDA, KMLD, KML, KM, K$], $CBJADLMK$

     (b) [QUEUE : $K, LDA, JMBLD, JMBL, CJMB, CJM, CJ, C$], $KADLBMJC$

8.79. (a) $G = [A:BLM; B:ACLM; C:BDJ; D:CK; J:CK; K:DJL; L:ABKM; M:ABL]$

     (b) [STACK : $C, JDB, KDB, LDB, MBAD, BAD, AD, D$], $CJKLMBAD$

     (c) [STACK : $K, LJD, MBAJD, BAJD, CAJD, JDA, DA, A$], $KLMBCJDA$

8.80. (a) [QUEUE : $C, JDB, MLAJD, KMLAJ, KMLA, KML, KM, K$], $CBDJALMK$

     (b) [QUEUE : $K, LJD, CLJ, CL, MBAC, MBA, MB, M$], $KDJLCABM$

CHAPTER 9

Directed Graphs

9.1 INTRODUCTION

Directed graphs are graphs in which the edges are one-way. Such graphs are frequently more useful in various dynamic systems such as digital computers or flow systems. However, this added feature makes it more difficult to determine certain properties of the graph. That is, processing such graphs may be similar to traveling in a city with many one-way streets.

This chapter gives the basic definitions and properties of directed graphs. Many of the definitions will be similar to those in the preceding chapter on (non-directed) graphs. However, for pedagogical reasons, this chapter will be mainly independent from the preceding chapter.

9.2 DIRECTED GRAPHS

A *directed graph G* or *digraph* (or simply *graph*) consists of two things:

   (i) A set *V* whose elements are called *vertices*, *nodes*, or *points*.

   (ii) A set *E* of *ordered* pairs (u, v) of vertices called *arcs* or *directed edges* or simply *edges*.

We will write $G(V, E)$ when we want to emphasize the two parts of *G*. We will also write $V(G)$ and $E(G)$ to denote, respectively, the set of vertices and the set of edges of a graph *G*. (If it is not explicitly stated, the context usually determines whether or not a graph *G* is a directed graph.)

Suppose $e = (u, v)$ is a directed edge in a digraph *G*. Then the following terminology is used:

   (a) *e begins* at *u* and *ends* at *v*.
   (b) *u* is the *origin* or *initial point* of *e*, and *v* is the *destination* or *terminal point* of *e*.
   (c) *v* is a *successor* of *u*.
   (d) *u* is *adjacent to v*, and *v* is *adjacent from u*.

If $u = v$, then *e* is called a *loop*.

The set of all successors of a vertex *u* is important; it is denoted and formally defined by

$$\text{succ}(u) = \{v \in V \mid \text{there exists an edge } (u, v) \in E\}$$

It is called the *successor list* or *adjacency list* of *u*.

A *picture* of a directed graph G is a representation of G in the plane. That is, each vertex u of G is represented by a dot (or small circle), and each (directed) edge $e = (u, v)$ is represented by an arrow or directed curve from the initial point u of e to the terminal point v. One usually presents a digraph G by its picture rather than explicitly listing its vertices and edges.

If the edges and/or vertices of a directed graph G are labeled with some type of data, then G is called a *labeled directed graph*.

A directed graph (V, E) is said to be *finite* if its set V of vertices and its set E of edges are finite.

EXAMPLE 9.1

(a) Consider the directed graph G pictured in Fig. 9-1(a). It consists of four vertices, A, B, C, D, that is, $V(G) = \{A, B, C, D\}$ and the seven following edges:

$$E(G) = \{e_1, e_2, \ldots, e_7\} = \{(A, D), (B, A), (B, A), (D, B), (B, C), (D, C), (B, B)\}$$

The edges e_2 and e_3 are said to be *parallel* since they both begin at B and end at A. The edge e_7 is a *loop* since it begins and ends at B.

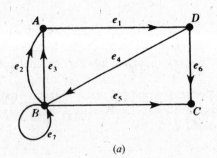 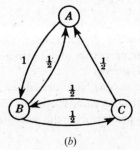

(a)   (b)

Fig. 9-1

(b) Suppose three boys, A, B, C, are throwing a ball to each other such that A always throws the ball to B, but B and C are just as likely to throw the ball to A as they are to each other. This dynamic system is pictured in Fig. 9-1(b) where edges are labeled with the respective probabilities, that is, A throws the ball to B with probability 1, B throws the ball to A and C each with probability 1/2, and C throws the ball to A and B each with probability 1/2.

Subgraphs

Let $G = G(V, E)$ be a directed graph, and let V' be a subset of the set V of vertices of G. Suppose E' is a subset of E such that the endpoints of the edges in E' belong to V'. Then $H(V', E')$ is a directed graph, and it is called a *subgraph* of G. In particular, if E' contains all the edges in E whose endpoints belong to V', then $H(V', E')$ is called the subgraph of G *generated* or *determined* by V'. For example, for the graph $G = G(V, E)$ in Fig. 9-1(a), $H(V', E')$ is the subgraph of G determine by the vertex set V' where

$$V' = \{B, C, D\} \quad \text{and} \quad E' = \{e_4, e_5, e_6, e_7\} = \{(D, B), (B, C), (D, C), (B, B)\}$$

9.3 BASIC DEFINITIONS

This section discusses the questions of degrees of vertices, paths, and connectivity in a directed graph.

Degrees

Suppose G is a directed graph. The *outdegree* of a vertex v of G, written outdeg(v), is the number of edges beginning at v, and the *indegree* of v, written indeg(v), is the number of edges ending at v. Since each edge begins and ends at a vertex we immediately obtain the following theorem.

Theorem 9.1: The sum of the outdegrees of the vertices of a digraph G equals the sum of the indegrees of the vertices, which equals the number of edges in G.

A vertex v with zero indegree is called a *source*, and a vertex v with zero outdegree is called a *sink*.

EXAMPLE 9.2 Consider the graph G in Fig. 9-1(a). We have:

$$\text{outdeg}\,(A) = 1, \quad \text{outdeg}\,(B) = 4, \quad \text{outdeg}\,(C) = 0, \quad \text{outdeg}\,(D) = 2,$$
$$\text{indeg}\,(A) = 2, \quad \text{indeg}\,(B) = 2, \quad \text{indeg}\,(C) = 2, \quad \text{indeg}\,(D) = 1.$$

As expected, the sum of the outdegrees equals the sum of the indegrees, which equals the number 7 of edges. The vertex C is a sink since no edge begins at C. The graph has no sources.

Paths

Let G be a directed graph. The concepts of path, simple path, trail, and cycle carry over from nondirected graphs to the directed graph G except that the directions of the edges must agree with the direction of the path. Specifically:

  (i) A (*directed*) *path* P in G is an alternating sequence of vertices and directed edges, say,

$$P = \left(v_0,\ e_1,\ v_1,\ e_2,\ v_2, \ldots, e_n,\ v_n\right)$$

     such that each edge e_i begins at v_{i-1} and ends at v_i. If there is no ambiguity, we denote P by its sequence of vertices or its sequence of edges.

 (ii) The *length* of the path P is n, its number of edges.

(iii) A *simple path* is a path with distinct vertices. A *trail* is a path with distinct edges.

(iv) A *closed path* has the same first and last vertices.

 (v) A *spanning path* contains all the vertices of G.

(vi) A *cycle* (or *circuit*) is a closed path with distinct vertices (except the first and last).

(vii) A *semipath* is the same as a path except the edge e_i may begin at v_{i-1} or v_i and end at the other vertex. *Semitrails* and *semisimple paths* are analogously defined.

A vertex v is *reachable* from a vertex u if there is a path from u to v. If v is reachable from u, then (by eliminating redundant edges) there must be a simple path from u to v.

EXAMPLE 9.3 Consider the graph G in Fig. 9-1(a).

(a) The sequence $P_1 = (D, C, B, A)$ is a semipath but not a path since (C, B) is not an edge; that is, the direction of $e_5 = (C, B)$ does not agree with the direction of P_1.

(b) The sequence $P_2 = (D, B, A)$ is a path from D to A since (D, B) and (B, A) are edges. Thus A is reachable from D.

Connectivity

There are three types of connectivity in a directed graph G:

(i) G is *strongly connected* or *strong* if, for any pair of vertices u and v in G, there is a path from u to v and a path from v to u, that is, each is reachable from the other.

(ii) G is *unilaterally connected* or *unilateral* if, for any pair of vertices u and v in G, there is a path from u to v or a path from v to u, that is, one of them is reachable from the other.

(iii) G is *weakly connected* or *weak* if there is a semipath between any pair of vertices u and v in G.

Let G' be the (nondirected) graph obtained from a directed graph G by allowing all edges in G to be nondirected. Clearly, G is weakly connected if and only if the graph G' is connected.

Observe that strongly connected implies unilaterally connected which implies weakly connected. We say that G is *strictly unilateral* if it is unilateral but not strong, and we say that G is *strictly weak* if it is weak but not unilateral.

Connectivity can be characterized in terms of spanning paths as follows:

Theorem 9.2: Let G be a finite directed graph. Then:

(i) G is strong if and only if G has a closed spanning path.

(ii) G is unilateral if and only if G has a spanning path.

(iii) G is weak if and only if G has a spanning semipath.

EXAMPLE 9.4 Consider the graph G in Fig. 9-1(a). It is weakly connected since the underlying nondirected graph is connected. There is no path from C to any other vertex, that is, C is a sink, so G is not strongly connected. However, $P = (B, A, D, C)$ is a spanning path, so G is unilaterally connected.

Graphs with sources and sinks appear in many applications (such as flow diagrams and networks). A sufficient condition for such vertices to exist follows.

Theorem 9.3: Suppose a finite directed graph G is cycle-free, that is, contains no (directed) cycles. Then G contains a source and a sink.

Proof: Let $P = (v_0, v_1, \ldots, v_n)$ be a simple path of maximum length, which exists since G is finite. Then the last vertex v_n is a sink; otherwise an edge (v_n, u) will either extend P or form a cycle if $u = v_i$, for some i. Similarly, the first vertex v_0 is a source.

9.4 ROOTED TREES

Recall that a tree graph is a connected cycle-free graph, that is, a connected graph without any cycles. A *rooted tree* T is a tree graph with a designated vertex r called the *root* of the tree. Since there is a unique simple path from the root r to any other vertex v in T, this determines a direction to the edges of T. Thus T may be viewed as a directed graph. We note that any tree may be made into a rooted tree by simply selecting one of the vertices as the root.

Consider a rooted tree T with root r. The length of the path from the root r to any vertex v is called the *level* (or *depth*) of v, and the maximum vertex level is called the *depth* of the tree. Those vertices with degree 1, other than the root r, are called the *leaves* of T, and a directed path from a vertex to a leaf is called a *branch*.

One usually draws a picture of a rooted tree T with the root at the top of the tree. Figure 9-2(a) shows a rooted tree T with root r and 10 other vertices. The tree has five leaves, d, f, h, i, and j. Observe that: $level(a) = 1$, $level(f) = 2$, $level(j) = 3$. Furthermore, the depth of the tree is 3.

The fact that a rooted tree T gives a direction to the edges means that we can give a precedence relationship between the vertices. Specifically, we will say that a vertex u *precedes* a vertex v or that v *follows* u if there is

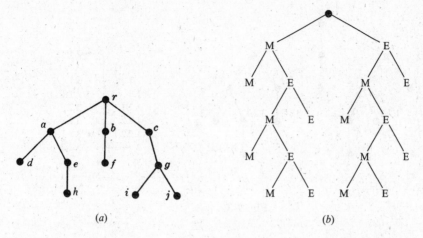

Fig. 9-2

a (directed) path from v to u. In particular, we say that v *immediately follows* u if (u, v) is an edge, that is, if v follows u and v is adjacent to u. We note that every vertex v, other than the root, immediately follows a unique vertex, but that v can be immediately followed by more than one vertex. For example, in Fig. 9-2(a), the vertex j follows c but immediately follows g. Also, both i and j immediately follow g.

A rooted tree T is also a useful device to enumerate all the logical possibilities of a sequence of events where each event can occur in a finite number of ways. This is illustrated in the following example.

EXAMPLE 9.5 Suppose Marc and Erik are playing a tennis tournament such that the first person to win two games in a row or who wins a total of three games wins the tournament. Find the number of ways the tournament can proceed.

The rooted tree in Fig. 9-2(b) shows the various ways that the tournament could proceed. There are 10 leaves which correspond to the 10 ways that the tournament can occur:

MM, MEMM, MEMEM, MEMEE, MEE, EMM, EMEMM, EMEME, EMEE, EE

Specifically, the path from the root to the leaf describes who won which games in the particular tournament.

Ordered Rooted Trees

Consider a rooted tree T in which the edges leaving each vertex are ordered. Then we have the concept of an *ordered rooted tree*. One can systematically label (or *address*) the vertices of such a tree as follows: We first assign 0 to the root r. We next assign $1, 2, 3, \ldots$ to the vertices immediately following r according as the edges were ordered. We then label the remaining vertices in the following way. If a is the label of a vertex v, then $a.1, a.2, \ldots$ are assigned to the vertices immediately following v according as the edges were ordered. We illustrate this address system in Fig. 9-3(a), where edges are pictured from left to right according to their order. Observe that the number of decimal points in any label is one less than the level of the vertex. We will refer to this labeling system as the *universal address system* for an ordered rooted tree.

The universal address system gives us an important way of linearly describing (or storing) an ordered rooted tree. Specifically, given addresses a and b, we let $a < b$ if $b = a.c$, (that is, a is an *initial segment* of b), or if there exist positive integers m and n with $m < n$ such that

$$a = \text{r.m.s} \quad \text{and} \quad b = \text{r.n.t}$$

This order is called the *lexicographic order* since it is similar to the way words are arranged in a dictionary. For example, the addresses in Fig. 9-3(a) are linearly ordered as pictured in Fig. 9-3(b). This lexicographic order is

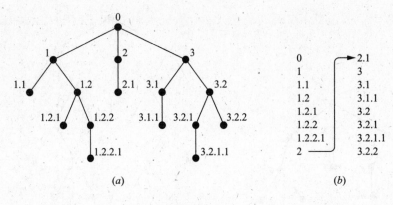

Fig. 9-3

identical to the order obtained by moving down the leftmost branch of the tree, then the next branch to the right, then the second branch to the right, and so on.

9.5   SEQUENTIAL REPRESENTATION OF DIRECTED GRAPHS

There are two main ways of maintaining a directed graph G in the memory of a computer. One way, called the *sequential representation* of G, is by means of its adjacency matrix A. The other way, called the *linked representation* of G, is by means of linked lists of neighbors. This section covers the first representation. The linked representation will be covered in Section 9.7.

Suppose a graph G has m vertices (nodes) and n edges. We say G is *dense* if $m = O(n^2)$ and *sparse* if $m = O(n)$ or even if $m = O(n \log n)$. The matrix representation of G is usually used when G is dense, and linked lists are usually used when G is sparse. Regardless of the way one maintains a graph G in memory, the graph G is normally input into the computer by its formal definition, that is, as a collection of vertices and a collection of edges (pairs of vertices).

Remark: In order to avoid special cases of our results, we assume, unless otherwise stated, that $m > 1$ where m is the number of vertices in our graph G. Therefore, G is not connected if G has no edges.

Digraphs and Relations, Adjacency Matrix

Let $G(V, E)$ be a *simple* directed graph, that is, a graph without parallel edges. Then E is simply a subset of $V \times V$, and hence E is a relation on V. Conversely, if R is a relation on a set V, then $G(V, R)$ is a simple directed graph. Thus the concepts of relations on a set and simple directed graphs are one and the same. In fact, in Chapter 2, we already introduced the directed graph corresponding to a relation on a set.

Suppose G is a simple directed graph with m vertices, and suppose the vertices of G have been ordered and are called $v_1, v_2, \ldots, v_m$. Then the *adjacency matrix* $A = [a_{ij}]$ of G is the $m \times m$ matrix defined as follows:

$$a_{ij} = \begin{cases} 1 & \text{if there is an edge } (v_i, v_j) \\ 0 & \text{otherwise} \end{cases}$$

Such a matrix A, which contains entries of only 0 or 1, is called a *bit matrix* or a *Boolean matrix*. (Although the adjacency matrix of an undirected graph is symmetric, this is not true here for a directed graph.)

The adjacency matrix A of the graph G does depend on the ordering of the vertices of G. However, the matrices resulting from two different orderings are closely related in that one can be obtained from the other by simply interchanging rows and columns. Unless otherwise stated, we assume that the vertices of our matrix have a fixed ordering.

Remark: The adjacency matrix $A = [a_{ij}]$ may be extended to directed graphs with parallel edges by setting:

$$a_{ij} = \text{the number of edges beginning at } v_i \text{ and ending at } v_j$$

Then the entries of A will be nonnegative integers. Conversely, every $m \times m$ matrix A with nonnegative integer entries uniquely defines a directed graph with m vertices.

EXAMPLE 9.6 Let G be the directed graph in Fig. 9-4(a) with vertices v_1, v_2, v_3, v_4. Then the adjacency matrix A of G appears in Fig. 9-4(b). Note that the number of 1's in A is equal to the number (eight) of edges.

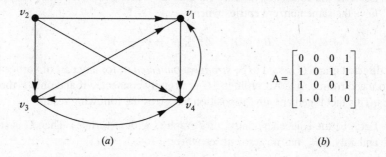

$$A = \begin{bmatrix} 0 & 0 & 0 & 1 \\ 1 & 0 & 1 & 1 \\ 1 & 0 & 0 & 1 \\ 1 & 0 & 1 & 0 \end{bmatrix}$$

(a) $\qquad\qquad\qquad\qquad\qquad\qquad$ (b)

Fig. 9-4

Consider the powers A, A^2, A^3,... of the adjacency matrix $A = [a_{ij}]$ of a graph G. Let

$$a_K(i, j) = \text{the } ij \text{ entry in the matrix } A^K$$

Note that $a_1(i, j) = a_{ij}$ gives the number of paths of length 1 from vertex v_i to vertex v_j. One can show that $a_2(i, j)$ gives the number of paths of length 2 from v_i to v_j. In fact, we prove in Problem 9.17 the following general result.

Proposition 9.4:   Let A be the adjacency matrix of a graph G. Then $a_K(i, j)$, the ij entry in the matrix A^K, gives the number of paths of length K from v_i to v_j.

EXAMPLE 9.7 Consider again the graph G and its adjacency matrix A appearing in Fig. 9-4. The powers A^2, A^3, and A^4 of A follow:

$$A^2 = \begin{bmatrix} 1 & 0 & 1 & 0 \\ 2 & 0 & 1 & 2 \\ 1 & 0 & 1 & 1 \\ 1 & 0 & 0 & 2 \end{bmatrix}, \quad A^3 = \begin{bmatrix} 1 & 0 & 0 & 2 \\ 3 & 0 & 2 & 3 \\ 2 & 0 & 1 & 2 \\ 2 & 0 & 2 & 1 \end{bmatrix}, \quad A^4 = \begin{bmatrix} 2 & 0 & 2 & 1 \\ 5 & 0 & 3 & 5 \\ 3 & 0 & 2 & 3 \\ 3 & 0 & 1 & 4 \end{bmatrix}$$

Observe that $a_2(4, 1) = 1$, so there is a path of length 2 from v_4 to v_1. Also, $a_3(2, 3) = 2$, so there are two paths of length 3 from v_2 to v_3; and $a_4(2, 4) = 5$, so there are five paths of length 4 from v_2 to v_4.

Remark: Let A be the adjacency matrix of a graph G, and let B_r be the matrix defined by:

$$B_r = A + A^2 + A^3 + \cdots + A^r$$

Then the ij entry of the matrix B_r gives the number of paths of length r or less from vertex v_i to vertex v_j.

Path Matrix

Let $G = G(V, E)$ be a simple directed graph with m vertices v_1, v_2, ..., v_m. The *path matrix* or *reachability matrix* of G is the m-square matrix $P = [p_{ij}]$ defined as follows:

$$p_{ij} = \begin{cases} 1 & \text{if there is a path from to } v_i \text{ to } v_j \\ 0 & \text{otherwise} \end{cases}$$

(The path matrix P may be viewed as the transitive closure of the relation E on V.)

Suppose now that there is a path from vertex v_i to vertex v_j in a graph G with m vertices. Then there must be a simple path from v_i to v_j when $v_i \neq v_j$, or there must be a cycle from v_i to v_j when $v_i = v_j$. Since G has m vertices, such a simple path must have length $m - 1$ or less, or such a cycle must have length m or less. This means that there is a nonzero ij entry in the matrix B_m (defined above) where A is the adjacency matrix of G. Accordingly, the path matrix P and B_m have the same nonzero entries. We state this result formally.

Proposition 9.5: Let A be the adjacency matrix of a graph G with m vertices. Then the path matrix P and B_m have the same nonzero entries where

$$B_m = A + A^2 + A^3 + \cdots + A^m$$

Recall that a directed graph G is said to be *strongly connected* if, for any pair of vertices u and v in G, there is a path from u to v and from v to u. Accordingly, G is strongly connected if and only if the path matrix P of G has no zero entries. This fact together with Proposition 9.5 gives the following result.

Proposition 9.6: Let A be the adjacency matrix of a graph G with m vertices. Then G is strongly connected if and only if B_m has no zero entries where

$$B_m = A + A^2 + A^3 + \cdots + A^m$$

EXAMPLE 9.8 Consider the graph G and its adjacency matrix A appearing in Fig. 9-4. Here G has $m = 4$ vertices. Adding the matrix A and matrices A^2, A^3, A^4 in Example 9.7, we obtain the following matrix B_4 and also path (reachability) matrix P by replacing the nonzero entries in B_4 by 1:

$$B_4 = \begin{bmatrix} 4 & 0 & 3 & 4 \\ 11 & 0 & 7 & 11 \\ 7 & 0 & 4 & 7 \\ 7 & 0 & 4 & 7 \end{bmatrix} \quad \text{and} \quad P = \begin{bmatrix} 1 & 0 & 1 & 1 \\ 1 & 0 & 1 & 1 \\ 1 & 0 & 1 & 1 \\ 1 & 0 & 1 & 1 \end{bmatrix}$$

Examining the matrix B_4 or P, we see zero entries; hence G is not strongly connected. In particular, we see that the vertex v_2 is not reachable from any of the other vertices.

Remark: The adjacency matrix A and the path matrix P of a graph G may be viewed as logical (Boolean) matrices where 0 represents "false" and 1 represents "true." Thus the logical operations of $\wedge$ (AND) and $\vee$ (OR) may be applied to the entries of A and P where these operations, used in the next section, are defined in Fig. 9-5.

$\wedge$	0	1
0	0	0
1	0	1

(a) AND.

$\vee$	0	1
0	0	1
1	1	1

(b) OR.

Fig. 9-5

Transitive Closure and the Path Matrix

Let R be a relation on a finite set V with m elements. As noted above, the relation R may be identified with the simple directed graph $G = G(V, R)$. We note that the composition relation $R^2 = R \times R$ consists of all pairs (u, v) such that there is a path of length 2 from u to v. Similarly:

$$R^K = \{(u, v) \mid \text{there is a path of length } K \text{ from } u \text{ to } v\}.$$

The transitive closure R^* of the relation R on V may now be viewed as the set of ordered pairs (u, v) such that there is a path from u to v in the graph G. Furthermore, by the above discussion, we need only look at simple paths of length $m - 1$ or less and cycles of length m or less. Accordingly, we have the following result which characterizes the transitive closure R^* of R.

Theorem 9.7: Let R be a relation on a set V with m elements. Then:

      (i) $R^* = R \cup R^2 \cup \ldots \cup R^m$ is the transitive closure of R.

      (ii) The path matrix P of $G(V, R)$ is the adjacency matrix of $G'(V, R^*)$.

9.6   WARSHALL'S ALGORITHM, SHORTEST PATHS

Let G be a directed graph with m vertices, $v_1, v_2, \ldots, v_m$. Suppose we want to find the path matrix P of the graph G. Warshall gave an algorithm which is much more efficient than calculating the powers of the adjacency matrix A. This algorithm is defined in this section, and a similar algorithm is used to find shortest paths in G when G is weighted.

Warshall's Algorithm

First we define m-square Boolean matrices $P_0, P_1, \ldots, P_m$ where $P_k[i, j]$ denotes the ij entry of the matrix P_k:

$$P_k[i, j] = \begin{cases} 1 & \text{if there is a simple path from } v_i \text{ to } v_j \text{ which does not use any} \\ & \text{other vertices except possibly } v_1, v_2, \ldots, v_k, \\ 0 & \text{otherwise.} \end{cases}$$

For example,

$$P_3[i, j] = 1 \quad \text{if there is a simple path from } v_i \text{ to } v_j \text{ which does not use any}$$
$$\text{other vertices except possibly } v_1, v_2, v_3.$$

Observe that the first matrix $P_0 = A$, the adjacency matrix of G. Furthermore, since G has only m vertices, the last matrix $P_m = P$, the path matrix of G.

Warshall observed that $P_k[i, j] = 1$ can occur only if one of the following two cases occurs:

(1) There is a simple path from v_i to v_j which does not use any other vertices except possibly $v_1, v_2, \ldots, v_{k-1}$; hence

$$P_{k-1}[i, j] = 1$$

(2) There is a simple path from v_i to v_k and a simple path from v_k to v_j where each simple path does not use any other vertices except possibly $v_1, v_2, \ldots, v_{k-1}$; hence

$$P_{k-1}[i, k] = 1 \quad \text{and} \quad P_{k-1}[k, j] = 1$$

These two cases are pictured as follows:

$$(1) \ v_i \rightarrow \cdots \rightarrow v_j; \qquad (2) \ v_i \rightarrow \cdots \rightarrow v_k \rightarrow \cdots v_j$$

where $\rightarrow \cdots \rightarrow$ denotes part of a simple path which does not use any other vertices except possibly $v_1, v_2, \ldots, v_{k-1}$. Accordingly, the elements of P_k can be obtained by:

$$P_k[i, j] = P_{k-1}[i, j] \ \vee \ (P_{k-1}[i, k] \ \wedge \ P_{k-1}[k, j])$$

where we use the logical operations of a $\wedge$ (AND) and $\vee$ (OR). In other words we can obtain each entry in the matrix P_k by looking at only three entries in the matrix P_{k-1}. Warshall's algorithm appears in Fig. 9-6.

Algorithm 9.1 (Warshall's Algorithm):   A directed graph G with M vertices is maintained in memory by its adjacency matrix A. This algorithm finds the (Boolean) path matrix P of the graph G.

Step 1.   Repeat for $I, J = 1, 2, \ldots, M$: [Initializes P.]
              If $A[I, J] = 0$, then: Set $P[I, J]: = 0$;
              Else: Set $P[I, J]: = 1$.
         [End of loop.]

Step 2.   Repeat Steps 3 and 4 for $K = 1, 2, \ldots, M$: [Updates P.]

Step 3.   Repeat Step 4 for $I = 1, 2, \ldots, M$:

Step 4.   Repeat for $J = 1, 2, \ldots, M$:
                        Set $P[I, J]:= P[I, J] \lor (P[I, K] \land P[K, J])$.
                  [End of loop.]
               [End of Step 3 loop.]
            [End of Step 2 loop.]

Step 5.   Exit.

Fig. 9-6

Shortest-path Algorithm

Let G be a simple directed graph with m vertices, $v_1, v_2, \ldots, v_m$. Suppose G is weighted; that is, suppose each edge e of G is assigned a nonnegative number $w(e)$ called the *weight* or *length* of e. Then G may be maintained in memory by its weight matrix $W = [w_{ij}]$ defined as follows:

$$w_{ij} = \begin{cases} w(e) & \text{if there is an edge } e \text{ from } v_i \text{ to } v_j \\ 0 & \text{if there is no edge from } v_i \text{ to } v_j \end{cases}$$

The path matrix P tells us whether or not there are paths between the vertices. Now we want to find a matrix Q which tells us the lengths of the shortest paths between the vertices or, more exactly, a matrix $Q = [q_{ij}]$ where

$$q_{ij} = \text{length of the shortest path from } v_i \text{ to } v_j$$

Next we describe a modification of Warshall's algorithm which efficiently finds us the matrix Q.

Here we define a sequence of matrices $Q_0, Q_1, \ldots, Q_m$ (analogous to the above matrices $P_0, P_1, \ldots, P_m$) where $Q_k[i, j]$, the ij entry of Q_k, is defined as follows:

$Q_k[i, j] = $ the smaller of the length of the preceding path from v_i to v_j or the sum of the lengths of the preceding paths from v_i to v_k and from v_k to v_j.

More exactly,

$$Q_k[i, j] = \text{MIN}(Q_{k-1}[i, j], \ Q_{k-1}[i, k] + Q_{k-1}[k, j])$$

The initial matrix Q_0 is the same as the weight matrix W except that each 0 in w is replaced by ∞ (or a very, very large number). The final matrix Q_m will be the desired matrix Q.

EXAMPLE 9.9 Figure 9-7 shows a weighted graph G and its weight matrix W where we assume that $v_1 = R$, $v_2 = S$, $v_3 = T$, $v_4 = U$.

Suppose we apply the modified Warshall's algorithm to our weighted graph G in Fig. 9-7. We will obtain the matrices Q_0, Q_1, Q_3, and Q_4 in Fig. 9-8. (To the right of each matrix Q_k in Fig. 9-8, we show the matrix of

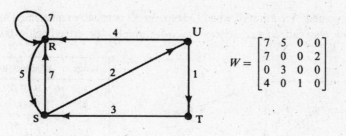

$$W = \begin{bmatrix} 7 & 5 & 0 & 0 \\ 7 & 0 & 0 & 2 \\ 0 & 3 & 0 & 0 \\ 4 & 0 & 1 & 0 \end{bmatrix}$$

Fig. 9-7

paths which correspond to the lengths in the matrix Q_k.) The entries in the matrix Q_0 are the same as the weight matrix W except each 0 in W is replaced by ∞ (a very large number). We indicate how the circled entries are obtained:

$$Q_1[4, 2] = \text{MIN}(Q_0[4, 2], \quad Q_0[4, 1] + Q_0[1, 2]) = \text{MIN}(\infty, 4 + 5) = 9$$
$$Q_2[1, 3] = \text{MIN}(Q_1[1, 3], \quad Q_1[1, 2] + Q_1[2, 3]) = \text{MIN}(\infty, 5 + \infty) = \infty$$
$$Q_3[4, 2] = \text{MIN}(Q_2[4, 2], \quad Q_2[4, 3] + Q_2[3, 2]) = \text{MIN}(9, 3 + 1) = 4$$
$$Q_4[3, 1] = \text{MIN}(Q_3[3, 1], \quad Q_3[3, 4] + Q_3[4, 1]) = \text{MIN}(10, 5 + 4) = 9$$

The last matrix $Q_4 = Q$, the desired shortest-path matrix.

$$Q_0 = \begin{bmatrix} 7 & 5 & \infty & \infty \\ 7 & \infty & \infty & 2 \\ \infty & 3 & \infty & \infty \\ 4 & \infty & 1 & \infty \end{bmatrix} \qquad \begin{bmatrix} RR & RS & — & — \\ SR & — & — & SU \\ — & TS & — & — \\ UR & — & UT & — \end{bmatrix}$$

$$Q_1 = \begin{bmatrix} 7 & 5 & \infty & \infty \\ 7 & 12 & \infty & 2 \\ \infty & 3 & \infty & \infty \\ 4 & ⑨ & 1 & \infty \end{bmatrix} \qquad \begin{bmatrix} RR & RS & — & — \\ SR & SRS & — & SU \\ — & TS & — & — \\ UR & URS & UT & — \end{bmatrix}$$

$$Q_2 = \begin{bmatrix} 7 & 5 & ⊗ & 7 \\ 7 & 12 & \infty & 2 \\ 10 & 3 & \infty & 5 \\ 4 & 9 & 1 & 11 \end{bmatrix} \qquad \begin{bmatrix} RR & RS & — & RSU \\ SR & SRS & — & SU \\ TSR & TS & — & TSU \\ UR & URS & UT & URS \end{bmatrix}$$

$$\begin{bmatrix} 7 & 5 & \infty & 7 \\ & 12 & \infty & 2 \\ & 3 & \infty & 5 \\ & & 1 & 6 \end{bmatrix} \qquad \begin{bmatrix} RR & RS & — & RSU \\ SR & SRS & — & SU \\ TSR & TS & — & TSU \\ UR & UTS & UT & UTSU \end{bmatrix}$$

$$\begin{bmatrix} & & & 7 \end{bmatrix} \qquad \begin{bmatrix} RR & RS & RSUT & RSU \\ SR & SURS & SUT & SU \\ TSUR & TS & TSUT & TSU \\ UR & UTS & UT & UTSU \end{bmatrix}$$

Fig. 9-8

9.7 LINKED REPRESENTATION OF DIRECTED GRAPHS

Let G be a directed graph with m vertices. Suppose the number of edges of G is $O(m)$ or even $O(m \log m)$, that is, suppose G is sparse. Then the adjacency matrix A of G will contain many zeros; hence a great deal of memory space will be wasted. Accordingly, when G is sparse, G is usually represented in memory by some type of *linked representation*, also called an *adjacency structure*, which is described below by means of an example.

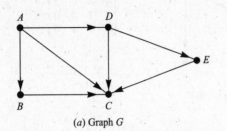

Vertex	Adjacency list
A	B, C, D
B	C
C	$\varnothing$
D	C, E
E	C

(a) Graph G (b) Adjacency lists of G

Fig. 9-9

Consider the directed graph G in Fig. 9-9(a). Observe that G may be equivalently defined by the table in Fig. 9-9(b), which shows each vertex in G followed by its *adjacency list*, also called its *successors* or *neighbors*. Here the symbol $\varnothing$ denotes an empty list. Observe that each edge of G corresponds to a unique vertex in an adjacency list and vice versa. Here G has seven edges and there are seven vertices in the adjacency lists. This table may also be presented in the following compact form where a colon ":" separates a vertex from its list of neighbors, and a semicolon ";" separates the different lists:

$$G = [A : B, C, D; \quad B : C; \quad C : \varnothing; \quad D : C, E; \quad E : C]$$

The *linked representation* of a directed graph G maintains G in memory by using linked lists for its adjacency lists. Specifically, the linked representation will normally contain two files (sets of records), one called the Vertex File and the other called the Edge File, as follows.

(a) *Vertex File:* The Vertex File will contain the list of vertices of the graph G usually maintained by an array or by a linked list. Each record of the Vertex File will have the form

VERTEX	NEXT-V	PTR	

Here VERTEX will be the name of the vertex, NEXT-V points to the next vertex in the list of vertices in the Vertex File, and PTR will point to the first element in the adjacency list of the vertex appearing in the Edge File. The shaded area indicates that there may be other information in the record corresponding to the vertex.

(b) *Edge File:* The Edge File contains the edges of G and also contains all the adjacency lists of G where each list is maintained in memory by a linked list. Each record of the Edge File will represent a unique edge in G and hence will correspond to a unique vertex in an adjacency list. The record will usually have the form

EDGE	BEG-V	END-V	NEXT-E	

Here:

(1) EDGE will be the name of the edge (if it has a name).
(2) BEG-V points to location in the Vertex File of the initial (beginning) vertex of the edge.
(3) END-V points to the location in the Vertex File of the terminal (ending) vertex of the edge. The adjacency lists appear in this field.
(4) NEXT-E points to the location in the Edge File of the next vertex in the adjacency list.

We emphasize that the adjacency lists consist of terminal vertices and hence are maintained by the END-V field. The shaded area indicates that there may be other information in the record corresponding to the edge. We note that the order of the vertices in any adjacency list does depend on the order in which the edges (pairs of vertices) appear in the input.

Figure 9-10 shows how the graph G in Fig. 9-9(a) may appear in memory. Here the vertices of G are maintained in memory by a linked list using the variable START to point to the first vertex. (Alternatively, one could use a linear array for the list of vertices, and then NEXT-V would not be required.) The choice of eight locations for the Vertex File and 10 locations for the Edge File is arbitrary. The additional space in the files will be used if additional vertices or edges are inserted in the graph. Figure 9-10 also shows, with arrows, the adjacency list [B, C, D] of the vertex A.

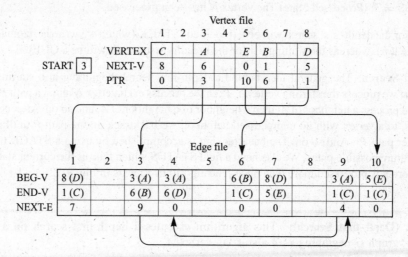

Fig. 9-10

9.8   GRAPH ALGORITHMS: DEPTH-FIRST AND BREADTH-FIRST SEARCHES

This section discusses two important graph algorithms for a given graph G. Any particular graph algorithm may depend on the way G is maintained in memory. Here we assume G is maintained in memory by its adjacency structure. Our test graph G with its adjacency structure appears in Fig. 9-11.

Many applications of graphs require one to systematically examine the vertices and edges of a graph G. There are two standard ways that this is done. One way is called a *depth-first search* (DFS) and the other is called a *breadth-first search* (BFS). (These algorithms are essentially identical to the analogous algorithms for undirected graphs in Chapter 8.)

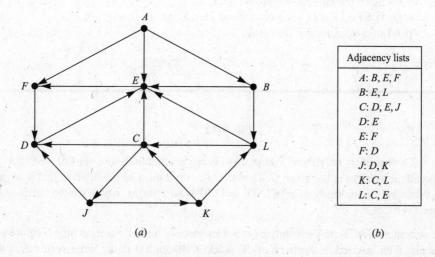

Fig. 9-11

During the execution of our algorithms, each vertex (node) N of G will be in one of three states, called the *status* of N, as follows:

STATUS = 1: (Ready State) The initial state of the vertex N.

STATUS = 2: (Waiting State) The vertex N is on a (waiting) list, waiting to be processed.

STATUS = 3: (Processed State) The vertex N has been processed.

The waiting list for the depth-first search will be a (modified) STACK (which we write horizontally with the TOP of STACK on the left), whereas the waiting list for the breadth-first search will be a QUEUE.

(a) **Depth-first Search:** The general idea behind a depth-first search beginning at a starting vertex A is as follows. First we process the starting vertex A. Then we process each vertex N along a path P which begins at A; that is, we process a neighbor of A, then a neighbor of a neighbor of A, and so on. After coming to a "dead end," that is, to a vertex with no unprocessed neighbor, we backtrack on the path P until we can continue along another path P'. And so on. The backtracking is accomplished by using a STACK to hold the initial vertices of future possible paths. We also need a field STATUS which tells us the current status of any vertex so that no vertex is processed more than once. The algorithm appears in Fig. 9-12.

Algorithm 9.2 (Depth-first Search):   This algorithm executes a depth-first search on a directed graph G beginning with a starting vertex A.

Step 1.   Initialize all vertices to the ready state (STATUS = 1).

Step 2.   Push the starting vertex A onto STACK and change the status of A to the waiting state (STATUS = 2).

Step 3.   Repeat Steps 4 and 5 until STACK is empty.

Step 4.   Pop the top vertex N of STACK. Process N, and set STATUS(N) = 3, the processed state.

Step 5.   Examine each neighbor J of N.
   (a)   If   STATUS (J) = 1 (ready state), push J onto STACK and reset STATUS (J) = 2 (waiting state).
   (b)   If STATUS (J) = 2 (waiting state), delete the previous J from the STACK and push the current J onto STACK.
   (c)   If STATUS (J) = 3 (processed state), ignore the vertex J.
   [End of Step 3 loop.]

Step 6.   Exit.

Fig. 9-12

Algorithm 9.2 will process only those vertices which are reachable from a starting vertex A. Suppose one wants to process all the vertices in the graph G. Then the algorithm must be modified so that it begins again with another vertex that is still in the ready state (STATE = 1). This new vertex, say B, can be obtained by traversing through the list of vertices.

Remark: The structure STACK in Algorithm 9.2 is not technically a stack since, in Step 5(b), we allow a vertex J to be deleted and then inserted in the front of the stack. (Although it is the same vertex J, it will represent a different edge.) If we do not delete the previous J in Step 5(b), then we obtain an alternative traversal algorithm.

EXAMPLE 9.10 Consider our test graph G in Fig. 9-11. Suppose we want to find and print all the vertices reachable from the vertex J (including J itself). One way to do this is to use a depth-first search of G starting at the vertex J.

Applying Algorithm 9.2, the vertices will be processed and printed in the following order:

$$J, \quad K, \quad L, \quad E, \quad F, \quad D, \quad C$$

Specifically, Fig. 9-13(a) shows the sequence of waiting lists in STACK and the vertices being processed. (The slash / indicates that vertex is deleted from the waiting list.) We emphasize that each vertex, excluding J, comes from an adjacency list and hence it is the terminal vertex of a unique edge of the graph. We have indicated the edge by labeling the terminal vertex with the initial vertex of the edge as a subscript. For example,

$$D_j$$

means D is in the adjacency list of J, and hence D is the terminal vertex of an edge beginning at J. These edges form a rooted tree T with J as the root, which is pictured in Fig. 9-13(b). (The numbers indicate the order the edges are added to the tree.) This tree T spans the subgraph G' of G consisting of vertices reachable from J.

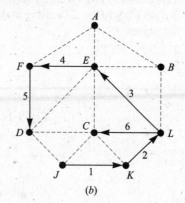

STACK	Vertex
J	J
K_J, D_J	K_J
L_K, C_K, D_J	L_K
$E_L, C_L, \cancel{C}_K, D_J$	E_L
F_E, C_L, D_J	F_E
$D_F, C_L, \cancel{D}_J$	D_F
C_L	C_L
$\varnothing$	

(a)

(b)

Fig. 9-13

(b) **Breath-first Search:** The general idea behind a breadth-first search beginning at a starting vertex A is as follows. First we process the starting vertex A. Then we process all the neighbors of A. Then we process all the neighbors of neighbors of A. And so on. Naturally we need to keep track of the neighbors of a vertex, and we need to guarantee that no vertex is processed twice. This is accomplished by using a QUEUE to hold vertices that are waiting to be processed; and by a field STATUS which tells us the current status of a vertex. The algorithm appears in Fig. 9-14.

Algorithm 9.3 will process only those vertices which are reachable from a starting vertex A. Suppose one wants to process all the vertices in the graph G. Then the algorithm must be modified so that it begins again with another vertex that is still in the ready state (STATE = 1). This new vertex, say B, can be obtained by traversing through the list of vertices.

EXAMPLE 9.11 Consider our test graph G in Fig. 9-11. Suppose G represents the daily flights between cities, and suppose we want to fly from city A to city J with the minimum number of stops. That is, we want to find a shortest path P from A to J (where each edge has weight 1). One way to do this is to use a breadth-first search of G starting at the vertex A, and stop as soon as J is encountered.

Figure 9-15(a) shows the sequence of waiting lists in QUEUE and the vertices being processed up to the time vertex J is encountered. We then work backwards from J to obtain the following desired path which is pictured in Fig. 9-15(b):

$$J_C \leftarrow C_L \leftarrow L_B \leftarrow B_A \leftarrow A \quad \text{or} \quad A \rightarrow B \rightarrow L \rightarrow C \rightarrow J$$

Thus a flight from city A to city J will make three intermediate stops at B, L, and C. Note that the path does not include all of the vertices processed by the algorithm.

Algorithm 9.3 (Breadth-first Search):   This algorithm executes a breadth-first search on a directed graph G beginning with a starting vertex A.

Step 1. Initialize all vertices to the ready state (STATUS = 1).

Step 2. Put the starting vertex A in QUEUE and change the status of A to the waiting state (STATUS = 2).

Step 3. Repeat Steps 4 and 5 until QUEUE is empty.

Step 4.   Remove the first vertex N of QUEUE. Process N, and set STATUS $(N) = 3$, the processed state.

Step 5.   Examine each neighbor J of N.
   (a)   If STATUS $(J) = 1$ (ready state), add J to the rear of QUEUE and reset STATUS $(J) = 2$ (waiting state).
   (b)   If STATUS $(J) = 2$ (waiting state) or STATUS $(J) = 3$ (processed state), ignore the vertex J.
   [End of Step 3 loop.]

Step 6. Exit.

Fig. 9-14

QUEUE	Vertex
A	A
F_A, E_A, B_A	B_A
L_B, F_A, E_A	E_A
L_B, F_A	F_A
D_F, L_B	L_B
C_L, D_F	D_F
C_L	C_L
J_C	J_C

(a)

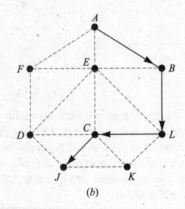

(b)

Fig. 9-15

9.9   DIRECTED CYCLE-FREE GRAPHS, TOPOLOGICAL SORT

Let S be a directed graph with the following two properties:

(1) Each vertex v_i of S represents a task.

(2) Each (directed) edge (u, v) of S means that task u must be completed before beginning task v.

We note that such a graph S cannot contain a cycle, such as $P = (u, v, w, u)$, since, otherwise, we would have to complete u before beginning v, complete v before beginning w, and complete w before beginning u. That is, we cannot begin any of the three tasks in the cycle.

Such a graph S, which represents tasks and a prerequisite relation and which cannot have any cycles, is said to be *cycle-free* or *acyclic*. A directed acyclic (cycle-free) graph is called a *dag* for short. Figure 9-16 is an example of such a graph.

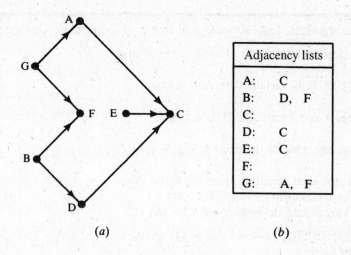

Fig. 9-16

A fundamental operation on a dag S is to process the vertices one after the other so that the vertex u is always processed before vertex v whenever (u, v) is an edge. Such a linear ordering T of the vertices of S, which may not be unique, is called a *topological sort*.

Figure 9-17 shows two topological sorts of the graph S in Fig. 9-16. We have included the edges of S in Fig. 9-17 to show that they agree with the direction of the linear ordering.

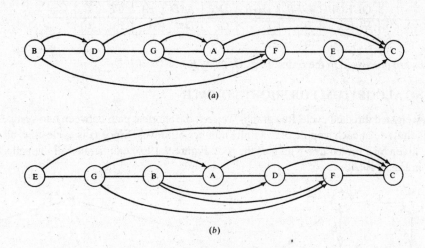

Fig. 9-17   Two topological sorts

The following is the main theoretical result in this section.

Theorem 9.8: Let S be a finite directed cycle-free graph. Then there exists a topological sort T of the graph S.

Note that the theorem states only that a topological sort exists. We now give an algorithm which will find a topological sort. The main idea of the algorithm is that any vertex (node) N with zero indegree may be chosen as the first element in the sort T. The algorithm essentially repeats the following two steps until S is empty:

(1) Find a vertex N with zero indegree.
(2) Delete N and its edges from the graph S.

We use an auxiliary QUEUE to temporarily hold all vertices with zero degree. The algorithm appears in Fig. 9-18.

Algorithm 9.4:   The algorithm finds a topological sort T of a directed cycle-free graph S.

Step 1. Find the indegree INDEG(N) of each vertex N of S.

Step 2. Insert in QUEUE all vertices with zero degree.

Step 3. Repeat Steps 4 and 5 until QUEUE is empty.

Step 4.   Remove and process the front vertex N of QUEUE.

Step 5.   Repeat for each neighbor M of the vertex N.
   (a)   Set INDEG(M) : = INDEG(M) − 1.
         [This deletes the edge from N to M.]

   (b)   If INDEG(M) = 0, add M to QUEUE.
         [End of loop.]
      [End of Step 3 loop.]

Step 6. Exit.

Fig. 9-18

EXAMPLE 9.12 Suppose Algorithm 9.4 is applied to the graph S in Fig. 9-16. We obtain the following sequence of the elements of QUEUE and sequence of vertices being processed:

QUEUE	GEB	DGE	DG	FAD	FA	CF	C	∅
Vertex	B	E	G	D	A	F	C	

Thus the vertices are processed in the order: B, E, G, D, A, F.

9.10 PRUNING ALGORITHM FOR SHORTEST PATH

Let G be a weighted directed cycle-free graph. We seek the shortest path between two vertices, say u and w. We assume G is finite so at each step there is a finite number of moves. Since G is cycle-free, all paths between u and w can be given by a rooted tree with u as the root. Figure 9-19(b) enumerates all the paths between u and w in the graph in Fig. 9-19(a).

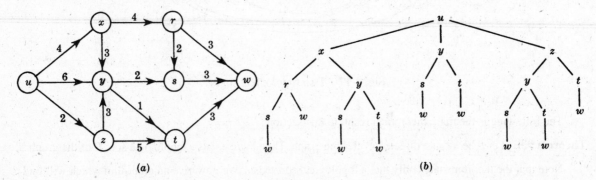

Fig. 9-19

One way to find the shortest path between u and w is simply to compute the lengths of all the paths of the corresponding rooted tree. On the other hand, suppose two partial paths lead to an intermediate vertex v. From then on, we need only consider the shorter partial path; that is, we prune the tree at the vertex corresponding to the longer partial path. This pruning algorithm is described below.

Pruning Algorithm

This algorithm finds the shortest path between a vertex u and a vertex w in a weighted directed cycle-free graph G. The algorithm has the following properties:

(a) During the algorithm each vertex v' of G is assigned two things:

 (1) A number $\ell(v')$ denoting the current minimal length of a path from u to v'.

 (2) A path $p(v')$ from u to v' of length $\ell(v')$.

(b) Initially, we set $\ell(u) = 0$ and $p(u) = u$. Every other vertex v is initially assigned $\ell(v) = \infty$ and $p(v) = \varnothing$.

(c) Each step of the algorithm examines an edge $e = (v', v)$ from v' to v with, say, length k. We calculate $\ell(v') + k$.

 (1) Suppose $\ell(v') + k < \ell(v)$. Then we have found a shorter path from u to v. Thus we update:

$$\ell(v) = \ell(v') + k \quad \text{and} \quad p(v) = p(v')v$$

 (This is always true when $\ell(v) = \infty$, that is, when we first enter the vertex v.)

 (2) Otherwise, we do not change $\ell(v)$ and $p(v)$.

 If no other unexamined edges enter v, we will say that $p(v)$ has been determined.

(d) The algorithm ends when $p(w)$ has been determined.

Remark: The edge $e = (v', v)$ in (c) can only be chosen if v' has been previously visited, that is, if $p(v')$ is not empty. Furthermore, it is usually best to examine an edge which begins at a vertex v' whose path $p(v')$ has been determined.

EXAMPLE 9.13 We apply the pruning algorithm to the graph G in Fig. 9-19(a).

From u: The successive vertices are x, y, and z, which are all entered for the first time. Thus:

 (1)   set $\ell(x) = 4$, $p(x) = ux$.

 (2)   set $\ell(y) = 6$, $p(y) = uy$.

 (3)   set $\ell(z) = 2$, $p(z) = uz$.

 Note that $p(x)$ and $p(z)$ have been determined.

From x: The successive vertices are r, entered for the first time, and y. Thus:

 (1) Set $\ell(r) = 4 + 4 = 8$ and $p(r) = p(x)r = uxr$.

 (2) We calculate:

$$\ell(x) + k = 4 + 3 = 7 \quad \text{which is not less than} \quad \ell(y) = 6.$$

 Thus we leave $\ell(y)$ and $p(y)$ alone.

 Note that $p(r)$ has been determined.

From z: The successive vertices are t, entered for the first time, and y. Thus:

 (1) Set $\ell(t) = \ell(z) + k = 2 + 5 = 7$ and $p(t) = p(z)t = urt$.

 (2) We calculate:

$$\ell(z) + k = 2 + 3 = 5 \quad \text{which is less than} \quad \ell(y) = 6.$$

 We have found a shorter path to y, and so we update $\ell(y)$ and $p(y)$; set:

$$\ell(y) = \ell(z) + k = 5 \quad \text{and} \quad p(y) = p(z)y = uzy$$

 Now $p(y)$ has been determined.

From y: The successive vertices are s, entered for the first time, and t. Thus:

   (1) Set $\ell(s) = \ell(y) + k = 5 + 2 = 7$ and $p(s) = p(y)s = uzys$.

   (2) We calculate:

$$\ell(y) + k = 5 + 1 = 6 \quad \text{which is less than} \quad \ell(t) = 7.$$

Thus we change $\ell(t)$ and $p(t)$ to read:

$$\ell(t) = \ell(y) + 1 = 6 \quad \text{and} \quad p(t) = p(y)t = uzyt.$$

Now $p(t)$ has been determined.

From r: The successive vertices are w, entered for the first time, and s. Thus:

   (1) Set $\ell(w) = \ell(r) + 3 = 11$ and $p(w) = p(r)w = uxrw$.

   (2) We calculate:

$$\ell(r) + k = 8 + 2 = 10 \quad \text{which is less than} \quad \ell(s) = 7.$$

Thus we leave $\ell(s)$ and $p(s)$ alone.

Note that $p(s)$ has been determined.

From s: The successive vertex is w. We calculate:

$$\ell(s) + k = 7 + 3 = 10 \quad \text{which is less than} \quad \ell(w) = 11.$$

Thus we change, $\ell(w)$ and $p(w)$ to read:

$$\ell(w) = \ell(s) + 3 = 10 \quad \text{and} \quad p(w) = p(s)w = uzysw.$$

From t: The successive vertex is w. We calculate:

$$\ell(t) + k = 6 + 3 = 9 \quad \text{which is less than} \quad \ell(w) = 10.$$

Thus we undate $\ell(w)$ and $p(w)$ as follows:

$$\ell(w) = \ell(t) + 3 = 9 \quad \text{and} \quad p(w) = p(t) = uzytw$$

Now $p(w)$ has been determined.
The algorithm is finished since $p(w)$ has been determined. Thus $p(w) = uzytw$ is the shortest path from u to w and $\ell(w) = 9$.

The edges which were examined in the above example form the rooted tree in Fig. 9-20. This is the tree in Fig. 9-19(b) which has been pruned at vertices belonging to longer partial paths. Observe that only 13 of the original 23 edges of the tree had to be examined.

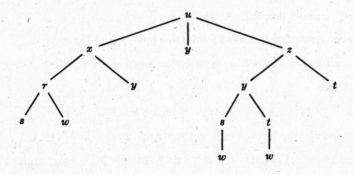

Fig. 9-20

Solved Problems

GRAPH TERMINOLOGY

9.1. Let G be the directed graph in Fig. 9-21(a).

(a) Describe G formally.   (d) Find all cycles in G.

(b) Find all simple paths from X to Z.   (e) Is G unilaterally connected?

(c) Find all simple paths from Y to Z.   (f) Is G strongly connected?

(a) The vertex set V has four vertices and the edge set E has seven (directed) edges as follows:

$$V = \{X, Y, Z, W\} \quad \text{and} \quad E = \{(X, Y), (X, Z), (X, W), (Y, W), (Z, Y), (Z, W), (W, Z)\}$$

(b) There are three simple paths from X to Z, which are (X, Z), (X, W, Z), and (X, Y, W, Z).

(c) There is only one simple path from Y to Z, which is (Y, W, Z).

(d) There is only one cycle in G, which is (Y, W, Z, Y).

(e) G is unilaterally connected since (X, Y, W, Z) is a spanning path.

(f) G is not strongly connected since there is no closed spanning path.

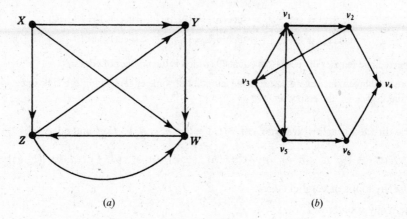

(a)   (b)

Fig. 9-21

9.2. Let G be the directed graph in Fig. 9-21(a).

(a) Find the indegree and outdegree of each vertex of G.

(b) Find the successor list of each vertex of G.

(c) Are there any sources or sinks?

(d) Find the subgraph H of G determined by the vertex set $V' = X, Y, Z$.

(a) Count the number of edges ending and beginning at a vertex v to obtain, respectively, indeg(v) and outdeg(v). This yields the data:

$$\text{indeg}(X) = 0, \quad \text{indeg}(Y) = 2, \quad \text{indeg}(Z) = 2, \quad \text{indeg}(W) = 3$$
$$\text{outdeg}(X) = 3, \quad \text{outdeg}(Y) = 1, \quad \text{outdeg}(Z) = 2, \quad \text{outdeg}(W) = 1$$

(As expected, the sum of the indegrees and the sum of the outdegrees each equal 7, the number of edges.)

(b) Add vertex v to the successor list succ(u) of u for each edge (u, v) in G. This yields:

$$\text{succ}(X) = [Y, Z, W], \quad \text{succ}(Y) = [W], \quad \text{succ}(Z) = [Y, W], \quad \text{succ}(W) = [Z]$$

(c) X is a source no edge enters X, that is, indeg$(X) = 0$. There are no sinks since every vertex is the initial point of an edge, that is, has nonzero outdegree.

(d) Let E' consists of all edges of G whose endpoints lie in V'. This yield $E' = \{(X, Y), (X, Z), (Z, Y)\}$. Then $H = H(V', E')$

9.3 Let G be the directed graph in Fig. 9-21(b).

(a) Find two simple paths from v_1 to v_6. Is $\alpha = (v_1, v_2, v_4, v_6)$ such a simple path?

(b) Find all cycles in G which include v_3.

(c) Is G unilaterally connected? Strongly connected?

(d) Find the successor list of each vertex of G.

(e) Are there any sources in G? Any sinks?

(a) A simple path is a path where all vertices are distinct. Thus (v_1, v_5, v_6) and $(v_1, v_2, v_3, v_5, v_6)$ are two simple paths from v_1 to v_6. The sequence is not even a path since the edge joining v_4 to v_6 does not begin at v_4.

(b) There are two such cycles: (v_3, v_1, v_2, v_3) and $(v_3, v_5, v_6, v_1, v_2, v_3)$.

(c) G is unilaterally connected since $(v_1, v_2, v_3, v_5, v_6, v_4)$ is a spanning path. G is not strongly connected since there is no closed spanning path.

(d) Add vertex v to the successor list succ(u) of u for each edge (u, v) in G. This yields:

$$\text{succ}(v_1) = [v_2, v_5], \quad \text{succ}(v_2) = [v_3, v_4], \quad \text{succ}(v_3) = [v_1, v_5]$$
$$\text{succ}(v_4) = \varnothing, \quad \text{succ}(v_5) = [v_6], \quad \text{succ}(v_6) = [v_1, v_4]$$

(As expected, the number of successors equals 9, which is the number of edges.)

(e) There are no sources since every vertex is the endpoint of some edge. Only v_4 is a sink since no edge begins at v_4, that is, succ$(v_4) = \varnothing$, the empty set.

9.4. Let G be the directed graph with vertex set $V(G) = (a, b, c, d, e, f, g\}$ and edge set:

$$E(G) = \{(a, a), (b, e), (a, e), (e, b), (g, c), (a, e), (d, f), (d, b), (g, g)\}$$

(a) Identify any loops or parallel edges.

(b) Are there any sources in G?

(c) Are there any sinks in G?

(d) Find the subgraph H of G determined by the vertex set $V' = \{a, b, c, d\}$.

(a) A loop is an edge with the same initial and terminal points; hence (a, a) and (g, g) are loops. Two edges are parallel if they have the same initial and terminal points. Thus (a, e) and (a, e) are parallel edges.

(b) The vertex d is a source since no edge ends in d, that is, d does not appear as the second element in any edge. There are no other sources.

(c) Both c and f are sinks since no edge begins at c or at f, that is, neither c nor f appear as the first element in any edge. There are no other sinks.

(d) Let E' consist of all edges of G whose endpoints lie in $V' = \{a, b, c, d\}$. This yields $E' = \{(a, a), (d, b)\}$. Then $H = H(V', E')$.

ROOTED TREES, ORDERED ROOTED TREES

9.5 Let T be the rooted tree in Fig. 9-22.

(a) Identify the path α from the root R to each of the following vertices, and find the level number n of the vertex: (i) H; (ii) F; (iii) M.

(b) Find the siblings of E.

(c) Find the leaves of T.

(a) List the vertices while proceeding from R down the tree to the vertex. The number of vertices, other than R, is the level number:

   (i) $\alpha = (R, A, C, H)$, $n = 3$;   (ii) $\alpha = (R, B, F)$, $n = 2$;   (iii) $\alpha = (R, B, G, L, M)$, $n = 4$.

(b) The siblings of E are F and G since they all have the same parent B.

(c) The leaves are vertices with no children, that is, H, D, I, J, K, M, N.

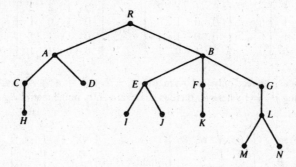

Fig. 9-22

9.6. Let T be the ordered rooted tree in Fig. 9-23 whose vertices are labeled using the universal address system. Find the lexicographic order of the addresses of the tree T.

An ordered rooted tree T is usually drawn so the edges are ordered from left to right as in Fig. 9-23. The lexicographic order can be obtained by reading down the leftmost branch, then the second branch from the left, and so forth.

   Reading down the leftmost branch of T we obtain:

$$0,\quad 1,\quad 1.1,\quad 1.1.1$$

The next branch is 1.2, 1.2.1, 1.2.1.1, so we add this branch to the list to obtain

$$0,\quad 1,\quad 1.1,\quad 1.1.1,\quad 1.2\quad 1.2.1,\quad 1.2.1.1$$

Proceeding in this manner, we finally obtain

$$0,\quad 1,\quad 1.1,\quad 1.1.1,\quad 1.2,\quad 1.2.1,\quad 1.2.1.1,\quad 1.2.2,\quad 1.3,\quad 2,\quad 2.1,\quad 2.2.1$$

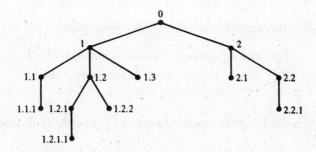

Fig. 9-23

SEQUENTIAL REPRESENTATION OF GRAPHS

9.7. Consider the graph G in Fig. 9-21(a), and suppose the vertices are stored in memery in the array:

$$\text{DATA: } X, Y, Z, W$$

(a) Find the adjacency matrix A of the graph G and the powers A^2, A^3, A^4.

(b) Find the path matrix P of G using the powers of A. Is G strongly connected?

(a) The vertices are normally ordered according to the way they appear in memory; that is, we assume $v_1 = X$, $v_2 = Y$, $v_3 = Z$, $v_4 = W$. The adjacency matrix $A = [a_{ij}]$ is obtained by setting $a_{ij} = 1$ if there is an edge from v_i to v_j; and 0 otherwise. The matrix A and its powers follow:

$$A = \begin{bmatrix} 0 & 1 & 1 & 1 \\ 0 & 0 & 0 & 1 \\ 0 & 1 & 0 & 1 \\ 0 & 0 & 1 & 0 \end{bmatrix}, \quad A^2 = \begin{bmatrix} 0 & 1 & 1 & 2 \\ 0 & 0 & 1 & 0 \\ 0 & 0 & 1 & 1 \\ 0 & 1 & 0 & 1 \end{bmatrix}, \quad A^3 = \begin{bmatrix} 0 & 1 & 2 & 2 \\ 0 & 1 & 0 & 1 \\ 0 & 1 & 1 & 1 \\ 0 & 0 & 1 & 1 \end{bmatrix}, \quad A^4 = \begin{bmatrix} 0 & 2 & 2 & 3 \\ 0 & 0 & 1 & 1 \\ 0 & 1 & 1 & 2 \\ 0 & 1 & 1 & 1 \end{bmatrix}$$

(b) Since G has 4 vertices, we need only find the matrix $B_4 = A + A^2 + A^3 + A^4$, and then the path matrix $P = [p_{ij}]$ is obtained by setting $p_{ij} = 1$ whenever there is a nonzero entry in the matrix B_4, and 0 otherwise. The matrices B_4 and P follow:

$$B_4 = \begin{bmatrix} 0 & 5 & 6 & 8 \\ 0 & 1 & 2 & 3 \\ 0 & 3 & 3 & 5 \\ 0 & 2 & 3 & 5 \end{bmatrix} \quad \text{and} \quad P = \begin{bmatrix} 0 & 1 & 1 & 1 \\ 0 & 1 & 1 & 1 \\ 0 & 1 & 1 & 1 \\ 0 & 1 & 1 & 1 \end{bmatrix}$$

The path matrix P shows there is no path from any node to v_1. Hence G is not strongly connected.

9.8. Consider the adjacency matrix A of the graph G in Fig. 9-19(a) obtained in Problem 9.7. Find the path matrix P of G using Warshall's algorithm rather than the powers of A.

Initially set $P_0 = A$. Then, P_1, P_2, P_3, P_4 are obtained recursively by setting

$$P_k[i, j] = P_{k-1}[i, j] \vee (P_{k-1}[i, k] \wedge P_{k-1}[k, j])$$

where $P_k[i, j]$ denotes the ij-entry in the matrix P_k. That is, by setting

$$P_k[i, j] = 1 \quad \text{if} \quad P_{k-1}[i, j] = 1 \quad \text{or if both} \quad P_{k-1}[i, k] = 1 \quad \text{and} \quad P_{k-1}[k, j] = 1$$

Then matrices P_1, P_2, P_3, P_4 follow:

$$P_1 = \begin{bmatrix} 0 & 1 & 1 & 1 \\ 0 & 0 & 0 & 1 \\ 0 & 1 & 0 & 1 \\ 0 & 0 & 1 & 0 \end{bmatrix}, \quad P_2 = \begin{bmatrix} 0 & 1 & 1 & 1 \\ 0 & 0 & 0 & 1 \\ 0 & 1 & 0 & 1 \\ 0 & 0 & 1 & 0 \end{bmatrix}, \quad P_3 = \begin{bmatrix} 0 & 1 & 1 & 1 \\ 0 & 0 & 0 & 1 \\ 0 & 1 & 0 & 1 \\ 0 & 1 & 1 & 1 \end{bmatrix}, \quad P_4 = \begin{bmatrix} 0 & 1 & 1 & 1 \\ 0 & 1 & 1 & 1 \\ 0 & 1 & 1 & 1 \\ 0 & 1 & 1 & 1 \end{bmatrix}$$

Observe that $P_1 = P_2 = A$. The changes in P_3 occur for the following reasons:

$$P_3[4, 2] = 1 \quad \text{because} \quad P_2[4, 3] = 1 \text{ and } P_2[3, 2] = 1$$
$$P_3[4, 4] = 1 \quad \text{because} \quad P_2[4, 3] = 1 \text{ and } P_2[3, 4] = 1$$

9.9. Draw a picture of the weighted graph G which is maintained in memory by the following vertex array DATA and weight matrix W:

$$\text{DATA: } X, Y, S, T; \quad W = \begin{bmatrix} 0 & 0 & 3 & 0 \\ 5 & 0 & 1 & 7 \\ 2 & 0 & 0 & 4 \\ 0 & 6 & 8 & 0 \end{bmatrix}$$

The picture appears in Fig. 9-24(a). The vertices are labeled by the entries in DATA.

Assuming $v_1 = X$, $v_2 = Y$, $v_3 = S$, $v_4 = T$, the order the vertices appear in the array DATA, we draw an edge from v_i to v_j with weight w_{ij} when $w_{ij} \neq 0$.

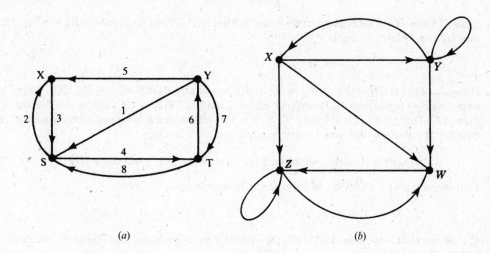

Fig. 9-24

LINKED REPRESENTATION OF GRAPHS

9.10. Let G be the graph presented by the following table:

$$G = [X : Y, Z, W; \quad Y : X, Y, W; \quad Z : Z, W; \quad W : Z]$$

(a) Find the number of vertices and edges in G.

(b) Draw the graph of G.

(c) Are there any sources or sinks?

(a) The table tells us that there are four vertices, X, Y, Z, W. The outdegrees of the vertices are 3, 3, 2, 1, respectively. Thus there are $3 + 3 + 2 + 1 = 9$ edges.

(b) Using the adjacency lists, draw the graph in Fig. 9-24(b).

(c) No vertex has zero outdegree, so there are no sinks. Also, no vertex has zero indegree, that is, each vertex is a successor; hence there are no sources.

9.11. A weighted graph G with six vertices, $A, B, \ldots, F$, is stored in memory using a linked representation with a vertex file and an edge file as in Fig. 9-25(a).

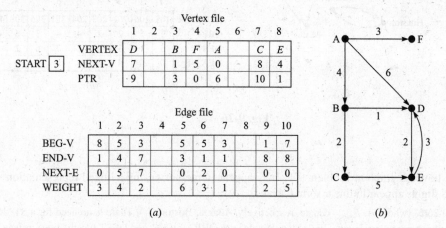

Vertex file

	1	2	3	4	5	6	7	8
VERTEX	D		B	F	A		C	E
NEXT-V	7		1	5	0	-	8	4
PTR	9		3	0	6		10	1

START 3

Edge file

	1	2	3	4	5	6	7	8	9	10
BEG-V	8	5	3		5	5	3		1	7
END-V	1	4	7		3	1	1		8	8
NEXT-E	0	5	7		0	2	0		0	0
WEIGHT	3	4	2		6	3	1		2	5

(a)

(b)

Fig. 9-25

(a) List the vertices in the order they appear in memory.

(b) Find the successor list succ(v) of each vertex v.

(c) Draw the graph of G.

(a) Since START = 3, the list begins with the vertex B. Then NEXT-V tells us to go to $1(D)$, then $7(C)$, then $8(E)$, then $4(F)$, and then $5(A)$; that is,

$$B, \quad D, \quad C, \quad E, \quad F, \quad A$$

(b) Here succ$(A) = [1(D), \; 4(F), \; 3(B)] = [D, F, B]$. Specifically, PTR$[5(A)] = 6$ and END-V$[6] = 1(D)$ tells us that succ(A) begins with D. Then NEXT-E$[6] = 2$ and END-V$[2] = 4(F)$ tells us that F is the next vertex in succ(A). Then NEXT-E$[2] = 5$ and END-V$[5] = 3(B)$ tells us that B is the next vertex in succ(A). However, NEXT-E$[5] = 0$ tells us that there are no more successors of A. Similarly,

$$\text{succ}(B) = [C, D], \quad \text{succ}(C) = [E], \quad \text{succ}(D) = [E], \quad \text{succ}(E) = [D]$$

Furthermore succ$(F) = \varnothing$, since PTR$[4(F)] = 0$. In other words,

$$G = [A:D, F, B; \quad B:C, D; \quad C:E; \quad D:E; \quad E:D; \quad F:\varnothing]$$

(c) Use the successor lists obtained in (b) and the weights of the edges in the Edge File in Fig. 9-25(a) to draw the graph in Fig. 9-25(b).

9.12. Suppose Friendly Airways has nine daily flights as follows:

103	Atlanta to Houston	203	Boston to Denver	305	Chicago to Miami
106	Houston to Atlanta	204	Denver to Boston	308	Miami to Boston
201	Boston to Chicago	301	Denver to Reno	401	Reno to Chicago

Describe the data by means of a labeled directed graph G.

The data are described by the graph in Fig. 9-26(a) (where the flight numbers have been omitted for notational convenience.)

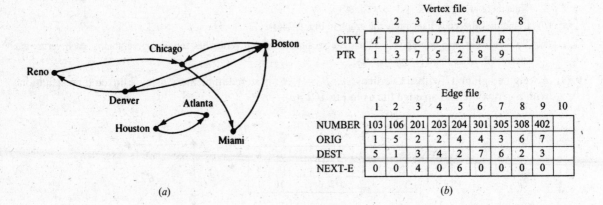

Fig. 9-26

9.13. Describe how the graph in Problem 9.12 may appear in memory using a linked representation where the cities and flights appear in linear sorted arrays.

See Fig. 9-26(b) (where A, B, ... denote, respectively, Atlanta, Boston,...). There is no need for a START variable since the cities form an array, not a linked list. We also use ORIG (origin) and DEST (destination) instead of BEG-V and END-V.

9.14. Clearly, the data in Problem 9.12 may be efficiently stored in a file where each record contains only three fields:

Flight Number, City of Origin, City of Destination

However, when there are many, many flights, such a representation does not easily answer the following natural questions:

   (i) Is there a direct flight from city X to city Y?
   (ii) Can one fly from city X to city Y?
   (iii) What is the most direct route (minimum number of stops) from city X to city Y?

Show how the answer, say, to (ii) may be more readily available if the data is stored in memory using the linked representation of the graph as in Fig. 9-26(b).

   One way to answer (ii) is to use a breadth-first or depth-first search algorithm to decide whether city Y is reachable from city X. Such algorithms require the adjacency lists, which can easily be obtained from the linked representation of a graph, but not from the above representation which uses only three fields.

MISCELLANEOUS PROBLEMS

9.15. Let $A = \begin{bmatrix} 0 & 2 & 0 & 1 \\ 0 & 0 & 1 & 1 \\ 2 & 1 & 1 & 0 \\ 0 & 0 & 1 & 1 \end{bmatrix}$ be the adjacency matrix of a multigraph G. Draw a picture of G.

Since A is a 4×4 matrix, G has four vertices, v_1, v_2, v_3, v_4. For each entry a_{ij} in A, draw a_{ij} arcs (directed edges) from vertex v_i to vertex v_j to obtain the graph in Fig. 9-27(a).

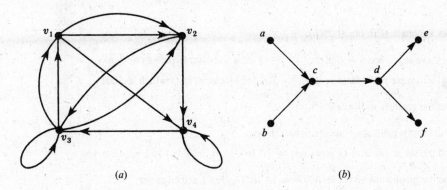

(a)   (b)

Fig. 9-27

9.16. Let S be the cycle-free graph in Fig. 9-27(b). Find all possible topological sorts of S.

There are four possible topological sorts of S. Specifically, each sort T must begin with either a or b, must end with e or f, and c and d must be the third and fourth elements, respectively. The four sorts follow:

$$T_1 = [a, b, c, d, e, f], \qquad T_2 = [b, a, c, d, e, f]$$
$$T_3 = [a, b, c, d, f, e], \qquad T_4 = [b, a, c, d, f, e]$$

9.17. Prove Proposition 9.4: Let A be the adjacency matrix of a graph G. Then $a_K[i, j]$, the ij-entry in the matrix A^K, gives the number of paths of length K from v_i to v_j.

The proof is by induction on K. A path of length 1 from v_i to v_j is precisely an edge (v_i, v_j). By definition of the adjacency matrix A, $a_1[i, j] = a_{ij}$ gives the number of edges from v_i to v_j. Hence the proposition is true for $K = 1$.

   Suppose $K > 1$. (Assume G has m nodes.) Since $A^K = A^{K-1}A$,

$$a_K[i, j] = \sum_{s=1}^{m} a_{K-1}[i, s]\, a_1[s, j]$$

By induction, $a_{K-1}[i, s]$ gives the number of paths of length $K - 1$ from v_i to v_s and $a_1[s, j]$ gives the number of paths of length 1 from v_s to v_j. Thus $a_{K-1}[i, s]a_1[s, j]$ gives the number of paths of length K from v_i to v_j where v_s is the next-to-last vertex. Thus all paths of length K from v_i to v_j can be obtained by summing up the product $a_{K-1}[i, s]a_1[s, j]$ for all s. Accordingly, $a_K[i, j]$ is the number of paths of length K from v_i to v_j. Thus the proposition is proved.

Supplementary Problems

GRAPH TERMINOLOGY

9.18. Consider the graph G in Fig. 9-28(a).

(a) Find the indegree and outdegree of each vertex.　(d) Find all cycles in G.

(b) Are there any sources or sinks?　(e) Find all paths of length 3 or less from v_1 to v_3.

(c) Find all simple paths from v_1 to v_4.　(f) Is G unilaterally or strongly connected?

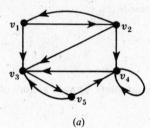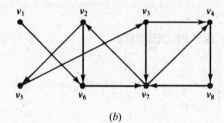

(a)　　　　　　　　　　　　(b)

Fig. 9-28

9.19. Consider the graph G in Fig. 9-28(b).

(a) Are there any sources or sinks?　(c) Find a non-simple path from v_1 to v_4.

(b) Find all simple paths from v_1 to v_4.　(d) Find all cycles in G which include v_4.

9.20. Consider the graph G in Fig. 9-28(b).

(a) Find: $\text{succ}(v_1)$, $\text{succ}(v_3)$, $\text{succ}(v_5)$, $\text{succ}(v_7)$.

(b) Find the subgraph H of G generated by: (i) $\{v_1, v_3, v_5, v_6\}$; (ii) $\{v_2, v_3, v_6, v_7\}$.

9.21. Let G be the graph with vertex set $V(G) = \{A, B, C, D, E\}$ and edge set

$$E(G) = \{(A, D),\quad (B, C),\quad (C, E),\quad (D, B),\quad (D, D),\quad (D, E),\quad (E, A)\}$$

(a) Express G by its adjacency table.

(b) Does G have any loops or parallel edges?

(c) Find all simple paths from D to E.

(d) Find all cycles in G.

(e) Is G unilaterally or strongly connected?

(f) Find the number of subgraphs of G with vertices C, D, E.

(g) Find the subgraph H of G generated by C, D, E.

9.22. Let G be the graph with vertex set $V(G) = \{a, b, c, d, e\}$ and the following successor lists:

$$\text{succ}(a) = [b, c],\quad \text{succ}(b) = \varnothing,\quad \text{succ}(c) = [d, e],\quad \text{succ}(d) = [a, b, e],\quad \text{succ}(e) = \varnothing$$

(a) List the edges of G. (b) Is G weakly, unilaterally, or strongly connected?

9.23. Let G be the graph in Fig. 9-29(a).

(a) Express G by its adjacency table.　(d) Find all cycles in G.

(b) Does G have any sources or sinks?　(e) Find a spanning path of G.

(c) Find all simple paths from A to E.　(f) Is G strongly connected?

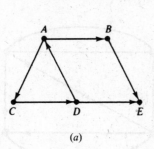

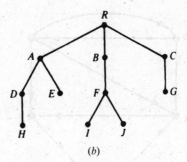

(a)   (b)

Fig. 9-29

ROOTED TREES, ORDERED ROOTED TREES

9.24. Let T be the rooted tree in Fig. 9-29(b).

(a) Identify the path α from the root R to each of the following vertices, and find the level number of the vertex: (i) D; (ii) J; (iii) G.

(b) Find the leaves of T.

(c) Assuming T is an ordered rooted tree, find the universal address of each leaf of T.

9.25. The following addresses are in random order:

2.1.1,   3.1,   2.1,   1,   2.2.1.2,   0,   3.2,   2.2,   1.1,   2,   3.1.1,   2.2.1,   3,   2.2.1.1

(a) Place the addresses in lexicographic order.

(b) Draw the corresponding rooted tree.

SEQUENTIAL REPRESENTATION OF GRAPHS

9.26. Let G be the graph in Fig. 9-30(a).

(a) Find the adjacency matrix A and the path matrix P for G.

(b) For all $k > 0$, find n_k where n_k denotes the number of paths of length k from v_1 to v_4.

(c) Is G weakly, unilaterally, or strongly connected?

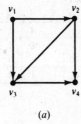

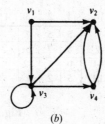

(a)   (b)

Fig. 9-30

9.27. Repeat Problem 9.26 for the graph G in Fig. 9-30(b).

9.28. Let P be the path matrix for a graph G. Describe P when G is: (a) strongly connected; (b) unilaterally connected.

9.29. Let G be the graph in Fig. 9-31(a) where the vertices are maintained in memory by the array DATA: X, Y, Z, S, T. (a) Find the adjacency matrix A and the path matrix P of G. (b) Find all cycles in G. (c) Is G unilaterally connected? Strongly connected?

9.30. Let G be the weighted graph in Fig. 9-31(b) where the vertices are maintained in memory by the array DATA: X, Y, S, T.

(a) Find the weight matrix W of G.

(b) Find the matrix Q of shortest paths using Warshall's algorithm.

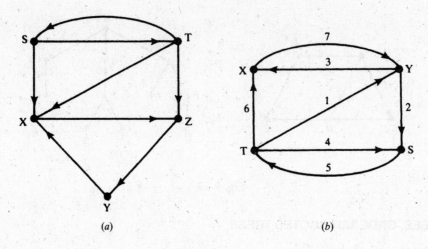

Fig. 9-31

LINKED REPRESENTATION OF GRAPHS

9.31. A weighted graph G with six vertices, A, B, ..., F, is stored in memory using a linked representation with a vertex file and an edge file as in Fig. 9-32.

    (a) List the vertices in the order they appear in memory.

    (b) Find the successor list $\text{succ}(v)$ of each vertex v in G.

    (c) Draw a picture of G.

Vertex file

	1	2	3	4	5	6	7	8
VERTEX	D		B	F	A		C	E
NEXT-V	3		8	1	0		4	5
PTR	7		5	9	2		3	0

START $\boxed{7}$

Edge file

	1	2	3	4	5	6	7	8	9	10	11	12
BEG-V	5	5	7		3	7	1		4	1	4	7
END-V	8	7	5		1	1	5		8	4	3	8
NEXT-E	0	1	12		0	0	10		11	0	0	6
WEIGHT	5	2	1		3	2	4		1	3	4	1

Fig. 9-32

9.32. Let G be the graph presented by the table: $G = [A:B,C; \ B:C,D; \ C:C; \ D:B; \ E:\varnothing]$.

    (a) Find the number of vertices and edges in G.   (c) Are there any sources or sinks?

    (b) Draw a picture of G.   (d) Is G weakly, unilaterally, or strongly connected?

9.33. Repeat Problem 9.32 for the table: $G = [A:D; \ B:C; \ C:E; \ D:B,D,E; \ E:A]$.

9.34. Repeat Problem 9.32 for the table: $G = [A:B,C,D,K; \ B:J; \ C:\varnothing; \ D:\varnothing; \ J:B,D,L; \ K:D,L; \ L:D]$.

9.35. Suppose Friendly Airways has eight daily flights serving the seven cities: Atlanta, Boston, Chicago, Denver, Houston, Philadelphia, Washington. Suppose the data on the flights are stored in memory as in Fig. 9-33; that is, using a linked representation where the cities and flights appear in linear sorted arrays. Draw a labeled directed graph G describing the data.

Vertex file

	1	2	3	4	5	6	7	8
CITY	A	B	C	D	H	P	W	
PTR	1	2	3	8	9	5	7	

Edge file

	1	2	3	4	5	6	7	8	9	10
NUMBER	101	102	201	202	203	301	302	401	402	
ORIG	1	2	3	1	6	6	7	4	5	
DEST	2	3	6	7	3	1	6	5	4	
NEXT-E	4	0	0	0	6	0	0	0	0	

Fig. 9-33

9.36. Using the data in Fig. 9-33, write a procedure with input CITY X and CITY Y which finds the flight number of a direct flight from city X to city Y, if it exists. Test the procedure using:

(a) X = Atlanta, Y = Philadelphia;　　(c) X = Houston, Y = Chicago;

(b) X = Philadelphia, Y = Atlanta;　　(d) X = Washington, Y = Chicago.

9.37. Using the data in Fig. 9-33, write a procedure with input CITY X and CITY Y which finds the most direct route (minimum number of stops) from city X to city Y, if it exists. Test the procedure using the input in Problem 9.36.

MISCELLANEOUS PROBLEMS

9.38. Use the pruning algorithm to find the shortest path from s to t in Fig. 9-34.

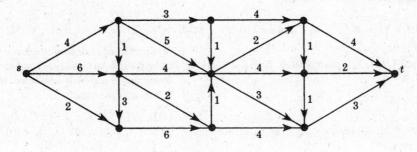

Fig. 9-34

9.39. Find a topological sort T of each of the following graphs:

(a) $G = [A:Z;\quad B:T;\quad C:B;\quad D:\emptyset;\quad X:D;\quad Y:X;\quad Z:B, X;\quad S:C, Z;\quad T:\emptyset]$

(b) $G = [A:X, Z;\quad B:A;\quad C:S, T;\quad D:Y;\quad X:S, T;\quad Y:B;\quad Z:\emptyset;\quad S:Y;\quad T:\emptyset]$

(c) $G = [A:C, S;\quad B:T, Z;\quad C:\emptyset;\quad D:Z;\quad X:A;\quad Y:A;\quad Z:X, Y;\quad S:\emptyset;\quad T:Y]$

9.40. Draw a labeled graph G that represents the following situation. Three sisters, Barbara, Rose, and Susan, each regularly telephone their mother Gertrude, though Gertrude calls only Rose. Susan will not call Rose, though Rose continues to call Susan. Barbara and Susan will call each other, and Barbara and Rose will call each other.

9.41. Let R be the relation (directed graph) on $V = \{2, 3, 4, 9, 15\}$ defined by "x is less than and relatively prime to y." (a) Draw the diagram of the graph R. (b) Is R weakly connected? Unilaterally connected? Strongly connected?

9.42. A directed graph G is *complete* if, for each pair of distinct vertices u and v, either (u, v) is an arc or (v, u) is an arc. Show that a finite complete directed graph G has a path which includes all vertices. (This obviously holds for non-directed complete graphs.) Thus G is unilaterally connected.

9.43. Suppose a graph G is input by means of an integer M, representing the vertices $1, 2, \ldots, M$, and a list of N ordered pairs of integers representing the edges of G. Write a procedure for each of the following:

(a) Finds the $M \times M$ adjacency matrix A of the graph G.

(b) Uses A and Warshall's algorithm to find the path matrix P of G.

Test the above procedure using the following data:

(i) $M = 5$; $N = 8$; (3, 4), (5, 3), (2, 4), (1, 5), (3, 2), (4, 2), (3, 1), (5, 1)

(ii) $M = 6$; $N = 10$; (1, 6), (2, 1), (2, 3), (3, 5), (4, 5), (4, 2), (2, 6), (5, 3), (4, 3), (6, 4)

9.44. Suppose a graph G is input by means of an integer M, representing the vertices $1, 2, \ldots, M$, and a list of N ordered triples (a_i, b_i, w_i) of integers such that (a_i, b_i) is an edge of G and w_i is its weight. Write a procedure for each of the following:

(a) Finds the $M \times M$ weight matrix W of the graph G.

(b) Uses W and Warshall's algorithm to find the matrix Q of shortest paths between the vertices of G.

Test the above procedure using the following data:

(i) $M = 4$; $N = 7$; (1, 2, 5), (2, 4, 2), (3, 2, 3), (1, 1, 7), (4, 1, 4), (4, 3, 1)

(ii) $M = 5$; $N = 8$; (3, 5, 3), (4, 1, 2), (5, 2, 2), (1, 5, 5), (1, 3, 1), (2, 4, 1), (3, 4, 4), (5, 4, 4)

9.45. Consider the graph G in Fig. 9-11. Show the sequence of waiting lists in STACK and the sequence of vertices processed while doing a depth-first search (DFS) of G beginning at the vertex: (a) B; (b) E; (c) K.

9.46. Consider the graph in Fig. 9-11. As in Example 9.11, find the shortest path from K to F using a breadth-first search of G. In particular, show the sequence of waiting lists in QUEUE during the search.

Answers to Supplementary Problems

Notation: $M = [R_1; \ R_2; \ \ldots; \ R_n]$ denotes a matrix with rows $R_1, R_2, \ldots, R_n$.

9.18. (a) Indegrees: 1, 1, 4, 3, 1; outdegrees: 2, 3, 1, 2, 2.

(b) None.

(c) (v_1, v_2, v_4), (v_1, v_3, v_5, v_4), $(v_1, v_2, v_3, v_5, v_4)$

(d) (v_3, v_5, v_4, v_3)

(e) (v_1, v_3), (v_1, v_2, v_3), (v_1, v_2, v_4, v_3), (v_1, v_2, v_1, v_3) (v_1, v_3, v_5, v_7)

(f) Unilaterally, but not strongly.

9.19. (a) Sources: v_1

(b) (v_1, v_6, v_7, v_4), $(v_1, v_6, v_7, v_2, v_5, v_3, v_4)$

(c) $(v_1, v_6, v_7, v_2, v_6, v_7, v_4)$

(d) (v_4, v_8, v_7, v_4), $(v_4, v_8, v_7, v_2, v_5, v_3, v_4)$

9.20. (a) succ(1) = [6], succ(3) = [4, 7], succ(5) = [3], succ(7) = [2, 4].

(b) (i) (1, 6), (5, 3); (ii) (2, 6), (6, 7), (7, 2), (3, 7).

9.21. (a) $G = [A : D; \ B : C; \ C : E; \ D : B, D, E; \ E : A]$

(b) Loop: (D, D)

(c) (D, E), (D, B, C, E)

(d) (A, D, E, A), (A, D, B, C, E, A)

(e) Unilaterally, and strongly.

(f) And (g) H has three edges: (C, E), (D, E), (D, D). There are $8 = 2^3$ ways of choosing some of the three edges; and each choice gives a subgraph.

9.22. (a) $(a, b), (a, c), (c, d), (c, e), (d, a), (d, b), (d, e)$

(b) Since b and e are sinks, there is no path from b to e or from e to b, so G is neither unilaterally nor strongly connected. G is weakly connected since, cc, a, b, d, e is a spanning semipath.

9.23. (a) $G = [A : B, C : B : E; \ C : D; \ E : \varnothing]$; (b) Sink: E; (c) (A, B, E), (A, C, D, E); (d) (A, C, D, A); (e) $C, D, A, B, E)$; (f) No.

9.24. (a) (i) (R, A, D), 2; (ii) (R, B, F, J), 3; (iii) (R, C, G), 2.

(b) H, E, I, J, G

(c) $H : 1.1.1, E : 1.2, I : 2.1.1, J : 2.1.2, G : 3.1$

9.25. (a) 0, 1, 1.1, 2, 2.1, 2.1.1, 2.2, 2.2.1, 2.2.1.1, 2.2.1.2, 3, 3.1, 3.1.1, 3.2. (b) Fig. 9-35(a).

9.26. (a) $A = [0, 1, 1, 0; 0, 0, 1, 1; 0, 0, 0, 1; 0, 0, 0, 0]$; $P = [0, 1, 1, 1; 0, 0, 1, 1; 0, 0, 0, 1; 0, 0, 0, 0]$;

(b) 0, 2, 1, 0, 0, $\ldots$; (c) Weakly and unilaterally.

9.27. (a) $A = [0, 1, 1, 0; 0, 0, 0, 0; 0, 1, 1, 1; 0, 2, 0, 0]$; $P = [0, 1, 1, 1; 0, 0, 0, 0; 0, 1, 1, 1; 0, 1, 0, 0]$;

(b) 0, 1, 1, 1, $\ldots$; (c) Weakly and unilaterally.

9.28. Let $P = [p_{ij}]$. For $i \ne j$: (a) $p_{ij} \ne 0$; (b) either $p_{ij} \ne 0$ or $p_{ji} \ne 0$.

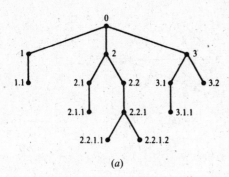

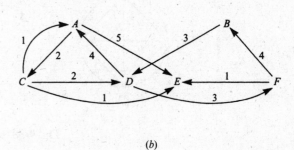

(a)   (b)

Fig. 9-35

9.29. (a) $A = [0, 0, 1, 0, 0; 1, 0, 0, 0, 0; 0, 1, 0, 0, 0;$
$1, 0, 0, 0, 1; 1, 0, 1, 1, 0];$
$P = [1, 1, 1, 0, 0; 1, 1, 1, 0, 0; 1, 1, 1, 0, 0;$
$1, 1, 1, 1, 1; 1, 1, 1, 1, 1; 1, 1, 1, 1, 1]$
(b) $(X, Z, Y, X); (S, T, S)$ (c) Unilaterally.

9.30. (a) $A = [0, 7, 0, 0; 3, 0, 2, 0; 0, 0, 0, 5; 6, 1, 4, 0]$
(b) $Q = [XYX, XY, XYS, XYST; YX, YSTY, YS, YST;$
$STYX, STY, STYS, ST; TX, TY, TYS, TYST]$

9.31. (a) $C, F, D, B, E, A;$ (b) $[A : C, E; B : D;$
$C : D, E, A; D : A, F; E : \varnothing; F : B, E];$ (c)
See Fig. 9-35(b).

9.32. (a) 5, 6; (b) source: A; (c) See Fig. 9-36(a); none.

9.33. (a) 5, 1; (b) none; (c) See Fig. 9-36(b); (d) all three.

9.34. (a) 7, 11; (b) source: A; sinks: C, D; (c) See
Fig. 9-36(c); (d) only weakly.

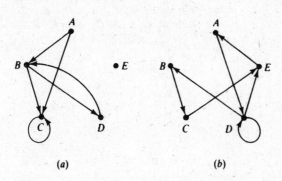

(a)   (b)   (c)

Fig. 9-36

9.35. See Fig. 9-37.

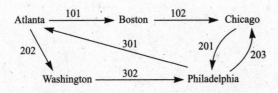

Fig. 9-37

9.36. (a) No; (b) yes; (c) no; (d) no.

9.37. (a) AWP; (b) PA; (c) none; (d) WPC.

9.38. $(s, 4, 1, 2, 1, 2, 1, 2, t)$

9.39. *Hint:* First draw the graph. (a) $ASYCZBXTD$; (b) none,
the graph is not cycle-free, e.g., $YBAXSY$ is a cycle; (c)
$BTYXACSDZ$.

9.40. See Fig. 9-38(a).

9.41. (a) See Fig. 9-38(b). (b) Only weakly connected.

9.42. *Hint:* Suppose $(\alpha = v_1, \ldots, v_m)$ is a longest path in G
and does not include vertex u. If $(u_1 v_1)$ is an arc, then
$\beta = (u, \alpha)$ extends α. Hence (v_1, u) is an arc. If (u, v_2)
is also an arc, then $\beta = (v_1, u, v_2, \ldots, v_m)$ extends α;
hence (v_2, u) is an arc. Similarly $(v_3, u), \ldots, (v_m, u)$
are arcs. Hence $\beta = (\alpha, u)$ extends α. This contradicts
the maximality of α.

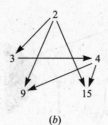

(a)   (b)

Fig. 9-38

9.43. (i) $A = [0, 0, 0, 0, 0; 0, 0, 0, 1, 0; 1, 1, 0, 1, 0;$
$0, 1, 0, 0, 0; 1, 0, 1, 0, 0]$
$P = [1, 1, 1, 1, 1; 0, 1, 0, 1, 0; 1, 1, 1, 1, 1;$
$0, 1, 0, 1, 0; 1, 1, 1, 1, 1]$

(ii) $A = [0, 0, 0, 0, 0, 1; 1, 0, 1, 0, 0, 1; 0, 0, 0, 0, 1, 0;$
$0, 1, 0, 1, 0; 0, 0, 1, 0, 0, 0; 0, 0, 0, 1, 0, 0]$
$P = [1, 1, 1, 1, 1, 1; 1, 1, 1, 1, 1, 1; 0, 0, 1, 0, 1, 0;$
$1, 1, 1, 1, 1, 1; 0, 0, 1, 0, 1, 0; 1, 1, 1, 1, 1, 1]$

9.44. (i) $W = [7, 5, 0, 0; 0, 0, 0, 2; 0, 3, 0, 0; 4, 0, 1, 0];$
$Q = [AA, AB, ABCD, ABD; BDA, BDCB, BDC, BD;$
$CBDA, CB, CBDC, CBD; DA, DCB, DC, DCBD],$
where A, B, C, D are the vertices.

(ii) $W = [0, 0, 1, 0, 5; 0, 0, 0, 1, 0; 0, 0, 0, 4, 3;$
$2, 0, 0, 0, 0; 0, 2, 0, 4, 0];$

$Q = [ACDA, ACEB, AC, ACD, ACE; BDA, BDACEB,$
$BDAC, BD, BDACE; CDA, CEB, CDAC, CD, CE;$
$DA, DACEB, DAC, DACD, DACEB; EDA, EB,$
$EDAC, ED, EDACE]$ where A, B, C, D, E are the
vertices.

9.45. (a) STACK: $B, L_B E_B, E_L C_L E_B, F_E C_L, D_F C_L, C_L,$
$J_C, K_J, \varnothing$; Vertex: $B, L_B, E_L, F_E, D_F, C_L, J_C, K_J$

(b) STACK: $E, F_E, D_F, \varnothing$; Vertex: E, F_E, D_F

(c) STACK: $K, L_K C_K, E_L C_L, C_K, C_L, D_F C_L,$
$C_L J_C, \varnothing$; Vertex: $K, L_K, E_L, F_E, D_F, C_L, J_C$

9.46. QUEUE: $K, L_K C_K, J_C E_C D_C L_K, J_C E_C D_C, J_C E_C,$
F_E; Vertex: $K, C_K, L_K, D_C, E_C, J_C, F_E$; Minimal
Path: $F_E \leftarrow E_C \leftarrow C_K \leftarrow$ or $K \rightarrow C_K \rightarrow E_C \rightarrow F_E$.

CHAPTER 10

Binary Trees

10.1 INTRODUCTION

The binary tree is a fundamental structure in mathematics and computer science. Some of the terminology of rooted trees, such as, edge, path, branch, leaf, depth, and level number, will also be used for binary trees. However, we will use the term node, rather than vertex, with binary trees. We emphasize that a binary tree is not a special case of a rooted tree; they are different mathematical objects.

10.2 BINARY TREES

A *binary tree T* is defined as a finite set of elements, called *nodes*, such that:

    (1) T is empty (called the *null tree* or *empty tree*), or

    (2) T contains a distinguished node R, called the *root* of T, and the remaining nodes of T form an ordered pair of disjoint binary trees T_1 and T_2.

If T does contain a root R, then the two trees T_1 and T_2 are called, respectively, the left and right subtrees of R. If T_1 is nonempty, then its root is called the *left successor* of R; similarly, if T_2 is nonempty, then its root is called the *right successor* of R.

The above definition of a binary tree T is recursive since T is defined in terms of the binary subtrees T_1 and T_2. This means, in particular, that every node N of T contains a left and a right subtree, and either subtree or both subtrees may be empty. Thus every node N in T has 0, 1, or 2 successors. A node with no successors is called a *terminal* node. Thus both subtrees of a terminal node are empty.

Picture of a Binary Tree

A binary tree T is frequently presented by a diagram in the plane called a *picture* of T. Specifically, the diagram in Fig. 10-1(a) represents a binary tree as follows:

    (i) T consists of 11 nodes, represented by the letters A through L, excluding I.

    (ii)   The root of T is the node A at the top of the diagram.

    (iii)   A left-downward slanted line at a node N indicates a left successor of N; and a right-downward slanted line at N indicates a right successor of N.

235

Accordingly, in Fig. 10-1(a):

(a) B is a left successor and C is a right successor of the root A.

(b) The left subtree of the root A consists of the nodes B, D, E, and F, and the right subtree of A consists of the nodes C, G, H, J, K, and L.

(c) The nodes A, B, C, and H have two successors, the nodes E and J have only one successor, and the nodes D, F, G, L, and K have no successors, i.e., they are terminal nodes.

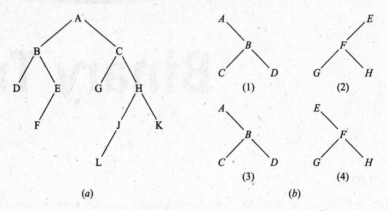

(a)   (b)

Fig. 10-1

Similar Binary Trees

Binary trees T and T' are said to be *similar* if they have the same structure or, in other words, if they have the same shape. The trees are said to be *copies* if they are similar and if they have the same contents at corresponding nodes.

EXAMPLE 10.1 Consider the four binary trees in Fig. 10-1(b). The three trees (1), (3), and (4) are similar. In particular the trees (1) and (3) are copies since they also have the same data at corresponding nodes. The tree (2) is neither similar nor a copy of the tree (4) because, in a binary tree, we distinguish between a left successor and a right successor even when there is only one successor.

Terminology

Terminology describing family relationships is frequently used to describe relationships between the nodes of a tree T. Specifically, suppose N is a node in T with left successor S_1 and right successor S_2. Then N is called the *parent* (or *father*) of S_1 and S_2. Analogously, S_1 is called the *left child* (or *left son*) of N, and S_2 is called the *right child* (or *right son*) of N. Furthermore, S_1 and S_2 are said to be *siblings* (or *brothers*). Every node N in a binary tree T, except the root, has a unique parent, called the *predecessor* of N.

The terms descendant and ancestor have their usual meaning. That is, a node L is called a *descendant* of a node N (and N is called an *ancestor* of L) if there is a succession of children from N to L. In particular, L is called a *left* or *right descendant* of N according to whether L belongs to the left or right subtree of N.

Terminology from graph theory and horticulture are also used with a binary tree T. Specifically, the line drawn from a node N of T to a successor is called an *edge*, and a sequence of consecutive edges is called a *path*. A terminal node is called a *leaf*, and a path ending in a leaf is called a *branch*.

Each node in a binary tree T is assigned a *level number*, as follows. The root R of the tree T is assigned the level number 0, and every other node is assigned a level number which is 1 more than the level number of its parent. Furthermore, those nodes with the same level number are said to belong to the same *generation*.

The *depth* (or *height*) of a tree T is the maximum number of nodes in a branch of T. This turns out to be 1 more than the largest level number of T. The tree T in Fig. 10-1(a) has depth 5.

10.3   COMPLETE AND EXTENDED BINARY TREES

This section considers two special kinds of binary trees.

Complete Binary Trees

Consider any binary tree T. Each node of T can have at most two children. Accordingly, one can show that level r of T can have at most 2^r nodes. The tree T is said to be *complete* if all its levels, except possibly the last, have the maximum number of possible nodes, and if all the nodes at the last level appear as far left as possible. Thus there is a unique complete tree T_n with exactly n nodes (where we ignore the contents of the nodes). The complete tree T_{26} with 26 nodes appears in Fig. 10-2.

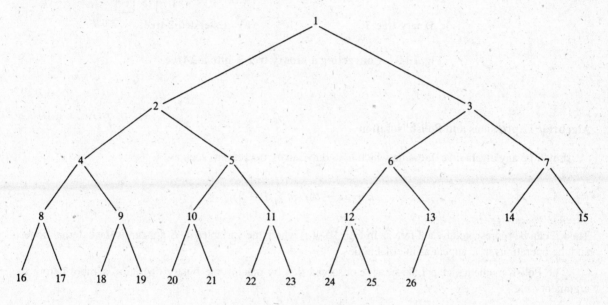

Fig. 10-2   Complete tree T_{26}

The nodes of the complete binary tree T_{26} in Fig. 10-2 have been purposely labeled by the integers $1, 2, \ldots, 26$, from left to right, generation by generation. With this labeling, one can easily determine the children and parent of any node K in any complete tree T_n. Specifically, the left and right children of the node K are, respectively, $2*K$ and $2*K + 1$, and the parent of K is the node $[K/2]$. For example, the children of node 9 are the nodes 18 and 19, and its parent is the node $[9/2] = 4$. The depth d_n of the complete tree T_n with n nodes is given by

$$d_n = \lfloor \log_2 n + 1 \rfloor$$

This is a relatively small number. For example, if the complete tree T_n has $n = 1\,000\,000$ nodes, then its depth $d_n = 21$.

Extended Binary Trees: 2-Trees

A binary tree tree T is said to be a *2-tree* or an *extended binary tree* if each node N has either 0 or 2 children. In such a case, the nodes with two children are called *internal nodes*, and the nodes with 0 children are called *external nodes*. Sometimes the nodes are distinguished in diagrams by using circles for internal nodes and squares for external nodes.

The term "extended binary tree" comes from the following operation. Consider any binary tree T, such as the tree in Fig. 10-3(a). Then T may be "converted" into a 2-tree by replacing each empty subtree by a new node, as pictured in Fig. 10-3(b). Observe that the new tree is, indeed, a 2-tree. Furthermore, the nodes in the original tree T are now the internal nodes in the extended tree, and the new nodes are the external nodes in the extended tree. We note that if a 2-tree has n internal nodes, then it will have $n + 1$ external nodes.

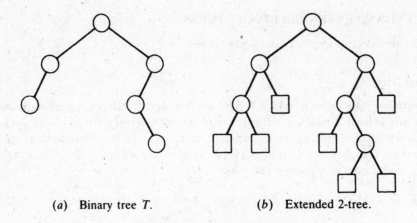

(a)   Binary tree T.   (b)   Extended 2-tree.

Fig. 10-3   Converting a binary tree T into a 2-tree

Algebraic Expressions and Polish Notation

Let E be any algebraic expression which uses only binary operations, such as,

$$E = (a - b)/((c \times d) + e)$$

Then E can be represented by a 2-tree as in Fig. 10-4(a) where the variables in E appear as the external nodes, and the operations in E appear as internal nodes.

The Polish mathematician Lukasiewicz observed that by placing the binary operation symbol before its arguments, e.g.,

$$+ab \text{ instead of } a + b \quad \text{and} \quad /cd \text{ instead of } c/d$$

one does not need to use any parentheses. This notation is called *Polish notation in prefix form*. (Analogously, one can place the symbol after its arguments, called *Polish notation in postfix form*.) Rewriting E in prefix form we obtain:

$$E = /-ab + \times cde$$

Observe that this is precisely the lexicographic order of the vertices in its 2-tree which can be obtained by scanning the tree as in Fig. 10-4(b).

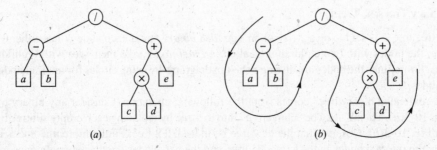

(a)   (b)

Fig. 10-4

10.4 REPRESENTING BINARY TREES IN MEMORY

Let T be a binary tree. This section discusses two ways of representing T in memory. The first and usual way is called the *linked representation* of T and is analogous to the way linked lists are represented in memory. The second way, which uses a single array, is called the *sequential representation* of T. The main requirement of any representation of T is that one should have direct access to the root R of T and, given any node N of T, one should have direct access to the children of N.

Linked Representation of Binary Trees

Consider a binary tree T. Unless otherwise stated or implied, T will be maintained in memory by means of a linked representation which uses three parallel arrays, INFO, LEFT, and RIGHT, and a pointer variable ROOT as follows. First of all, each node N of T will correspond to a location K such that:

(1) INFO[K] contains the data at the node N.
(2) LEFT[K] contains the location of the left child of node N.
(3) RIGHT[K] contains the location of the right child of node N.

Furthermore, ROOT will contain the location of the root R of T. If any subtree is empty, then the corresponding pointer will contain the null value; if the tree T itself is empty, then ROOT will contain the null value.

Remark 1: Most of our examples will show a single item of information at each node N of a binary tree T. In actual practice, an entire record may be stored at the node N. In other words, INFO may actually be a linear array of records or a collection of parallel arrays.

Remark 2: Any invalid address may be chosen for the null pointer denoted by NULL. In actual practice, 0 or a negative number is used for NULL.

EXAMPLE 10.2 Consider the binary tree T in Fig. 10-1(a). The linked representation of T appears in Fig. 10-5 where we have written the linear arrays vertical rather than horizontal for notational convenience. Note that ROOT = 5 points to INFO[5] = A since A is the root of T. Also, note that LEFT[5] = 10 points to INFO[10] = B since B is the left child of A, and RIGHT[5] = 2 points to INFO[2] = C since C is the right child of A. And so on. The choice of 18 elements for the arrays is arbitrary.

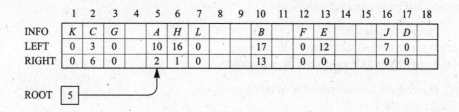

Fig. 10-5

Sequential Representation of Binary Trees

Suppose T is a binary tree that is complete or nearly complete. Then there is an efficient way of maintaining T in memory called the *sequential representation* of T. This representation uses only a single linear array TREE together with a pointer variable END as follows:

(1) The root R of T is stored in TREE[1].
(2) If a node N occupies TREE[K], then its left child is stored in TREE[2*K] and its right child is stored in TREE[2*K + 1].
(3) END contains the location of the last node of T.

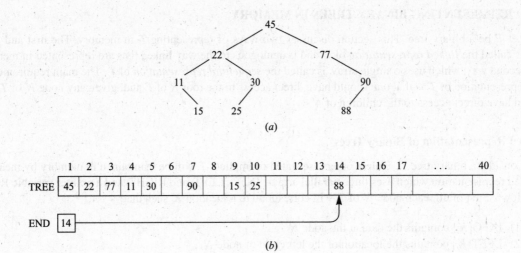

Fig. 10-6

Furthermore, the node N at TREE[K] contains an empty left subtree or an empty right subtree according as $2*K$ or $2*K + 1$ exceeds END or according as TREE[$2*K$] or TREE[$2*K + 1$] contains the NULL value.

The sequential representation of the binary tree T in Fig. 10-6(a) appears in Fig. 10-6(b). Observe that we require 14 locations in the array TREE even though T has only 9 nodes. Generally speaking, the sequential representation of a tree with depth d will require an array with approximately 2^d elements. Accordingly, this sequential representation is usually inefficient unless, as stated above, the binary tree T is complete or nearly complete. For example, the tree T in Fig. 10-1(a) has 11 nodes and depth 5, which means it would require an array with approximately $2^5 = 32$ elements.

10.5 TRAVERSING BINARY TREES

There are three standard ways of traversing a binary tree T with root R. These three algorithms, called *preorder*, *inorder*, and *postorder*, are as follows:

Preorder: (1) Process the root R.
(2) Traverse the left subtree of R in preorder.
(3) Traverse the right subtree of R in preorder.

Inorder: (1) Traverse the left subtree of R in inorder.
(2) Process the root R.
(3) Traverse the right subtree of R in inorder.

Postorder: (1) Traverse the left subtree of R in postorder.
(2) Traverse the right subtree of R in postorder.
(3) Process the root R.

Observe that each algorithm contains the same three steps, and that the left subtree of R is always traversed before the right subtree. The difference between the algorithms is the time at which the root R is processed. Specifically, in the "pre" algorithm, the root R is processed before the subtrees are traversed; in the "in" algorithm, the root R is processed between the traversals of the subtrees; and in the "post" algorithm, the root R is processed after the subtrees are traversed.

The three algorithms are sometimes called, respectively, the node-left-right (NLR) traversal, the left-node-right (LNR) traversal, and the left-right-node (LRN) traversal.

EXAMPLE 10.3 Consider the binary tree T in Fig. 10-7(a). Observe that A is the root of T, that the left subtree L_T of T consists of nodes B, D, and E, and the right subtree R_T of T consists of nodes C and F.

(a) The preorder traversal of T processes A, traverses L_T, and traverses R_T. However, the preorder traversal of L_T processes the root B, and then D and E; and the preorder traversal of R_T processes the root C and then F. Thus $ABDECF$ is the preorder traversal of T.

(b) The inorder traversal of T traverses L_T processes A, and traverses R_T. However, the inorder traversal of L_T processes D, B, and then E; and the inorder traversal of R_T processes C and then F. Thus $DBEACF$ is the inorder traversal of T.

(c) The postorder traversal of T traverses L_T, traverses R_T, and processes A. However, the postorder traversal of L_T, processes D, E, and then B, and the postorder traversal of RT processes F and then C. Accordingly, $DEBFCA$ is the postorder traversal of T.

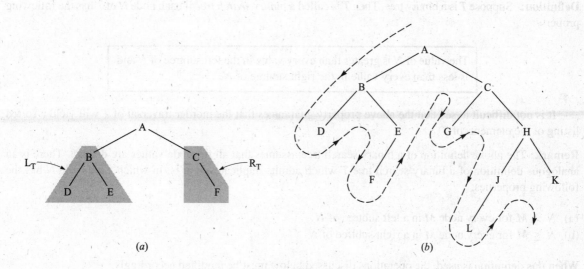

$$(a) \qquad\qquad\qquad\qquad\qquad\qquad (b)$$

Fig. 10-7

EXAMPLE 10.4 Let T be the binary tree in Fig. 10-7(b). The preorder traversal is as follows:

$$\text{(Preorder)} \quad A\ B\ D\ E\ F\ C\ G\ H\ J\ L\ K$$

This order is the same as the one obtained by scanning the tree from the left as indicated by the path in Fig. 10-7(b). That is, one "travels" down the leftmost branch until meeting a terminal node, then one backtracks to the next branch, and so on. In the preorder traversal, the rightmost terminal node, node K, is the last node scanned. Observe that the left subtree of the root A is traversed before the right subtree, and both are traversed after A. The same is true for any other node having subtrees, which is the underlying property of a preorder traversal.

The reader can verify by inspection that the other two ways of traversing the tree T in Fig. 10-7(b) are as follows:

$$\text{(Inorder)} \quad D\ B\ F\ E\ A\ G\ C\ L\ J\ H\ K$$
$$\text{(Postorder)} \quad D\ F\ E\ B\ G\ L\ J\ K\ H\ C\ A$$

Remark: Observe that the terminal nodes, D, F, G, L, and K, of the binary tree in Fig. 10-7(b) are traversed in the same order, from left to right, in all three traversals. We emphasize that this is true for any binary tree T.

10.6 BINARY SEARCH TREES

This section discusses one of the most important data structures in computer science, a binary search tree. This structure enables us to search for and find an element with an average running time $f(n) = 0(\log_2 n)$, where n is the number of data items. It also enables us to easily insert and delete elements. This structure contrasts with the following structures:

(a) *Sorted linear array:* Here one can search for and find an element with running time $f(n) = 0(\log_2 n)$. However, inserting and deleting elements is expensive since, on the average, it involves moving $0(n)$ elements.

(b) *Linked list:* Here one can easily insert and delete elements. However, it is expensive to search and find an element, since one must use a linear search with running time $f(n) = 0(n)$.

Although each node in a binary search tree may contain an entire record of data, the definition of the tree depends on a given field whose values are distinct and may be ordered.

Definition: Suppose T is a binary tree. Then T is called a *binary search tree* if each node N of T has the following property:

> The value of N is greater than every value in the left subtree of N and is less than every value in the right subtree of N.

It is not difficult to see that the above property guarantees that the inorder traversal of T will yield a sorted listing of the elements of T.

Remark: The above definition of a binary search tree assumes that all the node values are distinct. There is an analogous definition of a binary search tree T which admits duplicates, that is, in which each node N has the following properties:

(a) $N > M$ for every node M in a left subtree of N.
(b) $N \leq M$ for every node M in a right subtree of N.

When this definition is used, the operations discussed below must be modified accordingly.

EXAMPLE 10.5 The binary tree T in Fig. 10-8(a) is a binary search tree. That is, every node N in T exceeds every number in its left subtree and is less than every number in its right subtree. Suppose the 23 were replaced by 35. Then T would still be a binary search tree. On the other hand, suppose the 23 were replaced by 40. Then T would not be a binary search tree, since 40 would be in the left subtree of 38 but 40 > 38.

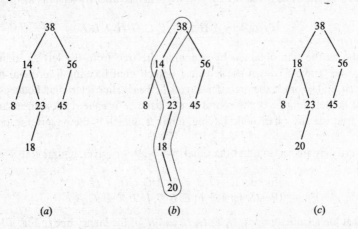

(a)	(b)	(c)

Fig. 10-8

Searching and Inserting in a Binary Search Tree

A search and insertion algorithm in a binary search tree T appears in Fig. 10-9.

Algorithm 10.1:   A binary search tree T and an ITEM of information is given. The algorithm finds
the location of ITEM in T, or inserts ITEM as a new node in the tree.

Step 1. Compare ITEM with the root N of the tree.
    (a) If ITEM $< N$, proceed to the left child of N.
    (b) If ITEM $> N$, proceed to the right child of N.

Step 2. Repeat Step 1 until one of the following occurs:
    (a) We meet a node N such that ITEM $= N$. In this case the search is successful.
    (b) We meet an empty subtree, which indicates the search is unsuccessful. Insert ITEM in
        place of the empty subtree.

Step 3. Exit.

Fig. 10-9

EXAMPLE 10.6 Consider the binary search tree T in Fig. 10-8(a). Suppose ITEM $= 20$ is given, and we want
to find or insert ITEM into T. Simulating Algorithm 10-1, we obtain the following steps:

    (1) Compare ITEM $= 20$ with root $R = 38$. Since $20 < 38$, proceed to the left child of 38, which is 14.
    (2) Compare ITEM $= 20$ with 14. Since $20 > 14$, proceed to the right child of 14, which is 23.
    (3) Compare ITEM $= 20$ with 23. Since $20 < 23$, proceed to the left child of 23, which is 18.
    (4) Compare ITEM $= 20$ with 18. Since $20 > 18$ and 18 has no right child, insert 20 as the right child of 18.

Figure 10-11(b) shows the new tree with ITEM $= 20$ inserted. The path down the tree during the algorithm has
been ringed.

Deleting in a Binary Search Tree

An algorithm which deletes a given ITEM from a binary search tree T appears in Fig. 10-10. It uses
Algorithm 10.1 in Fig. 10-9 to find the location of ITEM in T.

Remark: Observe that case (iii) in Step 2(c) is more complicated than the first two cases. The inorder successor
$S(N)$ of N is found as follows. From the node N, move right to the right child of N, and then successively move
left until meeting a node M with no left child. The node M is the inorder successor $S(N)$ of N.

EXAMPLE 10.7 Consider the binary tree T in Fig. 10-8(b). Suppose we want to delete ITEM $= 14$ from T.
First find the node N such that $N = 14$. Note $N = 14$ has two children. Moving right and then left, we find the
inorder successor $S(N) = 18$ of N. We delete $S(N) = 18$ by replacing it by its only child 20, then we replace
$N = 14$ by $S(N) = 18$. This yields the tree in Fig. 10-8(c).

Complexity of the Binary Search Tree Algorithms

Let T be a binary tree with n nodes and depth d, and let $f(n)$ denote the running time of either of the above
algorithms. Algorithm 10.1 tells us to proceed from the root R down through the tree T until finding ITEM in T
or inserting ITEM as a terminal node. Algorithm 10.2 tells us to proceed from the root R down through the tree
T to find ITEM and then to continue down the tree T to find the inorder successor of ITEM. In either case, the

---

Algorithm 10.2: A binary search tree T and an ITEM of information is given. $P(N)$ denotes the parent of a node N, and $S(N)$ denotes the inorder successor of N. The algorithm deletes ITEM from T.

Step 1. Use Algorithm 10.1 to find the location of the node N which contains ITEM and keep track of the location of the parent node $P(N)$. (If ITEM is not in T, then STOP and Exit.)

Step 2. Determine the number of children of N. There are three cases:
   - (a) **N has no children.** N is deleted from T by simply replacing the location of N in the parent node $P(N)$ by the NULL pointer.
   - (b) **N has exactly one child M.** N is deleted from T by replacing the location of N in the parent node $P(N)$ by the location of M. (This replaces N by M.)
   - (c) **N has two children.**
      - (i)   Find the inorder successor $S(N)$ of N. (Then $S(N)$ has no left child.)
      - (ii) Delete $S(N)$ from T using (a) or (b).
      - (iii) Replace N by $S(N)$ in T.

Step 3. Exit.

Fig. 10-10

number of moves cannot exceed the depth d of the tree. Thus the running time $f(n)$ of either algorithm depends on the depth d of the tree T.

Now suppose T has the property that, for any node N of T, the depths of the subtrees of N differ by at most 1. Then the tree T is said to be balanced, and $d \approx \log_2 n$. Therefore, the running time $f(n)$ of either algorithm in a balanced tree is very fast; specifically, $f(n) = 0(\log_2 n)$. On the other hand, as items are added into a binary search tree T, there is no guarantee that T remains balanced. It could even happen that $d \approx n$. In this case, $f(n)$ would be relatively slow; specifically, $f(n) = 0(n)$. Fortunately, there are techniques for rebalancing a binary search tree T as elements are added to T. Such techniques, however, lie beyond the scope of this text.

10.7 PRIORITY QUEUES, HEAPS

Let S be a priority queue. That is, S is a set where elements may be periodically inserted and deleted, but where the current largest element (element with highest priority) is always deleted. One may maintain S in memory as follows:

(a) **Linear array:** Here one can easily insert an element by simply adding it to the end of the array. However, it is expensive to search for and find the largest element, since one must use a linear search with running time $f(n) = 0(n)$.

(b) **Sorted linear array:** Here the largest element is either first or last, and so it is easily deleted. However, inserting and deleting elements is expensive since, on the average, it involves moving $0(n)$ elements.

This section introduces a discrete structure which can efficiently implement a priority queue S.

Heaps

Suppose H is a complete binary tree with n elements. We assume H is maintained in memory using its sequential representation, not a linked representation. (See Section 10.4.)

Definition 10.1: Suppose H is a complete binary tree. Then H is called a *heap*, or a *maxheap*, if each node N has the following property.

The value of N is greater than or equal to the value at each of the children of N.

Accordingly, in a heap, the value of N exceeds the value of every one of its descendants. In particular, the root of H is a largest value of H.

A *minheap* is defined analogously: The value of N is less than or equal to the value at each of its children.

EXAMPLE 10.8 Consider the complete binary tree H in Fig.10-11(a). Observe that H is a heap. This means, in particular, that the largest element of H appears at the "top" of the heap. Fig. 10-11(b) shows the sequential representation of H by the array TREE and the variable END. Accordingly:

(a) TREE[1] is the root R of H.

(b) TREE[$2K$] and TREE[$2K + 1$] are the left and right children of TREE[K].

(c) The variable END $= 20$ points to the last element in H.

(d) The parent of any nonroot node TREE(J) is the node TREE[$J \div 2$] (where $J \div 2$ means integer division).

Observe that the nodes of H on the same level appear one after another in the array TREE. The choice of 30 locations for TREE is arbitrary.

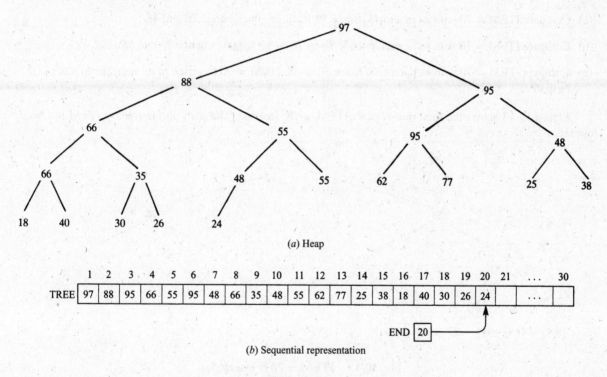

(a) Heap

(b) Sequential representation

Fig. 10-11

Inserting into a Heap

Figure 10-12 gives an algorithm which inserts a given data ITEM into a heap H.

Remark: One must verify that Algorithm 10.3 does always yield a heap as a final tree. This is not difficult to see, and we leave this verification to the reader.

EXAMPLE 10.9 Consider the heap H in Fig. 10-11. Suppose we want to insert ITEM $= 70$ into H. Simulating Algorithm 10.3, we first adjoin ITEM as the last element of the complete tree; that is, as the right child of 48.

Algorithm 10.3:   A heap H and a new ITEM are given. The algorithm inserts ITEM into H.

Step 1.   Adjoin ITEM at the end of H so that H is still a complete tree, but not necessarily a heap.

Step 2.   (Reheap)   Let ITEM rise to its "appropriate place" in H so that H is a heap. That is:
  (a)   Compare ITEM with its parent P(ITEM). If ITEM $> P$(ITEM), then interchange ITEM and P(ITEM).
  (b)   Repeat (a) until ITEM $\leq P$(ITEM).

Step 3.   Exit.

Fig. 10-12

In other words, we set TREE[21] $= 70$ and END $= 21$. Then we reheap, i.e., we let ITEM rise to its appropriate place as follows:

(a) Compare ITEM $= 70$ with its parent 48. Since $70 > 48$, we interchange 70 and 48.

(b) Compare ITEM $= 70$ with its new parent 55. Since $70 > 55$, we interchange 70 and 55.

(c) Compare ITEM $= 70$ with its parent 88. Since $70 < 88$, ITEM $= 70$ has risen to its appropriate place in the heap H.

Figure 10-13 shows the final tree H with ITEM $= 70$ inserted. The path up the tree by ITEM has been circled.

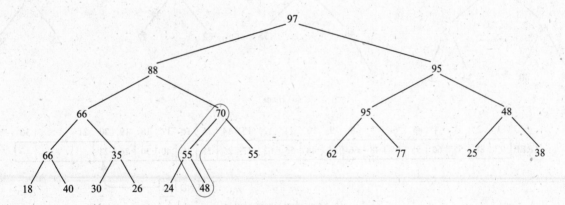

Fig. 10-13   ITEM = 70 is inserted

Deleting the Root from a Heap

Figure 10-14 gives an algorithm which deletes the root R from a heap H.

Remark: As with inserting in a heap, one must verify that Algorithm 10.4 does always yield a heap as a final tree. Again, we leave this verification to the reader. We also note that Step 3 may not end until the node L reaches the bottom of the tree, i.e., until L has no children.

EXAMPLE 10.10 Consider the heap H in Fig. 10-15(a), where $R = 95$ is the root and $L = 22$ is the last node of H. Suppose we want to delete $R = 95$ from the heap H. Simulating Algorithm 10.4, we first "delete" $R = 95$ by assigning ITEM $= 95$, and then we replace $R = 95$ by $L = 22$. This yields the complete tree in Fig. 10-15(b)

Algorithm 10.4: The algorithm deletes the root R from a given heap H.

Step 1. Assign the root R to some variable ITEM.

Step 2. Replace the deleted root R by the last node of L of H so that H is still a complete binary tree, but not necessarily a heap. [That is, set TREE[1] := TREE[END] and then set END := END−1.]

Step 3. (Reheap)   Let L sink to its "appropriate place" in H so that H is a heap. That is:
    (a) Find the larger child LARGE(L) of L. If $L <$ LARGE(L), then interchange L and LARGE(L).
    (b) Repeat (a) until $L \geq$ LARGE(L).

Step 4. Exit.

Fig. 10-14

which is not a heap. (Note that both subtrees of 22 are still heaps.) Then we reheap, that is, we let $L = 22$ sink to its appropriate place as follows:

(a) The children of $L = 22$ are 85 and 70. The larger is 85. Since $22 < 85$, interchange 22 and 85. This yields the tree in Fig. 10-15(c).

(b) The children of $L = 22$ are now 33 and 55. The larger is 55. Since $22 < 55$, interchange 22 and 55. This yields the tree in Fig. 10-15(d).

(c) The children of $L = 22$ are now 15 and 11. The larger is 15. Since $22 > 15$, the node $L = 22$ has sunk to its appropriate place in the heap.

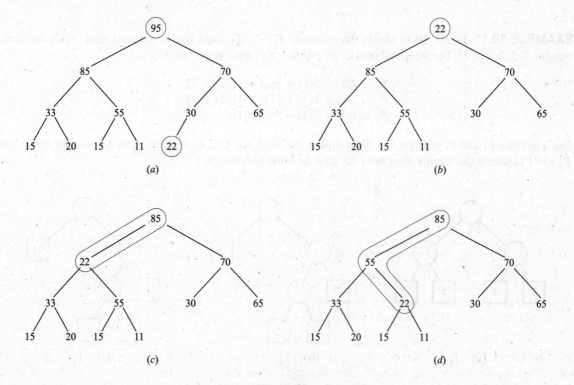

Fig. 10-15

Thus Fig. 10-15(d) is the required heap H without its original root $R = 95$. Observe that we ringed the paths as $L = 22$ made its way down the tree.

Complexity of the Heap Algorithms

Let H be a heap with n nodes. Since H is a complete tree, $d \approx \log_2 n$ where d is the depth of H. Algorithm 10.3 tells us to let the new ITEM proceed up the tree, from level to level, until finding its appropriate place in H. Algorithm 10.4 tells us to let the original last node L proceed down the tree, level by level, until finding its appropriate place in H. In either case, the number of moves cannot exceed the depth d of H. Thus the running time $f(n)$ of either algorithm is very fast; specifically, $f(n) = 0(\log_2 n)$. Accordingly, the heap is a much more efficient way to implement a priority queue S than either the linear array or sorted linear array mentioned at the beginning of the section.

10.8 PATH LENGTHS, HUFFMAN'S ALGORITHM

Let T be an extended binary tree or 2-tree (Section 10.3). Recall that if T has n external nodes, then T has $n - 1$ internal nodes. Figure 10-3(b) shows a 2-tree with seven external nodes and hence $7 - 1 = 6$ internal nodes.

Weighted Path Lengths

Suppose T is a 2-tree with n external nodes, and suppose each external node is assigned a (nonnegative) weight. The weighted path length (or simply path length) P of the tree T is defined to be the sum

$$P = W_1 L_1 + W_2 L_2 + \cdots + W_n L_n$$

where W_1 is the weight at an external node N_i, and L_i is the length of the path from the root R to the node L_i. (The path length P exists even for nonweighted 2-trees where one simply assumes the weight 1 at each external node.)

EXAMPLE 10.11 Figure 10-16 shows three 2-trees, T_1, T_2, T_3, each having external nodes with the same weights 2, 3, 5, and 11. The weighted path lengths of the three trees are as follows:

$$P_1 = 2(2) + 3(2) + 5(2) + 11(2) = 42$$
$$P_2 = 2(1) + 3(3) + 5(3) + 11(2) = 48$$
$$P_3 = 2(3) + 3(3) + 5(2) + 11(1) = 36$$

The quantities P_1 and P_3 indicate that the complete tree need not give a minimum path, and that the quantities P_2 and P_3 indicate that similar trees need not give the same path length.

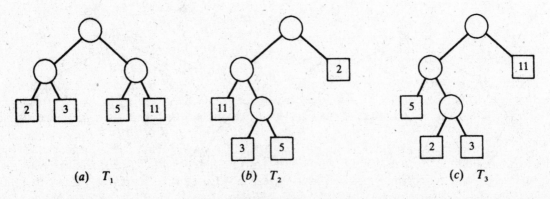

(a) T_1 (b) T_2 (c) T_3

Fig. 10-16

Huffman's Algorithm

The general problem we want to solve is the following. Suppose a list of n weights is given:

$$W_1, W_2, \ldots, W_n$$

Among all the 2-trees with n external nodes and with the given n weights, find a tree T with a minimum weighted path length. (Such a tree T is seldom unique.) Huffman gave an algorithm to find such a tree T.

Huffman's algorithm, which appears in Fig. 10-17, is recursively defined in terms of the number n of weights. In practice, we use an equivalent iterated form of the Huffman algorithm which constructs the desired tree T from the bottom up rather than from the top down.

Algorithm 10.5 (Huffman):   The algorithm recursively finds a weighted 2-tree T with n given
          weights $w_1, w_2, \ldots, w_n$ which has a minimum weighted path length.

Step 1. Suppose $n = 1$. Let T be the tree with one node N with weight w_1, and then Exit.

Step 2. Suppose $n > 1$.
   (a) Find two minimum weights, say w_i and w_j, among the n given weights.
   (b) Replace w_i and w_j in the list by $w_i + w_j$, so the list has $n - 1$ weights.
   (c) Find a tree T' which gives a minimum weighted path length for the $n - 1$ weights
   (d) In the tree T', replace the external node

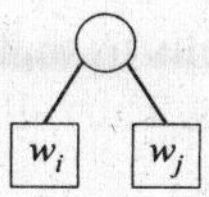

   (e) Exit.

Fig. 10-17

EXAMPLE 10.12 Let A, B, C, D, E, F, G, H be eight data items with the following assigned weights:

Data item:	A	B	C	D	E	F	G	H
Weight:	22	5	11	19	2	11	25	5

Construct a 2-tree T with a minimum weighted path length P using the above data as external nodes.

Apply the Huffman algorithm. That is, repeatedly combine the two subtrees with minimum weights into a single subtree as shown in Fig. 10-18(a). For clarity, the original weights are underlined, and a circled number indicates the root of a new subtree. The tree T is drawn from Step (8) backward yielding Fig. 10-18(b). (When splitting a node into two parts, we have drawn the smaller node on the left.) The path length P follows:

$$P = 22(2) + 11(3) + 11(3) + 25(2) + 5(4) + 2(5) + 5(5) + 19(3) = 280$$

Computer Implementation of Huffman's Algorithm

Consider again the data in Example 10.12. Suppose we want to implement the algorithm using the computer. Since some of the nodes in our binary tree are weighted, our tree can be maintained by four parallel arrays: INFO, WT, LEFT, and RIGHT. The first eight columns in Fig. 10-19 show how the data may be stored in the computer initially.

Each step in Huffman's algorithm assigns values to WT, LEFT, and RIGHT in columns 9 through 15, which correspond, respectively, to steps (2) through (8) in Fig. 10-18. Specifically, each step finds the current two minimal weights and their locations, and then enters the sum in WT and their locations in LEFT and RIGHT. For example, the current minimal weights after assigning values to column 11, which corresponds to step (4), are 12 and 19 appearing in WT[10] and WT[4]. Accordingly, we assign WT[12] = 12 + 19 = 31 and LEFT[12] = 10

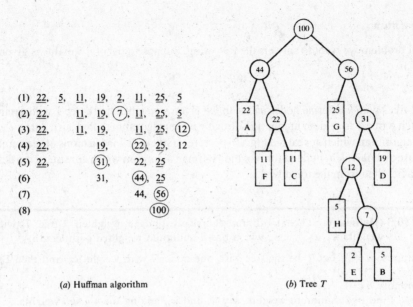

(1) 22, 5, 11, 19, 2, 11, 25, 5

(2) 22,   11, 19, (7), 11, 25, 5

(3) 22,   11, 19,   11, 25, (12)

(4) 22,   19,   (22), 25, 12

(5) 22,   (31),   22, 25

(6)   31,   (44), 25

(7)   44, (56)

(8)   (100)

(a) Huffman algorithm   (b) Tree T

Fig. 10-18

and RIGHT[12] = 4. The last step tells us that ROOT = 15, or we use that fact that ROOT = $2n - 1$, where $n = 8$ is the number of external nodes. Thus all of Fig. 10-19 gives the required tree T.

	1	2	3	4	5	6	7	8		9	10	11	12	13	14	15	16
INFO	A	B	C	D	E	F	G	H									
WT	22	5	11	19	2	11	25	5		7	12	22	31	44	56	100	
LEFT	0	0	0	0	0	0	0	0		5	8	6	10	1	7	13	
RIGHT	0	0	0	0	0	0	0	0		2	9	3	4	11	12	14	

(2) (3) (4) (5) (6) (7) (8)

ROOT $\boxed{15}$

Fig. 10-19

Remark: During the execution of Huffman's algorithm we need to keep track of the current weights and to find two of the minimal weights. This may be efficiently accomplished by maintaining an auxiliary minheap, where each node contains a weight and its location in the tree. We use a minheap rather than a maxheap since we want the node with the lowest weight to be on the top of the heap.

Application to Coding

Suppose a collection of n data items $A_1, A_2, \ldots, A_n$ are to be coded by means of strings of bits. Furthermore, suppose the data items do not occur with the same probability. Then memory space and time may be conserved by using variable-length strings, where items which occur frequently are assigned shorter strings and items which occur infrequently are assigned longer strings. For example, country telephone codes use this principle. The country code for United States is simply 1, for France is 33, and for Finland is 358. This section discusses a coding using variable length that is based on the *Huffman tree T* for weighted data items, that is, a 2-tree T with minimum path length P.

Huffman Code: Let T be the Huffman tree for the n weighted data items $A_1, A_2, \ldots A_n$. Each edge in T is assigned 0 or 1 according as the edge points to a left child or to a right child. The Huffman code assigns to each external node A_i, the sequence of bits from the root R of the tree T to the node A. The above Huffman code has the "prefix" property, that is, the code of any item is not an initial substring of the code of any other item. This means that there cannot be any ambiguity in decoding any message using a Huffman code.

EXAMPLE 10.13 Consider again the eight data items A, B, C, D, E, F, G, H in Example 10-12. Suppose the weights represent the percentage probabilities that the items will occur. Assigning, as above, bit labels to the edges in the Huffman tree in Fig. 10-18(b), that is, assigned 0 or 1 according as the edge points to a left child or to a right child, we obtain the following code for the data:

$$A : 00, \qquad B : 11011, \qquad C : 011, \qquad D : 111,$$
$$E : 11010, \qquad F : 010, \qquad G : 10, \qquad H : 1100.$$

For example, to get to E from the root, the path consists of a right edge, right edge, left edge, right edge, and left edge, yielding the code 11010 for E.

10.9 GENERAL (ORDERED ROOTED) TREES REVISITED

Let T be an ordered rooted tree (Section 9.4), which is also called a *general tree*. T may be formally defined as a nonempty set of elements, called nodes, such that:

(1) T contains a distinguished element R, called the *root* of T.
(2) The remaining elements of T form an ordered collection of zero or more disjoint trees, $T_1, T_2, \ldots, T_n$.

The trees $T_1, T_2, \ldots, T_n$ are called *subtrees* of the root R, and the roots of $T_1, T_2, \ldots, T_n$ are called *successors* of R.

Terminology from family relationships, graph theory, and horticulture is used for general trees in the same way as for binary trees. In particular, if N is a node with successors $S_1, S_2, \ldots, S_n$ then N is called the *parent* of the S_i, the S_i are called children of N, and the S_i are called siblings of each other.

EXAMPLE 10.14 Figure 10-20(a) is a picture of a general tree T with 13 nodes,

$$A, B, C, D, E, F, G, H, J, K, L, M, N$$

Unless otherwise stated, the root of a rooted tree T is the node at the top of the diagram, and the children of a node are ordered from left to right. Accordingly, A is the root of T, and A has three children; the first child B, the second child C, and the third child D. Observe that:

(a) C has three children.   (c) Each of D and H has only one child.

(b) Each of B and K has two children.   (d) Each of E, F, G, L, J, M and N has no children.

The last group of nodes, those with no children, are called *terminal nodes*.

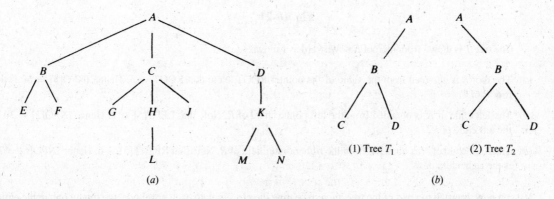

(a)   (1) Tree T_1 (2) Tree T_2

(b)

Fig. 10-20

Remark: A binary tree T is not a special case of a general tree T. They are two different objects. The two basic differences follow:

(1) A binary tree T' may be empty, but a general tree T is nonempty.
(2) Suppose a node N has only one child. Then the child is distinguished as a left child or right child in a binary tree T', but no such distinction exists in a general tree T.

The second difference is illustrated by the trees T_1 and T_2 in Fig. 10-20(b). Specifically, as binary trees, T_1 and T_2 are distinct trees, since B is the left child of A in the tree T_1 but B is the right child of A in the tree T_2. On the other hand, there is no difference between the trees T_1 and T_2 as general trees.

Forest

A *forest* F is defined to be an ordered collection of zero or more distinct general trees. Clearly, if we delete the root R from a general tree T, then we obtain the forest F consisting of the subtrees of R (which may be empty). Conversely, if F is a forest, then we may adjoin a node R to F to form a general tree T where R is the root of T and the subtrees of R consist of the original trees in F.

General Trees and Binary Trees

Suppose T is a general tree. Then we may assign a unique binary tree T' to T as follows. First of all, the nodes of the binary tree T' will be the same as the nodes of the general tree T, and the root of T' will be the root of T. Let N be an arbitrary node of the binary tree T'. Then the left child of N in T' will be the first child of the node N in the general tree T and the right child of N in T' will be the next sibling of N in the general tree T. This correspondence is illustrated in Problem 10.16.

Solved Problems

BINARY TREES

10.1. Suppose T is the binary tree stored in memory as in Fig. 10-21. Draw the diagram of T.

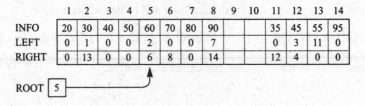

	1	2	3	4	5	6	7	8	9	10	11	12	13	14
INFO	20	30	40	50	60	70	80	90			35	45	55	95
LEFT	0	1	0	0	2	0	0	7			0	3	11	0
RIGHT	0	13	0	0	6	8	0	14			12	4	0	0

ROOT 5

Fig. 10-21

The tree T is drawn from its root R downward as follows:

(a) The root R is obtained from the value of the pointer ROOT. Note that ROOT = 5. Hence INFO[5] = 60 is the root R of T.

(b) The left child of R is obtained from the left pointer field of R. Note that LEFT[5] = 2. Hence INFO[2] = 30 is the left child of R.

(c) The right child of R is obtained from the right pointer field of R. Note that RIGHT[5] = 6. Hence INFO[6] = 70 is the right child of R.

We can now draw the top part of the tree, then, repeating the process with each new node, we finally obtain the entire tree T in Fig. 10-22(a)

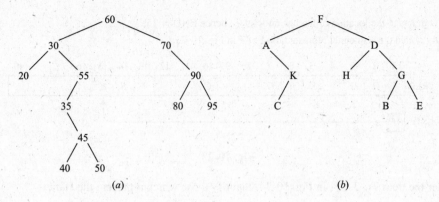

Fig. 10-22

10.2. Consider the binary tree T in Fig. 10-22(b)

    (a) Find the depth d of T.

    (b) Traverse T using the preorder algorithm.

    (c) Traverse T using the inorder algorithm.

    (d) Traverse T using the postorder algorithm.

    (e) Find the terminal nodes of T, and the order they are traversed in (b), (c), and (d).

    (a) The depth d is the number of nodes in a longest branch of T; hence $d = 4$.

    (b) The preorder traversal of T is a recursive NLR algorithm, that is, it first processes a node N, then its left subtree L, and finally its right subtree R. Letting $[A_1, \ldots, A_k]$ denote a subtree with nodes $A_1, \ldots, A_k$, the tree T is traversed as follows:

$$F-[A, K, C][D, H, G, B, E] \quad \text{or} \quad F-A-[K, C]-D-[H][G, B, E]$$

or, finally,

$$F-A-K-C-D-H-G-B-E$$

    (c) The inorder traversal of T is a recursive LNR algorithm, that is, it first processes a left subtree L, then its node N, and finally its right subtree R. Thus T is traversed as follows:

$$[A, K, C]-F-[D, H, G, B, E] \quad \text{or} \quad A-[K, C]-F-[H]-D-[G, B, E]$$

or finally,

$$A-K-C-F-H-D-B-G-E$$

    (d) The postorder traversal of T is a recursive LRN algorithm, that is, it first processes a left subtree L, then its right subtree R, and finally its node N. Thus T is traversed as follows:

$$[A, K, C][D, H, G, B, E]-F \quad \text{or} \quad [K, C]-A-[H][G, B, E]-D-F$$

or, finally,

$$C-K-A-H-B-E-G-D-F$$

    (e) The terminal nodes are the nodes without children. They are traversed in the same order in all three traversal algorithms: C, H, B, E.

10.3. Let T be the binary tree in Fig. 10-22(b). Find the sequential representation of T in memory.

    The sequential representation of T uses only a single array TREE and a variable pointer END.

    (a) The root R of T is stored in TREE[1]; hence $R = $ TREE[1] $= F$.

    (b) If node N occupies TREE[K], then its left and right children are stored in TREE[$2*K$] and TREE[$2*K + 1$], respectively. Thus TREE[2] $= A$ and TREE[3] $= D$ since A and D are the left and right children of F. And so on. Figure 10-23 contains the sequential representation of T. Note that TREE[10] $= C$ since C is the left child of K, which is stored in TREE[5]. Also, TREE[14] $= B$ and TREE[15] $= E$ since B and E are the left and right children of G, which is stored in TREE[7].

(c) END points to the location of the last node of T; hence END = 15.

We finally obtain the sequential representation of T in Fig. 10-23.

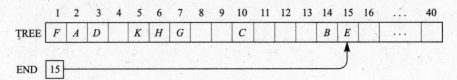

Fig. 10-23

10.4. Consider the trees T_1, T_2, T_3 in Fig. 10-24. Identify those which represent the same:

(a) rooted tree; (b) ordered rooted tree; (c) binary tree.

(a) They all represent the same rooted tree, that is, A is the root with children (immediate successors) B and C, and C has the single child D.

(b) Here T_1 and T_2 are the same ordered rooted tree but T_3 is different. Specifically, B is the first child of A in T_1 and T_2 but the second child of A in T_3.

(c) They all represent different binary trees. Specifically, T_1 and T_2 are different since we distinguish between left and right successors even when there is only one successor (which is not true for ordered rooted trees). That is, D is a left successor of C in T_1 but a right successor of C in T_2.

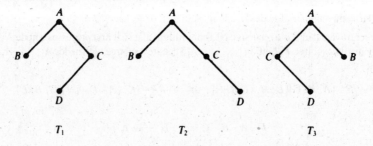

Fig. 10-24

10.5. A binary tree T has nine nodes. Draw a picture of T if the preorder and inorder traversal of T yield the following sequences of nodes:

$$\text{Preorder:} \quad G \quad B \quad Q \quad A \quad C \quad P \quad D \quad E \quad R$$
$$\text{Inorder:} \quad Q \quad B \quad C \quad A \quad G \quad P \quad E \quad D \quad R$$

The tree T is drawn from its root R downward as follows.

(a) The root of T is obtained by choosing the first node in its preorder. Thus G is the root of T.

(b) The left child of node G is obtained as follows. First use the inorder of T to find the nodes in the left subtree T_1 of G. Thus T_1 consists of the nodes Q, B, C, A, which are to the left of G in the inorder of T. Then the left child of G is obtained by choosing the first node (root) in the preorder of T_1 which appears in the preorder of T. Thus B is the left child of G.

(c) Similarly, the right subtree T_2 of G consists of the nodes P, E, D, R; and P is the root of T_2, that is, P is the right child of G.

Repeating the above process with each new node, we finally obtain the required tree T in Fig. 10-25(a).

10.6. Consider the algebraic expression $E = (2x + y)(5a - b)^3$.

(a) Draw the corresponding 2-tree. (b) Use T to write E in Polish prefix form.

(a) Use an arrow (↑) for exponentiation, an asterisk (*) for multiplication, and a slash (/) for division to obtain the tree in Fig. 10-25(b).

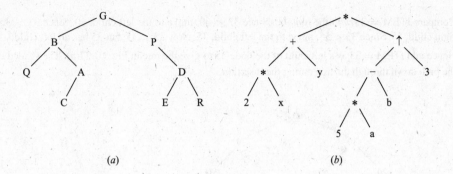

(a)   (b)

Fig. 10-25

(b) Scan the tree from the left as in Fig. 10-4(b) to obtain

$$* + * 2\, x\ y\ \uparrow\, - * 5\, a\, b\, 3$$

10.7. Draw all possible nonsimilar: (a) binary trees T with three nodes; (b) 2-trees T' with four external nodes.

(a) There are five such trees T, which are pictured in Fig. 10-26(a).

(b) Each 2-tree T' with four external nodes is determined by a binary tree T with three nodes, that is, by a tree T in part (a). Thus there are five such 2-trees T' which are pictured in Fig. 10-26(b).

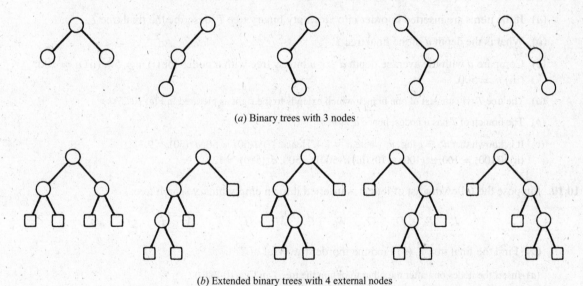

(a) Binary trees with 3 nodes

(b) Extended binary trees with 4 external nodes

Fig. 10-26

BINARY SEARCH TREES, HEAPS

10.8. Consider the binary tree T in Fig.10-22(a).

(a) Why is T a binary search tree?

(b) Suppose ITEM = 33 is added to the tree. Find the new tree T.

(a) T is a binary, search tree since each node N is greater than the values in its left subtree and less than the values in its right subtree.

(b) Compare ITEM = 33 with the root, 60. Since $33 < 60$, move to the left child, 30. Since $33 > 30$, move to the right child, 55. Since $33 < 55$, move to the left child, 35. Now $33 < 35$, but 35 has no left child.

Hence add ITEM = 33 as a left child of the node 35 to give the tree in Fig. 10-27(a). The shaded edges indicate the path down through the tree during the insertion.

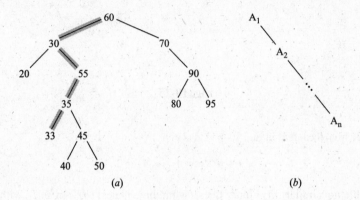

(a)   (b)

Fig. 10-27

10.9. Suppose n data items $A_1, A_2, \ldots, A_N$ are already sorted, i.e., $A_1 < A_2 < \cdots < A_N$.

(a) If the items are inserted in order into an empty binary tree T, describe the final tree T.

(b) What is the depth d of the final tree T.

(c) Compare d with the average depth d^* of a binary tree with n nodes for (i) $n = 50$; (ii) $n = 100$; (iii) $n = 500$.

(a) The tree T will consist of one branch which extends to the right as pictured in Fig. 10-27(b)

(b) The branch of T has n nodes; hence $d = n$.

(c) It is known that $d^* = c \log_2 n$, where $c \approx 1.4$. Hence: (i) $d(50) = 50, d^*(50) \approx 9$; (ii) $d(100) = 100, d^*(100) \approx 10$; (iii) $d(500) = 500, d^*(500) \approx 12$.

10.10. Suppose the following list of letters is inserted into an empty binary search tree:

$$J, \quad R, \quad D, \quad G, \quad W, \quad E, \quad M, \quad H, \quad P, \quad A, \quad F, \quad Q$$

(a) Find the final tree T. (b) Find the inorder traversal of T.

(a) Insert the nodes one after the other to obtain the tree T in Fig. 10-28(a)

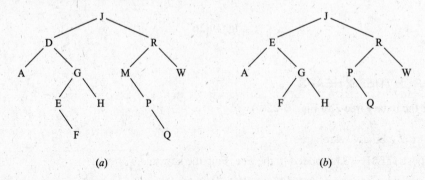

(a)   (b)

Fig. 10-28

(b) The inorder traversal of T follows:

$$A, \quad D, \quad E, \quad F, \quad G, \quad H, \quad J, \quad M, \quad P, \quad Q, \quad R, \quad W$$

Observe that this is the alphabetical listing of the letters. (The inorder traversal of any binary search tree T yields a sorted list of the nodes.)

10.11. Consider the binary tree T in Fig. 10-28(a). Describe the tree T after: (a) the node M and (b) the node D are deleted.

(a) The node M has only one child, P. Hence delete M and let P become the left child of R in place of M.

(b) The node D has two children. Find the inorder successor of D, which is the node E. First delete E from the tree, and then replace D by the node E.

Fig. 10-28(b) shows the updated tree T.

10.12. Let H be the minheap in Fig. 10-29(a). (H is a *minheap* since the smaller elements are on top of the heap, rather than the larger elements.) Describe the heap after ITEM = 11 is inserted into H.

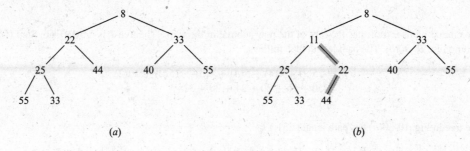

(a)   (b)

Fig. 10-29

First insert ITEM as the next node in the complete tree, that is, as the left child of node 44. Then repeatedly compare ITEM with its PARENT, and interchange ITEM and PARENT as long as ITEM < PARENT. Since 11 < 44, interchange 11 and 44. Since 11 < 22, interchange 11 and 22. Since 11 > 8, ITEM = 11 has found its appropriate place in the heap H. Figure 10-29(b) shows the final heap H. The shaded edges indicate the path of ITEM as it moves up the tree.

PATH LENGTHS, HUFFMAN'S ALGORITHM

10.13. Let T be the weighted 2-tree in Fig. 10-30(a). Find the weighted path length P of the tree T.

Multiply each weight W_i by the length L_i of the path from the root of T to the node containing the weight, and then sum all such products to obtain P. Thus:

$$\begin{aligned}
P &= 4(2) + 15(4) + 25(4) + 5(3) + 8(2) + 16(2) \\
&= 8 + 60 + 100 + 15 + 16 + 32 \\
&= 231
\end{aligned}$$

10.14. Suppose the six weights 4, 15, 25, 5, 8, 16 are given. Find a 2-tree T with the given weights and with a minimum path length P. (Compare T with the tree in Fig. 10-30(a).)

Use Huffman's algorithm. That is, repeatedly combine the two subtrees with minimum weights into a single subtree as follows:

(a) 4,   15,   25,   5,   8,   16;   (d) 25,   17,   ㉛;
(b)   15,   25,   ⑨,   8,   16;   (e)   ㊷   31;
(c)   15,   25,   ⑰   16;   (f)   ㊲

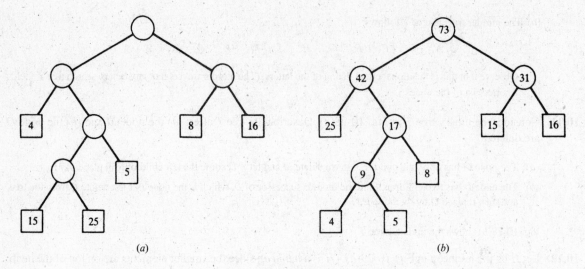

Fig. 10-30

(The circled number indicates the root of the new subtree in the step.) The tree T is drawn from Step (f) backward, yielding Fig. 10-30(b). The path length of T follows:

$$P = 25(2) + 4(4) + 5(4) + 8(3) + 15(2) + 16(2)$$
$$= 50 + 16 + 20 + 24 + 30 + 32$$
$$= 172$$

(The tree in Fig. 10-30(a) has path length 231.)

10.15. Suppose data items A, B, C, D, E, F, G occur with the following probability distribution:

Data item:	A	B	C	D	E	F	G
Probability:	10	30	5	15	20	15	5

Find a Huffman code for the data items.

As in Fig. 10-31(a), apply the Huffman algorithm to find a 2-tree with a minimum weighted path length P. (Again, the circled number indicates the root of the new subtree in the step.) The tree T is drawn from Step (g) backward, yielding Fig. 10-31(b). Assign bit labels to the edges of the tree T, 0 to a left edge and 1 to a right edge, as in Fig. 10-31(b). The tree T yields the following Huffman code:

$$A:000; \quad B:11; \quad C:0010; \quad D:100; \quad E:01; \quad F:101; \quad G:0011$$

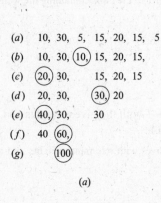

(a)

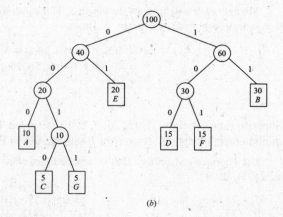

(b)

Fig. 10-31

GENERAL TREES

10.16. Let T be the general tree in Fig. 10-32(a). Find the corresponding binary tree T'.

The nodes of T' will be the same as the nodes of the general tree T. In particular, the root of T' will be the same as the root of T. Furthermore, if N is a node in the binary tree T', then its left child is the first child of N in T and its right child is the next sibling of N in T. Constructing T' from the root down we obtain the tree in Fig. 10-32(b).

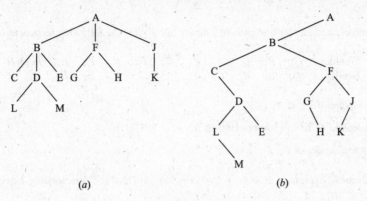

(a)   (b)

Fig. 10-32

Supplementary Problems

10.17. Consider the binary tree T in Fig. 10-33(a).

(a) Find: (i) depth d of T; (ii) descendants of B.

(b) Traverse T in: (i) preorder; (ii) inorder; (iii) postorder.

(c) Find the terminal nodes of T and the orders they are traversed in (b).

10.18. Repeat Problem 10.17 for the binary tree T in Fig. 10-33(b).

10.19. Repeat Problem 10.17 for the binary tree T in Fig. 10-33(c).

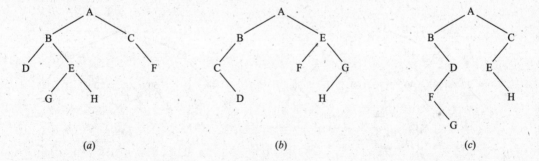

(a)   (b)   (c)

Fig. 10-33

10.20. Let T be the binary tree stored in memory as in Fig. 10-34 where ROOT = 14.

(a) Draw the diagram of T.

(b) Traverse T in: (i) preorder; (ii) inorder; (iii) postorder.

(c) Find the depth d of T.

(d) Find the minimum number of locations required for a linear array TREE if T were stored sequentially in TREE.

	1	2	3	4	5	6	7	8	9	10	11	12	13	14	15	16	17	18
INFO	H	R		P	B		E		C	F	Q	S		A	K	L		D
LEFT	4	0		0	18		1		0	15	0	0		5	2	0		0
RIGHT	11	0		0	7		0		10	16	12	0		9	0	0		0

Fig. 10-34

10.21. Suppose the preorder and inorder traversals of a binary tree T yield the following sequences of nodes:

$$\text{Preorder:} \quad G, \quad B, \quad Q, \quad A, \quad C, \quad K, \quad F, \quad P, \quad D, \quad E, \quad R, \quad H$$
$$\text{Inorder:} \quad Q, \quad B, \quad K, \quad C, \quad F, \quad A, \quad G, \quad P, \quad E, \quad D, \quad H, \quad R$$

(a) Draw the diagram of T.

(b) Find: (i) depth d of T;   (ii) descendants of B.

(c) List the terminal nodes of T.

10.22. Consider the algebraic expression $E = (x + 3y)^4(a - 2b)$. (a) Draw the corresponding 2-tree. (b) Write E in Polish prefix form.

BINARY SEARCH TREES, HEAPS

10.23. Find the final tree T if the following numbers are inserted into an empty binary search tree T:

$$50, \quad 33, \quad 44, \quad 22, \quad 77, \quad 35, \quad 60, \quad 40$$

10.24. Find the final heap H if the numbers in Problem 10-23 are inserted into an empty maxheap H.

10.25. Find the final heap H if the numbers in Problem 10-23 are inserted into an empty minheap H.

10.26. Let T be the binary search tree in Fig. 10-35(a). Suppose nodes 20, 55, 88 are added one after the other to T. Find the final tree T.

10.27. Let T be the binary search tree in Fig. 10-35(a). Suppose nodes 22, 25, 75 are deleted one after the other from T. Find the final tree T.

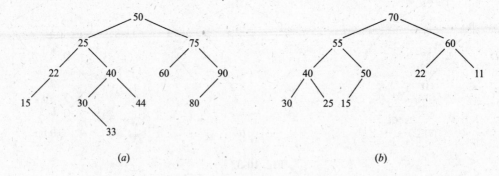

(a)   (b)

Fig. 10-35

10.28. Let H be the heap in Fig. 10-35(b). Find the final heap H if the numbers 65, 44, and 75 are inserted one after the other into H.

10.29. Let H be the heap in Fig. 10-35(b). Find the final heap H if the root and then the next root are deleted from H.

HUFFMAN ALGORITHM, GENERAL TREES

10.30. Consider the 2-tree T in Fig. 10-36(a) which contains the letters A, B, C, D, E, F, G as external nodes. Find the Huffman coding of the letters determined by the tree T.

10.31. Find the weighted path length P of the tree in Fig. 10-36(a) if the data items $A, B, \ldots, G$ are assigned the following weights:

$$(A, 13), \quad (B, 2), \quad (C, 19), \quad (D, 23), \quad (E, 29), \quad (F, 5), \quad (G, 9)$$

10.32. Using the data in Problem 10.31, find a Huffman coding for the seven letters using a 2-tree with a minimum path length P, and find P.

10.33. Let T be the general tree in Fig. 10-36(b). Find the corresponding binary tree T'.

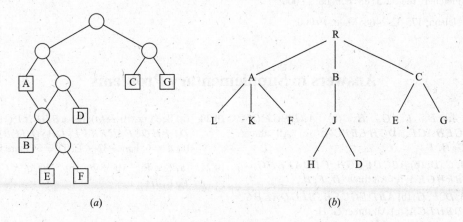

(a) (b)

Fig. 10-36

COMPUTER PROBLEMS

Problems 10.34 to 10.40 refer to Fig. 10-37 which is a list of employee records stored in memory. It is a binary search tree with respect to the NAME key. It uses a pointer HEAD where the number of employees is in SSN[HEAD], the total salary is in SALARY[HEAD], and the root of the tree is in LEFT[HEAD]. Also, to allow insertions, the available (empty) locations form a linked list with AVAIL pointing to the first element in the list, and the linking is maintained by the array LEFT.

		NAME	SSN	SEX	SALARY	LEFT	RIGHT
HEAD	1					0	
5	2	Davis	192-38-7282	Female	22 800	0	12
	3	Kelly	165-64-3351	Male	19 000	0	0
AVAIL	4	Green	175-56-2251	Male	27 200	2	0
8	5		0009		191 600	14	0
	6	Brown	178-52-1065	Female	14 700	0	0
	7	Lewis	181-58-9939	Female	16 400	3	10
	8					11	
	9	Cohen	177-44-4557	Male	19 000	6	4
	10	Rubin	135-46-6262	Female	15 500	0	0
	11					13	
	12	Evans	168-56-8113	Male	34 200	0	0
	13					1	
	14	Harris	208-56-1654	Female	22 800	9	7

Fig. 10-37

10.34. Draw a diagram of the binary search tree NAME.

10.35. Write a program which prints the list of employee records in alphabetical order. (*Hint*: Print the records in inorder.)

10.36. Write a program which reads the name *NNN* of an employee and prints the employee's record. Test the program using (a) Evans; (b) Smith; and (c) Lewis.

10.37. Write a program which reads the social security number *SSS* of an employee and prints the employee's record. Test the program using (a) 165-64-3351; (b) 135-46-626; and (c) 177-44-5555.

10.38. Write a program which reads an integer K and prints the name of each male employee when $K = 1$ or of each female employee when $K = 2$. Test the program using (a) $K = 2$; (b) $K = 5$; and (c) $K = 1$.

10.39. Write a program which reads the name *NNN* of an employee and deletes the employee's record from the structure. Test the program using (a) Davis; (b) Jones; and (c) Rubin.

10.40. Write a program which reads the record of a new employee and inserts the record in the file.
Test the program using:

(a) Fletcher; 168-52-3388; Female; 21 000;

(b) Nelson; 175-32-2468; Male; 19 000.

Answers to Supplementary Problems

10.17. (a) 4; *D*, *E*, *G*, *H*; (b) *ABDEGHCF*, *DBGEHACF*, *DGHEBFCA*; (c) All three: *D, G, H, F*.

10.18. (a) 4; *C, D*; (b) *ABCDEFGH, CDBAFEHG, DCBFHGEA*; (c) All three: *D, F, H*.

10.19. (a) 5; *D, F, G*; (b) *ABDFGCEH, BFGDAEHC, GFDBHECA*; (c) All three: *G, H*.

10.20. (a) See Fig. 10-38(a); (b) *ABDEHPQSCFKRL, DBPHQSEACRKFL, DPSQHEBRKLFCA*; (c) $d = 6$; hence $32 \le \text{END} = 64$; here END = 43.

10.21. (a) See Fig. 10-38(b); (b) 5; *QACKF*; (c) *Q, K, F, E, H*.

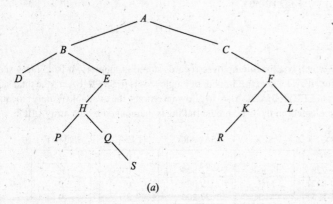

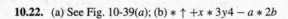

(a)   (b)

Fig. 10-38

10.22. (a) See Fig. 10-39(a); (b) $* \uparrow +x * 3y4 - a * 2b$ **10.23.** See Fig. 10-39(b).

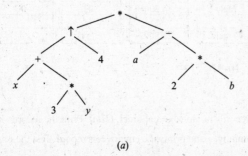   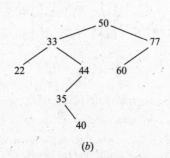

(a)   (b)

Fig. 10-39

10.24. Level by level: 77, 50, 60, 40, 33, 35, 44, 22.

10.25. Level by level: 22, 33, 35, 40, 77, 44, 60, 50.

10.26. See Fig. 10-40(*a*).

10.27. See Fig. 10-40(*b*).

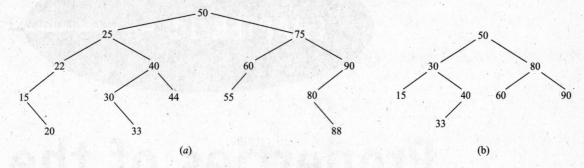

(*a*)

(*b*)

Fig. 10-40

10.28. Level by level: 75, 65, 70, 40, 55, 60, 11, 30, 25, 15, 50, 22, 44.

10.29. Level by level: 55, 50, 22, 40, 25, 15, 11, 30.

10.30. *A*: 00; *B*: 0100; *C*: 10; *D*: 011; *E*: 01010; *F*: 01011; *G*: 11.

10.31. *P* = 329.

10.32. *A*: 000; *B*: 00101; *C*: 10; *D*: 11; *E*: 01; *F*: 00100; *G*: 0011; *P* = 257.

10.33. See Fig. 10-41(*a*).

10.34. See Fig. 10-41(*b*) where only the first letter of each name is used.

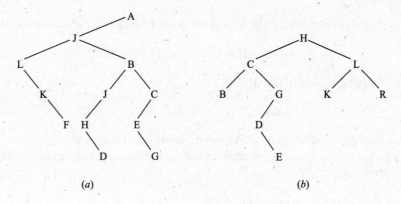

(*a*)

(*b*)

Fig. 10-41

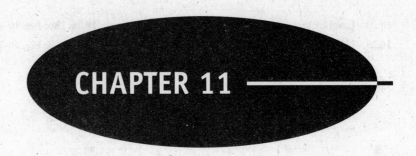

CHAPTER 11

Properties of the Integers

11.1 INTRODUCTION

This chapter investigates some basic properties of the *natural numbers* (or *positive integers*), that is, the set

$$\mathbf{N} = \{1, 2, 3, \ldots\}$$

and their "cousins," the integers, that is, the set

$$\mathbf{Z} = \{\ldots, -2, -1, 0, 1, 2, \ldots\}$$

(The letter **Z** comes from the word "Zahlen" which means numbers in German.)

The following simple rules concerning the addition and multiplication of these numbers are assumed (where a, b, c are arbitrary integers):

(a) Associative law for multiplication and addition:

$$(a + b) + c = a + (b + c) \quad \text{and} \quad (ab)c = a(bc)$$

(b) Commutative law for multiplication and addition:

$$a + b = b + a \quad \text{and} \quad ab = ba$$

(c) Distributive law:

$$a(b + c) = ab + ac$$

(d) Additive identity 0 and multiplicative identity 1:

$$a + 0 = 0 + a = a \quad \text{and} \quad a \cdot 1 = 1 \cdot a = a$$

(e) Additive inverse $-a$ for any integer a:

$$a + (-a) = (-a) + a = 0$$

264

Appendix **B** shows that other mathematical structures have the above properties. One fundamental property which distinguishes the integers **Z** from other structures is the Principle of Mathematical Induction (Section 1.8) which we rediscuss here. We also state and prove (Problem 11.30) the following theorem.

Fundamental Theorem of Arithmetic: Every positive integer $n > 1$ can be written uniquely as a product of prime numbers.

This theorem already appeared in Euclid's *Elements*. Here we also develop the concepts and methods which are used to prove this important theorem.

11.2   ORDER AND INEQUALITIES, ABSOLUTE VALUE

This section discusses the elementary properties of order and absolute value.

Order

Observe that we define order in **Z** in terms of the positive integers **N**. All the usual properties of this order relation are a consequence of the following two properties of **N**:

[P_1]   If a and b belong to **N**, then $a + b$ and ab belong to **N**.

[P_2]   For any integer a, either $a \in$ **N**, $a = 0$, or $-a \in$ **N**.

The following notation is also used:

$$a > b \text{ means } b < a; \qquad\qquad \text{read: } a \text{ is greater than } b.$$
$$a \leq b \text{ means } a < b \text{ or } a = b; \qquad \text{read: } a \text{ is less than or equal to } b.$$
$$a \geq b \text{ means } b \geq a; \qquad\qquad \text{read: } a \text{ is greater than or equal to } b.$$

The relations $<, >, \leq$ and $\geq$ are called *inequalities* in order to distinguish them from the relation $=$ of equality. The reader is certainly familiar with the representation of the integers as points on a straight line, called the *number line* **R**, as shown in Fig. 11-1.

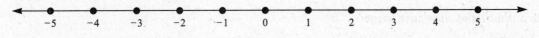

Fig. 11-1

We note that $a < b$ if and only if a lies to the left of b on the number line **R** in Fig. 11-1. For example,

$$2 < 5; \qquad -6 < -3; \qquad 4 \leq 4; \qquad 5 > -8; \qquad 6 \geq 0; \qquad -7 \leq 0$$

We also note that a is positive iff $a > 0$, and a is negative iff $a < 0$. (Recall "iff" means "if and only if.") Basic properties of the inequality relations follow.

Proposition 11.1: The relation $\leq$ in **Z** has the following properties:

(i) $a \leq a$, for any integer a.

(ii)   If $a \leq b$ and $b \leq a$, then $a = b$.

(iii)   If $a \leq b$ and $b \leq c$, then $a \leq c$.

Proposition 11.2 (Law of Trichotomy): For any integers a and b, exactly one of the following holds:

$$a < b, \quad a = b, \quad \text{or} \quad a > b$$

Proposition 11.3: Suppose $a \leq b$, and let c be any integer. Then:

(i) $a + c \leq b + c$.

(ii) $ac \leq bc$ when $c > 0$; but $ac \geq bc$ when $c < 0$.

(Problem 11.5 proves Proposition 11.3.)

Absolute Value

The *absolute value* of an integer a. written $|a|$, is formally defined by

$$|a| = \begin{cases} a & \text{if } a \geq 0 \\ -a & \text{if } a < 0 \end{cases}$$

Accordingly, $|a| > 0$ except when $a = 0$. Geometrically speaking, $|a|$ may be viewed as the distance between the points a and 0 on the number line $\mathbf{R}$. Also, $|a - b| = |b - a|$ may be viewed as the distance between the points a and b. For example:

(a) $|-3| = 3$; $|7| = 7$; $|-13| = 13$;   (b) $|2 - 7| = |-5| = 5$; $|7 - 2| = |5| = 5$

Some properties of the absolute value function follow. (Problems 11.6 and 11.7 prove (iii) and (iv).)

Proposition 11.4: Let a and b be any integers. Then:

(i) $|a| \geq 0$, and $|a| = 0$ iff $a = 0$ (iv) $|a \pm b| \leq |a| + |b|$
(ii) $-|a| \leq a \leq |a|$ (v) $||a| - |b|| \leq |a \pm b|$
(iii) $|ab| = |a||b|$

11.3   MATHEMATICAL INDUCTION

The principle of mathematical induction stated below essentially asserts that the positive integers $\mathbf{N}$ begin with the number 1 and the rest are obtained by successively adding 1. That is, we begin with 1, then $2 = 1 + 1$, then $3 = 2 + 1$, then $4 = 3 + 1$, and so on. The principle makes precise the vague phrase "and so on."

Principle of Mathematical Induction: Let S be a set of positive integers with the following two properties:

(i) 1 belongs to S.

(ii) If k belongs to S, then $k + 1$ belongs to S.

Then S is the set of all positive integers.

We shall not prove this principle. On the contrary, when the set $\mathbf{N}$ of positive integers (natural numbers) is developed axiomatically, this principle is given as one of the axioms.

There is an equivalent form of the above principle which is usually used when proving theorems:

Principle of Mathematical Induction: Let P be a proposition defined on the integers $n \geq 1$ such that:

(i) $P(1)$ is true.

(ii) $P(k + 1)$ is true whenever $P(k)$ is true.

Then P is true for every integer $n \geq 1$.

EXAMPLE 11.1

(a) Let P be the proposition that the sum of the first n odd numbers is n^2; that is:

$$P(n): 1 + 3 + 5 + \cdots + (2n - 1) = n^2$$

(The nth odd number is $2n - 1$ and the next odd number is $2n + 1$.)

Clearly, $P(n)$ is true for $n = 1$; that is:

$$P(1): 1 = 1^2$$

Suppose $P(k)$ is true. (This is called the inductive hypothesis.) Adding $2k + 1$ to both sides of $P(k)$ we obtain

$$1 + 3 + 5 + \cdots + (2k - 1) + (2k + 1) = k^2 + (2k + 1)$$
$$= (2k + 1)^2$$

which is $P(k + 1)$. We have shown that $P(k + 1)$ is true whenever $P(k)$ is true. By the principle of mathematical induction, P is true for all positive integers n.

(b) The symbol $n!$ (read: n factorial) is defined as the product of the first n positive integers; that is:

$$1! = 1, \quad 2! = 2 \cdot 1 = 2, \quad 3! = 3 \cdot 2 \cdot 1 = 6, \quad \text{and so on.}$$

This may be formally defined as follows:

$$1! = 1 \quad \text{and} \quad (n + 1)! = (n + 1)(n!), \quad \text{for} \quad n > 1.$$

Observe that if S is the set of positive integers for which ! is defined, then S satisfies the two properties of mathematical induction. Hence the above definition defines ! for every positive integer.

There is another form of the principle of mathematical induction (proved in Problem 11.13) which is sometimes more convenient to use. Namely:

Theorem 11.5 (Induction: Second Form): Let P be a proposition defined on the integers $n \geq 1$ such that:

   (i) $P(1)$ is true.

   (ii) $P(k)$ is true whenever $P(j)$ is true for all $1 \leq j < k$.

Then P is true for every integer $n \geq 1$.

Remark: The above theorem is true if we replace 1 by 0 or by any other integer a.

Well-Ordering Principle

A property of the positive integers which is equivalent to the principle of induction, although apparently very dissimilar, is the well-ordering principle (proved in Problem 11.12). Namely:

Theorem 11.6 (Well-Ordering Principle): Let S be a nonempty set of positive integers. Then S contains a *least element*; that is, S contains an element a such that $a \leq s$ for every s in S.

Generally speaking, an ordered set S is said to be *well-ordered* if every subset of S contains a first element. Thus Theorem 11.6 states that **N** is well ordered.

A set S of integers is said to be *bounded from below* if every element of S is greater than some integer m (which may be negative). (The number m is called a *lower bound* of S.) A simple corollary of the above theorem follows:

Corollary 11.7: Let S be a nonempty set of integers which is bounded from below. Then S contains a least element.

11.4 DIVISION ALGORITHM

The following fundamental property of arithmetic (proved in Problems 11.17 and 11.18) is essentially a restatement of the result of long division.

Theorem 11.8 (Division Algorithm): Let a and b be integers with $b \neq 0$. Then there exists integers q and r such that

$$a = bq + r \quad \text{and} \quad 0 \leq r < |b|$$

Also, the integers q and r are unique.

The number q in the above theorem is called the *quotient*, and r is called the *remainder*. We stress the fact that r must be non-negative. The theorem also states that

$$r = a - bq$$

This equation will be used subsequently

If a and b are positive, then q is non negative. If b is positive, then Fig. 11-2 gives a geometrical interpretation of this theorem. That is, the positive and negative multiples of b will be evenly distributed throughout the number line **R**, and a will fall between some multiples qb and $(q + 1)b$. The distance between qb and a is then the remainder r.

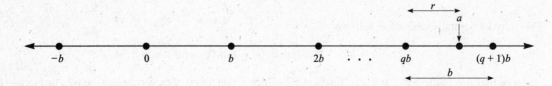

Fig. 11-2

Division Algorithm using a Calculator

Suppose a and b are both positive. Then one can find the quotient q and remainder r using a calculator as follows:

Step 1. Divide a by b using the calculator, that is, find a/b.

Step 2. Let q be the integer part of a/b, that is, let $q = INT(a/b)$.

Step 3. Let r be the difference between a and bq, that is, let $r = a - bq$.

EXAMPLE 11.2

(a) Let $a = 4461$ and $b = 16$. We can find that the quotient $q = 278$ and the remainder $r = 13$ by long division, Alternately, using a calculator, we obtain q and r as follows:

$$a/b = 278.8125\ldots, \quad q = 278, \quad r = 4461 - 16(278) = 13$$

As expected, $a = bq + r$, namely:
$$4461 = 16(278) + 13$$

(b) Let $a = -262$ and $b = 3$. First we divide $|a| = 262$ by $b = 3$. This yields a quotient $q' = 87$ and a remainder $r' = 1$. Thus
$$262 = 3(87) + 1$$

We need $a = -262$, so we multiply by -1 obtaining

$$-262 = 3(-87) - 1$$

However, -1 is negative and hence cannot be r. We correct this by adding and subtracting the value of b (which is 3) as follows:
$$-262 = 3(-87) - 3 + 3 - 1 = 3(-88) + 2$$

Therefore, $q = -88$ and $r = 2$.

(c) Let $b = 2$. Then any integer a can be written in the form

$$a = 2q + r \quad \text{where} \quad 0 \le r < 2$$

Thus r can only be 0 or 1. Thus every integer is of the form $2k$ or $2k + 1$. The integers of the form $2k$ are called *even* integers, while those of the form $2k + 1$ are called *odd* integers. (Usually, an even integer is defined as an integer divisible by 2, and all other integers are said to be odd. Thus the division algorithm proves that every odd integer has the form $2k + 1$.)

11.5   DIVISIBILITY, PRIMES

Let a and b be integers with $a \ne 0$. Suppose $ac = b$ for some integer c. We then say that a divides b or b is divisible by a, and we denote this by writing

$$a \mid b$$

We also say that b is a *multiple* of a or that a is a *factor* or *divisor* of b. If a does not divide b, we will write $a \nmid b$.

EXAMPLE 11.3

(a) Clearly, $3 \mid 6$ since $3 \cdot 2 = 6$, and $-4 \mid 28$ since $(-4)(-7) = 28$.

(b) The divisors of 4 are $\pm 1, \pm 2, \pm 4$ and the divisors of 9 are $\pm 1, \pm 3, \pm 9$.

(c) If $a \ne 0$, then $a \mid 0$ since $a \cdot 0 = 0$.

(d) Every integer a is divisible by ± 1 and $\pm a$. These are sometimes called the *trivial divisors* of a. The basic properties of divisibility is stated in the next theorem (proved in Problem 11.24).

Theorem 11.9:   Suppose a, b, c are integers.

  (i) If $a \mid b$ and $b \mid c$, then $a \mid c$.

  (ii) If $a \mid b$ then, for any integer x, $a \mid bx$.

  (iii) If $a \mid b$ and $a \mid c$, then $a \mid (b + c)$ and $a \mid (b - c)$.

  (iv) If $a \mid b$ and $b \ne 0$, then $a = \pm b$ or $|a| < |b|$.

  (v) If $a \mid b$ and $b \mid a$, then $|a| = |b|$, i.e., $a = \pm b$.

  (vi) If $a \mid 1$, then $a = \pm 1$

Putting (ii) and (iii) together, we obtain the following important result.

Corollary 11.10:   Suppose $a \mid b$ and $a \mid c$. Then, for any integers x and y, $a \mid (bx + cy)$. The expression $bx + cy$ will be called a *linear combination* of b and c.

Primes

A positive integer $p > 1$ is called a *prime number* or a *prime* if its only divisors are ± 1 and $\pm p$, that is, if p only has trivial divisors. If $n > 1$ is not prime, then n is said to be *composite*. We note (Problem 11.13) that if $n > 1$ is composite then $n = ab$ where $1 < a, b < n$.

EXAMPLE 11.4

(a) The integers 2 and 7 are primes, whereas $6 = 2 \cdot 3$ and $15 = 3 \cdot 5$ are composite.

(b) The primes less than 50 follow:

$$2, \quad 3, \quad 5, \quad 7, \quad 11, \quad 13, \quad 17, \quad 19, \quad 23, \quad 29, \quad 31, \quad 37, \quad 41, \quad 43, \quad 47$$

(c) Although 21, 24, and 1729 are not primes, each can be written as a product of primes:

$$21 = 3 \cdot 7; \qquad 24 = 2 \cdot 2 \cdot 2 \cdot 3 = 2^3 \cdot 3; \qquad 1729 = 7 \cdot 13 \cdot 19$$

The Fundamental Theorem of Arithmetic states that every integer $n > 1$ can be written as a product of primes in essentially one way; it is a deep and somewhat difficult theorem to prove. However, using induction, it is easy at this point to prove that such a product exists. Namely:

Theorem 11.11: Every integer $n > 1$ can be written as a product of primes.

Note that a product may consist of a single factor so that a prime p is itself a product of primes. We prove Theorem 11.11 here, since its proof is relatively simple.

Proof: The proof is by induction. Let $n = 2$. Since 2 is prime, n is a product of primes. Suppose $n > 2$, and the theorem holds for positive integers less than n. If n is prime, then n is a product of primes. If n is composite, then $n = ab$ where $a, b < n$. By induction, a and b are products of primes; hence $n = ab$ is also a product of primes.

Euclid, who proved the Fundamental Theorem of Arithmetic, also asked whether or not there was a largest prime. He answered the question thus:

Theorem 11.12: There is no largest prime, that is, there exists an infinite number of primes.

Proof: Suppose there is a finite number of primes, say $p_1, p_2, \ldots, p_m$. Consider the integer

$$n = p_1 p_2 \cdots p_m + 1$$

Since n is a product of primes (Theorem 11.11), it is divisible by one of the primes, say p_k. Note that p_k also divides the product $p_1 p_2 \ldots p_m$. Therefore p_k divides

$$n - p_1 p_2 \cdots p_m = 1$$

This is impossible, and so n is divisible by some other prime. This contradicts the assumption that $p_1, p_2, \ldots, p_m$ are the only primes. Thus the number of primes is infinite, and the theorem is proved.

11.6 GREATEST COMMON DIVISOR, EUCLIDEAN ALGORITHM

Suppose a and b are integers, not both 0. An integer d is called a *common divisor* of a and b if d divides both a and b, that is, if $d \mid a$ and $d \mid b$. Note that 1 is a positive common divisor of a and b, and that any common divisor of a and b cannot be greater than $|a|$ or $|b|$. Thus there exists a largest common divisor of a and b; it is denoted by

$$\gcd(a, b)$$

and it is called the *greatest common divisor* of a and b.

EXAMPLE 11.5

(a) The common divisors of 12 and 18 are $\pm 1, \pm 2, \pm 3, \pm 6$. Thus $\gcd(12, 18) = 6$. Similarly:

$$\gcd(12, -18) = 16, \qquad \gcd(12, -16) = 4, \qquad \gcd(29, 15) = 1, \qquad \gcd(14, 49) = 7$$

(b) For any integer a, we have $\gcd(1, a) = 1$.

(c) For any prime p, we have $\gcd(p, a) = p$ or $\gcd(p, a) = 1$ according as p does or does not divide a.

(d) Suppose a is positive. Then $a \mid b$ if and only if $\gcd(a, b) = a$.

The following theorem (proved in Problem 11.26) gives an alternative characterization of the greatest common divisor.

Theorem 11.13: Let d be the smallest positive integer of the form $ax + by$. Then

$$d = \gcd(a, b).$$

Corollary 11.14: Suppose $d = \gcd(a, b)$. Then there exist integers x and y such that $d = ax + by$.

Another way to characterize the greatest common divisor, without using the inequality relation, follows

Theorem 11.15: A positive integer $d = \gcd(a, b)$ if and only if d has the following two properties:

$\qquad$ (1) d divides both a and b.

$\qquad$ (2) If c divides both a and b, then $c \mid d$.

Simple properties of the greatest common divisor are:
(a) $\gcd(a, b) = \gcd(b, a)$. $\qquad\qquad\qquad$ (c) If $d = \gcd(a, b)$, then $\gcd(a/d, b/d) = 1$.
(b) If $x > 0$, then $\gcd(ax, bx) = x \cdot \gcd(a, b)$. $\quad$ (d) For any integer x, $\gcd(a, b) = \gcd(a, b + ax)$.

Euclidean Algorithm

Let a and b be integers, and let $d = \gcd(a, b)$. One can always find d by listing all the divisors of a and then all the divisors of b and then choosing the largest common divisor. The complexity of such an algorithm is $f(n) = 0(\sqrt{n})$ where $n = |a| + |b|$. Also, we have given no method to find the integers x and y such that $d = ax + by$.

This subsection gives a very efficient algorithm, called the Euclidean algorithm, with complexity $f(n) = O(\log n)$, for finding $d = \gcd(a, b)$ by applying the division algorithm to a and b and then repeatedly applying it to each new quotient and remainder until obtaining a nonzero remainder. The last nonzero remainder is $d = \gcd(a, b)$.

Then we give an "unraveling" algorithm which reverses the steps in the Euclidean algorithm to find the integers x and y such that $d = xa + yb$.

We illustrate the algorithms with an example.

EXAMPLE 11.6 Let $a = 540$ and $b = 168$. We apply the Euclidean algorithm to a and b. These steps, which repeatedly apply the division algorithm to each quotient and remainder until obtaining a zero remainder, are pictured in Fig. 11-3(a) using long division and also in Fig. 11-3(b) where the arrows indicate the quotient and remainder in the next step. The last nonzero remainder is 12. Thus

$$12 = \gcd(540, 168)$$

This follows from the fact that

$$\gcd(540, 168) = \gcd(168, 36) = \gcd(36, 24) = \gcd(24, 12) = 12$$

Next we find x and y such that $12 = 540x + 168y$ by "unraveling" the above steps in the Euclidean algorithm. Specifically, the first three quotients in Fig. 11-3 yield the following equations:

$$\text{(1) } 36 = 540 - 3(168), \qquad \text{(2) } 24 = 168 - 4(36), \qquad \text{(3) } 12 = 36 - 1(24)$$

Equation (3) tells us that $d = \gcd(a, b) = 12$ is a linear combination of 36 and 24. Now we use the preceding equations in reverse order to eliminate the other remainders. That is, first we use equation (2) to replace 24 in equation (3) so we can write 12 as a linear combination of 168 and 36 as follows:

$$\text{(4) } 12 = 36 - 1[168 - 4(36)] = 36 - 1(168) + 4(36) = 5(36) - 1(168)$$

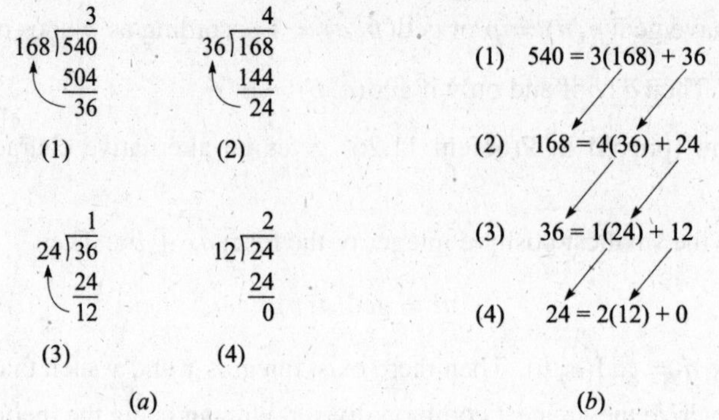

$$\begin{array}{r} 3 \\ 168\overline{)540} \\ \underline{504} \\ 36 \end{array}$$
(1)

$$\begin{array}{r} 4 \\ 36\overline{)168} \\ \underline{144} \\ 24 \end{array}$$
(2)

(1) $540 = 3(168) + 36$

(2) $168 = 4(36) + 24$

$$\begin{array}{r} 1 \\ 24\overline{)36} \\ \underline{24} \\ 12 \end{array}$$
(3)

$$\begin{array}{r} 2 \\ 12\overline{)24} \\ \underline{24} \\ 0 \end{array}$$
(4)

(3) $36 = 1(24) + 12$

(4) $24 = 2(12) + 0$

(a)

(b)

Fig. 11-3

Next we use equation (1) to replace 36 in (4) so we can write 12 as a linear combination of 168 and 540 as follows:

$$12 = 5[540 - 3(168)] - 1(168) = 5(54) - 15(168) - 1(168) = 5(540) - 16(168)$$

This is our desired linear combination. In other words, $x = 5$ and $y = -16$.

Least Common Multiple

Suppose a and b are nonzero integers. Note that $|ab|$ is a positive common multiple of a and b. Thus there exists a smallest positive common multiple of a and b; it is denoted by

$$\mathrm{lcm}(a, b)$$

and it is called the *least common multiple* of a and b.

EXAMPLE 11.7

(a) $\mathrm{lcm}(2, 3) = 6$; $\mathrm{lcm}(4, 6) = 12$; $\mathrm{lcm}(9, 10) = 90$.

(b) For any positive integer a, we have $\mathrm{lcm}(1, a) = a$.

(c) For any prime p and any positive integer a,

$$\mathrm{lcm}(p, a) = a \quad \text{or} \quad \mathrm{lcm}(p, a) = ap$$

according as p does or does not divide a.

(d) Suppose a and b are positive integers. Then $a \mid b$ if and only if $\mathrm{lcm}(a, b) = b$.

The next theorem gives an important relationship between the greatest common divisor and the least common multiple.

Theorem 11.16: Suppose a and b are nonzero integers. Then

$$\mathrm{lcm}(a,b) = \frac{|ab|}{\gcd(a, b)}$$

11.7 FUNDAMENTAL THEOREM OF ARITHMETIC

This section discusses the Fundamental Theorem of Arithmetic. First we define relatively prime integers.

Relatively Prime Integers

Two integers a and b are said to be *relatively prime* or *coprime* if $\gcd(a, b) = 1$. Accordingly, if a and b are relatively prime, then there exist integers x and y such that

$$ax + by = 1$$

Conversely, if $ax + by = 1$, then a and b are relatively prime.

EXAMPLE 11.8

(a) Observe that: $\gcd(12, 35) = 1$, $\gcd(49, 18) = 1$, $\gcd(21, 64) = 1$, $\gcd(-28, 45) = 1$

(b) If p and q are distinct primes, then $\gcd(p, q) = 1$.

(c) For any integer a, we have $\gcd(a, a + 1) = 1$, since any common factor of a and $a + 1$ must divide their difference $(a + 1) - a = 1$.

The relation of being relatively prime is particularly important because of the following results. The first theorem is proved in Problem 11.27, and we will prove the second theorem here.

Theorem 11.17: Suppose $\gcd(a, b) = 1$, and a and b both divide c. Then ab divides c.

Theorem 11.18: Suppose $a \mid bc$, and $\gcd(a, b) = 1$. Then $a \mid c$

Proof: Since $\gcd(a, b) = 1$, there exist x and y such that $ax + by = 1$. Multiplying by c yields:

$$acx + bcy = c$$

We have $a \mid acx$. Also, $a \mid bcy$ since, by hypothesis, $a \mid bc$. Hence a divides the sum $acx + bcy = c$.

Corollary 11.19: Suppose a prime p divides the product ab. Then $p \mid a$ or $p \mid b$.

This corollary (proved in Problem 11.28) dates back to Euclid; it is the basis of his proof of the Fundamental Theorem of Arithmetic.

Fundamental Theorem of Arithmetic

Theorem 11.11 asserts that every positive integer is a product of primes. Can different products of primes yield the same number? Clearly, we can rearrange the order of the prime factors, e.g.,

$$30 = 2 \cdot 3 \cdot 5 = 5 \cdot 2 \cdot 3 = 3 \cdot 2 \cdot 5$$

The Fundamental Theorem of Arithmetic (proved in Problem 11.30) says that this is the only way that two "different" products can give the same number. Namely:

Theorem 11.20 (Fundamental Theorem of Arithmetic): Every integer $n > 1$ can be expressed uniquely (except for order) as a product of primes.

The primes in the factorization of n need not be distinct. Frequently, it is useful to collect together all equal primes. Then n can be expressed uniquely in the form

$$n = p_1^{m_1} p_2^{m_2} \ldots p_r^{m_r}$$

where the m_i are positive and $p_1 < p_2 < \ldots < p_r$. This is called the *canonical factorization* of n.

EXAMPLE 11.9 Given $a = 2^4 \cdot 3^3 \cdot 7 \cdot 13$ and $b = 2^3 \cdot 3^2 \cdot 5^2 \cdot 11 \cdot 17$. Find $d = \gcd(a, b)$ and $m = \operatorname{lcm}(a, b)$.

(a) First we find $d = \gcd(a, b)$. Those primes p, which appear in both a and b, 2, 3, and 11, will also appear in d, and the exponent of p, in d will be the smaller of its exponents in a and b. Thus

$$d = \gcd(a, b) = 2^3 \cdot 3^2 \cdot 11 = 792$$

(b) Next we find $m = \mathrm{lcm}(a, b)$. Those primes p, which appear in either a or b, 2, 3, 5, 7, 11, 13, and 17, will also appear in m, and the exponent of p in m will be the larger of its exponents in a and b. Thus

$$m = \mathrm{lcm}(a, b) = 2^4 \cdot 3^3 \cdot 5^2 \cdot 11 \cdot 13 \cdot 17$$

We are so used to using numbers as if the Fundamental Theorem of Arithmetic were true that it may seem as if it needs no proof. It is a tribute to Euclid, who first proved the theorem, that he recognized that it does require proof. We emphasize the nontriviality of the theorem by giving an example of a system of numbers which does not satisfy this theorem.

EXAMPLE 11.10 Let F be the set of positive integers of the form $3x + 1$. Thus F consists of the numbers:

$$1, \ 4, \ 7, \ 10, \ 13, \ 16, \ 19, \ 22, \ \ldots$$

Note that the product of two numbers in F is again in F since:

$$(3x + 1)(3y + 1) = 9xy + 3x + 3y + 1 = 3(3xy + x + y) + 1$$

Our definition of primes makes perfectly good sense in F. Although $4 = 2 \cdot 2$, the number 2 is not in F. Thus 4 is prime in F since 4 has no factors except 1 and 4. Similarly $10, 22, 25, \ldots$ are primes in F. We list the first few primes in F:

$$4, \ 7, \ 10, \ 13, \ 19, \ 22, \ 25, \ \ldots$$

Note $100 = 3(33) + 1$ belongs to F. However, 100 has two essentially different factorizations into primes of F; namely,

$$100 = 4 \cdot 25 \qquad \text{and} \qquad 100 = 10 \cdot 10$$

Thus there is no unique factorization into primes in F.

11.8 CONGRUENCE RELATION

Let m be a positive integer. We say that a is *congruent* to b *modulo* m. written

$$a \equiv b \ (\text{modulo } m) \qquad \text{or simply} \qquad a \equiv b \ (\text{mod } m)$$

if m divides the difference $a - b$. The integer m is called the *modulus*. The negation of $a \equiv b \ (\text{mod } m)$ is written $a \not\equiv b \ (\text{mod } m)$. For example:

(i) $87 \equiv 23 \ (\text{mod } 4)$ since 4 divides $87 - 23 = 64$.

(ii) $67 \equiv 1 \ (\text{mod } 6)$ since 6 divides $67 - 1 = 66$.

(iii) $72 \equiv -5 \ (\text{mod } 7)$ since 7 divides $72 - (-5) = 77$.

(iv) $27 \not\equiv 8 \ (\text{mod } 9)$ since 9 does not divide $27 - 8 = 19$.

Our first theorem (proved in Problem 11.34) states that congruence modulo m is an equivalence relation.

Theorem 11.21: Let m be a positive integer. Then:

(i) For any integer a, we have $a \equiv a \ (\text{mod } m)$.

(ii) If $a \equiv b \ (\text{mod } m)$, then $b \equiv a \ (\text{mod } m)$.

(iii) If $a \equiv b \ (\text{mod } m)$ and $b \equiv c \ (\text{mod } m)$, then $a \equiv c \ (\text{mod } m)$.

Remark: Suppose m is positive, and a is any integer. By the Division Algorithm, there exist integers q and r with $0 = r \leq m$ such that $a = mq + r$. Hence

$$mq = a - r \quad \text{or} \quad m \mid (a - r) \quad \text{or} \quad a \equiv r \ (\text{mod } m)$$

Accordingly:

(1) Any integer a is congruent modulo m to a unique integer in the set

$$\{0, \ 1, \ 2, \ldots, m - 1\}$$

The uniqueness comes from the fact that m cannot divide the difference of two such integers.

(2) Any two integers a and b are congruent modulo m if and only if they have the same remainder when divided by m.

Residue Classes

Since congruence modulo m is an equivalence relation, it partitions the set $\mathbf{Z}$ of integers into disjoint equivalence classes called the *residue classes modulo m*. By the above remarks, a residue class consists of all those integers with the same remainder when divided by m. Therefore, there are m such residue classes and each residue class contains exactly one of the integers in the set of possible remainders, that is,

$$\{0, \ 1, \ 2, \ \ldots, m - 1\}$$

Generally speaking, a set of m integers $\{a_1, \ a_2, \ \ldots, a_m\}$ is said to be a *complete residue system modulo m* if each a_i comes from a distinct residue class. (In such a case, each a_i is called a *representative* of its equivalence class.)

Thus the integers from 0 to $m - 1$ form a complete residue system. In fact, any m consecutive integers form a complete residue system modulo m.

The notation $[x]_m$, or simply $[x]$ is used to denote the residue class (modulo m) containing an integer x, that is, those integers which are congruent to x. In other words,

$$[x] = \{a \in \mathbf{Z} \mid a \equiv x \ (\text{mod } m)\}$$

Accordingly, the residue classes can be denoted by

$$[0], \ [1], \ [2], \ldots, [m - 1]$$

or by using any other choice of integers in a complete residue system.

EXAMPLE 11.11 The residue classes modulo $m = 6$ follow:

$$[0] = \{\ldots, -18, -12, -6, 0, 6, 12, 18, \ldots\}, \quad |3| = \{\ldots, -15, -9, -3, 3, 9, 15, 21, \ldots\}$$
$$[1] = \{\ldots, -17, -11, -5, 1, 7, 13, 19, \ldots\}, \quad |4| = \{\ldots, -14, -8, -2, 4, 10, 16, 22, \ldots\}$$
$$[2] = \{\ldots, -16, -10, -4, 2, 8, 14, 20, \ldots\}, \quad |5| = \{\ldots, -13, -7, -1, 5, 11, 17, 23, \ldots\}$$

Note that $\{-2, -1, 0, 1, 2, 3\}$ is also a complete residue system modulo $m = 6$, and these representatives have minimal absolute values.

Congruence Arithmetic

The next theorem (proved in Problem 11.35) tells us that, under addition and multiplication, the congruence relation behaves very much like the relation of equality. Namely:

Theorem 11.22: Suppose $a \equiv c \ (\text{mod } m)$ and $b \equiv d \ (\text{mod } m)$. Then:
(i) $a + b \equiv c + d (\text{mod } m)$; (ii) $a \cdot b \equiv c \cdot d (\text{mod } m)$

Remark: Suppose $p(x)$ is a polynomial with integral coefficients. If $s \equiv t \ (\text{mod } m)$, then using Theorem 11.22 repeatedly we can show that $p(s) \equiv p(t) \ (\text{mod } m)$.

EXAMPLE 11.12 Observe that $2 \equiv 8 \pmod 6$ and $5 \equiv 41 \pmod 6$. Then:

(a) $2 + 5 \equiv 8 + 41 \pmod 6$ or $7 \equiv 49 \pmod 6$

(b) $2 \cdot 5 \equiv 8 \cdot 41 \pmod 6$ or $10 \equiv 328 \pmod 6$

(c) Suppose $p(x) = 3x^2 - 7x + 5$. Then

$$p(2) = 12 - 14 + 5 = 3 \qquad \text{and} \qquad p(8) = 192 - 56 + 5 = 141$$

Hence $3 \equiv 141 \pmod 6$.

Arithmetic of Residue Classes

Addition and multiplication are defined for our residue classes modulo m as follows:

$$[a] + [b] = [a + b] \qquad \text{and} \qquad [a] \cdot [b] = [ab]$$

For example, consider the residue classes modulo $m = 6$; that is,

$$[0], [1], [2], [3], [4], [5]$$

Then

$$[2] + [3] = [5], \quad [4] + [5] = [9] = [3], \quad [2] \cdot [2] = [4], \quad [2] \cdot [5] = [10] = [4]$$

The content of Theorem 11.22 tells us that the above definitions are well defined, that is, the sum and product of the residue classes do not depend on the choice of representative of the residue class.

There are only a finite number m of residue classes modulo m. Thus one can easily write down explicitly their addition and multiplication tables when m is small. Figure 11-4 shows the addition and multiplication tables for the residue classes modulo $m = 6$. For notational convenience, we have omitted brackets and simply denoted the residue classes by the numbers 0, 1, 2, 3, 4, 5.

+	0	1	2	3	4	5
0	0	1	2	3	4	5
1	1	2	3	4	5	0
2	2	3	4	5	0	1
3	3	4	5	0	1	2
4	4	5	0	1	2	3
5	5	0	1	2	3	4

×	0	1	2	3	4	5
0	0	0	0	0	0	0
1	0	1	2	3	4	5
2	0	2	4	0	2	4
3	0	3	0	3	0	3
4	0	4	2	0	4	2
5	0	5	4	3	2	1

Fig. 11-4

Integers Modulo m, $\mathbf{Z_m}$

The *integers modulo m*, denoted by $\mathbf{Z_m}$, refers to the set

$$\mathbf{Z_m} = \{0, 1, 2, 3, \ldots, m - 1\}$$

where addition and multiplication are defined by the arithmetic modulo m or, in other words, the corresponding operations for the residue classes. For example, Fig. 11-4 may also be viewed as the addition and multiplication tables for $\mathbf{Z_6}$, This means:

> There is no essential difference between $\mathbf{Z_m}$ and the arithmetic of the residue classes modulo m, and so they will be used interchangeably.

Cancellation Laws for Congruences

Recall that the integers satisfy the following:

> Cancellation law: If $ab = ac$ and $a \neq 0$, then $b = c$.

The critical difference between ordinary arithmetic and arithmetic modulo m is that the above cancellation law is not true for congruences. For example,

$$3 \cdot 1 \equiv 3 \cdot 5 \ (\text{mod } 6) \quad \text{but} \quad 1 \not\equiv 5 \ (\text{mod } 6)$$

That is, we cannot cancel the 3 even though $3 \not\equiv 0 \ (\text{mod } 6)$. However, we do have the following *Modified Cancellation Law* for our congruence relations.

Theorem 11.23 (Modified Cancellation Law): Suppose $ab \equiv ac \ (\text{mod } m)$ and $\gcd(a, m) = 1$.

$$\text{Then } b \equiv c (\text{mod } m).$$

The above theorem is a consequence of the following more general result (proved in Problem 11.37).

Theorem 11.24: Suppose $ab \equiv ac \ (\text{mod } m)$ and $d = \gcd(a, m)$. Then $b \equiv c \ (\text{mod } m/d)$.

EXAMPLE 11.13 Consider the following congruence:

$$6 \equiv 36 \ (\text{mod } 10) \tag{11.1}$$

Since $\gcd(3, 10) = 1$ but $\gcd(6, 10) \neq 1$, we can divide both sides of (11.1) by 3 but not by 6. That is,

$$2 \equiv 12 \ (\text{mod } 10) \quad \text{but} \quad 1 \not\equiv 6 \ (\text{mod } 10)$$

However, by Theorem 11.24, we can divide both sides of (11.1) by 6 if we also divide the modulus by 2 which equals $\gcd(6, 10)$. That is,

$$1 \equiv 6 \ (\text{mod } 5)$$

Remark: Suppose p is a prime. Then the integers 1 through $p - 1$ are relatively prime to p. Thus the usual cancellation law does hold when the modulus is a prime p. That is:

> If $ab \equiv ac \ (\text{mod } p)$ and $a \not\equiv 0 \ (\text{mod } p)$, then $b \equiv c \ (\text{mod } p)$.

Thus $\mathbf{Z_p}$, the integers modulo a prime p, plays a very special role in number theory.

Reduced Residue Systems, Euler Phi Function

The modified cancellation law, Theorem 11.23, is indicative of the special role played by those integers which are relatively prime (coprime) to the modulus m. We note that a is coprime to m if and only if every element in the residue class $[a]$ is coprime to m. Thus we can speak of a residue class being coprime to m.

The number of residue classes relatively prime to m or, equivalently, the number of integers between 1 and m (inclusive) which are relatively prime to m is denoted by

$$\phi(m)$$

The function $\phi(m)$ is called the *Euler phi function*. The list of numbers between 1 and m which are coprime to m or, more generally, any list of $\phi(m)$ incongruent integers which are coprime to m, is called a *reduced residue system modulo m*.

EXAMPLE 11.14

(a) Consider the modulus $m = 15$. There are eight integers between 1 and 15 which are coprime to 15:

$$1, \quad 2, \quad 4, \quad 7, \quad 8, \quad 11, \quad 13, \quad 14$$

Thus $\phi(15) = 8$ and the above eight integers form a reduced residue system modulo 15.

(b) Consider any prime p. All the numbers $1, 2, \ldots, p-1$ are coprime to p; hence $\phi(p) = p - 1$.

A function f with domain the positive integers N is said to be *multiplicative* if, whenever a and b are relatively prime,

$$f(ab) = f(a)f(b)$$

The following theorem (proved in Problem 11.44) applies.

Theorem 11.25: Euler's phi function is multiplicative. That is, if a and b are relatively prime, then

$$\phi(ab) = \phi(a)\phi(b)$$

11.9 CONGRUENCE EQUATIONS

A *polynomial congruence equation* or, simply, a *congruence equation* (in one unknown x) is an equation of the form

$$a_n x^n + a_{n-1} x^{n-1} + \ldots + a_1 x + a_0 \equiv 0 \quad (\text{mod } m) \tag{11.2}$$

Such an equation is said to be of *degree n* if $a \not\equiv 0 \ (\text{mod } m)$.

Suppose $s \equiv t \ (\text{mod } m)$. Then s is a solution of (11.2) if and only if t is a solution of (11.2). Thus the *number of solutions* of (11.2) is defined to be the number of incongruent solutions or, equivalently, the number of solutions in the set

$$\{0, 1, 2, \ldots, m-1\}$$

Of course, these solutions can always be found by testing, that is, by substituting each of the m numbers into (11.2) to see if it does indeed satisfy the equation.

The *complete set of solutions* of (11.2) is a maximum set of incongruent solutions whereas the *general solution* of (11.2) is the set of all integral solutions of (11.2). The general solution of (11.2) can be found by adding all the multiples of the modulus m to any complete set of solutions.

EXAMPLE 11.15 Consider the equations:

(a) $x^2 + x + 1 \equiv 0 \ (\text{mod } 4)$,   (b) $x^2 + 3 \equiv 0 \ (\text{mod } 6)$,   (c) $x^2 - 1 \equiv 0 \ (\text{mod } 8)$

Here we find the solutions by testing.

(a) There are no solutions since 0, 1, 2, and 3 do not satisfy the equation.

(b) There is only one solution among 0, 1, ..., 5 which is 3. Thus the general solution consists of the integers $3 + 6k$ where $k \in \mathbf{Z}$.

(c) There are four solutions, 1, 3, 5, and 7. This shows that a congruence equation of degree n can have more than n solutions.

We emphasize that we are not only interested in studying congruence equations in order to find their solutions; this can always be found by testing. We are mainly interested in developing techniques to help us find such solutions, and a theory which tells us conditions under which solutions exist and the number of such solutions. Such a theory holds for linear congruence equations which we investigate below. We will also discuss the Chinese Remainder Theorem, which is essentially a system of linear congruence equations.

Remark 1: The coefficients of a congruence equation can always be reduced modulo m since an *equivalent* equation, that is, an equation with the same solutions, would result. For example, the following are equivalent equations since the coefficients are congruent modulo $m = 6$:

$$15x^2 + 28x + 14 \equiv 0 \ (\mathrm{mod}\ 6), \quad 3x^2 + 4x + 2 \equiv 0 \ (\mathrm{mod}\ 6), \quad 3x^2 - 2x + 2 \equiv 0 \ (\mathrm{mod}\ 6),$$

Usually we choose coefficients between 0 and $m - 1$ or between $-m/2$ and $m/2$

Remark 2: Since we are really looking for solutions of (11.2) among the residue classes modulo m rather than among the integers, we may view (11.2) as an equation over the integers modulo m, rather than an equation over $\mathbf{Z}$, the integers. In this context, the number of solutions of (11.2) is simply the number of solutions in $\mathbf{Z_m}$.

Linear Congruence Equation: $ax \equiv 1 \ (\mathrm{mod}\ m)$

First we consider the special linear congruence equation

$$ax \equiv 1 \ (\mathrm{mod}\ m) \tag{11.3}$$

where $a \not\equiv 0 \ (\mathrm{mod}\ m)$. The complete story of this equation is given in the following theorem (proved in Problem 11.57).

Theorem 11.26: If a and m are relatively prime, then $ax \equiv 1 \ (\mathrm{mod}\ m)$ has a unique solution; otherwise it has no solution.

EXAMPLE 11.16

(a) Consider the congruence equation $6x \equiv 1 \ (\mathrm{mod}\ 33)$. Since $\gcd(6, 33) = 3$, this equation has no solution.

(b) Consider the congruence equation $7x \equiv 1 \ (\mathrm{mod}\ 9)$. Since $\gcd(7, 9) = 1$, the equation has a unique solution. Testing the numbers 0, 1, ..., 8, we find that

$$7(4) = 28 \equiv 1 (\mathrm{mod}\ 9)$$

Thus $x = 4$ is our unique solution. (The general solution is $4 + 9k$ for $k \in \mathbf{Z}$.)

Suppose a solution of (11.3) does exist, that is, suppose $\gcd(a, m) = 1$. Furthermore, suppose the modulus m is large. Then the Euclidean algorithm can be used to find a solution of (11.3). Specifically, we use the Euclidean algorithm to find x_0 and y_0 such that

$$ax_0 + my_0 = 1$$

From this it follows that $ax_0 \equiv 1 \ (\mathrm{mod}\ m)$; that is, x_0 is a solution to (11.3).

EXAMPLE 11.17 Consider the following congruence equation:

$$81 \equiv 1 \ (\text{mod } 256)$$

By observation or by applying the Euclidean algorithm to 81 and 256, we find that $\gcd(81, 256) = 1$. Thus the equation has a unique solution. Testing may not be an efficient way to find this solution since the modulus $m = 256$ is relatively large. Hence, we apply the Euclidean algorithm to $a = 81$ and $m = 256$. Specifically, as in Example 11.6, we find $x_0 = -25$ and $y_0 = 7$ such that

$$81x_0 + 256y_0 = 1$$

This means that $x_0 = -25$ is a solution of the given congruence equation. Adding $m = 256$ to -25, we obtain the following unique solution between 0 and 256:

$$x = 231$$

Linear Congruence Equation: $ax \equiv b \ (\text{mod } m)$

Now we consider the more general linear congruence equation

$$ax \equiv b \ (\text{mod } m) \tag{11.4}$$

where $a \not\equiv 0 \ (\text{mod } m)$. We first consider the case (proved in Problem 11.58) where a and m are coprime.

Theorem 11.27: Suppose a and m are relatively prime. Then $ax \equiv b \ (\text{mod } m)$ has a unique solution. Moreover, if s is the unique solution to $ax \equiv 1 \ (\text{mod } m)$, then the unique solution to $ax \equiv b \ (\text{mod } m)$ is $x = bs$.

EXAMPLE 11.18

(a) Consider the congruence equation $3x \equiv 5 \ (\text{mod } 8)$. Since 3 and 8 are coprime, the equation has a unique solution. Testing the integers $0, 1, \ldots, 7$, we find that

$$3(7) = 21 \equiv 5 \ (\text{mod } 8)$$

Thus $x = 7$ is the unique solution of the equation.

(b) Consider the linear congruence equation

$$33x \equiv 38 \ (\text{mod } 280) \tag{11.5}$$

Since $\gcd(33, 280) = 1$, the equation has a unique solution. Testing may not be an efficient way to find this solution since the modulus $m = 280$ is relatively large. We apply the Euclidean algorithm to first find a solution to

$$33x \equiv 1 \ (\text{mod } 280) \tag{11.6}$$

That is, as in Example 11.6, we find $x_0 = 17$ and $y_0 = 2$ to be a solution of

$$33x_0 + 280y_0 = 1$$

This means that $s = 17$ is a solution of (11.6). Then

$$sb = 17(38) = 646$$

is a solution of (11.5). Dividing 646 by $m = 280$, we obtain the remainder

$$x = 86$$

which is the unique solution 11.5 between 0 and 280. (The general solution is $86 + 280k$ with $k \in \mathbf{Z}$.)

The complete story of the general case of (11.4) is contained in the following theorem (proved in Problem 11.59).

Theorem 11.28: Consider the equation $ax \equiv b \pmod{m}$ where $d = \gcd(a, m)$.

   (i) Suppose d does not divide b. Then $ax \equiv b \pmod{m}$ has no solution.

   (ii) Suppose d does divide b. Then $ax \equiv b \pmod{m}$ has d solutions which are all congruent modulo M to the unique solution of
$$Ax \equiv B \pmod{M} \quad \text{where} \quad A = a/d, \ B = b/d, \ M = m/d.$$

We emphasize that Theorem 11.27 applies to the equation $Ax \equiv B \pmod{M}$ in Theorem 11.28 since $\gcd(A, M) = 1$.

EXAMPLE 11.19 Solve each congruence equation: (a) $4x \equiv 9 \pmod{14}$; (b) $8x \equiv 12 \pmod{28}$.

(a) Note $\gcd(4, 14) = 2$. However, 2 does not divide 9. Hence the equation does not have a solution.

(b) Note that $d = \gcd(8, 28) = 4$, and $d = 4$ does divide 12. Thus the equation has $d = 4$ solutions. Dividing each term in the equation by $d = 4$ we obtain the congruence equation (11.7) which has a unique solution.

$$2x \equiv 3 \pmod{7} \tag{11.7}$$

Testing the integers $0, 1, \ldots, 6$, we find that 5 is the unique solution of (11.7). We now add $d - 1 = 3$ multiples of 7 to the solution 5 of (11.7) obtaining:

$$5 + 7 = 12, \quad 5 + 2(7) = 19, \quad 5 + 3(7) = 26$$

Accordingly, 5, 12, 19, 26 are the required $d = 4$ solutions of the original equation $8x \equiv 12 \pmod{28}$.

Remark: The solution of equation (11.7) in Example 11.19 was obtained by inspection. However, in case the modulus m is large, we can always use the Euclidean algorithm to find its unique solution as in Example 11.17.

Chinese Remainder Theorem

An old Chinese riddle asks the following question.

> Is there a positive integer x such that when x is divided by 3 it yields a remainder 2, when x is divided by 5 it yields a remainder 4, and when x is divided by 7 it yields a remainder 6?

In other words, we seek a common solution of the following three congruence equations:

$$x \equiv 2 \pmod{3}, \quad x \equiv 4 \pmod{5}, \quad x \equiv 6 \pmod{7}$$

Observe that the moduli 3, 5, and 7 are pairwise relatively prime. (Moduli is the plural of modulus.) Thus the following theorem (proved in Problem 11.60) applies; it tells us that there is a unique solution modulo $M = 3 \cdot 5 \cdot 7 = 105$.

Theorem 11.29 (Chinese Remainder Theorem): Consider the system

$$x \equiv r_1 \pmod{m_1}, \quad x \equiv r_2 \pmod{m_2}, \quad \cdots, \quad x \equiv r_k \pmod{m_k} \tag{11.8}$$

where the m_i are pairwise relatively prime. Then the system has a unique solution modulo $M = m_1 m_2 \cdots m_k$.

One can actually give an explicit formula for the solution of the system (11.8) in Theorem 11.29 which we state as a proposition.

Proposition 11.30: Consider the system (11.8) of congruence equations. Let $M = m_1 m_2 \ldots m_k$, and

$$M_1 = \frac{M}{m_1}, \quad M_2 = \frac{M}{m_2}, \quad \ldots, \quad M_k = \frac{M}{m_k}$$

(Then each pair M_i and m_i are co-prime.) Let $s_1, s_2, \ldots, s_k$ be the solutions respectively, of the congruence equations

$$M_1 x \equiv 1 \ (\text{mod } m_1), \ M_2 x \equiv 1 \ (\text{mod } m_2), \ldots, \ M_k x \equiv 1 \ (\text{mod } m_k)$$

Then the following is a solution of the system (11.8):

$$x_0 = M_1 s_1 r_1 + M_2 s_2 r_2 + \cdots + M_k s_k r_k \tag{11.9}$$

We now solve the original riddle in two ways.

Method 1: First we apply the Chinese Remainder Theorem (CRT) to the first two equations,

$$\text{(a) } x \equiv 2 \ (\text{mod } 3) \qquad \text{and} \qquad \text{(b) } x \equiv 4 \ (\text{mod } 5)$$

CRT tells us there is a unique solution modulo $M = 3 \cdot 5 = 15$. Adding multiples of the modulus $m = 5$ to the given solution $x = 4$ of the second equation (b), we obtain the following three solutions of (b) which are less than 15:

$$4, \quad 9, \quad 14$$

Testing each of these solutions in equation (a), we find that 14 is the only solution of both equations. Now we apply the same process to the two equations

$$\text{(c) } x \equiv 14 \ (\text{mod } 15) \quad \text{and} \quad \text{(d) } x \equiv 6 \ (\text{mod } 7)$$

CRT tells us there is a unique solution modulo $M = 15 \cdot 7 = 105$. Adding multiples of the modulus $m = 15$ to the given solution $x = 14$ of the first equation (c), we obtain the following seven solutions of (b) which are less than 105:

$$14, \quad 29, \quad 44, \quad 59, \quad 74, \quad 89, \quad 104$$

Testing each of these solutions of (c) in the second equation (d) we find that 104 is the only solution of both equations. Thus the smallest positive integer satisfying all three equations is

$$x = 104$$

This is the solution of the riddle.

Method 2: Using the above notation, we obtain

$$M = 3 \cdot 5 \cdot 7 = 105, \quad M_1 = 105/3 = 35, \quad M_2 = 105/5 = 21, \quad M_3 = 105/7 = 15$$

We now seek solutions to the equations

$$35x \equiv 1 \ (\text{mod } 3), \quad 21x \equiv 1 \ (\text{mod } 5), \quad 15x \equiv 1 \ (\text{mod } 7)$$

Reducing 35 modulo 3, reducing 21 modulo 5, and reducing 15 modulo 7, yields the system

$$2x \equiv 1 \ (\text{mod } 3), \quad x \equiv 1 \ (\text{mod } 5), \quad x \equiv 1 \ (\text{mod } 7)$$

The solutions of these three equations are, respectively,

$$s_1 = 2, \quad s_2 = 1, \quad s_3 = 1$$

We now substitute into the formula (11.9) to obtain the following solution of our original system:

$$x_0 = 35 \cdot 2 \cdot 2 + 21 \cdot 1 \cdot 4 + 15 \cdot 1 \cdot 6 = 314$$

Dividing this solution by the modulus $M = 105$, we obtain the remainder

$$x = 104$$

which is the unique solution of the riddle between 0 and 105.

Remark: The above solutions $s_1 = 2$, $s_2 = 1$, $s_3 = 1$ were obtained by inspection. If the moduli are large, we can always use the Euclidean algorithm to find such solutions as in Example 11.17.

Solved Problems

INEQUALITIES, ABSOLUTE VALUE

11.1. Insert the correct symbol, $<$, $>$, or $=$, between each pair of integers:

    (a) 4 _____ -7;   (b) -2 _____ -9;   (c)$(-3)^2$ _____ 9;   (d) -8 _____ 3,

         For each pair of integers, say a and b, determine their relative positions on the number line **R**; or, alternatively, compute $b - a$, and write $a < b, a > b$, or $a = b$ according as $b - a$ is positive, negative, or zero. Hence:

    (a) $4 > -7$;   (b) $-2 > -9$;   (c) $(-3)^2 = 9$;   (d) $-8 < 3$.

11.2. Evaluate: (a) $|2 - 5|$, $|-2 + 5|$, $|-2 - 5|$;   (b) $|5 - 8| + |2 - 4|$, $|4 - 3| - |3 - 9|$.
     Evaluate inside the absolute value sign first:

    (a) $|2 - 5| = |-3| = 3$, $|-2 + 5| = |3| = 3$, $|-2 - 5| = |-7| = 7$

    (b) $|5 - 8| + |2 - 4| = |-3| + |-2| = 3 + 2 = 5$; $|4 - 3| - |3 - 9| = |1| - |-6| = 1 - 6 = -5$

11.3. Find the distance d between each pair of integers:
    (a) 3 and –7; (b) –4 and 2; (c) 1 and 9; (d) –8 and –3; (e) –5 and –8.

         The distance d between a and b is given by $d = |a - b| = |b - a|$. Alternatively, as indicated by Fig. 11-5, $d = |a| + |b|$ when a and b have different signs, and $d = |a| - |b|$ when a and b have the same sign and $|a| > |b|$. Thus: (a) $d = 3 + 7 = 10$; (b) $d = 4 + 2 = 6$; (c) $d = 9 - 1 = 8$; (d) $d = 8 - 3 = 5$; (e) $d = 8 - 5 = 3$.

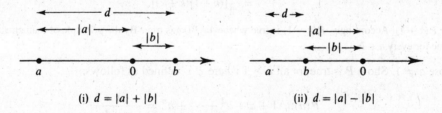

                  (i) $d = |a| + |b|$ (ii) $d = |a| - |b|$

Fig. 11-5

11.4. Find all integers n such that: (a) $1 < 2n - 6 < 14$; (b) $2 < 8 - 3n < 18$.

    (a) Add 6 to the "three sides" to get $7 < 2n < 20$. Then divide all sides by 2 (or multiply by 1/2) to get $3.5 < n < 10$. Hence $n = 4, 5, 6, 7, 8, 9$.

    (b) Add –8 to the three sides to get $-6 < -3n < 10$. Divide by –3 (or multiply by –1/3) and, since –3 is negative, change the direction of the inequality to get

$$2 > n > -3.3 \quad \text{or} \quad -3.3 < n < 2$$

     Hence $n = -3, -2, -1, 0, 1$.

11.5. Prove Proposition 11.3: Suppose $a \leq b$, and c is any integer. Then: (i) $a + c \leq b + c$,

(ii) $ac = bc$ when $c > 0$; but $ac = bc$ when $c < 0$.

The proposition is certainly true when $a = b$. Hence we need only consider the case when $a < b$, that is, when $b - a$ is positive.

(i) The following difference is positive: $(b + c) - (a + c) = b - a$. Hence $a + c < b + c$.

(ii) Suppose c is positive. By property $[\mathbf{P_1}]$ of the positive integers $\mathbf{N}$, the product $c(b - a)$ is also positive. Thus $ac < bc$.

Now suppose c is negative. Then $-c$ is positive, and the product $(-c)(b - a) = ac - bc$ is also positive. Accordingly, $bc < ac$, whence $ac > bc$.

11.6. Prove Proposition 11.4 (iii): $|ab| = |a||b|$.

The proof consists of analysing the following five cases: (a) $a = 0$ or $b = 0$; (b) $a > 0$ and $b > 0$; (c) $a > 0$ and $b < 0$; (d) $ba < 0$ and $b > 0$; (e) $ba < 0$ and $b < 0$. We only prove the third case here. (c) Since $a > 0$ and $b < 0$, $|a| = a$ and $|b| = -b$. Also $ab < 0$. Hence $|ab| = -(ab) = a(-b) = |a||b|$.

11.7 Prove Proposition 11.4 (iv): $|a \pm b| \leq |a| + |b|$.

Now $ab \leq |ab| = |a||b|$, and so $2ab \leq 2|a||b|$. Hence

$$(a + b)^2 = a^2 + 2ab + b^2 \leq |a|^2 + 2|a||b| + |b|^2 = (|a| + |b|)^2$$

But $\sqrt{(a + b)^2} = |a + b|$. Thus the square root of the above yields $|a + b| \leq |a| + |b|$. Also,

$$|a - b| = |a + (-b)| \leq |a| + |-b| = |a| + |b|$$

MATHEMATICAL INDUCTION, WELL-ORDERING PRINCIPLE

11.8. Prove the proposition that the sum of the first n positive integers is $n(n + 1)/2$; that is:

$$P(n): \quad 1 + 2 + \cdots + n = \frac{1}{2}n(n + 1)$$

$P(1)$ is true since $1 = \frac{1}{2}(1)(1 + 1)$. Assuming $P(k)$ is true, we add $k + 1$ to both sides of $P(k)$ obtaining

$$1 + 2 + 3 + \cdots + k + (k + 1) = \frac{1}{2}k(k + 1) + (k + 1) = \frac{1}{2}[k(k + 1) + 2(k + 1)]$$
$$= \frac{1}{2}[(k + 1)(k + 2)]$$

This is $P(k + 1)$. Accordingly, $P(k + 1)$ is true whenever $P(k)$ is true. By the principle of mathematical induction, P is true for every $n \in \mathbf{N}$.

11.9. Suppose $a \neq 1$. Show P is true for all $n \geq 1$ where P is defined as follows:

$$P(n): \quad 1 + a + a^2 + \cdots + a^n = \frac{a^{n+1} - 1}{a - 1}$$

$P(1)$ is true since

$$1 + a = \frac{a^2 - 1}{a - 1}$$

Assuming $P(k)$ is true, we add a^{k+1} to both sides of $P(k)$, obtaining

$$1 + a + a^2 + \ldots + a^k + a^{k+1} = \frac{a^{k+1} - 1}{a - 1} + a^{k+1} = \frac{a^{k+1} - 1 + (a - 1)a^{k+1}}{a - 1}$$
$$= \frac{a^{k+2} - 1}{a - 1}$$

This is $P(k + 1)$. Thus, $P(k + 1)$ is true whenever $P(k)$ is true. By the principle of mathematical induction, P is true for every $n \in \mathbf{N}$.

11.10. Suppose n is a positive integer. Prove $n \geq 1$. (This is not true for the rational numbers $\mathbf{Q}$.) In other words, if $P(n)$ is the statement that $n \geq 1$, then $P(n)$ is true for every $n \in \mathbf{N}$.

$P(n)$ holds for $n = 1$ since $1 \geq 1$. Assuming $P(k)$ is true, that is, $k \geq 1$, add 1 to both sides to obtain

$$k + 1 \geq 1 + 1 = 2 > 1$$

This is $P(k + 1)$. Thus $P(k + 1)$ is true whenever $P(k)$ is true. By the principle of mathematical induction, P is true for every $n \in \mathbf{N}$.

11.11. Suppose a and b are positive integers. Prove:

    (a) If $b \neq 1$, then $a < ab$.

    (b) If $ab = 1$, then $a = 1$ and $b = 1$.

    (c) If n is composite, then $n = ab$ where $1 < a, b < n$.

    (a) By Problem 11.10, $b > 1$. Hence $b - 1 > 0$, that is, $b - 1$ is positive. By the property $[\mathbf{P}_1]$ of the positive integers $\mathbf{N}$, the following product is also positive:

$$a(b - 1) = ab - a$$

Thus $a < ab$, as required.

    (b) Suppose $b \neq 1$. By (a), $a < ab = 1$. This contradicts Problem 11.10; hence $b = 1$. It then follows that $a = 1$.

    (c) If n is not prime, then n has a positive divisor a such that $a \neq 1$ and $a \neq n$. Then $n = ab$ where $b \neq 1$ and $b \neq n$. Thus, by Problem 11.10 and by part (a), $1 < a, b < ab = n$.

11.12. Prove Theorem 11.6 (Well-Ordering Principle): Let S be a nonempty set of positive integers. Then S contains a least element.

    Suppose S has no least element. Let M consist of those positive integers which are less than every element of S. Then $1 \in M$; otherwise, $1 \in S$ and 1 would be a least element of S. Suppose $k \in M$. Then k is less than every element of S. Therefore $k + 1 \in M$; otherwise $k + 1$ would be a least element of S.

    By the Principle of Mathematical Induction, M contains every positive integer. Thus S is empty which contradicts the hypothesis that S is nonempty. Accordingly, the original assumption that S has no least element cannot be true. Thus the theorem is true.

11.13. Prove Theorem 11.5 (Induction: Second Form): Let P be a proposition defined on the integers $n \geq 1$ such that: (i) $P(1)$ is true. (ii) $P(k)$ is true whenever $P(j)$ is true for all $1 \leq j < k$.

Then P is true for all $n \geq 1$.

    Let A be the set of integers $n \geq 1$ for which P is not true. Suppose A is not empty. By the Well-Ordering Principle, A contains at least element a_0 By (i), $a_0 \neq 1$.

    Since a_0 is the least element of A, P is true for every integer j where $1 \leq j < a_0$. By (ii), P is true for a_0. This contradicts the fact that $a_0 \in A$. Hence A is empty, and so P is true for every integer $n > 1$.

DIVISION ALGORITHM

11.14. For each pair of integers a and b, find integers q and r such that $a = bq + r$ and $0 < r < |b|$:

    (a) $a = 258$ and $b = 12$;   (b) $a = 573$ and $b = -16$.

    (a) Here a and b are positive. Simply divide a by b, that is, 258 by 12, say by long division, to obtain the quotient $q = 21$ and remainder $r = 6$. Alternately, using a calculator, we get

$$258/12 = 21.5, \quad q = INT(a/b) = 21, \quad r = a - bq = 258 - 12(21) = 6$$

(b) Here a is positive, but b is negative. Divide a by $|b|$, that is, 573 by 16, say with a calculator to obtain:

$$a/|b| = 573/16 = 35.8125, \quad q' = INT(a/|b|) = 35, \quad r' = 573 - 16(35) = 13$$

Then

$$573 = (16)(35) + 13 \quad \text{and} \quad 573 = (-16)(-35) + 13$$

Thus $q = -35$ and $r = 13$.

11.15. For each pair of integers a and b, find integers q and r such that $a = bq + r$ and $0 < r < |b|$:

(a) $a = -381$ and $b = 14$; (b) $a = -433$ and $b = -17$.

Here a is negative in each case, hence we have to make some adjustments to be sure that $0 < r < |b|$.

(a) Divide $|a| = 381$ by $b = 14$, say with a calculator, to obtain the quotient $q' = 27$ and remainder $r' = 3$. Then

$$381 = (14)(27) + 3 \quad \text{and so} \quad -381 = (14)(-27) - 3$$

But -3 is negative and cannot be the remainder r; hence we add and subtract $b = 14$ as follows:

$$-381 = (14)(-27) - 14 + 14 - 3 = (14)(-28) + 11$$

Thus $q = -28$ and $r = 11$.

(b) Divide $|a| = 433$ by $|b| = 17$, say by a calculator, to obtain the quotient $q' = 25$ and remainder $r' = 8$. Then:

$$433 = (17)(25) + 8 \quad \text{and so} \quad -433 = (-17)(25) - 8$$

But -8 is negative and cannot be the remainder r; we correct this by adding and subtracting $|b| = 17$ as follows:

$$-433 = (-17)(25) - 17 + 17 - 8 = (-17)(26) + 9$$

Thus $q = 26$ and $r = 9$.

11.16. Prove $\sqrt{2}$ is not rational, that is, $\sqrt{2} \neq a/b$ where a and b are integers.

Suppose $\sqrt{2}$ is rational and $\sqrt{2} = a/b$ where a and b are integers reduced to lowest terms, i.e., $\gcd(a, b) = 1$. Squaring both sides yields

$$2 = \frac{a^2}{b^2} \quad \text{or} \quad a^2 = 2b^2$$

Then 2 divides a^2. Since 2 is a prime, 2 also divides a. Say $a = 2c$. Then

$$2b^2 = a^2 = 4c^2 \quad \text{or} \quad b^2 = 2c^2$$

Then 2 divides b^2. Since 2 is a prime 2 also divides b. Thus 2 divides both a and b. This contradicts the assumption that $\gcd(a, b) = 1$. Therefore, $\sqrt{2}$ is not rational.

11.17. Prove Theorem 11.8 (Division Algorithm) for the case of positive integers. That is, assuming a and b are positive integers, prove there exist nonnegative integers q and r such that

$$a = bq + r \qquad \text{and} \qquad 0 \leq r < b \tag{11.10}$$

If $a < b$, choose $q = 0$ and $r = a$. If $a = b$, choose $q = 1$ and $r = 0$. In either case, q and r satisfy (11.10).

The proof is now by induction on a. If $a = 1$ then $a < b$ or $a = b$; hence the theorem holds when $a = 1$. Suppose $a > b$. Then $a - b$ is positive and $a - b < a$. By induction, the theorem holds for $a - b$. Thus there exists q' and r' such that

$$a - b = bq' + r' \qquad \text{and} \qquad 0 \leq r' < b$$

Then

$$a = bq' + b + r' = b(q' + 1) + r'$$

Choose $q = q' + 1$ and $r = r'$. Then q and r are nonnegative integers and satisfy (11.10). Thus the theorem is proved.

11.18. Prove Theorem 11.8 (Division Algorithm): Let a and b be integers with $b \neq 0$. Then there exists integers q and r such that $a = bq + r$ and $0 \leq r' < |b|$. Also, the integers q and r are unique.

Let M be the set of nonnegative integers of the form $a - xb$ for some integer x. If $x = -|a|b$ then $a - xb$ is nonnegative; hence M is nonempty. By the Well-Ordering Principle, M has a least element, say r. Since $r \in M$, we have

$$r \geq 0 \qquad \text{and} \qquad r = a - qb$$

for some integer q. We need only show that $r < |b|$. Suppose $r \geq |b|$. Let $r' = r - |b|$.

Then $r' \geq 0$ and also $r' < r$ because $b \neq 0$. Furthermore,

$$r' = r - |b| = a - qb - |b| = \begin{cases} a - (q+1)b, & \text{if } b < 0 \\ a - (q-1)b, & \text{if } b > 0 \end{cases}$$

In either case, r' belongs to M. This contradicts the fact that r is the least element of M Accordingly, $r < |b|$. Thus the existence of q and r is proved.

We now show that q and r are unique. Suppose there exist integers q and r and q' and r' such that

$$a = bq + r \quad \text{and} \quad a = bq' + r' \quad \text{where} \quad 0 < r, r' < |b|$$

Then $bq + r = bq' + r'$; hence

$$b(q - q') = r' - r$$

Thus b divides $r' - r$. But $|r' - r| < |b|$ since $0 < r, r' < |b|$. Accordingly, $r' - r = 0$. Since $b \neq 0$ this implies $q - q' = 0$. Consequently, $r' = r$ and $q' = q$; that is, q and r are uniquely determined by a and b.

DIVISIBILITY, PRIMES, GREATEST COMMON DIVISOR

11.19. Find all positive divisors of: (a) 18; (b) $256 = 2^8$; (c) $392 = 2^3 \cdot 7^2$.

(a) Since 18 is relatively small, we simply write down all positive integers (≤ 18) which divide 18. These are:

$$1, \quad 2, \quad 3, \quad 6, \quad 9, \quad 18$$

(b) Since 2 is a prime, the positive divisors of $256 = 2^8$ are simply the lower powers of 2, i.e.,

$$2^0, \quad 2^1, \quad 2^2, \quad 2^3, \quad 2^4, \quad 2^5, \quad 2^6, \quad 2^7, \quad 2^8$$

In other words, the positive divisors of 256 are:

$$1, \quad 2, \quad 4, \quad 8, \quad 16, \quad 32, \quad 64, \quad 128, \quad 256$$

(c) Since 2 and 7 are prime, the positive divisors of $392 = 2^3 \cdot 7^2$ are products of lower powers of 2 times lower powers of 7, i.e.,

$$2^0 \cdot 7^0, \quad 2^1 \cdot 7^0, \quad 2^2 \cdot 7^0, \quad 2^3 \cdot 7^0, \quad 2^0 \cdot 7^1, \quad 2^1 \cdot 7^1, \quad 2^2 \cdot 7^1, \quad 2^3 \cdot 7^1,$$
$$2^0 \cdot 7^2, \quad 2^1 \cdot 7^2, \quad 2^2 \cdot 7^2, \quad 2^3 \cdot 7^2$$

In other words, the positive powers of 392 are:

$$1, \quad 2, \quad 4, \quad 8, \quad 7, \quad 14, \quad 28, \quad 56, \quad 49, \quad 98, \quad 196, \quad 392.$$

(We have used the usual convention that $n^0 = 1$ for any nonzero number n.)

11.20. List all primes between 50 and 100.

Simply list all numbers p between 50 and 100 which cannot be written as a product of two positive integers, excluding 1 and p. This yields:

$$51, \quad 53, \quad 57, \quad 59, \quad 61, \quad 67, \quad 71, \quad 73, \quad 79, \quad 83, \quad 87, \quad 89, \quad 91, \quad 93, \quad 97$$

11.21. Let $a = 8316$ and $b = 10\,920$.

    (a) Find $d = \gcd(a, b)$, the greatest common divisor of a and b.

    (b) Find integers m and n such that $d = ma + nb$.

    (c) Find $\operatorname{lcm}(a, b)$, the least common multiple of a and b.

    (a) Apply the Euclidean algorithm to a and b. That is, apply the division algorithm to a and b and then repeatedly apply the division algorithm to each quotient and remainder until obtaining a zero remainder. These steps are pictured in Fig. 11-6(a) using long division and also in Fig. 11-6(b) where the arrows indicate the quotient and remainder in the next step. The last nonzero remainder is 84. Thus $84 = \gcd(8316, 10\,920)$.

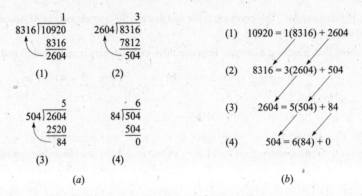

Fig. 11-6

    (b) Now find m and n such that $84 = 8316m + 1092n$ by "unraveling" the above steps in the Euclidean algorithm. Specifically, the first three quotients in Fig. 11-6 yields the equations:

$$(1)\ 2604 = 10\,920 - 1(8316); \quad (2)\ 504 = 8316 - 3(2604); \quad (3)\ 84 = 2604 - 5(504).$$

Equation (3) tells us that $d = 84$ is a linear combination of 2604 and 504. Use (2) to replace 504 in (3) so 84 can be written as a linear combination of 2604 and 8316 as follows:

$$(5)\ 84 = 2604 - 5[8316 - 3(2604)] = 2604 - 5(8316) + 15(2604)$$
$$= 16(2604) - 5(8316)$$

Now use (1) to replace 2604 in (5) so 84 can be written as a linear combination of 8316 and 10 290 as follows:

$$(6)\ 84 = 16[10\,920 - 1(8316)] - 5(8316) = 16(10\,920) - 16(8316) - 5(8316)$$
$$= -21(8316) + 16(10\,920)$$

This is our desired linear combination. In other words, $m = -21$ and $n = 16$.

    (c) By Theorem 11.16,

$$\operatorname{lcm}(a, b) = \frac{|ab|}{\gcd(a, b)} = \frac{(8316)(10\,920)}{84} = 1\,081\,080$$

11.22. Find the unique factorization of each number: (a) 135;   (b) 1330;   (c) 3105;   (d) 211.

    (a) $135 = 5 \cdot 27 = 5 \cdot 3 \cdot 3 \cdot 3$ or $135 = 3^3 \cdot 5$.

    (b) $1330 = 2 \cdot 665 = 2 \cdot 5 \cdot 133 = 2 \cdot 5 \cdot 7 \cdot 19$.

    (c) $3105 = 5 \cdot 621 = 5 \cdot 3 \cdot 207 = 5 \cdot 3 \cdot 3 \cdot 69 = 5 \cdot 3 \cdot 3 \cdot 3 \cdot 23$,   or $3105 = 3^3 \cdot 5 \cdot 23$.

    (d) None of the primes 2, 3, 5, 7, 11, 13 divides 211; hence 211 cannot be factored, that is, 211 is a prime.

    (**Remark**: We need only test those primes less than $\sqrt{211}$.)

11.23. Let $a = 2^3 \cdot 3^5 \cdot 5^4 \cdot 11^6 \cdot 17^3$ and $b = 2^5 \cdot 5^3 \cdot 7^2 \cdot 11^4 \cdot 13^2$. Find $\gcd(a, b)$ and $\text{lcm}(a, b)$.

Those primes p_i which appear in both a and b will also appear in $\gcd(a, b)$. Furthermore, the exponent of p_i in $\gcd(a, b)$ will be the smaller of its exponents in a and b. Hence

$$\gcd(a, b) = 2^3 \cdot 5^3 \cdot 11^4$$

Those primes p_i which appear in either a or b will also appear in $\text{lcm}(a, b)$. Also, the exponent of p_i in $\text{lcm}(a, b)$ will be the larger of its exponent in a and b. Hence

$$\text{lcm}(a, b) = 2^5 \cdot 3^5 \cdot 5^4 \cdot 7^2 \cdot 11^6 \cdot 13^2 \cdot 17^3$$

11.24. Prove Theorem 11.9: Suppose a, b, c are integers.

(i) If $a \mid b$ and $b \mid c$, then $a \mid c$.
(ii) If $a \mid b$ then, for any integer x, $a \mid bx$.
(iii) If $a \mid b$ and $a \mid c$, then $a \mid (b + c)$ and $a \mid (b - c)$.
(iv) If $a \mid b$ and $b \neq 0$, then $a = \pm b$ or $|a| < |b|$.
(v) If $a \mid b$ and $b \mid a$, then $|a| = |b|$, that is, $a = \pm b$.
(vi) If $a \mid 1$, then $a = \pm 1$.

(i) If $a \mid b$ and $b \mid c$, then there exist integers x and y such that $ax = b$ and $by = c$. Replacing b by ax, we obtain $axy = c$. Hence $a \mid c$.

(ii) If $a \mid b$, then there exists an integer c such that $ac = b$. Multiplying the equation by x, we obtain $acx = bx$. Hence $a \mid bx$.

(iii) If $a \mid b$ and $a \mid c$, then there exist integers x and y such that $ax = b$ and $ay = c$. Adding the equalities, we obtain

$$ax + ay = b + c \quad \text{and so} \quad a(x + y) = b + c$$

Hence $a \mid (b + c)$. Subtracting the equalities $ay = b$ and $by = c$, we obtain

$$ax - ay = b - c \quad \text{and so} \quad a(x - y) = b - c.$$

Thus $a \mid (b - c)$.

(iv) If $a \mid b$, then there exists c such that $ac = b$. Then

$$|b| = |ac| = |a||c|$$

Hence either $|c| = 1$ or $|a| < |a||c| = |b|$. If $|c| = 1$, then $c = \pm 1$; where $a = \pm b$, as required.

(v) If $a \mid b$, then $a = \pm b$ or $|a| < |b|$. If $|a| < |b|$ then $b \mid a$. Hence $a = \pm b$.

(vi) If $a \mid 1$, then $a = \pm 1$ or $|a| < |1| = 1$. By Problem 11.11, $|a| \geq 1$. Therefore, $a = \pm 1$.

11.25. A nonempty subset J of $\mathbf{Z}$ is called an *ideal* if J has the following two properties:

(1) If $a, b \in J$, then $a + b \in J$.   (2) If $a \in J$ and $n \in \mathbf{Z}$, then $na \in J$.

Let d be the least positive integer in an ideal $J \neq \{0\}$. Prove that d divides every element of J.

Since $J \neq \{0\}$, there exists $a \in J$ with $a \neq 0$. Then $-a = (-1)a \in J$. Thus J contains positive elements. By the Well-Ordering Principle, J contains a least positive integer, so d exists. Now let $b \in J$. Dividing b by d, the division algorithm tells us there exist q and r such that

$$b = qd + r \quad \text{and} \quad 0 \leq r < d$$

Now $b, d \in J$ and J is an ideal; hence $b + (-q)d = r$ also belongs to J. By the minimal property of d, we must have $r = 0$. Hence $d \mid b$, as required.

11.26. Prove Theorem 11.13: Let d be the smallest positive integer of the form $ax + by$. Then $d = \gcd(a, b)$.

Consider the set $J = \{ax + yb \mid x, y \in \mathbf{Z}\}$. Then

$$a = 1(a) + 0(b) \in J \quad \text{and} \quad b = 0(a) + 1(b) \in J$$

Also, suppose $s, t \in J$, say, $s = x_1 a + y_1 b$ and $t = x_2 a + y_2 b$. Then, for any $n \in \mathbf{Z}$, the following belong to J:

$$s + t = (x_1 + x_2)a + (y_1 + y_2)b \quad \text{and} \quad ns = (nx_1)a + (ny_1)b$$

Thus J is an ideal. Let d be the least positive element in J. We claim $d = \gcd(a, b)$.

By the preceding Problem 11.25, d divides every element of J. Thus, in particular, d divides a and b. Now suppose h divides both a and b. Then h divides $xa + yb$ for any x and y; that is, h divides every element of J. Thus h divides d, and so $h \leq d$. Accordingly, $d = \gcd(a, b)$.

11.27. Prove Theorem 11.17: Suppose $\gcd(a, b) = 1$, and a and b divide c. Then ab divides c.

Since $\gcd(a, b) = 1$, there exist x and y such that $ax + by = 1$. Since $a \mid c$ and $b \mid c$, there exist m and n such that $c = ma$ and $c = nb$. Multiplying $ax + by = 1$ by c yields

$$acx + bcy = c \quad \text{or} \quad a(nb)x + b(ma)y = c \quad \text{or} \quad ab(nx + my) = c$$

Thus ab divides c.

11.28. Prove Corollary 11.19: Suppose a prime p divides a product ab. Then $p \mid a$ or $p \mid b$.

Suppose p does not divide a. Then $\gcd(p, a) = 1$ since the only divisors of p are ± 1 and $\pm p$. Thus there exist integers m and n such that $1 = mp + nq$. Multiplying by b yields $b = mpb + nab$. By hypothesis, $p \mid ab$, say, $ab = cp$. Then:

$$b = mpb + nab = mpb + ncp = p(mb + nc).$$

Hence $p \mid b$, as required.

11.29. Prove: (a) Suppose $p \mid q$ where p and q are primes. Then $p = q$.
(b) Suppose $p \mid q_1 q_2 \cdots q_r$ where p and the q's are primes. Then p is equal to one of the q's.

(a) The only divisors of q are ± 1 and $\pm q$. Since $p > 1$, $p = q$.

(b) If $r = 1$, then $p = q_1$ by (a). Suppose $r > 1$. By Problem 11.28 (Corollary 11.19) $p \mid q_1$ or $p \mid (q_2 \cdots q_r)$. If $p \mid q_1$,
then $p = q_1$ by (a). If not, then $p \mid (q_2 \cdots q_r)$. We repeat the argument. That is, we get $p = p_2$ or $p \mid (q_3 \cdots q_r)$. Finally (or by induction) p must equal one of the q's.

11.30. Prove the Fundamental Theorem of Arithmetic (Theorem 11.20): Every integer $n > 1$ can be expressed uniquely (except for order) as a product of primes.

We already proved Theorem 11.11 that such a product of primes exists. Hence we need only show that such a product is unique (except for order). Suppose

$$n = p_1 p_2 \cdots p_k = q_1 q_2 \cdots q_r$$

where the p's and q's are primes. Note that $p_1 \mid (q_1 q_2 \cdots q_r)$. By the preceding Problem 11.29, p_1 equals one of the q's. We rearrange the q's so that $p_1 = q_1$. Then

$$p_1 p_2 \cdots p_k = p_1 q_2 \cdots q_r \quad \text{and so} \quad p_2 \cdots p_k = q_2 \cdots q_r$$

By the same argument, we can rearrange the remaining q's so that $p_2 = q_2$. And so on. Thus n can be expressed uniquely as a product of primes (except for order).

CONGRUENCES

11.31. Which of the following are true?

(a) $446 \equiv 278 \pmod{7}$, (c) $269 \equiv 413 \pmod{12}$, (e) $445 \equiv 536 \pmod{18}$

(b) $793 \equiv 682 \pmod{9}$, (d) $473 \equiv 369 \pmod{26}$, (f) $383 \equiv 126 \pmod{15}$

Recall $a \equiv b \pmod{m}$ if and only if m divides $a - b$.

(a) Find the difference $446 - 278 = 168$. Divide the difference 168 by the modulus $m = 7$. The remainder is 0; hence the statement is true.

(b) Divide the difference $793 - 682 = 111$ by the modulus $m = 9$. The remainder is not 0; hence the statement is false.

(c) True; since 12 divides $269 - 413 = -144$.

(d) True; since 26 divides $473 - 369 = 104$.

(e) False; since 18 does not divide $445 - 536 = -91$.

(f) False; since 15 does not divide $383 - 126 = 157$.

11.32. Find the smallest integer in absolute value which is congruent modulo $m = 7$ to each of the following numbers: (a) 386; (b) 257; (c) -192; (d) -466.

The integer should be in the set $\{-3, -2, -1, 0, 1, 2, 3\}$.

(a) Dividing 386 by $m = 7$ yields a remainder 1; hence $386 \equiv 1 \pmod 7$.

(b) Dividing 257 by $m = 7$ yields a remainder 5; hence $257 \equiv 5 \equiv -2 \pmod 7$. (We obtain -2 by subtracting the modulus $m = 7$ from 5.)

(c) Dividing 192 by $m = 7$ yields a remainder 3; hence $-192 \equiv -3 \pmod 7$.

(d) Dividing 466 by $m = 7$ yields a remainder 4; hence $-466 \equiv -4 \equiv 3 \pmod 7$. (We obtain 3 by adding the modulus $m = 7$ to -4.)

11.33. Find all numbers between -50 and 50 which are congruent to 21 modulo $m = 12$, that is, find all x such that $-50 \le x \le 50$ and $x \equiv 21 \pmod{12}$.

Add and subtract multiples of the modulus $m = 12$ to the given number 21 to obtain:

$$21 + 0 = 21, \quad 21 + 12 = 33, \quad 33 + 12 = 46, \quad 21 - 12 = 9$$
$$9 - 12 = -3, \quad -3 - 12 = -15, \quad -15 - 12 = -27, \quad -27 - 12 = -39$$

That is: $-39, -27, -15, -3, 9, 21, 33, 46$.

11.34. Prove Theorem 11.21: Let m be a positive integer. Then:

(i) For any integer a, we have $a \equiv a \pmod m$.

(ii) If $a \equiv b \pmod m$, then $b \equiv a \pmod m$.

(iii) If $a \equiv b \pmod m$ and $b \equiv c \pmod m$, then $a \equiv c \pmod m$.

(i) The difference $a - a = 0$ is divisible by m; hence $a \equiv a \pmod m$.

(ii) If $a \equiv b \pmod m$, then $m \mid (a - b)$. Hence m divides $-(a - b) = b - a$. Therefore, $b \equiv a \pmod m$.

(iii) We are given $m \mid (a - b)$ and $m \mid (b - c)$. Hence m divides the sum $(a - b) + (b - c) = a - c$. Therefore, $a \equiv c \pmod m$.

11.35. Prove Theorem 11.22: Suppose $a \equiv c \pmod m$ and $b \equiv d \pmod m$. Then:

(i) $a + b \equiv c + d \pmod m$.   (ii) $a \cdot b \equiv c \cdot d \pmod m$.

We are given that $m \mid (a - c)$ and $m \mid (b - d)$.

(i) Then m divides the sum $(a - c) + (b - d) = (a + b) - (c + d)$. Hence $a + b \equiv c + d \pmod m$.

(ii) Then m divides $b(a - c) = ab - bc$ and m divides $c(b - d) = bc - cd$. Thus m divides the sum $(ab - bc) + (bc - cd) = ab - cd$. Thus $ab \equiv cd \pmod m$.

11.36. Let $d = \gcd(a, b)$. Show that a/d and b/d are relatively prime.

There exists x and y such that $d = xa + yb$. Dividing the equation by d, we get $1 = x(a/d) + y(b/d)$. Hence a/d and b/d are relatively prime.

11.37. Prove Theorem 11.24: Suppose $ab \equiv ac \pmod m$ and $d = \gcd(a, m)$. Then $b \equiv c \pmod{m/d}$.

By hypothesis, m divides $ab - ac = a(b - c)$. Hence, there is an integer x such that $a(b - c) = mx$. Dividing by d yields $(a/d)(b - c) = (m/d)x$. Thus m/d divides $(a/d)(b - c)$. Since m/d and a/d are relatively prime, m/d divides $b - c$. That is, $b \equiv c \pmod{m/d}$, as required.

RESIDUE SYSTEMS, EULER PHI FUNCTION ϕ

11.38. For each modulo m, exhibit two complete residue systems, one consisting of the smallest nonnegative integers, and the other consisting of the integers with the smallest absolute values: (a) $m = 9$; (b) $m = 12$.

In the first case choose $\{0, 1, 2, \ldots, m - 1\}$, and in the second case choose

$$\{-(m-1)/2, \ldots, -1, 0, 1, \ldots, (m-1)/2\} \quad \text{or} \quad \{-(m-2)/2, \ldots, -1, 0, 1, \ldots, m/2\}$$

according as m is even or odd:

(a) $\{0, 1, 2, 3, 4, 5, 6, 7, 8\}$ and $\{-4, -3, -2, -1, 0, 1, 2, 3, 4\}$

(b) $\{0, 1, 2, 3, 4, 5, 6, 7, 8, 9, 10, 11\}$ and $\{-5, -4, -3, -2, -1, 0, 1, 2, 3, 4, 5, 6\}$.

11.39. Find a reduced residue system modulo m and $\phi(m)$ where: (a) $m = 9$; (b) $m = 16$; (c) $m = 7$.

Choose those positive numbers less than m and relatively prime to m. The number of such numbers is $\phi(m)$.

(a) $\{1, 2, 4, 5, 7, 8\}$; hence $\phi(9) = 6$.

(b) $\{1, 3, 5, 7, 9, 11, 13, 15\}$; hence $\phi(16) = 8$.

(c) $\{1, 2, 3, 4, 5, 6\}$; hence $\phi(7) = 6$. (This is expected since $\phi(p) = p - 1$ for any prime p.)

11.40. Recall $S_m = 0, 1, 2, \ldots, m - 1$ is a complete residue system modulo m. Prove:

(a) Any m consecutive integers is a complete residue system modulo m.

(b) If $\gcd(a, m) = 1$, then $aS_m = \{0, a, 2a, 3a, \ldots, (m-1)a\}$ is a complete residue system modulo m.

(a) Consider any other sequence of m integers, say $\{a, a + 1, a + 2, \ldots, a + (m - 1)\}$. The absolute value of the difference s of any two of the integers is less than m. Thus m does not divide s, and so the numbers are incongruent modulo m.

(b) Suppose $ax \equiv ay \pmod{m}$ where $x, y \in S_m$. Since $\gcd(a, m) = 1$, the modified cancellation law Theorem 11.24 tells us $x \equiv y \pmod{m}$. Since $x, y \in S_m$, we must have $x = y$. That is, aS_m is a complete residue system modulo m.

11.41. Exhibit a complete residue system modulo $m = 8$ consisting entirely of multiples of 3.

By Problem 11.40(b), $3S_8 = \{0, 3, 6, 9, 12, 15, 18, 21\}$ is a complete residue system modulo $m = 8$.

11.42. Show that if p is a prime, then $\phi(p^n) = p^n - p^{n-1} = p^{n-1}(p - 1)$.

Clearly, $\gcd(a, p^n) \neq 1$ if and only if p divides a. Thus the only numbers between 1 and p^n which are not relatively prime to p^n are the multiples of p, that is, $p, 2p, 3p, \ldots, p^{n-1}(p)$. There are p^{n-1} such multiples of p. All the other numbers between 1 and p^n are relatively prime to p^n. Thus, as claimed:

$$\phi(p^n) = p^n - p^{n-1} = p^{n-1}(p - 1).$$

11.43. Find: (a) $\phi(81)$, $\phi(7^6)$; (b) $\phi(72)$, $\phi(3000)$.

(a) By Problem 11.42,

$$\phi(81) = \phi(3^4) = 3^3(3 - 1) = 27(2) = 54 \quad \text{and} \quad \phi(7^6) = 7^5(7 - 1) = 6(7^5)$$

(b) Use Theorem 11.24 that ϕ is multiplicative:

$$\phi(72) = \phi(3^2 \cdot 2^3) = \phi(3^2)\phi(2^3) = 3(3 - 1) \cdot 2^2(2 - 1) = 24$$
$$\phi(3000) = \phi(3 \cdot 2^2 \cdot 5^3) = \phi(3)\phi(2^2)\phi(5^3) = 2 \cdot 2 \cdot 5^2(5 - 1) = 400$$

11.44. Prove Theorem 11.25: If a and b are relatively prime, then $\phi(ab) = \phi(a)\phi(b)$.

Let a and b be coprime (relatively prime) positive integers, and let S be the set of numbers from 1 to ab arranged in an array as in Fig. 11-7. That is, the first row of S is the list of numbers from 1 to a, the second row is the list of numbers from $a + 1$ to $2a$, and so on. Since a and b are coprime, any integer x is coprime to ab if and only if it is coprime to both a and b. We find the number of such integers x in the array S.

Since $na + k \equiv k \pmod{a}$, each column in S belongs to the same residue class modulo a. Therefore, any integer x in S is coprime to a if and only if x belongs to a column headed by some integer k which is coprime to a. On the other hand, there are $\phi(a)$ such columns since the first row is a residue system modulo a.

$$
\begin{array}{cccccccc}
1 & 2 & 3 & \dots & k & \dots & a \\
a+1 & a+2 & a+3 & \dots & a+k & \dots & 2a \\
2a+1 & 2a+2 & 2a+3 & \dots & 2a+k & \dots & 3a \\
\hline
(b-1)a+1 & & \dots\dots\dots & & (b-1)a+k & \dots & ba
\end{array}
$$

Fig. 11-7

Now let us consider an arbitrary column in the array S which consists of the numbers:

$$k, \quad a+k, \quad 2a+k, \quad 3a+k, \dots, (b-1)a+k \tag{11.11}$$

By Problem 11.10, these b integers form a residue system modulo b, that is, no two of the integers are congruent modulo b. Therefore, (11.11) contains exactly $\phi(b)$ integers which are coprime to b. We have shown that the array S contains $\phi(a)$ columns consisting of those integers which are coprime to a, and each such column contains $\phi(b)$ integers which are coprime to b. Thus there are $\phi(a)\,\phi(b)$ integers in the array S which are coprime to both a and b and hence are coprime to ab. Accordingly, as required

$$\phi(ab) = \phi(a)\phi(b)$$

ARITHMETIC MODULO m, $\mathbf{Z}_m$

11.45. Exhibit the addition and multiplication tables for: (a) $\mathbf{Z}_4$;   (b) $\mathbf{Z}_7$

(a) See Fig. 11-8. (b) See Fig. 11-9.

+	0	1	2	3
0	0	1	2	3
1	1	2	3	0
2	2	3	0	1
3	3	0	1	2

×	0	1	2	3
0	0	0	0	0
1	0	1	2	3
2	0	2	0	2
3	0	3	2	1

Fig. 11-8

+	0	1	2	3	4	5	6
0	0	1	2	3	4	5	6
1	1	2	3	4	5	6	0
2	2	3	4	5	6	0	1
3	3	4	5	6	0	1	2
4	4	5	6	0	1	2	3
5	5	6	0	1	2	3	4
6	6	0	1	2	3	4	5

×	0	1	2	3	4	5	6
0	0	0	0	0	0	0	0
1	0	1	2	3	4	5	6
2	0	2	4	6	1	3	5
3	0	3	6	2	5	1	4
4	0	4	1	5	2	6	3
5	0	5	3	1	6	4	2
6	0	6	5	4	3	2	1

Fig. 11-9

11.46. In $\mathbf{Z}_{11}$, find: (a) $-2, -5, -9, -10$; (b) $2/7, 3/7, 5/7, 8/7, 10/7, 1/7$.

(a) Note $-a = m - a$ since $(m - a) + a = 0$. Therefore:

$$-2 = 11 - 2 = 9, \quad -5 = 11 - 5 = 6, \quad -9 = 11 - 9 = 2, \quad -10 = 11 - 10 = 1$$

(b) By definition a/b is the integer c such that $bc = a$. Since we are dividing by 7, first compute the multiplication table for 7 in $\mathbf{Z}_{11}$ as in Fig. 11-10. Now find the number inside the table, and the answer will be above this number. Thus:

×	0	1	2	3	4	5	6	7	8	9	10
7	0	7	3	10	6	2	9	5	1	8	4

Fig. 11-10

$$2/7 = 5, \quad 3/7 = 2, \quad 5/7 = 7, \quad 8/7 = 9, \quad 10/7 = 3, \quad 1/7 = 8$$

Note that $7^{-1} = 8$ since $7(8) = 8(7) = 1$.

11.47. Consider $\mathbf{Z_p}$ where p is a prime. Prove:

(a) If $ab = ac$ and $a \neq 0$, then $b = c$;

(b) If $ab = 0$, then $a = 0$ or $b = 0$.

(a) If $ab = ac$ in $\mathbf{Z}_p$, then $ab \equiv ac \pmod{p}$. Since $a \neq 0$, $\gcd(a, p) = 1$. By Theorem 11.23 we can cancel the a's to obtain $b \equiv c \pmod{p}$. Therefore $b = c$ in $\mathbf{Z_p}$.

(b) If $ab = 0$ in $\mathbf{Z_p}$, then $ab = 0 \pmod{p}$. Therefore, p divides the product ab. Since p is a prime, $p \mid a$ or $p \mid b$; that is, $a \equiv 0 \pmod{p}$ or $b \equiv 0 \pmod{p}$. Thus $a = 0$ or $b = 0$ in $\mathbf{Z_p}$.

11.48. Consider $a \neq 0$ in $\mathbf{Z}_m$ where $\gcd(a, m) = 1$. Show that a has a multiplicative inverse in $\mathbf{Z}_m$.

Since $a \neq 0$ and $\gcd(a, m) = 1$, there exists integers x and y such that $ax + my = 1$ or $ax - 1 = my$. Thus m divides $ax - 1$ and hence $ax = 1 \pmod{m}$. Reduce x modulo m to an element x' in $\mathbf{Z}_m$. Then $ax' = 1$ in $\mathbf{Z}_m$.

11.49. Find a^{-1} in $\mathbf{Z}_m$ where: (a) $a = 37$ and $m = 249$;   (b) $a = 15$ and $m = 234$.

(a) First find $d = \gcd(37, 249)$ obtaining $d = 1$. Then, as in Example 11.6, find x and y such that $ax + my = 1$. This yields $x = -74$ and $y = 14$. That is,

$$-74(37) + 11(249) = 1 \quad \text{so} \quad -74(37) \equiv 1 \pmod{249}$$

Add $m = 249$ to -74 to obtain $-74 + 249 = 175$. Thus $(175)(37) \equiv 1 \pmod{249}$.
Accordingly, $a^{-1} = 175$ in $\mathbf{Z}_{249}$.

(b) First find $d = \gcd(15, 234)$ obtaining $d = 3$. Thus $d \neq 1$, and hence 15 has no multiplicative inverse in $\mathbf{Z}_{234}$.

11.50. For the following polynomials over $\mathbf{Z}_7$ find: (a) $f(x) + g(x)$ and (b) $f(x)h(x)$.

$$f(x) = 6x^3 - 5x^2 + 2x - 4, \quad g(x) = 5x^3 + 2x^2 + 6x - 1, \quad h(x) = 3x^2 - 2x - 5$$

Perform the operations as if the polynomials were over the integers $\mathbf{Z}$, and then reduce the coefficients modulo 7.

(a) We get: $f(x) + g(x) = 11x^3 - 3x^2 + 8x - 5 = 4x^3 - 3x^2 + x - 5 = 4x^3 + 4x^2 + x + 2$

(b) First find the product $f(x)h(x)$ as in Fig. 11-11. Then, reducing modulo 7, we obtain:

$$f(x)h(x) = 4x^5 - 6x^4 + 2x^2 - 2x + 6 = 4x^5 + x^4 + 2x^2 + 5x + 6$$

$$6x^3 - 5x^2 + 2x - 4$$
$$\underline{3x^2 - 2x - 5}$$
$$18x^5 - 15x^4 + 6x^3 - 12x^2$$
$$-12x^4 + 10x^3 - 4x^2 + 8x$$
$$\underline{-30x^3 + 25x^2 - 10x + 20}$$
$$18x^5 - 27x^4 - 14x^3 + 9x^2 - 2x + 20$$

Fig. 11-11

CONGRUENCE EQUATIONS

11.51. Solve the congruence equation $f(x) = 4x^4 - 3x^3 + 2x^2 + 5x - 4 \equiv 0 \pmod 6$.

Since the equation is not linear, we solve the equation by testing the numbers in a complete residue system modulo 6, say, $\{0, 1, 2, 3, 4, 5\}$. We have:

$$f(0) = -4 \not\equiv 0 \pmod 6, \quad f(2) = 54 \equiv 0 \pmod 6, \qquad f(4) = 880 \equiv 4 \not\equiv 0 \pmod 6$$
$$f(1) = 4 \not\equiv 0 \pmod 6, \quad\ f(3) = 272 \equiv 2 \not\equiv 0 \pmod 6, \quad f(5) = 2196 \equiv 0 \pmod 6$$

Thus only 2 and 5 are roots of $f(x)$ modulo 6. That is, $\{2, 5\}$ is a complete set of solutions.

11.52. Solve the congruence equation $f(x) = 26x^4 - 31x^3 + 46x^2 - 76x + 57 \equiv 0 \pmod 8$.

First we reduce the coefficients of $f(x)$ modulo 8 to obtain the equivalent congruence equation

$$g(x) = 2x^4 - 7x^3 + 6x^2 - 4x + 1 \equiv 0 \pmod 8$$

Since $7 \equiv -1 \pmod 8$ and $6 = -2 \pmod 8$, we can further simplify our original equation to obtain the equivalent congruence equation

$$h(x) = 2x^4 + x^3 - 2x^2 - 4x + 1 \equiv 0 \pmod 8$$

We test the numbers in a complete residue system modulo 8 and, in order to keep our arithmetic as simple as possible, we choose $\{-3, -2, -1, 0, 1, 2, 3, 4\}$. (That is, we chose those numbers whose absolute value is minimal.) Substituting these numbers in $h(x)$ we obtain:

$$h(-3) = 130 \equiv 2 \pmod 8, \quad h(0) = 1 \equiv 1 \pmod 8, \qquad h(3) = 160 \equiv 0 \pmod 8$$
$$h(-2) = 9 \equiv 1 \pmod 8, \quad\ h(1) = -2 \equiv 6 \pmod 8, \quad h(4) = 529 \equiv 1 \pmod 8$$
$$h(-1) = 4 \equiv 4 \pmod 8, \quad\ h(2) = 25 \equiv 1 \pmod 8,$$

Thus 3 is the only solution of $f(x) \pmod 8$.

11.53. Solve each linear congruence equation:

(a) $3x \equiv 2 \pmod 8$; (b) $6x \equiv 5 \pmod 9$; (c) $4x \equiv 6 \pmod{10}$

Since the moduli are relatively small, we find all the solutions by testing. Recall $ax \equiv b \pmod m$ has exactly $d = \gcd(a, m)$ solution providing d divides b.

(a) Here $\gcd(3, 8) = 1$, hence the equation has a unique solution. Testing $0, 1, 2, \ldots, 7$, we find that $3(6) = 18 \equiv 2 \pmod 8$. Thus 6 is the unique solution.

(b) Here $\gcd(6, 9) = 3$, but 3 does not divide 5. Hence the system has no solution.

(c) Here $\gcd(4, 10) = 2$ and 2 divides 6; hence the system has two solutions. Testing $0, 1, 2, 3, \ldots, 9$, we see that

$$4(4) = 16 \equiv 6 \pmod{10} \quad \text{and} \quad 4(9) = 36 \equiv 6 \pmod{10}$$

Hence 4 and 9 are our two solutions.

11.54. Solve the congruence equation $1092x \equiv 213 \pmod{2295}$.

Testing is not an efficient way to solve this equation since the modulus $m = 2295$ is large. First use the Euclidean algorithm to find $d = \gcd(1092, 2295) = 3$. Dividing 213 by $d = 3$ yields 0 as a remainder; that is, 3 does divide 213. Thus the equation will have three (incongruent) solutions.

Divide the equation and the modulus $m = 2295$ by $d = 3$ to obtain the congruence equation

$$364x \equiv 71 \ (\text{mod} \ 765) \tag{11.12}$$

We know that 364 and 796 are relatively prime since we divided by $d = \gcd(1092, 2295) = 3$; hence the equation (11.12) has unique solution modulo 765. We solve (11.12) by first finding the solution of the equation

$$364x \equiv 1 \ (\text{mod} \ 765) \tag{11.13}$$

This solution is obtained by finding s and t such that

$$364s + 765t = 1$$

Using the Euclidean algorithm and "unraveling" as in Example 11.6 and Problem 11.21, we obtain $s = 124$ and $t = -59$.

Accordingly, $s = 124$ is the unique solution of (11.13). Multiplying this solution $s = 124$ by 71 and reducing modulo 765 we obtain

$$124(71) = 8804 \equiv 389 \ (\text{mod} \ 765)$$

This is the unique solution of (11.12).

Lastly, we add the new modulus $m = 765$ to the solution $x_1 = 389$ two times to obtain the other two solutions of the given equation:

$$x_2 = 389 + 765 = 1154, \quad x_3 = 1154 + 765 = 1919$$

In other words, $x_1 = 389$, $x_2 = 1154$, $x_3 = 1919$ form a complete set of solutions of the given congruence equation $1092x \equiv 213 \ (\text{mod} \ 2295)$.

11.55. Solve the congruence equation $455x \equiv 204 \ (\text{mod} \ 469)$.

First use the Euclidean algorithm to find $d = \gcd(455, 469) = 7$. Dividing 204 by $d = 7$ yields 1 as a remainder; that is, 7 does not divide 204. Thus the equation has no solution.

11.56. Find the smallest positive integer x such that when x is divided by 3 it yields a remainder 2, when x is divided by 7 it yields a remainder 4, and when x is divided by 10 it yields a remainder 6.

We seek the smallest positive common solution of the following three congruence equations:

$$(a) \ x \equiv 2 (\text{mod} \ 3); \quad (b) \ x \equiv 4 \ (\text{mod} \ 7); \quad (c) \ x \equiv 6 \ (\text{mod} \ 10)$$

Observe that the moduli 3, 7, and 10 are pairwise relatively prime. (Moduli is the plural of modulus.) The Chinese Remainder Theorem (CRT), Theorem 11.29, tells us that there is a unique solution modulo the product $m = 3(7)(10) = 210$. We solve the problems in two ways.

Method 1: First we apply CRT to the first two equations,

$$(a) \ x \equiv 2 \ (\text{mod} \ 3) \quad \text{and} \quad (b) \ x \equiv 4 \ (\text{mod} \ 7)$$

We know there is a unique solution modulo $M = 3 \cdot 7 = 21$. Adding multiples of the modulus $m = 7$ to the given solution $x = 4$ of the second equation (b), we obtain the following three solutions of (b) which are less than 21:

$$4, \quad 11, \quad 18$$

Testing each of these solutions of (b) in the first equation (a), we find that 11 is the only solution of both equations.

Now we apply the same process to the two equations

$$(c) \ x \equiv 6 \ (\text{mod} \ 10) \quad \text{and} \quad (d) \ x \equiv 11 \ (\text{mod} \ 21)$$

CRT tells us there is a unique solution modulo $M = 21 \cdot 10 = 210$. Adding multiples of the modulus $m = 21$ to the given solution $x = 11$ of the equation (d), we obtain the following 10 solutions of (d) which are less than 210:

$$11, \ 32, \ 53, \ 74, \ 95, \ 116, \ 137, \ 158, \ 179, \ 210$$

Testing each of these solutions of (d) in the equation (c), we find that $x = 116$ is the only solution of equation (c). Accordingly, $x = 116$ is the smallest positive integer satisfying all of the three given equations (a), (b), and (c).

Method 2: Using the notation of Proposition 11.30, we obtain

$$M = 3 \cdot 7 \cdot 10 = 210, \quad M_1 = 210/3 = 70, \quad M_2 = 210/7 = 30, \quad M_3 = 210/10 = 21$$

We now seek solutions to the equations

$$70x \equiv 1 \ (\text{mod} \ 3), \quad 30x \equiv 1 \ (\text{mod} \ 7), \quad 21x \equiv 1 \ (\text{mod} \ 10)$$

Reducing 70 modulo 3, reducing 30 modulo 7, and reducing 21 modulo 10, we obtain the equivalent system

$$x \equiv 1 \ (\text{mod} \ 3), \quad 2x \equiv 1 \ (\text{mod} \ 7), \quad x \equiv 1 \ (\text{mod} \ 10)$$

The solutions of these three equations are, respectively,

$$s_1 = 1, \quad s_2 = 4, \quad s_3 = 1$$

Substituting into the formula

$$x_0 = M_1 s_1 r_1 + M_2 s_2 r_2 + \cdots + M_k s_k r_k$$

we obtain the following solution of our original system:

$$x_0 = 70 \cdot 1 \cdot 2 + 30 \cdot 4 \cdot 4 + 21 \cdot 1 \cdot 6 = 746$$

Dividing this solution by the modulus $M = 210$, we obtain the remainder $x = 116$ which is the unique solution of the original system between 0 and 210.

11.57. Prove Theorem 11.26: If a and m are relatively prime, then $ax \equiv 1$ (mod m) has a unique solution; otherwise it has no solution.

Suppose x_0 is a solution. Then m divides $ax_0 - 1$, and hence there exists y_0 such that $my_0 = ax_0 - 1$. Therefore

$$ax_0 + my_0 = 1 \tag{11.14}$$

and a and m are coprime (relatively prime). Conversely, if a and m are coprime, then there exist x_0 and y_0 satisfying (11.14), in which case x_0 is a solution of $ax \equiv 1$ (mod m).

It remains to prove that x_0 is a unique solution modulo m. Suppose x_1 is another solution. Then

$$ax_0 \equiv 1 \equiv ax_1 \, (\text{mod} \ m)$$

Since a and m are coprime, the Modified Cancellation Law holds here, so

$$x_0 \equiv x_1 \ (\text{mod} \ m)$$

Thus the theorem is proved.

11.58. Prove Theorem 11.27: Suppose a and m are relatively prime. Then $ax \equiv b$ (mod m) has a unique solution. Moreover, if s is the unique solution of $ax \equiv 1$ (mod m), then $x = bs$ is the unique solution to $ax \equiv b$ (mod m).

By Theorem 11.26 (proved in Problem 11.57), a unique solution s of $ax \equiv 1$ (mod m) exists. Hence $as \equiv 1$ (mod m) and so

$$a(bs) = (as)b \equiv 1 \cdot b = b \ (\text{mod} \ m)$$

That is, $x = bs$ is a solution of $ax \equiv b$ (mod m). Suppose x_0 and x_1 are two such solutions. Then

$$ax_0 \equiv b \equiv ax_1 \ (\text{mod} \ m)$$

Since a and m are coprime, the Modified Cancellation Law tells us that $x_0 \equiv x_1$ (mod m). That is, $ax \equiv b$ (mod m) has a unique solution modulo m.

11.59. Prove Theorem 11.28: Consider the following equation where $d = \gcd(a, m)$:

$$ax \equiv b \pmod{m} \qquad\qquad (11.15)$$

(i) Suppose d does not divide b. Then (11.15) has no solution.

(ii) Suppose d does divide b. Then (11.15) has d solutions which are all congruent modulo M to the unique solution of the following equation where $A = a/d$, $B = b/d$, $M = m/d$:

$$Ax \equiv B \pmod{M} \qquad\qquad (11.16)$$

(i) Suppose x_0 is a solution of (11.15). Then $ax_0 \equiv b \pmod{m}$, and so m divides $ax_0 - b$. Thus there exists an integer y_0 such that $my_0 = ax_0 - b$ or $my_0 + ax_0 = b$. But $d = \gcd(a, m)$, and so d divides $my_0 + ax_0$. That is, d divides b. Accordingly, if d does not divide b, then no solution exists.

(ii) Suppose x_0 is a solution of (11.15). Then, as above,

$$my_0 + ax_0 = b$$

Dividing through by d we get (11.16). Hence M divides $Ax_0 - B$ and so x_0 is a solution of (11.16). Conversely, suppose x_1 is a solution of (11.16). Then, as above, there exists an integer y_1 such that

$$My_1 + Ax_1 = B$$

Multiplying through by d yields

$$dMy_1 + dAx_1 = dB \quad \text{or} \quad my_1 + ax_1 = b$$

Therefore m divides $ax_1 - b$ whence x_1 is a solution of (11.15). Thus (11.16) has the same integer solution. Let x_0 be the unique smallest positive solutions of (11.16). Since $m = dM$,

$$x_0, \quad x_0 + M, \quad x_0 + 2M, \quad x_0 + 3M, \quad \dots, \quad x_0 + (d-1)M$$

are precisely the solution of (11.16) and (11.15) between 0 and m. Thus (11.15) has d solutions modulo m, and all are congruent to x_0 modulo M.

11.60. Prove the Chinese Remainder Theorem (Theorem 11.29:) Given the system:

$$x \equiv r_1 \pmod{m_1}, \quad x \equiv r_2 \pmod{m_2}, \quad \dots, \quad x \equiv r_k \pmod{m_k} \qquad (11.17)$$

where the m_i are pairwise relatively prime. Then the system has a unique solution modulo $M = m_1 m_2 \cdots m_k$.

Consider the integer

$$x_0 = M_1 s_1 r_1 + M_2 s_2 r_2 + \cdots + M_k s_k r_k$$

where $M_i = M/m_i$ and s_i is the unique solution of $M_i x \equiv 1 \pmod{m_i}$. Let j be given.

For $i \neq j$, we have $m_j | M_i$ and hence

$$M_i s_i r_i \equiv 0 \pmod{m_j}$$

On the other hand, $M_j S_j \equiv 1 \pmod{m_j}$; and hence

$$M_j s_j r_j \equiv r_j \pmod{m_j}$$

Accordingly,

$$x_0 \equiv 0 + \cdots + 0 + r_j + 0 + \cdots + 0 \equiv r_j \pmod{m_j}$$

In other words, x_0 is a solution of each of the equations in (11.17).

It remains to show that x_0 is the unique solution of the system (11.17) modulo M.

Suppose x_1 is another solution of all the equations in (11.17). Then:

$$x_0 \equiv x_1 \pmod{m_1}, \quad x_0 \equiv x_1 \pmod{m_2}, \quad \cdots, \quad x_0 \equiv x_1 \pmod{m_k}$$

Hence $m_i | (x_0 - x_1)$, for each i. Since the m_i are relatively prime, $M = \operatorname{lcm}(m_1, m_2, \dots, m_k)$ and so $M | (x_0 - x_1)$. That is, $x_0 = x_1 \pmod{M}$. Thus the theorem is proved.

Supplementary Problems

ORDER AND INEQUALITIES, ABSOLUTE VALUE

11.61. Insert the correct symbol, $<$, $>$, or $=$, between each pair of integers:

    (a) 2 ___ -6;   (c) -7 ___ 3;   (e) 2^3 ___ 11;   (g) -2 ___ -7;

    (b) -3 ___ -5;   (d) -8 ___ -1;   (f) 2^3 ___ -9;   (h) 4 ___ -9.

11.62. Evaluate: (a) $|3 - 7|, |-3 + 7|, |-3 - 7|$; (b) $|2 - 5| + |3 + 7|, |1 - 4| - |2 - 9|$;

    (c) $|5 - 9| + |2 - 3|, |-6 - 2| - |2 - 6|$.

11.63. Find the distance d between each pair of integers: (a) 2 and -5; (b) -6 and 3; (c) 2 and 8; (d) -7 and -1;

    (e) 3 and -3; (f) -7 and -9.

11.64. Find all integers n such that: (a) $3 < 2n - 4 < 10$;   (b) $1 < 6 - 3n < 13$.

11.65. Prove Proposition 11.1: (i) $a \leq a$, for any integer a; (ii) If $a \leq b$ and $b \leq a$, then $a = b$.

11.66. Prove Proposition 11.2: For any integers a and b, exactly one of the following holds: $a < b, a = b,$ or $a > b$.

11.67. Prove: (a) $2ab \leq a^2 + b^2$;   (b) $ab + ac + bc \leq a^2 + b^2 + c^2$.

11.68. Proposition 11.4:   (i) $|a| \geq 0$, and $|a| = 0$ iff $a = 0$;   (ii) $-|a| \leq a \leq |a|$;   (iii) $||a| - |b|| \leq |a \pm b|$.

11.69. Show that $a - xb \geq 0$ if $b \neq 0$, and $x = -|a|b$.

MATHEMATICAL INDUCTION, WELL-ORDERING PRINCIPLE

11.70. Prove the proposition that the sum of the first n even positive integers is $n(n + 1)$; that is,

$$P(n): 2 + 4 + 6 + \cdots + 2n- = n(n + 1)$$

11.71. Prove that the sum of the first n cubes is equal to the square of the sum of the first n positive integers:

$$P(n): 1^3 + 2^3 + 3^3 + \cdots + n^3 = (1 + 2 + \cdots + n)^2$$

11.72. Prove: $1 + 4 + 7 + \cdots + (3n - 2) = n(3n - 1)/2$

11.73. Prove: (a) $a^n a^m = a^{n+m}$; (b) $(a^n)^m = a^{nm}$; (c) $(ab)^n = a^n b^n$

11.74. Prove: $\frac{1}{1 \cdot 2} + \frac{1}{2 \cdot 3} + \frac{1}{3 \cdot 4} + \cdots + \frac{1}{n(n+1)} = \frac{n}{n+1}$

11.75. Prove: $\frac{1}{1 \cdot 3} + \frac{1}{3 \cdot 5} + \frac{1}{5 \cdot 7} + \cdots + \frac{1}{(2n-1)(2n+1)} = \frac{n}{2n+1}$

11.76. Prove: $\frac{1^2}{1 \cdot 3} + \frac{2^2}{3 \cdot 5} + \frac{3^2}{5 \cdot 7} + \cdots + \frac{n^2}{(2n-1)(2n+1)} = \frac{n(n+1)}{2(2n+1)}$

11.77. Prove: $x^{n+1} - y^{n+1} = (x - y)(x^n + x^{n-1}y + x^{n-2}y^2 + \cdots + y^n)$

11.78. Prove: $|P(A)| = 2^n$ where $|A| = n$. (Here $P(A)$ is the power set of the set A with n elements.)

DIVISION ALGORITHM

11.79. For each pair of integers a and b, find integers q and r such that $a = bq + r$ and $0 \leq r < |b|$:

    (a) $a = 608$ and $b = -17$;   (b) $a = -278$ and $b = 12$;   (c) $a = -417$ and $b = -8$.

11.80. Prove each of the following statements:

    (a) Any integer a is of the form $5k, 5k + 1, 5k + 2, 5k + 3,$ or $5k + 4$.

    (b) One of five consecutive integers is a multiple of 5.

11.81. Prove each of the following statements:

    (a) The product of any three consecutive integers is divisible by 6.

    (b) The product of any four consecutive integers is divisible by 24.

11.82. Show that each of the following numbers is not rational: (a) $\sqrt{3}$; (b) $\sqrt[3]{2}$.

11.83. Show that $\sqrt{p}$ is not rational, where p is any prime number.

DIVISIBILITY, GREATEST COMMON DIVISORS, PRIMES

11.84. Find all possible divisors of: (a) 24;   (b) $19683 = 3^9$;   (c) $432 = 2^4 \cdot 3^3$.

11.85. List all prime numbers between 100 and 150.

11.86. Express as a product of prime numbers: (a) 2940;   (b) 1485;   (c) 8712;   (d) 319 410.

11.87. For each pair of integers a and b, find $d = \gcd(a, b)$ and find m and n such that $d = ma + nb$:

   (a) $a = 356, b = 48$;   (b) $a = 1287, b = 165$;   (c) $a = 2310, b = 168$;   (d) $a = 195, b = 968$;
   (e) $a = 249, b = 37$.

11.88. Find: (a) lcm$(5, 7)$;   (b) lcm$(3, 33)$;   (c) lcm $(12, 28)$.

11.89. Suppose $a = 5880$ and $b = 8316$. (a) Express a and b as products of primes.
   (b) Find $\gcd(a, b)$ and lcm(a, b). (c) Verify that lcm$(a, b) = |ab| / \gcd(a, b)$.

11.90. Prove: (a) If $a|b$, then (i) $a| - b$, (ii) $-a|b$, (iii) $-a| - b$; (b) If $ab|ac$, then $b|c$.

11.91. Prove: (a) If $n > 1$ is composite, then n has a positive divisor d such that $d \le \sqrt{n}$.   (b) If $n > 1$ is not divisible
   by a prime $p \le \sqrt{n}$, then n is a prime.

11.92. Prove: (a) If $am + bn = 1$, then $\gcd(a, b) = 1$;   (b) If $a = bq + r$, then $\gcd(a, b) = \gcd(b, r)$.

11.93. Prove: (a) $\gcd(a, a + k)$ divides k;   (b) $\gcd(a, a + 2)$ equals 1 or 2.

11.94. Prove: (a) If $a > 2$ and $k > 1$, then $a^k - 1$ is composite.   (b) If $n > 0$ and $2^n - 1$ is prime, then n is prime.

11.95. Let n be a positive integer. Prove:

   (a) 3 divides n if and only if 3 divides the sum of the digits of n.

   (b) 9 divides n if and only if 9 divides the sum of the digits of n.

   (c) 8 divides n if and only if 8 divides the integer formed by the last three digits of n.

11.96. Extend the definition of gcd and lcm to any finite set of integers, that is, for integers $a_1, a_2, \ldots, a_k$, define:
   (a) $\gcd(a_1, a_2, \ldots, a_k)$;   (b) lcm$(a_1, a_2, \ldots, a_k)$.

11.97. Prove: If $a_1|n, a_2|n, \ldots, a_k|n$, then $m|n$ where $m = $ lcm$(a_1, a_2, \ldots, a_k)$.

11.98. Prove: There are arbitrarily large gaps between prime numbers, that is, for any positive integer k, there exist k
   consecutive composite (nonprime) integers.

CONGRUENCES

11.99. Which of the following are true?
   (a) $224 \equiv 762 \pmod 8$;   (b) $582 \equiv 263 \pmod{11}$;   (c) $156 \equiv 369 \pmod 7$;   (d) $-238 \equiv 483 \pmod{13}$.

11.100. Find the smallest nonnegative integer which is congruent modulo $m = 9$ to each of the following numbers:
   (a) 457; (b) 1578; (c) -366; (d) -3288. (The integer should be in the set $\{0, 1, 2, \ldots, 7, 8\}$).

11.101. Find the smallest integer in absolute value which is congruent modulo $m = 9$ to each of the following numbers:
   (a) 511; (b) 1329; (c) -625; (d) -2717. (The integer should be in the set $\{-4, -3, -2, -1, 0, 1, 2, 3, 4\}$).

11.102. Find all numbers between 1 and 100 which are congruent to 4 modulo $m = 11$.

11.103. Find all numbers between -50 and 50 which are congruent to 12 modulo $m = 9$.

RESIDUE SYSTEMS, EULER PHI FUNCTION ϕ

11.104. For each modulo m, exhibit two complete residue systems, one consisting of the smallest nonnegative integers, and
   the other consisting of the integers with the smallest absolute values: (a) $m = 11$; (b) $m = 14$.

11.105. Exhibit a reduced residue system modulo m and find $\phi(m)$ where: (a) $m = 4$;   (b) $m = 11$;   (c) $m = 14$;
   (d) $m = 15$.

11.106. Exhibit a complete residue system modulo $m = 8$ consisting entirely of: (a) multiples of 5; (b) powers of 3.

11.107. Show that $\{1^2, 2^2, 3^2, \ldots, m^2\}$ is not a complete residue system modulo m for $m > 2$.

11.108. Find: (a) $\phi(10)$;   (b) $\phi(12)$;   (c) $\phi(15)$;   (d) $\phi(3^7)$;   (e) $\phi(5^6)$;   (f) $\phi(2^4 \cdot 7^6 \cdot 13^3)$.

11.109. Find the number s of positive integers less than 3200 which are coprime to 8000.

11.110. Consider an arbitrary column in the array S in Fig. 11-7 which consists of the numbers:

$$k, a + k, 2a + k, 3a + k, \ldots, (b - 1)a + k$$

Show that these b integers form a residue system modulo b.

ARITHMETIC MODULO m, Z_m

11.111. Exhibit the addition and multiplication tables for: (a) $\mathbf{Z}_2$; (b) $\mathbf{Z}_8$.

11.112. In. $\mathbf{Z}_{13}$, find: (a) $-2, -3, -5, -9, -10, -11$; (b) 2/9, 4/9, 5/9, 7/9, 8/9.

11.113. In $\mathbf{Z}_{17}$, find: (a) $-3, -5, -6, -8, -13, -15, -16$; (b) 3/8, 5/8, 7/8, 13/8, 15/8.

11.114. Find a^{-1} in $\mathbf{Z}_m$ Where: (a) $a = 15, m = 127$;   (b) $a = 61, m = 124$;   (c) $a = 12, m = 111$.

11.115. Find the product $f(x)g(x)$ for the following polynomials over $\mathbf{Z}_5$:

$$f(x) = 4x^3 - 2x^2 + 3x - 1, g(x) = 3x^2 - x - 4$$

CONGRUENCE EQUATIONS

11.116. Solve each congruence equation:

(a) $f(x) = 2x^3 - x^2 + 3x + 1 \equiv 0 \pmod 5$

(b) $g(x) = 3x^4 - 2x^3 + 5x^2 + x + 2 \equiv 0 \pmod 7$

(c) $h(x) = 45x^3 - 37x^2 + 26x + 312 \equiv 0 \pmod 6$

11.117. Solve each linear congruence equation:

(a) $7x \equiv 3 \pmod 9$;   (b) $4x \equiv 6 \pmod{14}$;   (c) $6x \equiv 4 \pmod 9$.

11.118. Solve each linear congruence equation:

(a) $5x \equiv 3 \pmod 8$;   (b) $6x \equiv 9 \pmod{16}$;   (c) $9x \equiv 12 \pmod{21}$.

11.119. Solve each linear congruence equation: (a) $37x \equiv 1 \pmod{249}$;   (b) $195x \equiv 23 \pmod{968}$.

11.120. Solve each linear congruence equation: (a) $132x \equiv 169 \pmod{735}$;   (b) $48x \equiv 284 \pmod{356}$

11.121. A puppet theater has only 60 seats. The admission to the theater is $2.25 per adult and $1.00 per child. Suppose $117.25 was collected. Find the number of adults and children attending the performance.

11.122. A boy sells apples for 12 cents each and pears for 7 cents each. Suppose the boy collected $3.21. Find the number of apples and pears that he sold.

11.123. Find the smallest positive solution of each system of congruence equations:

(a) $x \equiv 2 \pmod 3$, $x \equiv 3 \pmod 5$, $x \equiv 4 \pmod{11}$
(b) $x \equiv 3 \pmod 5$, $x \equiv 4 \pmod 7$, $x \equiv 6 \pmod 9$
(c) $x \equiv 5 \pmod{45}$, $x \equiv 6 \pmod{49}$, $x \equiv 7 \pmod{52}$

Answers to Supplementary Problems

11.61. (a) $2 > -6$;   (b) $-3 > -5$;   (c) $-7 < 3$;
(d) $-8 < -1$;   (e) $2^3 < 11$;   (f) $2^3 > -9$;
(g) $-2 > -7$;   (h) $4 > -9$

11.62. (a) 4, 4, 10;   (b) $3 + 10 = 13, 3 - 7 = -4$;
(c) $4 + 1 = 5, 8 - 4 = 4$.

11.63. (a) 7;   (b) 9;   (c) 6;   (d) 6;   (e) 6;   (f) 2.

11.64. (a) 4, 5, 6;   (b) $-2, -1, 0, 1$.

11.79. (a) $q = -15, r = 13$;   (b) $q = -24, r = 10$.
(c) $q = 53, r = 7$

11.81. (a) One is divisible by 2 and one is divisible by 3.
(b) One is divisible by 4, another is divisible by 2, and one is divisible by 3.

11.84. (a) 1, 2, 3, 4, 6, 8, 12, 24; (b) 3^n for $n = 0$ to 9;
(c) $2^r 3^s$ for $r = 0$ to 4 and $s = 0$ to 3.

11.85. 101, 103, 107, 109, 113, 127, 131, 137, 139, 149.

11.86. (a) $2940 = 2^2 \cdot 3 \cdot 5 \cdot 7^2$;   (b) $1485 = 3^3 \cdot 5 \cdot 11$;
(c) $8712 = 2^3 \cdot 3^2 \cdot 11^2$; (d) $319410 = 2 \cdot 3^3 \cdot 5 \cdot 7 \cdot 13^2$.

11.87. (a) $d = 4 = 5(356) - 37(48)$; (b) $d = 33 = 8(165) - 1(1287)$; (c) $d = 42 = 14(168) - 1(2310)$; (d) $d = 1 = 139(195) - 28(968)$; (e) $11(249) - 74(37)$.

11.88. (a) 35; (b) 33; (c) 84.

11.89. (a) $a = 2^3 \cdot 3 \cdot 5 \cdot 7^2$, $b = 2^2 \cdot 3^3 \cdot 7 \cdot 11$; (b) $\gcd(a, b) = 2^2 \cdot 3 \cdot 7$, $\mathrm{lcm}(a, b) = 2^3 \cdot 3^3 \cdot 5 \cdot 7^2 \cdot 11 = 1,164,240$.

11.94. (a) Hint: $a^k - 1 = (a-1)(1 + a + a^2 + \cdots + a^{k-1})$; (b) Hint: If $n = ab$, then $2^n - 1 = (2^a)^b - 1$.

11.98. $(k + 1)! + 2$, $(k + 1)! + 3$, $(k + 1)! + 4, \ldots,$ $(k+1)! + (k + 1)$ are divisible by $2, 3, 4, \ldots, k + 1$, respectively.

11.99. (a) False; (b) true; (c) false; (d) false.

11.100. (a) 7; (b) 3; (c) 3; (d) 6.

11.101. (a) -2; (b) -3; (c) -4; (d) 1.

11.102. 4, 15, 26, 37, 48, 59, 70, 81, 92.

11.103. $-42, -33, -24, -15, -6, 3, 12, 21, 30, 39, 48$.

11.104. (a) $\{0, 1, \ldots, 10\}$ and $\{-5, -4, \ldots, -1, 0, 1, \ldots, 4, 5\}$.
(b) $\{0, 1, \ldots, 13\}$ and $\{-6, -5, \ldots, -1, 0, 1, \ldots, 6, 7\}$.

11.105. (a) $\{1, 3\}$; (b) $\{1, 2, \ldots, 10\}$; (c) $\{1, 3, 5, 9, 11, 13\}$; (d) $\{1, 2, 4, 7, 8, 11, 13, 14\}$.

11.106. (a) $\{5, 10, 15, 20, 25, 30, 35, 40\}$; (b) $\{3, 9, 27, 81, 243, 729, 2187, 6561\}$.

11.107. $m - 1 \equiv -1 \pmod{m}$ and so $(m - 1)^2 \equiv 1 \pmod{m}$.

11.108. (a) 4; (b) 4; (c) $= 8$; (d) $2(3^6)$; (e) $4(5^5)$; (f) $(2^3)(6 \cdot 7^5)(12 \cdot 13^2)$

11.109. $\phi(8000) = \phi(2^5 \cdot 5^2) = 2^4 \cdot 4 \cdot 5 = 320$. Hence $s = 4(320) = 1280$.

11.112. (a) 11, 10, 8, 4, 3, 2; (b) 6, 12, 2, 8, 11.

11.113. (a) 14, 12, 11, 9, 4, 2; (b) 11, 7, 3, 8, 4.

11.114. (a) 17; (b) 61; (c) a^{-1} does not exist.

11.115. $2x^5 + 2x^2 - x + 4$

11.116. (a) 1, 3, 4; (b) 2, -2; (c) 0, 2, 3, -1.

11.117. (a) 3; (b) 5, 12; (c) no solution.

11.118. (a) 7; (b) no solution; (c) 6, 13, 20.

11.119. (a) 175; (b) 293.

11.120. (a) No solution; (b) 43, 132, 221, 310.

11.121. 49 adults, 7 children.

11.122. 25 apples, 3 pears; 18 apples, 15 pears; 11 apples, 27 pears; or 4 apples, 39 pears.

11.123. (a) 158; (b) 1(123); (c) 31 415.

CHAPTER 12

Languages, Automata, Grammars

12.1 INTRODUCTION

This chapter discusses three topics, *languages*, *automata*, and *grammars*. These three topics are closely related to each other. Our languages will use the letters $a, b, \ldots$ to code data rather than the digits 0 and 1 used by some other texts.

12.2 ALPHABET, WORDS, FREE SEMIGROUP

Consider a nonempty set A of symbols. *A word* or *string* w on the set A is a finite sequence of its elements. For example, suppose $A = \{a, b, c\}$. Then the following sequences are words on A:

$$u = ababb \quad \text{and} \quad v = accbaaa$$

When discussing words on A, we frequently call A the *alphabet*, and its elements are called *letters*. We will also abbreviate our notation and write a^2 for aa, a^3 for aaa, and so on. Thus, for the above words, $u = abab^2$ and $v = ac^2ba^3$.

The empty sequence of letters, denoted by λ (Greek letter lambda) or ϵ (Greek letter epsilon), or 1, is also considered to be a word on A, called the *empty word*. The set of all words on A is denoted by A^* (read: "A star").

The *length* of a word u, written $|u|$ or $l(u)$, is the number of elements in its sequence of letters. For the above words u and v, we have $l(u) = 5$ and $l(v) = 7$. Also, $l(\lambda) = 0$, where λ is the empty word.

Remark: Unless otherwise stated, the alphabet A will be finite, the symbols u, v, w will be reserved for words on A, and the elements of A will come from the letters a, b, c.

Concatenation

Consider two words u and v on the alphabet A. The *concatenation* of u and v, written uv, is the word obtained by writing down the letters of u followed by the letters of v. For example, for the above words u and v, we have

$$uv = ababbaccbaaa = abab^2 ac^2 ba^3$$

As with letters, for any word u, we define $u^2 = uu$, $u^3 = uuu$, and, in general, $u^{n+1} = uu^n$.

Clearly, for any words u, v, w, the words $(uv)w$ and $u(vw)$ are identical, they simply consist of the letters of u, v, w written down one after the other. Also, adjoining the empty word before or after a word u does not change the word u. That is:

Theorem 12.1: The concatenation operation for words on an alphabet A is associative. The empty word λ is an identity element for the operation.

(Generally speaking, the operation is not commutative, e.g., $uv \neq vu$ for the above words u and v.)

Subwords, Initial Segments

Consider any word $u = a_1 a_2 \ldots a_n$ on an alphabet A. Any sequence $w = a_j a_{j+1} \ldots a_k$ is called a *subword* of u. In particular, the subword $w = a_1 a_2 \ldots a_k$ beginning with the first letter of u, is called an *initial segment* of u. In other words, w is a subword of u if $u = v_1 w v_2$ and w is an initial segment of u if $u = wv$. Observe that λ and u are both subwords of uv since $u = \lambda u$.

Consider the word $u = abca$. The subwords and initial segments of u follow:

    (1) Subwords: $\lambda, a, b, c, ab, bc, ca, abc, bca, abca = u$

    (2) Initial segments: $\lambda, a, ab, abc, abca = u$.

Observe that the subword $w = a$ appears in two places in u. The word ac is not a subword of u even though all its letters belong to u.

Free Semigroup, Free Monoid

Let F denote the set of all nonempty words from an alphabet A with the operation of concatenation. As noted above, the operation is associative. Thus F is a semigroup; it is called the *free semigroup over A* or the *free semigroup generated by A*. One can easily show that F satisfies the right and left cancellation laws. However, F is not commutative when A has more than one element. We will write F_A for the free semigroup over A when we want to specify the set A.

Now let $M = A^*$ be the set of all words from A including the empty word λ. Since λ is an identity element for the operation of concatenation, M is a monoid, called the *free monoid* over A.

12.3 LANGUAGES

A *language L* over an alphabet A is a collection of words on A. Recall that A^* denotes the set of all words on A. Thus a language L is simply a subset of A^*.

EXAMPLE 12.1 Let $A = \{a, b\}$. The following are languages over A.

    (a) $L_1 = \{a, ab, ab^2, \ldots\}$ (c) $L_3 = \{a^m b^m \mid m > 0\}$

    (b) $L_2 = \{a^m b^n \mid m > 0, n > 0\}$ (d) $L_4 = b^m a b^n \mid m \geq 0, n \geq 0\}$

One may verbally describe these languages as follows.

(a) L_1 consists of all words beginning with an a and followed by zero or more b's.

(b) L_2 consists of all words beginning with one or more as followed by one or more b's.

(c) L_3 consists of all words beginning with one or more as and followed by the same number of b's.

(d) L_4 consists of all words with exactly one a.

Operations on Languages

Suppose L and M are languages over an alphabet A. Then the "concatenation" of L and M, denoted by LM, is the language defined as follows:

$$LM = \{uv \mid u \in L, \ v \in V\}$$

That is, LM denotes the set of all words which come from the concatenation of a word from L with a word from M. For example, suppose

$$L_1 = \{a, b^2\}, \quad L_2 = \{a^2, ab, b^3\}, \quad L_3 = \{a^2, a^4, a^6, \ldots\}$$

Then:

$$L_1 L_1 = \{a^2, ab^2, b^2 a, b^4\}, \quad L_1 L_2 = \{a^3, a^2 b, ab^3, b^2 a^2, b^2 ab, b^5\}$$
$$L_1 L_3 = \{a^3, a^5, a^7, \ldots, b^2 a^2, b^2 a^4, b^2 a^6, \ldots\}$$

Clearly, the concatenation of languages is associative since the concatenation of words is associative.

Powers of a language L are defined as follows:

$$L^0 = \{\lambda\}, \quad L^1 = L, \quad L^2 = LL, \quad L^{m+1} = L^m L \quad \text{for} \quad m > 1.$$

The unary operation $L*$ (read "L star") of a language L, called the *Kleene* closure of L because Kleene proved Theorem 12.2, is defined as the infinite union:

$$L^* = L^0 \cup L^1 \cup L^2 \cup \cdots = \bigcup_{k=0}^{\infty} L^k$$

The definition of L^* agrees with the notation A^* which consists of all words over A. Some texts define L^+ to be the union of $L^1, L^2, \ldots$ that is, L^+ is the same as L^* but without the empty word λ.

12.4 REGULAR EXPRESSIONS, REGULAR LANGUAGES

Let A be a (nonempty) alphabet. This section defines a regular expression r over A and a language $L(r)$ over A associated with the regular expression r. The expression r and its corresponding language $L(r)$ are defined inductively as follows.

Definition 12.1: Each of the following is a regular expression over an alphabet A.

(1) The symbol "λ" (empty word) and the pair "$(\ )$" (empty expression) are regular expressions.

(2) Each letter a in A is a regular expression.

(3) If r is a regular expression, then (r^*) is a regular expression.

(4) If r_1 and r_2 are regular expressions, then $(r_1 \vee r_2)$ is a regular expression.

(5) If r_1 and r_2 are regular expressions, then $(r_1 r_2)$ is a regular expression.

All regular expressions are formed in this way.

Observe that a regular expression r is a special kind of a word (string) which uses the letters of A and the five symbols:

$$(\quad)\quad *\quad \vee\quad \lambda$$

We emphasize that no other symbols are used for regular expressions.

Definition 12.2: The language $L(r)$ over A defined by a regular expression r over A is as follows:

(1) $L(\lambda) = \{\lambda\}$ and $L((\ )) = \varnothing$, the empty set.

(2) $L(a) = \{a\}$, where a is a letter in A.

(3) $L(r^*) = (L(r))^*$ (the Kleene closure of $L(r)$).

(4) $L(r_1 \vee r_2) = L(r_1) \cup L(r_2)$ (the union of the languages).

(5) $L(r_1 r_2) = L(r_1) L(r_2)$ (the concatenation of the languages).

Remark: Parentheses will be omitted from regular expressions when possible. Since the concatenation of languages and the union of languages are associative, many of the parentheses may be omitted. Also, by adopting the convention that "*" takes precedence over concatenation, and concatenation takes precedence over "∨," other parentheses may be omitted.

Definition 12.3: Let L be a language over A. Then L is called a *regular language* over A if there exists a regular expression r over A such that $L = L(r)$.

EXAMPLE 12.2 Let $A = \{a, b\}$. Each of the following is an expression r and its corresponding language $L(r)$:

(a) Let $r = a^*$. Then $L(r)$ consists of all powers of a including the empty word λ.

(b) Let $r = aa^*$. Then $L(r)$ consists of all positive powers of a excluding the empty word.

(c) Let $r = a \vee b^*$. Then $L(r)$ consists of a or any word in b, that is, $L(r) = \{a, \lambda, b, b^2, \cdots\}$.

(d) Let $r = (a \vee b)^*$. Note $L(a \vee b) = \{a\} \cup \{b\} = A$; hence $L(r) = A^*$, all words over A.

(e) Let $r = (a \vee b)^* bb$. Then $L(r)$ consists of the concatenation of any word in A with bb, that is, all words ending in b^2.

(f) Let $r = a \wedge b^*$. $L(r)$ does not exist since r is not a regular expression. (Specifically, $\wedge$ is not one of the symbols used for regular expressions.)

EXAMPLE 12.3 Consider the following languages over $A = \{a, b\}$:

(a) $L_1 = \{a^m b^n \mid m > 0, n > 0\}$;   (b) $L_2 = \{b^m ab^n \mid m > 0, n > 0\}$;   (c) $L_3 = \{a^m b^m \mid m > 0\}$.

Find a regular expression r over $A = \{a, b\}$ such that $L_i = L(r)$ for $i = 1, 2, 3$.

(a) L_1 consists of those words beginning with one or more a's followed by one or more b's. Thus we can set $r = aa^* bb^*$. Note that r is not unique; for example, $r = a^* abb^*$ is another solution.

(b) L_2 consists of all words which begin with one or more b's followed by a single a which is then followed by one or more b's, that is, all words with exactly one a which is neither the first nor the last letter. Hence $r = bb^* abb^*$ is a solution.

(c) L_3 consists of all words beginning with one or more a's followed by the same number of b's. There exists no regular expression r such that $L_3 = L(r)$; that is, L_3 is not a regular language. The proof of this fact appears in Example 12.8.

12.5 FINITE STATE AUTOMATA

A *finite state automaton* (FSA) or, simply, an *automaton M*, consists of five parts:

(1) A finite set (alphabet) A of inputs.

(2) A finite set S of (internal) states.

(3) A subset Y of S (called accepting or "yes" states).

(4) An initial state s_0 in S.

(5) A next-state function F from $S \times A$ into S.

Such an automaton M is denoted by $M = (A, S, Y, s_0, F)$ when we want to indicate its five parts. (The plural of automaton is *automata*.)

Some texts define the next-state function $F : S \times A \to S$ in (5) by means of a collection of functions $f_a : S \to S$, one for each $a \in A$. Setting $F(s, a) = f_a(s)$ shows that both definitions are equivalent.

EXAMPLE 12.4 The following defines an automaton M with two input symbols and three states:

(1) $A = \{a, b\}$, input symbols.

(2) $S = \{s_0, s_1, s_2\}$, internal states.

(3) $Y = \{s_0, s_1\}$, "yes" states.

(4) s_0, initial state.

(5) Next-state function $F : S \times A \to S$ defined explicitly in Fig. 12-1(a) or by the table in Fig. 12-1(b).

$$F(s_0, a) = s_0, \quad F(s_1, a) = s_0, \quad F(s_2, a) = s_2$$
$$F(s_0, b) = s_1, \quad F(s_1, b) = s_2, \quad F(s_2, b) = s_2$$

F	a	b
s_0	s_0	s_1
s_1	s_0	s_2
s_2	s_2	s_2

(a)   (b)

Fig. 12-1

State Diagram of an Automaton M

An automaton M is usually defined by means of its state diagram $D = D(M)$ rather than by listing its five parts. The state diagram $D = D(M)$ is a labeled directed graph as follows.

(1) The vertices of $D(M)$ are the states in S and an accepting state is denoted by means of a double circle.

(2) There is an arrow (directed edge) in $D(M)$ from state s_j to state s_k labeled by an input a if $F(s_j, a) = s_k$ or, equivalently, if $f_a(s_j) = s_k$.

(3) The initial state s_0 is indicated by means of a special arrow which terminates at s_0 but has no initial vertex.

For each vertex s_j and each letter a in the alphabet A, there will be an arrow leaving s_j which is labeled by a; hence the outdegree of each vertex is equal to number of elements in A. For notational convenience, we label a single arrow by all the inputs which cause the same change of state rather than having an arrow for each such input.

The state diagram $D = D(M)$ of the automaton M in Example 12.4 appears in Fig. 12-2. Note that both a and b label the arrow from s_2 to s_2 since $F(s_2, a) = s_2$ and $F(s_2, b) = s_2$. Note also that the outdegree of each vertex is 2, the number of elements in A.

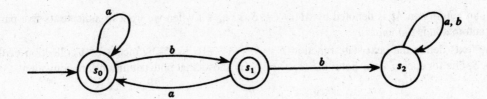

Fig. 12-2

Language $L(M)$ Determined by an Automaton M

Each automaton M with input alphabet A defines a language over A, denoted by $L(M)$, as follows.

Let $w = a_1 a_2 \cdots a_m$ be a word on A. Then w determines the following path in the state diagram graph $D(M)$ where s_0 is the initial state and $F(s_{i-1}, a_i) = s_i$ for $i \geq 1$:

$$P = (s_0, a_1, s_1, a_2, s_2, \cdots, a_m, s_m)$$

We say that M *recognizes* the word w if the final state s_m is an accepting state in Y. The language $L(M)$ of M is the collection of all words from A which are accepted by M.

EXAMPLE 12.5 Determine whether or not the automaton M in Fig. 12-2 accepts the words:

$$w_1 = ababba; \quad w_2 = baab; \quad w_3 = \lambda \text{ the empty word.}$$

Use Fig. 12-2 and the words w_1 and w_2 to obtain the following respective paths:

$$P_1 = s_0 \xrightarrow{a} s_0 \xrightarrow{b} s_1 \xrightarrow{a} s_0 \xrightarrow{b} s_1 \xrightarrow{b} s_2 \xrightarrow{a} s_2 \quad \text{and} \quad P_2 = s_0 \xrightarrow{b} s_1 \xrightarrow{a} s_0 \xrightarrow{a} s_0 \xrightarrow{b} s_1$$

The final state in P_1 is s_2 which is not in Y; hence w_1 is not accepted by M. On the other hand, the final state in P_2 is s_1 which is in Y; hence w_2 is accepted by M. The final state determined by w_3 is the initial state s_0 since $w_3 = \lambda$ is the empty word. Thus w_3 is accepted by M since $s_0 \in Y$.

EXAMPLE 12.6 Describe the language $L(M)$ of the automaton M in Fig. 12-2.

$L(M)$ will consist of all words w on A which do not have two successive b's. This comes from the following facts:

(1) We can enter the state s_2 if and only if there are two successive b's.

(2) We can never leave s_2.

(3) The state s_2 is the only rejecting (nonaccepting) state.

The fundamental relationship between regular languages and automata is contained in the following theorem (whose proof lies beyond the scope of this text).

Theorem 12.2 (Kleene): A language L over an alphabet A is regular if and only if there is a finite state automaton M such that $L = L(M)$.

The star operation L^* on a language L is sometimes called the Kleene closure of L since Kleene first proved the above basic result.

EXAMPLE 12.7 Let $A = \{a, b\}$. Construct an automaton M which will accept precisely those words from A which end in two b's.

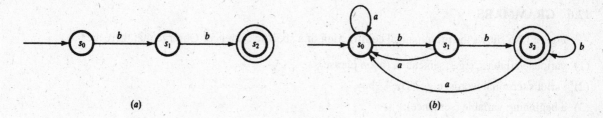

Fig. 12-3

Since b^2 is accepted, but not λ or b, we need three states, s_0, the initial state, and s_1 and s_2 with an arrow labeled b going from s_0 to s_1 and one from s_1 to s_2. Also, s_2 is an accepting state, but not s_0 nor s_1. This gives the graph in Fig. 12-3(a). On the other hand, if there is an a, then we want to go back to s_0, and if we are in s_2 and there is a b, then we want to stay in s_2. These additional conditions give the required automaton M which is shown in Fig. 12-3(b).

Pumping Lemma

Let M be an automaton over A with k states. Suppose $w = a_1 a_2 \cdots a_n$ is a word over A accepted by M and suppose $|w| = n > k$, the number of states. Let

$$P = (s_0, s_1, \ldots, s_n)$$

be the corresponding sequence of states determined by the word w. Since $n > k$, two of the states in P must be equal, say $s_i = s_j$ where $i < j$. Let w be divided into subwords x, y, z as follows:

$$x = a_1 a_2 \cdots a_i, \quad y = a_{i+1} \cdots a_j, \quad z = a_{j+1} \cdots a_n$$

As shown in Fig. 12-4, xy ends in $s_i = s_j$; hence xy^m also ends in s_i. Thus, for every m, $w_m = xy^m z$ ends in s_n, which is an accepting state.

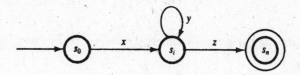

Fig. 12-4

The above discussion proves the following important result.

Theorem 12.3 (Pumping Lemma): Suppose M is an automaton over A such that:
(i) M has k states. (ii) M accepts a word w from A where $|w| > k$.
Then $w = xyz$ where, for every positive m, $w_m = xy^m z$ is accepted by M.
The next example gives an application of the Pumping Lemma.

EXAMPLE 12.8 Show that the language $L = \{a^m b^m \mid m$ is positive$\}$ is not regular.

Suppose L is regular. Then, by Theorem 12.2, there exists a finite state automaton M which accepts L. Suppose M has k states. Let $w = a^k b^k$. Then $|w| > k$. By the Pumping Lemma (Theorem 12.3), $w = xyz$ where y is not empty and $w_2 = xy^2 z$ is also accepted by M. If y consists of only a's or only b's, then w^2 will not have the same number of a's as b's. If y contains both a's and b's, then w_2 will have a's following b's. In either case w_2 does not belong to L, which is a contradiction. Thus L is not regular.

12.6 GRAMMARS

Figure 12-5 shows the grammatical construction of a specific sentence. Observe that there are:

(1) various variables, e.g., (sentence), (noun phrase), $\cdots$;

(2) various terminal words, e.g., "The", "boy," $\cdots$;

(3) a beginning variable (sentence);

(4) various substitutions or productions, e.g.

$$
\begin{aligned}
\langle sentence \rangle &\rightarrow \langle noun\ phrase \rangle\ \langle verb\ phrase \rangle \\
\langle object\ phrase \rangle &\rightarrow \langle article \rangle\ \langle noun \rangle \\
\langle noun \rangle &\rightarrow apple
\end{aligned}
$$

The final sentence only contains terminals, although both variables and terminals appear in its construction by the productions. This intuitive description is given in order to motivate the following definition of a grammar and the language it generates.

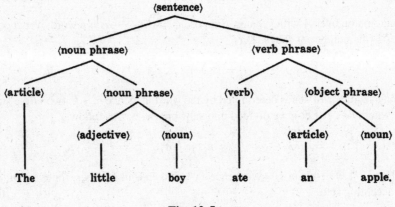

Fig. 12-5

Definition 12.4: A *phrase structure grammar* or, simply, a *grammar G* consists of four parts:

(1) A finite set (vocabulary) V.

(2) A subset T of V whose elements are called *terminals*; the elements of $N = V \setminus T$ are called *non-terminals* or *variables*.

(3) A nonterminal symbol S called the *start* symbol.

(4) A finite set P of productions. (A production is an ordered pair (α, β), usually written $\alpha \rightarrow \beta$, where α and β are words in V, and the production must contain at least one nonterminal on its left side α.)

Such a grammar G is denoted by $G = G(V, T, S, P)$ when we want to indicate its four parts.

The following notation, unless otherwise stated or implied, will be used for our grammars. Terminals will be denoted by italic lower case Latin letters, $a, b, c, \cdots$, and nonterminals will be denoted by italic capital Latin letters, $A, B, C, \cdots$, with S as the start symbol. Also, Greek letters, $\alpha, \beta, \cdots$, will denote words in V, that is, words in terminals and nonterminals. Furthermore, we will write

$$\alpha \rightarrow (\beta_1, \beta_2, \cdots, \beta_k) \quad \text{instead of} \quad \alpha \rightarrow \beta_1, \alpha \rightarrow \beta_2, \cdots, \alpha \rightarrow \beta_k$$

Remark: Frequently, we will define a grammar G by only giving its productions, assuming implicitly that S is the start symbol and that the terminals and nonterminals of G are only those appearing in the productions.

EXAMPLE 12.9 The following defines a grammar G with S as the start symbol:

$$V = \{A, B, S, a, b\}, \quad T = \{a, b\}, \quad P = \{S \overset{1}{\to} AB, A \overset{2}{\to} Aa, B \overset{3}{\to} Bb, A \overset{4}{\to} a, B \overset{5}{\to} b\}$$

The productions may be abbreviated as follows: $S \to AB, A \to (Aa, a), B \to (Bb, b)$

Language $L(G)$ of a Grammar G

Suppose w and w' are words over the vocabulary set V of a grammar G. We write

$$w \Rightarrow w'$$

if w' can be obtained from w by using one of the productions; that is, if there exists words u and v such that $w = u\alpha v$ and $w' = u\beta v$ and there is a production $\alpha \to \beta$. Furthermore, we write

$$w \Rightarrow\Rightarrow w' \quad \text{or} \quad w \overset{*}{\Rightarrow} w'$$

if w' can be obtained from w using a finite number of productions.

Now let G be a grammar with terminal set T. The language of G, denoted by $L(G)$, consists of all words in T that can be obtained from the start symbol S by the above process; that is,

$$L(G) = \{w \in T^* \mid S \Rightarrow\Rightarrow w\}$$

EXAMPLE 12.10 Consider the grammar G in Example 12.9. Observe that $w = a^2b^4$ can be obtained from the start symbol S as follows:

$$S \Rightarrow AB \Rightarrow AaB \Rightarrow aaB \Rightarrow aaBb \Rightarrow aaBbb \Rightarrow aaBbbb \Rightarrow aabbbb = a^2b^4$$

Here we used the productions 1, 2, 4, 3, 3, 3, 5, respectively. Thus we can write $S \Rightarrow\Rightarrow a^2b^4$. Hence $w = a^2b^4$ belongs to $L(G)$. More generally, the production sequence:

$$1, \ 2 \ (r \ \text{times}), \ 4, \ 3 \ (s \ \text{times}), \ 5$$

will produce the word $w = a^r ab^s b$ where r and s are nonnegative integers. On the other hand, no sequence of productions can produce an a after a b. Accordingly,

$$L(G) = \{a^m b^n \mid m \text{ and } n \text{ are positive integers}\}$$

That is, the language $L(G)$ of the grammar G consists of all words which begin with one or more a's followed by one or more b's.

EXAMPLE 12.11 Find the language $L(G)$ over $\{a, b, c\}$ generated by the grammar G:

$$S \to aSb, \quad aS \to Aa, \quad Aab \to c$$

First we must apply the first production one or more times to obtain the word $w = a^n Sb^n$ where $n > 0$. To eliminate S, we must apply the second production to obtain the word $w' = a^m Aabb^m$ where $m = n - 1 \geq 0$. Now we can only apply the third production to finally obtain the word $w' = a^m cb^m$ where $m \geq 0$. Accordingly,

$$L(G) = \{a^m cb^m \mid m \text{ nonnegative}\}$$

That is, $L(G)$ consists of all words with the same nonnegative number of a's and b's separated by a c.

Types of Grammars

Grammars are classified according to the kinds of production which are allowed. The following grammar classification is due to Noam Chomsky.

A Type 0 grammar has no restrictions on its productions. Types 1, 2, and 3 are defined as follows:

(1) A grammar G is said to be of Type 1 if every production is of the form $\alpha \to \beta$ where $|\alpha| \le |\beta|$ or of the form $\alpha \to \lambda$.

(2) A grammar G is said to be of Type 2 if every production is of the form $A \to \beta$ where the left side A is a nonterminal.

(3) A grammar G is said to be of Type 3 if every production is of the form $A \to a$ or $A \to aB$, that is, where the left side A is a single nonterminal and the right side is a single terminal or a terminal followed by a nonterminal, or of the form $S \to \lambda$.

Observe that the grammars form a hierarchy; that is, every Type 3 grammar is a Type 2 grammar, every Type 2 grammar is a Type 1 grammar, and every Type 1 grammar is a Type 0 grammar.

Grammars are also classified in terms of context-sensitive, context-free, and regular as follows.

(a) A grammar G is said to be *context-sensitive* if the productions are of the form

$$\alpha A \alpha' \to \alpha \beta \alpha'$$

The name "context-sensitive" comes from the fact that we can replace the variable A by β in a word only when A lies between α and $\alpha\prime$.

(b) A grammar G is said to be *context-free* if the productions are of the form

$$A \to \beta$$

The name "context-free" comes from the fact that we can now replace the variable A by β regardless of where A appears.

(c) A grammar G is said to be *regular* if the productions are of the form

$$A \to a, \quad A \to aB, \quad S \to \lambda$$

Observe that a context-free grammar is the same as a Type 2 grammar, and a regular grammar is the same as a Type 3 grammar.

A fundamental relationship between regular grammars and finite automata follows.

Theorem 12.4: A language L can be generated by a Type 3 (regular) grammar G, if and only if there exists a finite automaton M which accepts L.

Thus a language L is regular iff $L = L(r)$ where r is a regular expression iff $L = L(M)$ where M is a finite automaton iff $L = L(G)$ where G is a regular grammar. (Recall that "iff" is an abbreviation for "if and only if.")

EXAMPLE 12.12 Consider the language $L = \{a^n b^n \mid n > 0\}$.

(a) Find a context-free grammar G which generates L.

Clearly the grammar G with the following productions will generate L:

$$S \to ab, \quad S \to aSb$$

Note that G is context-free

(b) Find a regular grammar G which generates L.

By Example 12.8, L is not a regular language. Thus L cannot be generated by a regular grammar.

Derivation Trees of Context-Free Grammars

Consider a context-free (Type 2) grammar G. Any derivation of a word w in $L(G)$ can be represented graphically by means of an ordered, rooted tree T, called a *derivation tree* or *parse tree*. We illustrate such a derivation tree below.

Let G be the context-free grammar with the following productions:

$$S \rightarrow aAB, \quad A \rightarrow Bba, \quad B \rightarrow bB, \quad B \rightarrow c$$

The word $w = acbabc$ can be derived from S as follows:

$$S \Rightarrow aAB \Rightarrow a(Bba)B \Rightarrow acbaB \Rightarrow acba(bB) \Rightarrow acbabc$$

One can draw a derivation tree T of the word w as indicated by Fig. 12-6. Specifically, we begin with S as the root and then add branches to the tree according to the production used in the derivation of w. This yields the completed tree T which is shown in Fig. 12-6(e). The sequence of leaves from left to right in T is the derived word w. Also, any non-leaf in T is a variable, say A, and the immediate successors (children) of A form a word α where $A \rightarrow \alpha$ is the production of G used in the derivation of w.

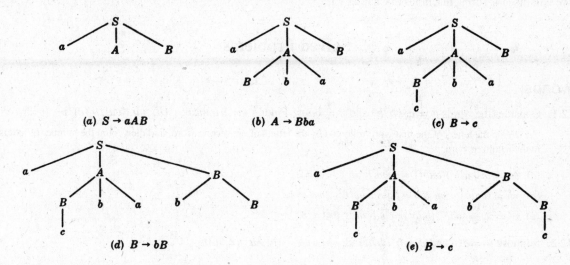

Fig. 12-6

Backus-Naur Form

There is another notation, called the Backus-Naur form, which is sometimes used for describing the productions of a context-free (Type 2) grammar. Specifically:

(i) "::=" is used instead of "→."

(ii) Every nonterminal is enclosed in brackets ⟨ ⟩.

(iii) All productions with the same nonterminal left-hand side are combined into one statement with all the right-hand sides listed on the right of :: = separated by vertical bars.

For instance, the productions $A \rightarrow aB$, $A \rightarrow b$, $A \rightarrow BC$ are combined into the one statement:

$$\langle A \rangle ::= a \langle B \rangle |b| \langle B \rangle \langle C \rangle$$

Machines and Grammars

Theorem 12.4 tells us that the regular languages correspond to the finite state automata (FSA). There are also machines, more powerful than the FSA, which correspond to the other grammars.

(a) **Pushdown Automata:** A pushdown automaton P is similar to a FSA except that P has an auxiliary stack which provides an unlimited amount of memory for P. A language L is recognized by a pushdown automaton P if and only if L is context-free.

(b) **Linear Bounded Automata:** A linear bounded automaton B is more powerful than a pushdown automaton. Such an automaton B uses a tape which is linearly bounded by the length of the input word w. A language L is recognized by a linear bounded automaton B if and only if L is context-sensitive.

(c) **Turing Machine:** A Turing machine M, named after the British mathematician Alan Turing, uses an infinite tape; it is able to recognize every language L that can be generated by any phase-structure grammar G. In fact, a Turing machine M is one of a number of equivalent ways to define the notion of a "computable" function.

The discussion of the pushdown automata and the linear bounded automata lies beyond the scope of this text. We will discuss Turing machines in Chapter 13.

Solved Problems

WORDS

12.1. Consider the words $u = a^2ba^3b^2$ and $v = bab^2$. Find: (a) uv; $|uv|$; (b) vu, $|vu|$; (c) v^2, $|v^2|$.

   Write the letters of the first word followed by the letters of the second word, and then count the number of letters in the resulting word.

   (a) $uv = (a^2ba^3b^2)(bab^2) = a^2ba^3b^3ab^2$; $|uv| = 12$

   (b) $vu = (bab^2)(a^2ba^3b^2) = bab^2a^2ba^3b^2$; $|vu| = 12$

   (c) $v^2 = vv = (bab^2)(bab^2) = bab^3ab^2$; $|v^2| = 8$

12.2. Suppose $u = a^2b$ and $v = b^3ab$. Find: (a) uvu; (b) λu, $u\lambda$, $u\lambda v$.

   (a) Write down the letters in u, then v, and finally u to obtain $uvu = a^2b^4aba^2b$.

   (b) Since λ is the empty word, $\lambda u = u\lambda = u = a^2b$ and $u\lambda v = uv = a^2b^4ab$.

12.3. Let $w = abcd$. (a) Find all subwords of w. (b) Which of them are initial segments?

   (a) The subwords are: $\lambda, a, b, c, d, ab, bc, cd, abc, bcd, w = abcd$. (We emphasize that $v = acd$ is not a subword of w even though all its letters belong to w.)

   (b) The initial segments are $\lambda, a, ab, abc, w = abcd$.

12.4. For any words u and v, show that: (a) $|uv| = |u| + |v|$; (b) $|uv| = |vu|$.

   (a) Suppose $|u| = r$ and $|v| = s$. Then uv will consist of the r letters of u followed by the s letters of v; hence $|uv| = r + s = |u| + |v|$.

   (b) Using (a) yields $|uv| = |u| + |v| = |v| + |u| = |vu|$.

12.5. State the difference between the free semigroup on an alphabet A and the free monoid on A.

   The free semigroup on A is the set of all nonempty words in A under the operation of concatenation; it does not include the empty word λ. On the other hand, the free monoid on A does include the empty word λ.

LANGUAGES

12.6. Let $A = \{a, b\}$. Describe verbally the following languages over A (which are subsets of A^*):

(a) $L_1 = \{(ab)^m \mid m > 0\}$; (b) $L_2 = \{a^r b a^s b a^t \mid r, s, t \geq 0\}$; (c) $L_3 = \{a^2 b^m a^3 \mid m > 0\}$.

(a) L_1 consists of words $w = ababab \cdots ab$, that is, beginning with a, alternating with b, and ending with b.

(b) L_2 consists of all words with exactly two b's.

(c) L_3 consists of all words beginning with a^2 and ending with a^3 with one or more b's between them.

12.7. Let $K = \{a, ab, a^2\}$ and $L = \{b^2, aba\}$ be languages over $A = \{a, b\}$. Find: (a) KL; (b) LL.

(a) Concatenate words in K with words in L to obtain $KL = \{ab^2, a^2ba, ab^3, ababa, a^2b^2, a^3ba\}$.

(b) Concatenate words in L with words in L to obtain $LL = \{b^4, b^2aba, abab^2, aba^2ba\}$.

12.8. Consider the language $L = \{ab, c\}$ over $A = \{a, b, c\}$. Find: (a) L^0; (b) L^3; (c) L^{-2}.

(a) $L^0 = \{\lambda\}$, by definition.

(b) Form all three-word sequences from L to obtain:

$$L^3 = \{ababab, ababc, abcab, abc^2, cabab, cabc, c^2ab, c^3\}$$

(c) The negative power of a language is not defined.

12.9. Let $A = \{a, b, c\}$. Find L^* where: (a) $L = \{b^2\}$; (b) $L = \{a, b\}$; (c) $L = \{a, b, c^3\}$.

(a) L^* consists of all words b^n where n is even (including the empty word λ).

(b) L^* consists of words in a and b.

(c) L^* consists of all words from A with the property that the length of each maximal subword composed entirely of c's is divisible by 3.

12.10. Consider a countable alphabet $A = \{a_1, a_2, \ldots\}$. Let L_k be the language over A consisting of those words w such that the sum of the subscripts of the letters in w is equal to k. (For example, $w = a_2a_3a_3a_6a_4$ belongs to L_{18}.) (a) Find L_4. (b) Show that L_k is finite. (c) Show that A^* is countable. (d) Show that any language over A is countable.

(a) No word in L_4 can have more than four letters, and no letter a_n with $n > 4$ can be used. Thus we obtain the following list:

$$a_1a_1a_1a_1, \quad a_1a_1a_2, \quad a_1a_2a_1, \quad a_2a_1a_1, \quad a_1a_3, \quad a_3a_1, \quad a_2a_2, \quad a_4$$

(b) Only a finite number of the a's, that is, $a_1, a_2, \ldots, a_k$, can be used in L_k and no word in L_k can have more than k letters. Thus L_k is finite.

(c) A^* is the countable union of the finite sets L_k; hence A^* is countable.

(d) L is a subset of the countable set A^*; hence L is also countable.

REGULAR EXPRESSIONS, REGULAR LANGUAGES

12.11. Let $A = \{a, b\}$. Describe the language $L(r)$ where:

(a) $r = abb^*a$; (b) $r = b^*ab^*ab^*$; (c) $r = a^* \vee b^*$; (d) $r = ab^* \wedge a^*$.

(a) $L(r)$ consists of all words beginning and ending in a and enclosing one or more b's.

(b) $L(r)$ consists of all words with exactly two a's.

(c) $L(r)$ consists of all words only in a or only in b, that is, $L(r) = \{\lambda, a, a^2, \cdots, b, b^2, \cdots\}$

(d) Here r is not a regular expression since $\wedge$ is not one of the symbols used in forming regular expressions.

12.12. Let $A = \{a, b, c\}$ and let $w = abc$. State whether or not w belongs to $L(r)$ where:

    (a) $r = a^* \vee (b \vee c)^*$;   (b) $r = a^*(b \vee c)^*$.

    (a) No. Here $L(r)$ consists of word in a or words in b and c.

    (b) Yes, since $a \in L(a)^*$ and $bc \in (b \vee c)^*$.

12.13. Let $A = \{a, b\}$. Find a regular expression r such that $L(r)$ consists of all words w where:

    (a) w begins with a^2 and ends with b^2;   (b) w contains an even number of a's.

    (a Let $r = a^2(a \vee b)^*b^2$. (Note $(a \vee b)^*$ consists of all words on A.)

    (b) Note $s = b^*ab^*ab^*$ consists of all words with exactly two a's. Then let $r = s^* = (b^*ab^*ab^*)^*$.

FINITE AUTOMATA

12.14. Let M be the automaton with the following input set A, state set S with initial state s_0, and accepting ("yes") state set Y:

$$A = \{a, b\}, \quad S = \{s_0, s_1, s_2\}, \quad Y = \{s_2\}$$

Suppose next state function F of M is given by the table in Fig. 12-7(a).

    (a) Draw the state diagram $D = D(M)$ of M.

    (b) Describe the language $L = L(M)$ accepted by M.

    (a) The state diagram D appears in Fig. 12.7(b). The vertices of D are the states, and a double circle indicates an accepting state. If $F(s_j, x) = s_k$, then there is a directed edge from s_j to s_k labeled by the input symbol x. Also, there is a special arrow which terminates at the initial state s_0.

    (b) $L(M)$ consists of all words w with exactly one b. Specifically, if an input word w has no b's, then it terminates in s_0 and if w has two or more b's then it terminates in s_2. Otherwise w terminates in s_1, which is the only accepting state.

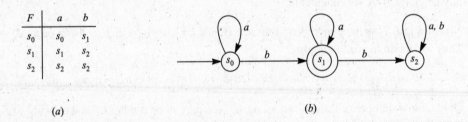

F	a	b
s_0	s_0	s_1
s_1	s_1	s_2
s_2	s_2	s_2

    (a)   (b)

Fig. 12-7

12.15. Let $A = \{a, b\}$. Construct an automaton M which will accept precisely those words from A which have an even number of a's. For example, $aababbab$, aa, bbb, $ababaa$ will be accepted by M, but $ababa$, aaa, $bbabb$ will be rejected by M.

    We need only two states, s_0 and s_1. We assume that M is in state s_0 or s_1 according as the number of a's up to the given step is even or odd. (Thus s_0 is an accepting state, but s_1 is a rejecting state.) Then only a will change the state. Also, s_0 is the initial state. The state diagram of M is shown in Fig. 12-8(a).

12.16. Let $A = \{a, b\}$. Construct an automaton M which will accept those words from A which begin with an a followed by (zero or more) b's.

    The automaton M appears in Fig. 12-8(b).

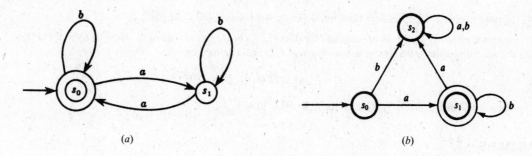

Fig. 12-8

12.17. Describe the words w in the language L accepted by the automaton M in Fig. 12-9(a).

The system can reach the accepting state s_2 only when there exists an a in w which follows a b.

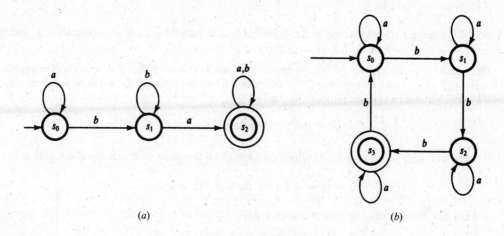

Fig. 12-9

12.18. Describe the words w in the language L accepted by the automaton M in Fig. 12-9(b).

Each a in w does not change the state of the system, whereas each b in w changes the state from R_i, to s_{i+1} (modulo 4). Thus w is accepted by M if the number n of b's in w is congruent to 3 modulo 4, that is, where $n = 3, 7, 11, \cdots$.

12.19. Suppose L is a language over A which is accepted by the automaton $M = (A, S, Y, s_0, F)$. Find an automaton N which accepts L^C, that is, those words from A which do not belong to L.

Simply interchange the accepting and rejecting states in M to obtain N. Then w will be accepted in the new machine N if and only if w is rejected in M, that is, if and only if w belongs to L^C. Formally, $N = (A, S, S \backslash Y, s_0, F)$.

12.20. Let $M = (A, S, Y, s_0, F)$ and $M' = (A, S', Y', S_0', F')$ be automata over the same alphabet A which accept the languages $L(M)$ and $L(M')$ over A, respectively. Construct an automaton N over A which accepts precisely $L(M) \cap L(M')$.

Let $S \times S'$ be the set of states of N. Let (s, s') be an accepting state of N if both s and s' are accepting states in M and M', respectively. Let (s_0, s_0') be the initial state of N. Let the next-state function of N, $G : (S \times S') \times A \to (S \times S')$, be defined by:

$$G((s, s'), a) = (F(s, a), F'(s', a))$$

Then N will accept precisely those words in $L(M) \cap L(M')$.

12.21. Repeat Problem 12.20 except now let N accept precisely $L(M) \cup L(M')$.

Again, let $S \times S'$ be the set of states of N and let (s_0, s_0) be the initial state of N. Now let $(S \times Y') \cup (Y \times S')$ be the accepting states in N. The next-state function G is again defined by

$$G((s, s'), a) = (F(s, a), F'(s', a))$$

Then N will accept precisely those words in $L(M) \cup L(M')$.

GRAMMARS

12.22. Define: (a) context-free grammar; (b) regular grammar.

(a) A context-free grammar is the same as a Type 2 grammar, that is, every production is of the form $A \rightarrow \beta$, that is, the left side is a single variable and the right side is a word in one or more symbols.

(b) A regular grammar is the same as a Type 3 grammar, that is, every production is of the form $A \rightarrow a$ or of the form $A \rightarrow aB$, that is, the left side is a single variable and the right side is either a single terminal or a terminal followed by a variable.

12.23. Find the language $L(G)$ generated by the grammar G with variables S, A, B, terminals a, b, and productions $S \rightarrow aB$, $B \rightarrow b$, $B \rightarrow bA$, $A \rightarrow aB$.

Observe that we can only use the first production once since the start symbol S does not appear anywhere else. Also, we can only obtain a terminal word by finally using the second production. Otherwise we alternately add a's and b's using the third and fourth productions. Accordingly,

$$L(G) = \{(ab)^n = ababab \cdots ab \mid n \in N\}$$

12.24. Let L be the language on $A = \{a, b\}$ which consists of all words w with exactly one b, that is,

$$L = \{b, a^r b, ba^s, a^r ba^s \mid r > 0, \ s > 0\}$$

(a) Find a regular expression r such that $L = L(r)$.

(b) Find a regular grammar G which generates the language L.

(a) Let $r = a^* ba^*$. Then $L(r) = L$.

(b) The regular grammar G with the following productions generates L:

$$S \rightarrow (b, aA), \quad A \rightarrow (b, aA, bB), \quad B \rightarrow (a, aB)$$

That is, the letter b can only appear once in any word derived from S. G is regular since it has the required form.

12.25. Let G be the regular grammar with productions: $S \rightarrow aA$, $A \rightarrow aB$, $B \rightarrow bB$, $B \rightarrow A$.

(a) Find the derivation tree of the word $w = aaba$.

(b) Describe all words w in the language L generated by G.

(a) Note first that w can be derived from S as follows:

$$S \Rightarrow aA \Rightarrow a(aB) \Rightarrow aa(bB) \Rightarrow aaba$$

Figure 12-10(a) shows the corresponding derivation tree.

(b) Using the production 1, then 2, then 3, r times, and finally 4 will derive the word $w = aab^r a$ where $r \geq 0$. No other word can be derived from S.

12.26. Figure 12-10(b) is the derivation tree of a word w in the language L of a context-free grammar G. (a) Find w. (b) Which terminals, variables, and productions must lie in G?

(a) The sequence of leaves from left to right yields the word $w = ababbbba$.

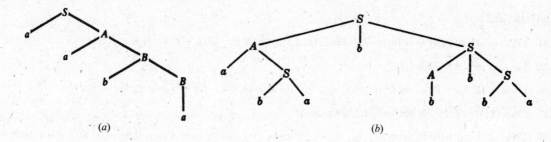

Fig. 12-10

(b) The leaves show that a and b must be terminals, and the internal vertices show that S and A must be variables with S the starting variable. The children of each variable show that $S \to AbS$, $A \to aS$, $S \to ba$, and $A \to b$ must be productions.

12.27. Does a derivation tree exist for any word w derived from the start symbol S in a grammar G?

No. Derivation trees only exist for Type 2 and 3 grammars, that is, for context-free and regular grammars.

12.28. Determine the type of grammar G which consists of the following productions:

(a) $S \to aA$, $A \to aAB$, $B \to b$, $A \to a$

(b) $S \to aAB$, $AB \to bB$, $B \to b$, $A \to aB$

(c) $S \to aAB$, $AB \to a$, $A \to b$, $B \to AB$

(d) $S \to aB$, $B \to bA$, $B \to b$, $B \to a$, $A \to aB$, $A \to a$

(a) Each production is of the form $A \to \alpha$; hence G is a context-free or Type 2 grammar.

(b) The length of the left side of each production does not exceed the length of the right side; hence G is a Type 1 grammar.

(c) The production $AB \to a$ means G is a Type 0 grammar.

(d) G is a regular or Type 3 grammar since each production has the form $A \to a$ or $A \to aB$.

12.29. Rewrite each grammar G in Problem 12.28 in Backus-Naur form.

The Backus-Naur form only applies to context-free grammars (which includes regular grammars). Thus only (a) and (d) can be written in Backus-Naur form. The form is obtained as follows:

(i) Replace $\to$ by $::=$.

(ii) Enclose nonterminals in brackets $\langle \rangle$.

(iii) All productions with the same left-hand side are combined in one statement with all the right-hand sides listed on the right of $::=$ separated by vertical bars.

Accordingly:

(a) $\langle S \rangle ::= a \langle A \rangle$, $\langle A \rangle ::= a \langle A \rangle \langle B \rangle \mid a$, $\langle B \rangle ::= b$

(b) $\langle S \rangle ::= a \langle B \rangle$, $\langle B \rangle ::= b \langle A \rangle \mid b \mid a$, $\langle A \rangle ::= a \langle B \rangle \mid a$

Supplementary Problems

WORDS

12.30. Consider the words $u = ab^2 a^3$ and $v = aba^2 b^2$. Find: (a) uv;   (b) vu;   (c) u^2;   (d) λu;   (e) $v \lambda v$.

12.31. For the words $u = ab^2 a^3$ and $v = aba^2 b^2$, find: $|u|$, $|v|$, $|uv|$, $|vu|$, and $|v^2|$.

12.32. Let $w = abcde$. (a) Find all subwords of w. (b) Which of them are initial segments?

12.33. Suppose $u = a_1 a_2 \cdots a_r$ and the a_k are distinct. Find the number n of subwords of u.

LANGUAGES

12.34. Let $L = \{a^2, ab\}$ and $K = \{a, ab, b^2\}$. Find: (a) LK;   (b) KL;   (c) $L \vee K$;   (d) $K \wedge L$.

12.35. Let $L = \{a^2, ab\}$. Find: (a) L^0;   (b) L^2;   (c) L^3.

12.36. Let $A = \{a, b, c\}$. Describe L^* if: (a) $L = \{a^2\}$;   (b) $L = \{a, b^2\}$;   (c) $L = \{a, b^2, c^3\}$.

12.37. Does $(L^2)^* = (L^*)^2$? If not, how are they related?

12.38. Consider a countable alphabet $A = \{a_1, a_2, \cdots\}$. Let L_k the language over A consisting of those words w such that the sum of the subscripts of the letters in w is equal to k. (See Problem 12.10.) Find: (a) L_3;   (b) L_5.

REGULAR EXPRESSIONS, REGULAR LANGUAGES

12.39. Let $A = \{a, b, c\}$. Describe the language $L(r)$ for each of the following regular expressions:

(a) $r = ab^*c$;   (b) $r = (ab \vee c)^*$;   (c) $r = ab \vee c^*$.

12.40. Let $A = \{a, b\}$. Find a regular expression r such that $L(r)$ consists of all words w where:

(a) w contains exactly three a's.

(b) The number of a's is divisible by 3.

12.41. Let $A = \{a, b, c\}$ and let $w = ac$. State whether or not w belongs to $L(r)$ where:

(a) $r = a^*bc^*$;   (b) $r = a^*b^*c$;   (c) $r = (ab \vee c)^*$

12.42. Let $A = \{a, b, c\}$ and let $w = abc$. State whether or not w belongs to $L(r)$ where:

(a) $r = ab^*(bc)^*$;   (b) $r = a^* \vee (b \vee c)^*$;   (c) $r = a^*b(bc \vee c^2)^*$.

FINITE AUTOMATA

12.43. Let $A = \{a, b\}$. Construct an automaton M such that $L(M)$ will consist of those words w where:
(a) the number of b's is divisible by 3. (b) w begins with a and ends in b.

12.44. Let $A = \{a, b\}$. Construct an automaton M which will accept the language:

(a) $L(M) = \{b^r ab^s \mid r > 0, s > 0\}$;   (b) $L(M) = \{a^r b^s \mid r > 0, s > 0\}$.

12.45. Let $A = \{a, b\}$. Construct an automaton M such that $L(M)$ will consist of those words where the number of a's is divisible by 2 and the number of b's is divisible by 3.

(*Hint*: Use Problems 12.15, 12.43(a), and 12.20.)

12.46. Find the language $L(M)$ accepted by the automaton M in Fig. 12-11.

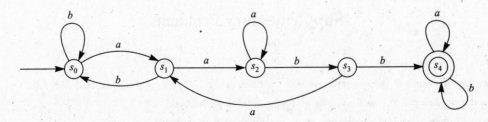

Fig. 12-11

GRAMMARS

12.47. Determine the type of grammar G which consists of the productions:

$$(a)\ S \to aAB; \quad S \to AB; \quad A \to a; \quad B \to b$$
$$(b)\ S \to aB; \quad B \to AB; \quad aA \to b; \quad A \to a; \quad B \to b$$
$$(c)\ S \to aB; \quad B \to bB; \quad B \to bA; \quad A \to a; \quad B \to b$$

12.48. Find a regular grammar G which generates the language L which consists of all words in a and b such that no two a's appear next to each other.

12.49. Find a context-free grammar G which generates the language L which consists of all words in a and b with twice as many a's as b's.

12.50. Find a grammar G which generates the language L which consists of all words in a and b with an even number of a's.

12.51. Find a grammar G which generates the language L which consists of all words of the form $a^n b a^n$ with $n \geq 0$.

12.52. Show that the language G in Problem 12.51 is not regular. (*Hint*: Use the Pumping Lemma.)

12.53. Describe the language $L = L(G)$ where G has the productions $S \to aA$, $A \to bbA$, $A \to c$.

12.54. Describe the language $L = L(G)$ where G has the productions $S \to aSb$, $Sb \to bA$, $abA \to c$.

12.55. Write each grammar G in Problem 12.47 in Backus-Naur form.

12.56. Let G be the context-free grammar with productions $S \to (a, aAS)$ and $A \to bS$.

(a) Write G in Backus-Naur form. (b) Find the derivation tree of the word $w = abaa$.

12.57. Figure 12-12 is the derivation tree of a word w in a language L of a context-free grammar G.

(a) Find w. (b) Which terminals, variables, and productions must belong to G?

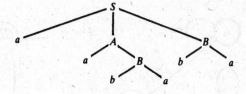

Fig. 12-12

Answers to Supplementary Problems

12.30. (a) $uv = ab^2 a^4 ba^2 b^2$; (b) $vu = aba^2 b^2 ab^2 a^3$;
(c) $u^2 = ab^2 a^4 b^2 a^3$; (d) $\lambda u = u$;
(e) $v\lambda v = v^2 = aba^2 b^2 aba^2 b^2$.

12.31. 6, 6, 12, 12, 12.

12.32. (a) λ, a, b, c, d, e, ab, bc, cd, de, abc, bcd, cde, $abcd$, $bcde$, $w = abcde$.

(b) λ, a, ab, abc, $abcd$, $w = abcde$.

12.33. If $\mathbf{u} = \lambda$ then $n = 1$; otherwise, $n = 1 + [r + (r-1) + \cdots + 2 + 1] = 1 + r(r+1)/2$.

12.34. (a) $LK = \{a^3,\ a^3 b,\ a^2 b^2,\ aba,\ abab,\ ab^3\}$;
(b) $KL = \{a^3,\ a^2 b,\ aba^2,\ abab,\ b^2 a^2,\ b^2 ab\}$;
(c) $L \vee K = \{a^2,\ ab,\ a,\ b^2\}$; (d) $K \wedge L$　not defined.

12.35. (a) $L^0 = \{\lambda\}$; (b) $L^2 = \{a^4,\ a^3 b,\ aba^2,\ abab\}$;
(c) $L^3 = \{a^6,\ a^5 b,\ a^3 ba^2,\ a^3 bab,\ aba^4,\ aba^3 b,\ ababa^2,\ ababab\}$.

12.36. (a) $L* = \{a^n \mid n \text{ is even}\}$. (b) All words w in a and b with only even powers of b.

(c) All words in a, b, c with each power of b even and each power of c a multiple of 3.

12.37. No. $(L^2)^* \subseteq (L^*)^2$.

12.38. (a) $a_1 a_1 a_1$, $a_1 a_2$, $a_2 a_1 a_3$ (b) $a_1 a_1 a_1 a_1 a_1$, $a_1 a_1 a_1 a_2$, $a_1 a_1 a_2 a_1$, $a_1 a_2 a_1 a_1$, $a_2 a_1 a_1 a_1$, $a_1 a_1 a_3$, $a_1 a_3 a_1$, $a_3 a_1 a_1$, $a_2 a_1 a_3$, $a_3 a_2$, $a_1 a_4$, $a_4 a_1$, a_5

12.39. (a) $L(r) = \{ab^n c \mid n \geq 0\}$. (b) All words in x and c where $x = ab$. (c) $L(r) = ab \cup \{c^n \mid n \geq 0\}$.

12.40. (a) $r = b^* ab^* ab^* ab^*$; (b) $r = (b^* ab^* ab^* ab^*)^*$.

12.41. (a) No; (b) yes; (c) no.

12.42. (a) Yes; (b) no; (c) no.

12.43. See: (a) Fig. 12-13(*a*); (b) Fig. 12-13(*b*).

12.44. See: (a) Fig. 12-14; (b) Fig. 12-15(*a*).

12.45. See: Fig. 12-15(*b*).

12.46. $L(M)$ consists of all words w which contain $aabb$ as a subword.

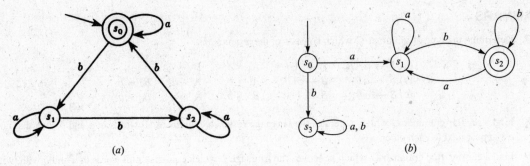

Fig. 12-13

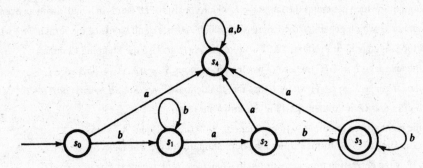

Fig. 12-14

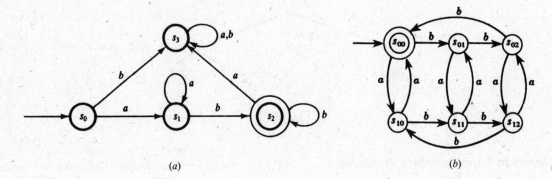

Fig. 12-15

12.47. (a) Type 2; (b) Type 0; (c) Type 3.

12.48. $S \rightarrow (a, b, aB, bA)$, $A \rightarrow (bA, ab, a, b)$, $B \rightarrow$ (b, bA).

12.49. $S \rightarrow (AAB, ABA, BAA)$, $A \rightarrow (a, BAAA, ABAA, AABA, AAAB)$,
$B \rightarrow (b, BBAA, BABA, aBAAB, ABAB, AABB)$.

12.50. $S \rightarrow (aA, bB)$, $A \rightarrow (aB, bA, a)$, $B \rightarrow$ (bB, aA, b)

12.51. $S \rightarrow (aSa, b)$.

12.53. $L = \{ab^{2n}c \mid n \geq 0\}$

12.54. $L = \{a^n cb^n \mid n > 0\}$

12.55. (a) $\langle S \rangle ::= a\langle A \rangle \langle B \rangle \mid \langle A \rangle \langle B \rangle$, $\langle A \rangle ::= a$, $\langle B \rangle ::= b$.
(b) Not defined for Type 0 language.

(c) $\langle S \rangle ::= a \langle B \rangle$, $\langle B \rangle ::= b \langle B \rangle \mid b \langle A \rangle$, $\langle A \rangle ::= a \mid b$.

12.56. (a) $\langle S \rangle ::= a \mid a \langle A \rangle \langle S \rangle$, $\langle A \rangle ::= b \langle S \rangle$; (b) See Fig. 12-16.

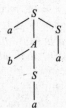

Fig. 12-16

12.57. (a) $w = aababa$; (b) $S \rightarrow aAB$, $A \rightarrow aB$, $B \rightarrow ba$.

CHAPTER 13

Finite State Machines and Turing Machines

13.1 INTRODUCTION

This chapter discusses two types of "machines." The first is a finite state machine (FSM) which is similar to a finite state automaton (FSA) except that the finite state machine "prints" an output using an output alphabet which may be distinct from the input alphabet. The second is the celebrated Turing machine which may be used to define computable functions.

13.2 FINITE STATE MACHINES

A finite state machine (or complete sequential machine) M consists of six parts:

(1) A finite set A of input symbols.   (4) An initial state s_0 in S.
(2) A finite set S of "internal" states.   (5) A next-state function f from $S \times A$ into S.
(3) A finite set Z of output symbols.   (6) An output function g from $S \times A$ into Z.

Such a machine M is denoted by $M = M(A, S, Z, s_0, f, g)$ when we want to indicate its six parts.

EXAMPLE 13.1 The following defines a finite state machine M with two input symbols, three internal states, and three output symbols:
(1) $A = \{a, b\}$,   (2) $S = \{s_0, s_1, s_2\}$,   (3) $Z = \{x, y, z\}$,   (4) Initial state s_0,

(5) Next-state function $f : S \times A \to S$ defined by:

$$f(s_0, a) = s_1, \quad f(s_1, a) = s_2, \quad f(s_2, a) = s_0$$
$$f(s_0, b) = s_2, \quad f(s_1, b) = s_1, \quad f(s_2, b) = s_1$$

(6) Output function $g : S \times A \to Z$ defined by:

$$g(s_0, a) = x, \quad g(s_1, a) = x, \quad g(s_2, a) = z$$
$$g(s_0, b) = y, \quad g(s_1, b) = z, \quad g(s_2, b) = y$$

State Table and State Diagram of a Finite State Machine

There are two ways of representing a finite state machine M in compact form. One way is by a table called the *state table* of the machine M, and the other way is by a labeled directed graph called the *state diagram* of the machine M.

The state table combines the next-state function f and the output function g into a single table which represent the function $F : S \times A \to S \times Z$ defined as follows:

$$F(s_i, a_j) = [f(s_i, a_j), g(s_i, a_j)]$$

For instance, the state table of the machine M in Example 13.1 is pictured in Fig. 13-1(a). The states are listed on the left of the table with the initial state first, and the input symbols are listed on the top of the table. The entry in the table is a pair (s_k, z_r) where $s_k = f(s_i, a_j)$ is the next state and $z_r = g(s_i, a_j)$ is the output symbol. One assumes that there are no output symbols other than those appearing in the table.

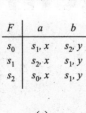

F	a	b
s_0	s_1, x	s_2, y
s_1	s_2, x	s_1, y
s_2	s_0, x	s_1, y

(a)

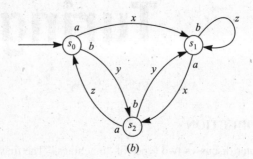

(b)

Fig. 13-1

The state diagram $D = D(M)$ of a finite state machine $M = M(A, S, Z, s_0, f, g)$ is a labeled directed graph. The vertices of D are the states of M. Moreover, if

$$F(s_i, a_j) = (s_k, z_r), \quad \text{or equivalently,} \quad f(s_i, a_j) = s_k \quad \text{and} \quad g(s_i, a_j) = z_r$$

then there is an arc (arrow) from s_i to s_k which is labeled with the pair a_j, z_r. We usually put the input symbol a_i near the base of the arrow (near s_i) and the output symbol z_r near the center of the arrow. We also label the initial state s_0 by drawing an extra arrow into s_0. For instance, the state diagram of the machine M in Example 13.1 appears in Fig. 13-1(b).

Input and Output Tapes

The above discussion of a finite state machine M does not show the dynamic quality of M. Suppose M is given a string (word) of input symbols, say

$$u = a_1 a_2 \ldots a_m$$

We visualize these symbols on an "input tape." The machine M "reads" these input symbols one by one and, simultaneously, changes through a sequence of states

$$v = s_0 s_1 s_2 \ldots s_m$$

where s_0 is the initial state, while printing a string (word) of output symbols

$$w = z_1 z_2 \ldots z_m$$

on an "output tape." Formally, the initial state s_0 and the input string u determine the strings v and w as follows, where $i = 1, 2, \ldots, m$:

$$s_i = f(s_{i-1}, a_i) \quad \text{and} \quad z_i = g(s_{i-1}, a_i)$$

EXAMPLE 13.2 Consider the machine M in Fig. 13-1, that is, Example 13.1. Suppose the input is the word

$$u = abaab$$

We calculate the sequence v of states and the output word w from the state diagram as follows. Beginning at the initial state s_0 we follow the arrows which are labeled by the input symbols as follows:

$$s_0 \xrightarrow{a,x} s_1 \xrightarrow{b,z} s_1 \xrightarrow{a,x} s_2 \xrightarrow{a,z} s_0 \xrightarrow{b,y} s_2$$

This yields the following sequence v of states and output word w:

$$v = s_0 s_1 s_1 s_2 s_0 s_2 \quad \text{and} \quad w = xzxzy$$

Binary Addition

This subsection describes a finite state machine M which can do binary addition. By adding 0's at the beginning of our numbers, we can assume that our numbers have the same number of digits. If the machine is given the input

$$1101011$$
$$\underline{+0111011}$$

then we want the output to be the binary sum 10100110. Specifically, the input is the string of pairs of digits to be added:

$$11, \quad 11, \quad 00, \quad 11, \quad 01, \quad 11, \quad 10, \quad b$$

where b denotes blank spaces, and the output should be the string:

$$0, \quad 1, \quad 1, \quad 0, \quad 0, \quad 1, \quad 0, \quad 1$$

We also want the machine to enter a state called "stop" when the machine finishes the addition.

The input symbols and output symbols are, respectively, as follows:

$$A = \{00, 01, 10, 11, b\} \quad \text{and} \quad Z = \{0, 1, b\}$$

The machine M that we "construct" will have three states:

$$S = \{\text{carry } (c), \text{ no carry } (n), \text{ stop } (s)\}$$

Here n is the initial state. The machine is shown in Fig. 13-2.

In order to show the limitations of our machines, we state the following theorem.

Theorem 13.1: There is no finite state machine M which can do binary multiplication.

If we limit the size of the numbers that we multiply, then such machines do exist. Computers are important examples of finite state machines which multiply numbers, but the numbers are limited as to their size.

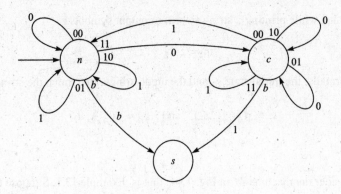

Fig. 13-2

13.3 GÖDEL NUMBERS

Recall (Section 11.5) that a positive integer $p > 1$ is called a prime number if its only positive divisors are 1 and p. We let $p_1, p_2, p_3, \ldots$ denote the successive prime numbers. Thus

$$p_1 = 2, \quad p_2 = 3, \quad p_3 = 5, \quad p_4 = 7, \quad p_5 = 11, \ldots$$

(By Theorem 11.12, there exists an infinite number of primes.) The Fundamental Theorem of Arithmetic (Theorem 11.20) states that any positive integer $n > 1$ can be written uniquely (except for order) as a product of prime numbers. The German logician Kurt Gödel used this result to code finite sequences of numbers and also to code words over a finite or countable alphabet. Each sequence or word is assigned a positive integer called its *Gödel number* as follows.

The Gödel number of the sequence $s = (n_1, n_2, \ldots, n_k)$ of nonnegative integers is the positive integer $c(s)$ where n_i is the exponent of p_i in the prime decomposition of $c(s)$, that is,

$$c(s) = p_1^{n_1} p_2^{n_2} \cdots p_k^{n_k}$$

For example,

$$s = (3, 1, 2, 0, 2) \quad \text{is coded by} \quad c(s) = 2^3 \cdot 3 \cdot 5^2 \cdot 7^0 \cdot 11^2 = 72\,600$$

The Gödel number of a word w on an alphabet $\{a_0, a_1, a_2, a_3, \ldots\}$ is the positive integer $c(w)$ where the subscript of the ith letter of w is the exponent of p_i in the prime decomposition of $c(w)$. For instance,

$$w = a_4 a_1 a_3 a_2 a_2 \quad \text{is coded by} \quad c(w) = 2^4 \cdot 3 \cdot 5^3 \cdot 7^2 \cdot 11^2$$

(Observe that both codes are essentially the same since we may view a word w as the sequence of the subscripts of its letters.)

The above coding is essentially the proof of the main result of this section:

Theorem 13.2: Suppose an alphabet A is countable. Then any language L over A is also countable.

Proof: The Gödel coding is a one-to-one mapping $c: L \to \mathbf{N}$. Thus L is countable.

13.4 TURING MACHINES

There are a number of equivalent ways to formally define a "computable" function. We do it by means of a Turing machine M. This section formally defines a Turing machine M, and the next section defines a computable function.

Our definition of a Turing machine uses an infinite two-way tape, quintuples, and three halt states. Other definitions use a one-way infinite tape and/or quadruples, and one halt state. However, all the definitions are equivalent.

Basic Definitions

A *Turing machine* M involves three disjoint nonempty sets:

(1) A finite *tape* set where $B = a_0$ is the "blank" symbol:

$$A = \{a_1, a_2, \ldots, a_m\} \cup \{B\}$$

(2) A finite *state* set where s_0 is the *initial state*:

$$S = \{s_1, s_2, \ldots, s_n\} \cup \{s_0\} \cup \{s_H, s_Y, s_N\}$$

Here s_H (HALT) is the halting state, s_Y (YES) is the accepting state, and s_N (NO) is the nonaccepting state.

(3) A *direction* set where L denotes "left" and R denotes "right:"

$$d = \{L, R\}$$

Definition 13.1: An *expression* is a finite (possibly empty) sequence of elements from $A \cup S \cup d$.

In other words, an expression is a word whose letters (symbols) come from the sets A, S, and d.

Definition 13.2: A *tape expression* is an expression using only elements from the tape set A.

The Turing machine M may be viewed as a read/write tape head which moves back and forth along an infinite tape. The tape is divided lengthwise into squares (cells), and each square may be blank or hold one tape symbol. At each step in time, the Turing machine M is in a certain internal state s_i scanning one of the tape symbols a_j on the tape. We assume that only a finite number of nonblank symbols appear on the tape.

Figure 13-3(a) is a picture of a Turing machine M in state s_2 scanning the second symbol where $a_1 a_3 B a_1 a_1$ is printed on the tape. (Note again that B is the blank symbol.) This picture may be represented by the expression $\alpha = a_1 s_2 a_3 B a_1 a_1$ where we write the state s_2 of M before the tape symbol a_3 that M is scanning. Observe that α is an expression using only the tape alphabet A except for the state symbol s_2 which is not at the end of the expression since it appears before the tape symbol a_3 that M is scanning. Figure 13-3 shows two other informal pictures and their corresponding picture expressions.

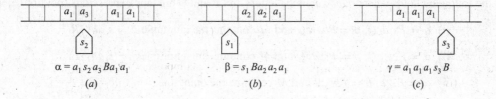

$$\alpha = a_1 s_2 a_3 B a_1 a_1 \qquad\qquad \beta = s_1 B a_2 a_2 a_1 \qquad\qquad \gamma = a_1 a_1 a_1 s_3 B$$
$$(a) \qquad\qquad\qquad\qquad (b) \qquad\qquad\qquad\qquad (c)$$

Fig. 13-3

We give formal definitions.

Definition 13.3: A *picture* α is an expression as follows where P and Q are tape expressions (possibly empty):

$$\alpha = P s_i a_k Q$$

Definition 13.4: Let $\alpha = P s_i a_k Q$ be a picture. We say that the Turing machine M *is in state s_i scanning the letter a_k* and that the *expression on the tape* is the expression $P a_k Q$, that is, without its state symbol s_i.

As mentioned above, at each step in time the Turing machine M is in a certain state s_i and is scanning a tape symbol a_k. The Turing machine M is able to do the following three things simultaneously:

 (i) M erases the scanned symbol a_k and writes in its place a tape symbol a_l (where we permit $a_l = a_k$).

(ii) M changes its internal states s_i to a state s_j (where we permit $s_j = s_j$).

(iii) M moves one square to the left or moves one square to the right.

The above action by M may be described by a five-letter expression called a *quintuple* which we define below.

Definition 13.5: A quintuple q is a five-letter expression of the following form:

$$q = \left(s_i, a_k, a_l, s_j, \begin{Bmatrix} L \\ R \end{Bmatrix} \right)$$

That is, the first letter of q is a state symbol, the second is a tape symbol, the third is a tape symbol, the fourth is a state symbol, and the last is a direction symbol L or R.

Next we give a formal definition of a Turing machine.

Definition 13.6: A Turing machine M is a finite set of quintuples such that:

 (i) No two quintuples begin with the same first two letters.

(ii) No quintuple begins with s_H, s_Y, or s_N.

Condition (i) in the definition guarantees that the machine M cannot do more than one thing at any given step, and condition (ii) guarantees that M halts in state s_H, s_Y, or s_N.

The following is an alternative equivalent definition.

Definition 13.6′: Turing machine M is a partial function from

$$S \backslash \{ s_H, s_Y \text{ or } s_N \} \times A \quad \text{into} \quad A \times S \times d$$

The term partial function simply means that domain of M is a subset of $S \backslash \{s_H, s_Y, \text{ or } s_N\} \times A$.

The action of the Turing machine described above can now be formally defined.

Definition 13.7: Let α and β be pictures. We write

$$\alpha \to \beta$$

if one of the following holds where a, b, c are tape letters and P and Q are tape expressions (possibly empty):

 (i) $\alpha = Ps_i acQ$, $\beta = Pbs_j cQ$ and M contains the quintuple $q = s_i abs_j R$.

(ii) $\alpha = Pcs_i aQ$, $\beta = Ps_j cbQ$ and M contains the quintuple $q = s_i abs_j L$.

(iii) $\alpha = Ps_i a$, $\beta = Pbs_j B$ and M contains the quintuple $q = s_i abs_j R$.

(iv) $\alpha = s_i aQ$, $\beta = s_j BbQ$ and M contains the quintuple $q = s_i abs_j L$.

Observe that, in all four cases, M replaces a on the tape by b (where we permit $b = a$), and M changes its state from s_i to s_j (where we permit $s_j = s_i$). Furthermore:

 (i) Here M moves to the right.

(ii) Here M moves to the left.

(iii) Here M moves to the right; however, since M is scanning the rightmost letter, it must add the blank symbol B on the right.

(iv) Here M moves to the left; however, since M is scanning the leftmost letter, it must add the blank symbol B on the left.

Definition 13.8: A picture α is said to be *terminal* if there is no picture β such that $\alpha \to \beta$.

In particular, any picture α in one of the three halt states must be terminal since no quintuple begins with s_H, s_Y or s_n.

Computing with a Turing Machine

The above is a static (one-step) description of a Turing machine M. Now we discuss its dynamics.

Definition 13.9: A *computation* of a Turing machine M is a sequence of pictures $\alpha_1, \alpha_2, \ldots, \alpha_m$ such that $\alpha_{i-1} \to \alpha_i$, for $i = 1, 2, \ldots, m$, and α_m is a terminal picture.

In other words, a computation is a sequence

$$\alpha_0 \to \alpha_1 \to \alpha_2 \to \ldots \to \alpha_m$$

which cannot be extended since α_m is terminal. We will let term(α) denote the final picture of a computation beginning with α. Thus term$(\alpha_0) = \alpha_m$ in the above computation.

Turing Machines with Input

The following definition applies.

Definition 13.10: An *input* for a Turing machine M is a tape expression W. The *initial picture* for an input W is $\alpha(W)$ where $\alpha(W) = s_0(W)$.

Observe that the initial picture $\alpha(W)$ of the input W is obtained by placing the initial state s_0 in front of the input tape expression W. In other words, the Turing machine M begins in its initial state s_0 and it is scanning the first letter of W.

Definition 13.11: Let M be a Turing machine and let W be an input. We say M halts on W if there is a computation beginning with the initial picture $\alpha(W)$.

That is, given an input W, we can form the initial picture $\alpha(W) = s_0(W)$ and apply M to obtain the sequence

$$\alpha(W) \to \alpha_1 \to \alpha_2 \to \ldots$$

Two things can happen:

(1) **M halts on W.** That is, the sequence ends with some terminal Picture α_r.

(2) **M does not halt on W.** That is, the sequence never ends.

Grammars and Turing Machines

Turing machines may be used to recognize languages. Specifically, suppose M is a Turing machine with tape set A. Let L be the set of words W in A such that M halts in the accepting state s_Y when W is the input. We will then write $L = L(M)$, and we will say that M recognizes the language L. Thus an input W does not belong to $L(M)$ if M does not halt on W or if M halts on W but not in the accepting state s_Y.

The following theorem is the main result of this subsection; its proof lies beyond the scope of this text.

Theorem 13.3: A language L is recognizable by a Turing machine M if and only if L is a type 0 language.

Remark: The reason for three halt states is that s_Y and s_N are used for recognizing languages, whereas s_H is used for computations discussed in the next section.

EXAMPLE 13.3 Suppose a Turing machine M with tape set $A = \{a, b, c\}$ contains the following four quintuples:

$$q_1 = s_0 a a s_0 R, \quad q_2 = s_0 b b s_0 R, \quad q_3 = s_0 B B s_N R, \quad q_4 = s_0 c c s_Y R$$

(a) Suppose $W = W(a, b, c)$ is an input Without any c's.

By the quintuples q_1 and q_2, M stays in state s_0 and moves to the right until it encounters a blank symbol B. Then M changes its state to the nonaccepting state s_N and halts.

(b) Suppose $W = W(a, b, c)$ is an input with at least one c symbol.

By the quintuple q_4, when M initially meets the first c in W it changes its state to the accepting state s_Y and halts.

Thus M recognizes the language L of all words W in a, b, c with at least one letter c. That is, $L = L(M)$.

13.5   COMPUTABLE FUNCTIONS

Computable functions are defined on the set of nonnegative integers. Some texts use $\mathbf{N}$ to denote this set. We use $\mathbf{N}$ to denote the set of positive integers, so we will use the notation

$$\mathbf{N_0} = \{0, 1, 2, 3, \ldots\}$$

Throughout this section, the terms number, integer, and nonnegative integer are used synonymously. The preceding section described the way a Turing machine M manipulates and recognizes character data. Here we show how M manipulates numerical data. First, however, we need to be able to represent our numbers by our tape set A. We will write 1 for the tape symbol a_1 and 1^n for $111\ldots1$, where 1 occurs n times.

Definition 13.12: Each number n will be represented by the tape expression $\langle n \rangle$ where $\langle n \rangle = 1^{n+1}$. Thus:

$$\langle 4 \rangle = 11111 = 1^5, \quad \langle 0 \rangle = 1, \quad \langle 2 \rangle = 111 = 1^3.$$

Definition 13.13: Let E be an expression. Then $[E]$ will denote the number of times 1 occurs in E. Thus

$$[11Bs_2a_3111Ba_4] = 5, \quad [a_4s_2Ba_2] = 0, \quad [\langle n \rangle] = n + 1.$$

Definition 13.14: A function $f: \mathbf{N_0} \to \mathbf{N_0}$ is computable if there exists a Turing machine M such that, for every integer n, M halts on $\langle n \rangle$ and

$$f(n) = [\text{term}(\alpha(\langle n \rangle))]$$

We then say that M computes f.

That is, given a function f and an integer n, we input $\langle n \rangle$ and apply M. If M always halts on $\langle n \rangle$ and the number of 1's in the final picture is equal to $f(n)$, then f is a computable function and we say that M computes f.

EXAMPLE 13.4 The function $f(n) = n + 3$ is computable. The input is $W = 1^{n+1}$. Thus we need only add two 1's to the input. A Turing machine M which computes f follows:

$$M = \{q_1, q_2, q_3\} = \{s_0 1 s_0 L, \quad s_0 B 1 s_1 L, \quad s_1 B 1 s_H L\}$$

Observe that:

(1) q_1 moves the machine M to the left.

(2) q_2 writes 1 in the blank square B, and moves M to the left.

(3) q_3 writes 1 in the blank square B, and halts M.

Accordingly, for any positive integer n,

$$s_0 1^{n+1} \to s_0 B 1^{n+1} \to s_1 B 1^{n+2} \to s_H B 1^{n+3}$$

Thus M computes $f(n) = n + 3$. It is clear that, for any positive integer k, the function $f(n) = n + k$ is computable.

The following theorem applies.

Theorem 13.4: Suppose $f : \mathbf{N_0} \to \mathbf{N_0}$ and $g : \mathbf{N_0} \to \mathbf{N_0}$ are computable. Then the composition function $h = g \circ f$ is computable.

We indicate the proof of this theorem here. Suppose M_f and M_g are the Turing machines which compute f and g, respectively. Given the input $\langle n \rangle$, we apply M_f to $\langle n \rangle$ to finally obtain an expression E with $[E] = f(n)$. We then arrange that $E = s_0 1^{f(n)}$. We next add 1 to E to obtain $E' = s_0 11^{f(n)}$ and apply M_g to E'. This yields E'' where $[E''] = g(f(n)) = (g \circ f)(n)$, as required.

Functions of Several Variables

This subsection defines a computable function $f(n_1, n_2, \ldots, n_k)$ of k variables. First we need to represent the list $m = (n_1, n_2, \ldots, n_k)$ in our alphabet A.

Definition 13.15: Each list $m = (n_1, n_2, \ldots, n_k)$ of k integers is represented by the tape expression

$$\langle m \rangle = \langle n_1 \rangle B \langle n_2 \rangle B \cdots B \langle n_k \rangle$$

For example, $\langle (2, 0, 4) \rangle = 111B1B11111 = 1^3 B 1^1 B 1^5$.

Definition 13.16: A function $f(n_1, n_2, \ldots, n_k)$ of k variables is computable if there is a Turing machine M such that, for every list $m = (n_1, n_2, \ldots, n_k)$, M halts on $\langle m \rangle$ and

$$f(m) = [\text{term}(\alpha(\langle m \rangle))]$$

We then say that M computes f.

The definition is analogous to Definition 13.14 for one variable.

EXAMPLE 13.5 The addition function $f(m, n) = m + n$ is computable. The input is $W = 1^{m+1} B 1^{n+1}$. Thus we need only erase two of the 1s. A Turing machine M which computes f follows:

$$M = \{q_1, q_2, q_3, q_4\} = \{s_0 1 B s_1 R, \ s_1 1 B s_H R, \ s_1 B B s_2 R, \ s_2 1 B s_H R\}$$

Observe that:

(1) q_1 erases the first 1 and moves M to the right.

(2) If $m \neq 0$, then q_2 erases the second 1 and halts M.

(3) If $m = 0$, q_3 moves M to the right past the blank space B.

(4) q_4 erases the 1 and halts M.

Accordingly, if $m \neq 0$ we have:

$$s_0 1^{m+1} B 1^{n+1} \to s_1 1^m B 1^{n+1} \to s_H 1^{m-1} B 1^{n+1}$$

but if $m = 0$ and $m + n = n$, we have

$$s_0 1 B 1^{n+1} \to s_1 B 1^{n+1} \to s_2 1^{n+1} \to s^H 1^n$$

Thus M computes $f(m, n) = m + n$.

Solved Problems

FINITE STATE MACHINES

13.1. Let M be the finite state machine with state table appearing in Fig. 13-4(a).

    (a) Find the input set A, the state set S, the output set Z, and the initial state.

    (b) Draw the state diagram $D = D(M)$ of M.

    (c) Suppose $w = aababaabbab$ is an input word (string). Find the corresponding output word v.

    (a) The input symbols are on the top of the table, the states are listed on the left, and the output symbols appear in the table. Thus:
$$A = \{a, b\}, \quad S = \{s_0, s_1, s_2, s_3\}, \quad Z = \{x, y, z\}$$
The state s_0 is the initial state since it is the first state listed in the table.

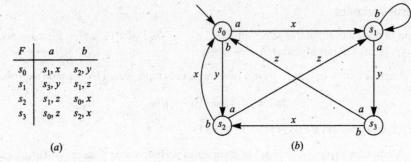

F	a	b
s_0	s_1, x	s_2, y
s_1	s_3, y	s_1, z
s_2	s_1, z	s_0, x
s_3	s_0, z	s_2, x

(a)

(b)

Fig. 13-4

(b) The state diagram $D = D(M)$ appears in Fig. 13-4(b). Note that the vertices of D are the states of M. Suppose $F(s_i, a_j) = (s_k, z_r)$. (That is, $f(s_i, a_j) = s_k$ and $g(s_i, a_j) = z_r$.)

Then there is a directed edge from s_i to s_k labeled by the pair a_j, z_r. Usually, the input symbol a_j is put near the base of the arrow (near s_i) and output symbol z_r is put near the center of the arrow.

(d) Starting at the initial state s_0, we move from state to state by the arrows which are labeled respectively, by the given input symbols as follows:

$$s_0 \xrightarrow{a} s_1 \xrightarrow{a} s_3 \xrightarrow{b} s_2 \xrightarrow{a} s_1 \xrightarrow{b} s_1 \xrightarrow{a} s_3 \xrightarrow{a} s_0 \xrightarrow{b} s_2 \xrightarrow{b} s_0 \xrightarrow{a} s_1 \xrightarrow{b} s_1$$

The output symbols on the above arrows yield the required output word $v = xyxzzyzxxz$.

13.2. Let M be the finite state machine with input set $A = \{a, b\}$, output set $Z = \{x, y, z\}$, and state diagram $D = D(M)$ in Fig. 13-5(a).

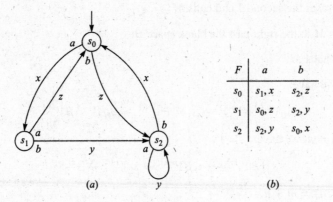

F	a	b
s_0	s_1, x	s_2, z
s_1	s_0, z	s_2, y
s_2	s_2, y	s_0, x

(a)

(b)

Fig. 13-5

(a) Construct the state table of M.

(b) Find the output word v if the input is the word: (i) $w = a^2 b^2 abab$; (ii) $w = abab^3 a^2$.

(a) The state table appears in Fig. 13-5(b). Since s_0 is the initial state, it is listed first.

Also, $F(s_i, a_j) = (s_k, z_r)$ if there is a directed edge from s_i to s_k labeled by the pair a_j, z_r.

(b) Move from state to state by the arrows which are labeled, respectively, by the given input symbols to obtain the following output: (i) $v = xz^2 x^2 y^2 x$; (ii) $v = xy^2 xzxx^2 z$.

TURING MACHINES

13.3. Let M be a Turing machine. Determine the picture α corresponding to each situation:

(a) M is in state s_3 and scanning the third letter of the tape expression $w = aabca$.

(b) M is in state s_2 and scanning the last letter of the tape expression $w = abca$.

(c) The input is the tape expression $w = 1^4 B 1^2$.

The picture α is obtained by placing the state symbol before the tape letter being scanned. Initially, M is in state s_0 scanning the first letter of an input. Thus:

(a) $\alpha = aas_3 bca$; (b) $\alpha = abcs_2 a$; (c) $\alpha = s_0 1111 B 11$.

13.4. Suppose $\alpha = aas_2 ba$ is a picture. Find β such that $\alpha \to \beta$ if the Turing machine M has the quintuple q where: (a) $q = s_2 bas_1 L$; (b) $q = s_2 bbs_3 R$; (c) $q = s_2 bas_2 R$; (d) $q = s_3 abs_1 L$.

(a) Here M erases b and writes a, changes its state to s_1, and moves left. Thus $\beta = as_1 aaa$.

(b) Here M does not change the scanned letter b, changes its state to s_3, and moves right. Thus $\beta = aabs_3 a$.

(c) Here M erases b and writes a, keeps its state s_2, and moves right. Thus $\beta = aaas_2 a$.

(d) Here q has no effect on α since q does not begin with $s_2 b$.

13.5. Let $A = \{a, b\}$ and let $L = \{a^r b^s \mid r > 0, s > 0\}$, that is, L consists of all words W beginning with one or more a's and followed by one or more b's. Find a Turing machine M which recognizes L.

The strategy is that we want M to: (1) move right over all the a's; (2) move right over the b's, and (3) halt in the accepting state s_Y when it meets the blank symbol B. The following quintuples do this:

$$q_1 = s_0 aas_1 R, \quad q_2 = s_1 aas_1 R, \quad q_3 = s_1 bbs_2 R, \quad q_4 = s_2 bbs_2 R, \quad q_5 = s_2 BBs_Y R,$$

Specifically, q_1 and q_2 do (1), q_3 and q_4 do (2), and q_5 does (3).

However, we also want M to non-accept an input word W which does not belong to L. Thus we also need the following quintuples:

$$q_6 = s_0 BBs_N R, \quad q_7 = s_0 bbs_N R, \quad q_8 = s_1 BBs_N R, \quad q_9 = s_2 aas_N R$$

Here q_6 is used if the input $W = \lambda = B$, the empty word; q_7 is used if the input W is an expression beginning with b; q_8 is used if the input W only contains a's; and q_9 is used if the input W contains the letter a following a letter b.

COMPUTABLE FUNCTIONS

13.6. Find $\langle m \rangle$ if: (a) $m = 5$; (b) $m = (4, 0, 3)$; (c) $m = (3, -2, 5)$.

Recall $\langle n \rangle = 1^{n+1} = 11^n$ and $\langle (n_1, n_2, \ldots n_r) \rangle = \langle n_1 \rangle B \langle n_2 \rangle B \cdots B \langle n_r \rangle$. Thus:

(a) $\langle m \rangle = 1^6 = 111111$

(b) $\langle m \rangle = 1^5 B 1^1 B 1^4 = 11111 B 1 B 1111$

(c) $\langle m \rangle$ is not defined for negative integers.

13.7. Find $[E]$ for the expressions:

(a) $E = alls_2 Bb111$; (c) $E = \langle m \rangle$ where $m = (4, 1, 2)$;

(b) $E = aas_3 bb$; (d) $E = \langle m \rangle$ where $m = (n_1, n_2, \ldots, n_r)$.

Recall that $[E]$ counts the number of 1s in E. Thus:

(a) $[E] = 5$; (b) $[E] = 0$; (c) $[E] = 10$ since $E = 1^5 B 1^2 B 1^3$;

(d) $[E] = n_1 + n_2 + \cdots + n_r + r$ since the number of 1s contributed by each n_k to E is $n_k + 1$.

13.8. Let f be the function $f(n) = n - 1$ when $n > 0$ and $f(0) = 0$. Show that f is computable.

We need to find a Turing machine M which computes f. Specifically, we want M to erase two of the 1s in the input $\langle n \rangle$ when $n > 0$, but only one 1 when $n = 0$. This is accomplished with the following quintuples:

$$q_1 = s_0 1Bs_1 R, \quad q_2 = s_1 BBs_H R, \quad q_3 = s_1 1Bs_H R$$

Here q_1 erases the first 1 and moves M right. If there is only one 1, then M is now scanning a blank symbol B and q_2 tells the computer to halt. Otherwise, q_3 erases the second 1 and halts M.

13.9. Let f be the function $f(x, y) = y$. Show that f is computable.

We need to find a Turing machine M which computes f. Specifically, we want M to erase all the 1s from $\langle x \rangle$ and one of the 1s from $\langle y \rangle$. This is accomplished with the following quintuples:

$$q_1 = s_0 1 B s_1 R, \quad q_2 = s_0 B B s_1 R, \quad q_3 = s_1 1 B s_H R$$

Here q_1 erases all the 1s in $\langle x \rangle$ while moving M to the right. When M scans the dividing blank B, q_2 changes the state of M from s_0 to s_1 and moves M right. Then $q3$ erases the first 1 in $\langle y \rangle$ and halts M.

Supplementary Problems

FINITE STATE MACHINES

13.10. Let M be the finite state machine with state table appearing in Fig. 13-6(a).

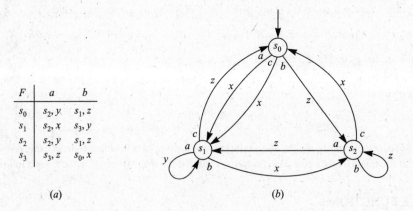

F	a	b
s_0	s_2, y	s_1, z
s_1	s_2, x	s_3, y
s_2	s_2, y	s_1, z
s_3	s_3, z	s_0, x

(a) (b)

Fig. 13-6

(a) Find the input set A, the state set S, the output set Z, and the initial state of M.

(b) Draw the state diagram $D = D(M)$ of M.

(c) Find the output word v if the input is the word: (i) $w = ab^3a^2ba^3b$; (ii) $w = a^2b^2ab^2a^2b$.

13.11. Let M be the finite state machine with input set $A = \{a, b, c\}$, output set $Z = \{x, y, z\}$ and state diagram $D = D(M)$ in Fig. 13-6(b).

(a) Construct the state table of M.

(b) Find the output word v if the input is the word: (i) $w = a^2c^2b^2cab^3$; (ii) $w = ca^2b^2ac^2ab$.

13.12. Let M be the finite state machine with input set $A = \{a, b\}$, output set $Z = \{x, y, z\}$, and state diagram $D = D(M)$ in Fig. 13-7(a). Find the output word v if the input is the word:

(a) $w = ab^3a^2ba^3b$; (b) $w = aba^2b^2ab^2a^2ba^2$.

13.13. Repeat Problem 13.12 for the state diagram $D = D(M)$ in Fig. 13-7(b).

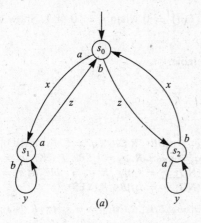

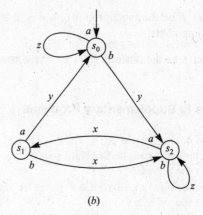

(a) (b)

Fig. 13-7

TURING MACHINES

13.14. Let M be a Turing machine. Determine the picture α corresponding to each situation:

  (a) M is in state s_2 and scanning the third letter of the tape expression $w = abbaa$.

  (b) M is in state s_3 and scanning the last letter of the tape expression $w = aabb$.

  (c) The input is the word $W = a^3 b^3$.

  (d) The input is the tape expression $W = \langle (3, 2) \rangle$.

13.15. Suppose $\alpha = abs_2aa$ is a picture. Find β such that $\alpha \to \beta$ if the Turing machine M has the quintuple q where:

  (a) $q = s_2abs_1R$;   (b) $q = s_2aas_3L$;   (c) $q = s_2abs_2R$;

  (d) $q = s_2abs_3L$;   (e) $q = s_3abs_2R$;   (f) $q = s_2aas_2L$.

13.16. Repeat Problem 13.15 for the picture $\alpha = s_2aBab$.

13.17. Find distinct pictures $\alpha_1, \alpha_2, \alpha_3, \alpha_4$ and a Turing machine M such that the following sequence does not terminate:

$$\alpha_1 \to \alpha_2 \to \alpha_3 \to \alpha_4 \to \alpha_1 \to \alpha_2 \to \cdots$$

13.18. Suppose $\alpha \to \beta_1$ and $\alpha \to \beta_2$ Must $\beta_1 \to \beta_2$?

13.19. Suppose $\alpha = \alpha(W)$ for some input W, and suppose $\alpha \to \beta \to \alpha$. Can M recognize W?

13.20. Let $A = \{a, b\}$. Find a Turing machine M which recognizes the language $L = \{ab^n \mid n > 0\}$, that is, where L consists of all words W beginning with one a and followed by one or more bs.

13.21. Let $A = \{a, b\}$. Find a Turing machine M which recognizes the finite language $L = \{a, a^2\}$, that is, where L consists of the first two nonzero powers of a.

COMPUTABLE FUNCTIONS

13.22. Find $\langle m \rangle$ if: (a) $m = 6$;   (b) $m = (5, 0, 3, 1)$;   (c) $m = (0, 0, 0)$;   (d) $m = (2, 3, -1)$.

13.23. Find $[E]$ for the expressions: (a) $E = 111s_2aa1B111$; (b) $E = a11bs_1Bb$; (c) $E = \langle m \rangle$ where $m = (2, 5, 4)$.

13.24. Let f be the function $f(n) = n - 2$ when $n > 1$ and $f(n) = 0$ when $n = 0$ or 1. Show that f is computable.

13.25. Let f be the function $f(x, y) = x$. Show that f is computable.

Answers to Supplementary Problems

13.10. (a) $A = (a, b)$, $S = \{s_0, s_1, s_2\}$, $Z = \{x, y, z\}$, and s_0 is the initial state. (b) See Fig. 13-8(a).
(c) $v = y^2 zyzxzxyz$.

13.11. (a) See Fig. 13-8. (b) (i) $v = xyz^2 x^2 zx^3 z^2$,
(ii) $v = xy^2 xz^3 xyx$.

13.12. (a) $xy^3 zxyzxz^2$; (b) $xyzxy^2 z^2 x^2 z^2 y^2$.

13.13. (a) $zyz^2 xy^2 xyzy$; (b) $zyxy^2 zx^2 zxy^2 xy$.

13.14. (a) $\alpha = abs_2 baa$; (b) $\alpha = aabs_3 b$; (c) $\alpha = s_0 aaabbb$;
(d) $\alpha = s_0 1111B111$.

13.15. (a) $\beta = abbs_3 a$; (b) $\beta = as_3 baa$; (c) $\beta = abbs_2 a$;
(d) $\beta = as_3 bba$; (e) α a is not changed by q;
(f) $\beta = as_2 baa$.

13.16. (a) $\beta = bs_1 Bab$; (b) $\beta = s_3 BaBab$; (c) $\beta = bs_2 Bab$;
(d) $\beta = s_3 BbBab$; (e) α is not changed by q;
(f) $\beta = s_2 BaBab$.

13.17. $\alpha_1 = s_0 ab$, $\alpha_2 = bs_1 b$, $\alpha_3 = s_2 bb$, $\alpha_4 = as_3 b$;
$q_1 = s_0 abs_1 R$, $q_2 = s_1 bbs_2 L$, $q_3 = s_2 bas_3 R$,
$q_4 = s_3 bbs_0 L$.

13.18. Yes.

13.19. No, since $\alpha \rightarrow \beta \rightarrow \alpha \rightarrow \beta \rightarrow \alpha \rightarrow \beta \rightarrow \cdots$ never ends.

13.20. $q_1 = s_0 BBs_N R$ (*NO*); $q_2 = s_0 bbs_N R$ (NO);
$q_3 = s_0 aas_1 R$; $q_4 = s_1 BBs_N R$ (NO); $q_5 = s_1 aas_N R$
(NO); $q_6 = s_1 bbs_N R$; $q_7 = s_2 bbs_2 R$; $q_8 = s_2 aas_N R$
(NO); $q_9 = s_2 BBs_Y R$ (YES).

13.21. $q_1 = s_0 BBs_N R$ (NO); $q_2 = s_0 bbs_N R$ (NO);
$q_3 = s_0 aas_1 R$; $q_4 = s_1 BBs_Y R$ (YES);
$q_5 = s_1 bbs_N R$ (NO); $q_6 = s_1 aas_2 R$; $q_7 = s_2 BBs_Y R$
(YES); $q_8 = s_2 aas_N R$ (NO); $q_9 = s_2 bbs_N R$ (NO).

13.22. (a) $\langle 6 \rangle = 1^7$; (b) $\langle m \rangle = 1^6 B1B1^4 B1^2$;
(c) $\langle m \rangle = 1B1B1$; (d) not defined.

13.23. (a) $[E] = 7$; (b) $[E] = 2$; (c) $[E] = 14$.

13.24. Strategy: Erase first three 1s
$$q_1 = S_0 1Bs_1 R; \quad q_2 = s_1 BBs_H R \text{ (HALT)}; \quad q_3 = S_1 1Bs_2 R; \quad q_4 = s_2 BBs_H R \text{ (HALT)}; \quad q_5 = s_2 BBs_H R \text{ (HALT)}.$$

13.25. Strategy: Erase first 1 and then all 1s after B.
$$q_1 = s_0 1Bs_1 R; \quad q_2 = S_1 11s_1 R; \quad q_3 = S_1 BBs_2 R; \quad q_4 = S_2 1Bs_3 R; \quad q_5 = S_3 1Bs_3 R; \quad q_6 = S_3 BBs_H R \text{ (HALT)}.$$

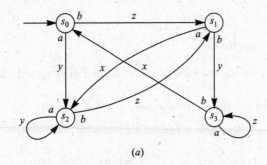

F	a	b	c
s_0	s_1, x	s_2, z	s_1, x
s_1	s_1, y	s_2, z	s_0, z
s_0	s_1, z	s_2, x	s_0, x

(a)   (b)

Fig. 13-8

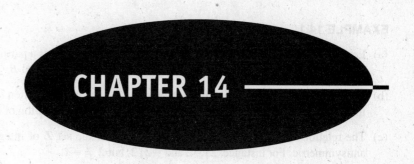

CHAPTER 14

Ordered Sets and Lattices

14.1 INTRODUCTION

Order and precedence relationships appear in many different places in mathematics and computer science. This chapter makes these notions precise. We also define a lattice, which is a special kind of an ordered set.

14.2 ORDERED SETS

Suppose R is a relation on a set S satisfying the following three properties:

[O$_1$] (Reflexive)   For any $a \in S$, we have aRa.

[O$_2$] (Antisymmetric)   If aRb and bRa, then $a = b$.

[O$_3$] (Transitive)   If aRb and bRc, then aRc.

Then R is called a *partial order* or, simply, an *order* relation, and R is said to define a *partial ordering* of S. The set S with the partial order is called a *partially ordered set* or, simply, an *ordered set* or *poset*. We write (S, R) when we want to specify the relation R.

The most familiar order relation, called the *usual order*, is the relation $\leq$ (read "less than or equal") on the positive integers **N** or, more generally, on any subset of the real numbers **R**. For this reason, a partial order relation is usually denoted by $\precsim$; and

$$a \precsim b$$

is read "a precedes b." In this case we also write:

$a \prec b$ means $a \precsim b$ and $a \neq b$;   read "a strictly precedes b."
$b \succsim a$ means $a \precsim b$;   read "b succeeds a."
$b \succ a$ means $a \prec b$;   read "b strictly succeeds a."
$\not\precsim, \not\prec, \not\succsim$, and $\not\succ$ are self-explanatory.

When there is no ambiguity, the symbols $\leq$, $<$, $>$, and $\geq$ are frequently used instead of $\precsim$, $\prec$, $\succ$, and $\succsim$, respectively.

EXAMPLE 14.1

(a) Let S be any collection of sets. The relation $\subseteq$ of set inclusion is a partial ordering of S. Specifically, $A \subseteq A$ for any set A; if $A \subseteq B$ and $B \subseteq A$ then $A = B$; and if $A \subseteq B$ and $B \subseteq C$ then $A \subseteq C$.

(b) Consider the set $\mathbf{N}$ of positive integers. We say "a divides b," written $a \mid b$, if there exists an integer c such that $ac = b$. For example, $2 \mid 4, 3 \mid 12, 7 \mid 21$, and so on. This relation of divisibility is a partial ordering of $\mathbf{N}$.

(c) The relation "$\mid$" of divisibility is not an ordering of the set $\mathbf{Z}$ of integers. Specifically, the relation is not antisymmetric. For instance, $2 \mid -2$ and $-2 \mid 2$, but $2 \neq -2$.

(d) Consider the set $\mathbf{Z}$ of integers. Define aRb if there is a positive integer r such that $b = a^r$. For instance, $2 \, R \, 8$ since $8 = 2^3$. Then R is a partial ordering of $\mathbf{Z}$.

Dual Order

Let $\precsim$ be any partial ordering of a set S. The relation $\succsim$, that is, a succeeds b, is also a partial ordering of S; it is called the *dual order*. Observe that $a \precsim b$ if and only if $b \succsim a$; hence the dual order $\succsim$ is the inverse of the relation $\precsim$, that is, $\succsim = \precsim^{-1}$.

Ordered Subsets

Let A be a subset of an ordered set S, and suppose $a, b \in A$. Define $a \precsim b$ as elements of A whenever $a \precsim b$ as elements of S. This defines a partial ordering of A called the *induced order* on A. The subset A with the induced order is called an *ordered subset* of S. Unless otherwise stated or implied, any subset of an ordered set S will be treated as an ordered subset of S.

Quasi-order

Suppose $\prec$ is a relation on a set S satisfying the following two properties:

[Q₁] (Irreflexive)   For any $a \in A$, we have $a \nprec a$.

[Q₂] (Transitive)   If $a \prec b$, and $b \prec c$, then $a \prec c$.

Then $\prec$ is called a *quasi-order* on S.

There is a close relationship between partial orders and quasi-orders. Specifically, if $\precsim$ is a partial order on a set S and we define $a \prec b$ to mean $a \precsim b$ but $a \neq b$, then $\prec$ is a quasi-order on S. Conversely, if $\prec$ is a quasi-order on a set S and we define $a \precsim b$ to mean $a \prec b$ or $a = b$, then $\precsim$ is a partial order on S. This allows us to switch back and forth between a partial order and its corresponding quasi-orders using whichever is more convenient.

Comparability, Linearly Ordered Sets

Suppose a and b are elements in a partially ordered set S. We say a and b are *comparable* if

$$a \precsim b \quad \text{or} \quad b \precsim a$$

that is, if one of them precedes the other. Thus a and b are *noncomparable*, written

$$a \parallel b$$

if neither $a \precsim b$ nor $b \precsim a$.

The word "partial" is used in defining a partially ordered set S since some of the elements of S need not be comparable. Suppose, on the other hand, that every pair of elements of S are comparable. Then S is said to be *totally ordered* or *linearly ordered*, and S is called a *chain*. Although an ordered set S may not be linearly ordered, it is still possible for a subset A of S to be linearly ordered. Clearly, every subset of a linearly ordered set S must also be linearly ordered.

EXAMPLE 14.2

(a) Consider the set $\mathbf{N}$ of positive integers ordered by divisibility. Then 21 and 7 are comparable since $7 \mid 21$. On the other hand, 3 and 5 are noncomparable since neither $3 \mid 5$ nor $5 \mid 3$. Thus $\mathbf{N}$ is not linearly ordered by divisibility. Observe that $A = \{2, 6, 12, 36\}$ is a linearly ordered subset of $\mathbf{N}$ since $2 \mid 6$, $6 \mid 12$ and $12 \mid 36$.

(b) The set $\mathbf{N}$ of positive integers with the usual order $\leq$ (less than or equal) is linearly ordered and hence every ordered subset of $\mathbf{N}$ is also linearly ordered.

(c) The power set $P(A)$ of a set A with two or more elements is not linearly ordered by set inclusion. For instance, suppose a and b belong to A. Then $\{a\}$ and $\{b\}$ are noncomparable. Observe that the empty set $\varnothing$, $\{a\}$, and A do form a linearly ordered subset of $P(A)$ since $\varnothing \subseteq \{a\} \subseteq A$. Similarly, $\varnothing$, $\{b\}$, and A form a linearly ordered subset of $P(A)$.

Product Sets and Order

There are a number of ways to define an order relation on the Cartesian product of given ordered sets. Two of these ways follow:

(a) **Product Order:** Suppose S and T are ordered sets. Then the following is an order relation on the product set $S \times T$, called the *product order*:

$$(a, b) \precsim (a', b') \quad \text{if} \quad a \leq a' \text{ and } b \leq b'$$

(b) **Lexicographical Order:** Suppose S and T are linearly ordered sets. Then the following is an order relation on the product set $S \times T$, called the *lexicographical* or *dictionary order*:

$$(a, b) \prec (a', b') \quad \text{if} \quad a < b \quad \text{or if} \quad a = a' \text{ and } b < b'$$

This order can be extended to $S_1 \times S_2 \times \cdots \times S_n$ as follows:

$$(a_1, a_2, \ldots, a_n) \prec (a_1', a_2', \ldots, a_n') \quad \text{if} \quad a_i = a_i' \text{ for } i = 1, 2, \ldots, k-1 \text{ and } a_k < a_k'$$

Note that the lexicographical order is also linear.

Kleene Closure and Order

Let A be a (nonempty) linearly ordered alphabet. Recall that A^*, called the Kleene closure of A, consists of all words w on A, and $|w|$ denotes the length of w. Then the following are two order relations on A^*.

(a) **Alphabetical (Lexicographical) Order:** The reader is no doubt familiar with the usual alphabetical ordering of A^*. That is:

  (i) $\lambda < w$, where λ is the empty word and w is any nonempty word.

  (ii) Suppose $u = au'$ and $v = bv'$ are distinct nonempty words where $a, b \in A$ and $u', v' \in A^*$. Then

$$u \prec v \quad \text{if } a < b \quad \text{or} \quad \text{if } a = b \text{ but } u' \prec v'$$

(b) **Short-lex Order:** Here A^* is ordered first by length, and then alphabetically. That is, for any distinct words u, v in A^*,

$$u \prec v \quad \text{if } |u| < |v| \quad \text{or} \quad \text{if } |u| = |v| \text{ but } u \text{ precedes } v \text{ alphabetically}$$

For example, "to" precedes "and" since $|\text{to}| = 2$ but $|\text{and}| = 3$. However, "an" precedes "to" since they have the same length, but "an" precedes "to" alphabetically. This order is also called the *free semigroup order*.

14.3   HASSE DIAGRAMS OF PARTIALLY ORDERED SETS

Let S be a partially ordered set, and suppose a, b belong to S. We say that a is an *immediate predecessor* of b, or that b is an *immediate successor* of a, or that b is a *cover* of a, written

$$a \ll b$$

if $a < b$ but no element in S lies between a and b, that is, there exists no element c in S such that $a < c < b$.

Suppose S is a finite partially ordered set. Then the order on S is completely known once we know all pairs a, b in S such that $a \ll b$, that is, once we know the relation $\ll$ on S. This follows from the fact that $x < y$ if and only if $x \ll y$ or there exist elements $a_1, a_2, \ldots, a_m$ in S such that

$$x \ll a_1 \ll a_2 \ll \cdots \ll a_m \ll y$$

The *Hasse diagram* of a finite partially ordered set S is the directed graph whose vertices are the elements of S and there is a directed edge from a to b whenever $a \ll b$ in S. (Instead of drawing an arrow from a to b, we sometimes place b higher than a and draw a line between them. It is then understood that movement upwards indicates succession.) In the diagram thus created, there is a directed edge from vertex x to vertex y if and only if $x \ll y$. Also, there can be no (directed) cycles in the diagram of S since the order relation is antisymmetric.

The Hasse diagram of a poset S is a picture of S; hence it is very useful in describing types of elements in S. Sometimes we define a partially ordered set by simply presenting its Hasse diagram. We note that the Hasse diagram of a poset S need not be connected.

Remark: The Hasse diagram of a finite poset S turns out to be a directed cycle-free graph (DAG) studied in Section 9.9. The investigation here is independent of the previous investigation. Here we mainly think of order in terms of "less than" or "greater than" rather than in terms of directed adjacency relations. Accordingly, there will be some overlap in the content.

EXAMPLE 14.3

(a) Let $A = \{1, 2, 3, 4, 6, 8, 9, 12, 18, 24\}$ be ordered by the relation "x divides y." The diagram of A is given in Fig. 14-1(a). (Unlike rooted trees, the direction of a line in the diagram of a poset is always upward.)

(b) Let $B = \{a, b, c, d, e\}$. The diagram in Fig. 14-1(b) defines a partial order on B in the natural way. That is, $d \leq b, d \leq a, e \leq c$ and so on.

(c) The diagram of a finite linearly ordered set, i.e., a finite chain, consists simply of one path. For example, Fig. 14-1(c) shows the diagram of a chain with five elements.

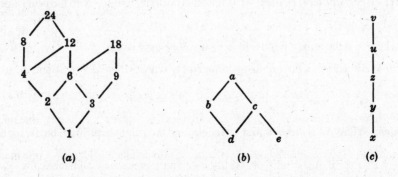

(a)   (b)   (c)

Fig. 14-1

EXAMPLE 14.4 A *partition* of a positive integer m is a set of positive integers whose sum is m. For instance, there are seven partitions of $m = 5$ as follows:

$$5, \quad 3 - 2, \quad 2 - 2 - 1, \quad 1 - 1 - 1 - 1 - 1, \quad 4 - 1, \quad 3 - 1 - 1, \quad 2 - 1 - 1 - 1$$

We order the partitions of an integer m as follows. A partition P_1 precedes a partition P_2 if the integers in P_1 can be added to obtain the integers in P_2 or, equivalently, if the integers in P_2 can be further subdivided to obtain the integers in P_1. For example,

$$2 - 2 - 1 \quad \text{precedes} \quad 3 - 2$$

since $2 + 1 = 3$. On the other hand, $3 - 1 - 1$ and $2 - 2 - 1$ are noncomparable.

Figure 14-2 gives the Hasse diagram of the partitions of $m = 5$.

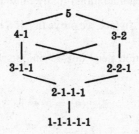

Fig. 14-2

Minimal and Maximal, and First and Last Elements

Let S be a partially ordered set. An element a in S is called a *minimal* element if no other element of S strictly precedes (is less than) a. Similarly, an element b in S is called a *maximal* element if no element of S strictly succeeds (is larger than) b. Geometrically speaking, a is a minimal element if no edge enters a (from below), and b is a maximal element if no edge leaves b (in the upward direction). We note that S can have more than one minimal and more than one maximal element.

If S is infinite, then S may have no minimal and no maximal element. For instance, the set $\mathbf{Z}$ of integers with the usual order $\leq$ has no minimal and no maximal element. On the other hand, if S is finite, then S must have at least one minimal element and at least one maximal element.

An element a is S is called a *first* element if for every element x in S,

$$a \precsim x$$

that is, if a precedes every other element in S. Similarly, an element b in S is called a *last* element if for every element y in S,

$$y \precsim b$$

that is, if b succeeds every other element in S. We note that S can have at most one first element, which must be a minimal element, and S can have at most one last element, which must be a maximal element. Generally speaking, S may have neither a first nor a last element, even when S is finite.

EXAMPLE 14.5

(a) Consider the three partially ordered sets in Example 14-3 whose Hasse diagrams appear in Fig. 14-1.

  (i) A has two maximal elements 18 and 24 and neither is a last element. A has only one minimal element, 1, which is also a first element.

  (ii) B has two minimal elements, d and e, and neither is a first element. B has only one maximal element, a, which is also a last element.

  (iii) The chain has one minimal element, x, which is a first element, and one maximal element, v, which is a last element.

(b) Let A be any nonempty set and let $P(A)$ be the power set of A ordered by set inclusion. Then the empty set $\varnothing$ is a first element of $P(A)$ since, for any set X, we have $\varnothing \subseteq X$. Moreover, A is a last element of $P(A)$ since every element Y of $P(A)$ is, by definition, a subset of A, that is, $Y \subseteq A$.

14.4 CONSISTENT ENUMERATION

Suppose S is a finite partially ordered set. Frequently we want to assign positive integers to the elements of S in such a way that the order is preserved. That is, we seek a function $f\colon S \to \mathbf{N}$ so that if $a \prec b$ then $f(a) < f(b)$. Such a function is called a *consistent enumeration* of S. The fact that this can always be done is the content of the following theorem.

Theorem 14.1: There exists a consistent enumeration for any finite poset A.

We prove this theorem in Problem 14.8. In fact, we prove that if S has n elements then there exists a consistent enumeration $f\colon S \to \{1, 2, \ldots, n\}$.

We emphasize that such an enumeration need not be unique. For example, the following are two such enumerations for the poset in Fig. 14-1(b):

(i) $f(d) = 1$, $f(e) = 2$, $f(b) = 3$, $f(c) = 4$, $f(a) = 5$.

(ii) $g(e) = 1$, $g(d) = 2$, $g(c) = 3$, $g(b) = 4$, $g(a) = 5$.

However the chain in Fig. 14-1(c) admits only one consistent enumeration if we map the set into $\{1, 2, 3, 4, 5\}$. Specifically, we must assign:

$$h(x) = 1, \quad h(y) = 2, \quad h(z) = 3, \quad h(u) = 4, \quad h(v) = 5$$

14.5 SUPREMUM AND INFIMUM

Let A be a subset of a partially ordered set S. An element M in S is called an *upper bound* of A if M succeeds every element of A, i.e., if, for every x in A, we have

$$x \precsim M$$

If an upper bound of A precedes every other upper bound of A, then it is called the *supremum* of A and is denoted by

$$\sup(A)$$

We also write $\sup(a_1, \ldots, a_n)$ instead of $\sup(A)$ if A consists of the elements $a_1, \ldots, a_n$. We emphasize that there can be at most one $\sup(A)$; however, $\sup(A)$ may not exist.

Analogously, an element m in a poset S is called a *lower bound* of a subset A of S if m precedes every element of A, i.e., if, for every y in A, we have

$$m \precsim y$$

If a lower bound of A succeeds every other lower bound of A, then it is called the *infimum* of A and is denoted by

$$\inf(A), \quad \text{or} \quad \inf(a_1, \ldots, a_n)$$

if A consists of the elements $a_1, \ldots, a_n$. There can be at most one $\inf(A)$ although $\inf(A)$ may not exist.

Some texts use the term *least upper bound* instead of supremum and then write $\operatorname{lub}(A)$ instead of $\sup(A)$, and use the term *greatest lower bound* instead of infimum and write $\operatorname{glb}(A)$ instead of $\inf(A)$.

If A has an upper bound we say A is *bounded above*, and if A has a lower bound we say A is *bounded below*. In particular, A is *bounded* if A has an upper and lower bound.

EXAMPLE 14.6

(a) Let $S = \{a, b, c, d, e, f\}$ be ordered as pictured in Fig. 14-3(a), and let $A = \{b, c, d\}$. The upper bounds of A are e and f since only e and f succeed every element in A. The lower bounds of A are a and b since only a and b precede every element of A. Note that e and f are noncomparable; hence sup(A) does not exist. However, b also succeeds a, hence inf$(A) = b$.

(b) Let $S = \{1, 2, 3, \ldots, 8\}$ be ordered as pictured in Fig. 14-3(b), and let $A = \{4, 5, 7\}$. The upper bounds of A are 1, 2, and 3, and the only lower bound is 8. Note that 7 is not a lower bound since 7 does not precede 4. Here sup$(A) = 3$ since 3 precedes the other upper bounds 1 and 2. Note that inf$(A) = 8$ since 8 is the only lower bound.

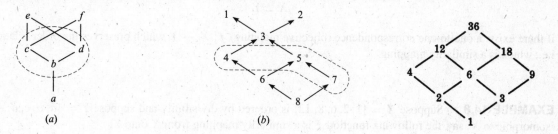

(a)   (b)

Fig. 14-3   **Fig. 14-4**

Generally speaking, sup(a, b) and inf(a, b) need not exist for every pair of elements a and b in a poset S. We now give two examples of partially ordered sets where sup(a, b) and inf(a, b) do exist for every a, b in the set.

EXAMPLE 14.7

(a) Let the set **N** of positive integers be ordered by divisibility. The *greatest common divisor* of a and b in **N**, denoted by

$$\gcd(a, b)$$

is the largest integer which divides a and b. The *least common multiple* of a and b, denoted by

$$\operatorname{lcm}(a, b)$$

is the smallest integer divisible by both a and b.

An important theorem in number theory says that every common divisor of a and b divides $\gcd(a, b)$. One can also prove that lcm(a, b) divides every multiple of a and b. Thus

$$\gcd(a, b) = \inf(a, b) \quad \text{and} \quad \operatorname{lcm}(a, b) = \sup(a, b)$$

In other words, inf(a, b) and sup(a, b) do exist for any pair of elements of **N** ordered by divisibility.

(b) For any positive integer m, we will let $\mathbf{D}_m$ denote the set of divisors of m ordered by divisibility. The Hasse diagram of

$$\mathbf{D}_{36} = \{1, 2, 3, 4, 6, 9, 12, 18, 36\}$$

appears in Fig. 14-4. Again, inf$(a, b) = \gcd(a, b)$ and sup$(a, b) = \operatorname{lcm}(a, b)$ exist for any pair a, b in $\mathbf{D}_m$.

14.6 ISOMORPHIC (SIMILAR) ORDERED SETS

Suppose X and Y are partially ordered sets. A one-to-one (injective) function $f: X \to Y$ is called a *similarity mapping* from X into Y if f preserves the order relation, that is, if the following two conditions hold for any pair a and a' in X:

(1) If $a \precsim a'$ then $f(a) \precsim f(a')$.

(2) If $a \parallel a'$ (noncomparable), then $f(a) \parallel f(a')$.

Accordingly, if A and B are linearly ordered, then only (1) is needed for f to be a similarity mapping.

Two ordered sets X and Y are said to be *isomorphic* or *similar*, written

$$X \simeq Y$$

if there exists a one-to-one correspondence (bijective mapping) $f: X \to Y$ which preserves the order relations, i.e., which is a similarity mapping.

EXAMPLE 14.8   Suppose $X = \{1, 2, 6, 8, 12\}$ is ordered by divisibility and suppose $Y = \{a, b, c, d, e\}$ is isomorphic to X; say, the following function f is a similarity mapping from X onto Y:

$$f = \{(1, e), (2, d), (6, b), (8, c), (12, a)\}$$

Draw the Hasse diagram of Y.

The similarity mapping preserves the order of the initial set X and is one-to-one and onto. Thus the mapping can be viewed simply as a relabeling of the vertices in the Hasse diagram of the initial set X. The Hasse diagrams for both X and Y appear in Fig. 14-5.

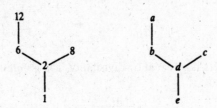

Fig. 14-5

14.7 WELL-ORDERED SETS

We begin with a definition.

Definition 14.1: An ordered set S is said to be *well-ordered* if every subset of S has a first element.

The classical example of a well-ordered set is the set **N** of positive integers with the usual order $\leq$. The following facts follow from the definition.

(1) A well-ordered set is linearly ordered. For if $a, b, \in S$, then $\{a, b\}$ has a first element; hence a and b are comparable.

(2) Every subset of a well-ordered set is well-ordered.

(3) If X is well-ordered and Y is isomorphic to X, then Y is well-ordered.

(4) All finite linearly ordered sets with the same number n of elements are well-ordered and are all isomorphic to each other. In fact, they are all isomorphic to $\{1, 2, \ldots, n\}$ with the usual order $\leq$.

(5) Every element $a \in S$, other than a last element, has an immediate successor. For, let $M(a)$ denote the set of elements which strictly succeed a. Then the first element of $M(a)$ is the immediate successor of a.

EXAMPLE 14.9

(a) The set $\mathbf{Z}$ of integers with the usual order $\leq$ is linearly ordered and every element has an immediate successor and an immediate predecessor, but $\mathbf{Z}$ is not well-ordered. For example, $\mathbf{Z}$ itself has no first element. However, any subset of $\mathbf{Z}$ which is bounded from below is well-ordered.

(b) The set $\mathbf{Q}$ of rational numbers with the usual order $\leq$ is linearly ordered, but no element in $\mathbf{Q}$ has an immediate successor or an immediate predecessor. For if $a, b \in Q$, say $a < b$, then $(a + b)/2 \in \mathbf{Q}$ and

$$a < \frac{a+b}{2} < b$$

(c) Consider the disjoint well-ordered sets

$$A = \{1, 3, 5, \ldots, \} \quad \text{and} \quad B = \{2, 4, 6, \ldots\}$$

Then the following ordered set

$$S = \{A; B\} = \{1, 3, 5, \ldots; 2, 4, 6, \ldots\}$$

is well-ordered. Note that, besides the first element 1, the element 2 does not have an immediate predecessor.

Notation: Here and subsequently, if A, B, ... are disjoint ordered sets, then $\{A; B; \ldots\}$ means the set $A \cup B \cup \ldots$ ordered positionwise from left to right; that is, the elements in the same set keep their order, and any element in a set on the left precedes any element in a set on its right. Thus every element in A precedes every element in B, and so on.

Transfinite Induction

First we restate the principle of mathematical induction. (See Section 1.8 and 11.3.)

Principle of Mathematical Induction: Let A be a subset of the set $\mathbf{N}$ of positive integers with the following two properties:

   (i) $1 \in A$.
   (ii) If $k \in A$, then $k + 1 \in A$.

Then $A = \mathbf{N}$.

The above principle is one of Peano's axioms for the natural numbers (positive integers) $\mathbf{N}$. There is another form which is sometimes more convenient to use. Namely:

Principle of Mathematical Induction (Second Form):   Let A be a subset of $\mathbf{N}$ with the following two properties:

   (i) $1 \in A$.
   (ii) If j belongs to A for $1 \leq j < k$, then $k \in A$.

Then $A = \mathbf{N}$.

The second form of induction is equivalent to the fact that $\mathbf{N}$ is well-ordered (Theorem 11.6). In fact, there is a somewhat similar statement which is true for every well-ordered set.

Principle of Transfinite Induction: Let A be a subset of a well-ordered set S with the following two properties:

(i) $a_0 \in A$.
(ii) If $s(a) \subseteq A$, then $a \in A$.

Then $A = S$.

Here a_0 is the first element of S, and $s(a)$, called the *initial segment* of a, is defined to be the set of all elements of S which strictly precede a.

Axiom of Choice, Well-Ordering Theorem

Let $\{A_i \mid i \in I\}$ be a collection of nonempty disjoint sets. We assume every $A_i \subseteq X$. A function $f: \{A_i\} \to X$ is called a *choice function* if $f(A_i) = a_i \in A_i$. In other words, f "chooses" a point $a_i \in A_i$ for each set A_i.

The axiom of choice lies at the foundations of mathematics and, in particular, the theory of sets. This "innocent looking" axiom, which follows, has as a consequence some of the most powerful and important results in mathematics.

Axiom of Choice: There exists a choice function for any nonempty collection of nonempty disjoint sets.

One of the consequences of the axiom of choice is the following theorem, which is attributed to Zermelo.

Well-Ordering Theorem: Every set S can be well-ordered.

The proof of this theorem lies beyond the scope of this text. Moreover, since all of our structures are finite or countable, we will not need to use this theorem. Ordinary mathematical induction suffices.

14.8   LATTICES

There are two ways to define a lattice L. One way is to define L in terms of a partially ordered set. Specifically, a lattice L may be defined as a partially ordered set in which $\inf(a, b)$ and $\sup(a, b)$ exist for any pair of elements $a, b \in L$. Another way is to define a lattice L axiomatically. This we do below.

Axioms Defining a Lattice

Let L be a nonempty set closed under two binary operations called *meet* and *join*, denoted respectively by $\wedge$ and $\vee$. Then L is called *lattice* if the following axioms hold where a, b, c are elements in L:

[L_1] Commutative law:
    (1a) $a \wedge b = b \wedge a$ (1b) $a \vee b = b \vee a$

[L_2] Associative law:
    (2a) $(a \wedge b) \wedge c = a \wedge (b \wedge c)$ (2b) $(a \vee b) \vee c = a \vee (b \vee c)$

[L_3] Absorption law:
    (3a) $a \wedge (a \vee b) = a$ (3b) $a \vee (a \wedge b) = a$

We will sometimes denote the lattice by $(L, \wedge, \vee)$ when we want to show which operations are involved.

Duality and the Idempotent Law

The *dual* of any statement in a lattice $(L, \wedge, \vee)$ is defined to be the statement that is obtained by interchanging $\wedge$ and $\vee$. For example, the dual of

$$a \wedge (b \vee a) = a \vee a \quad \text{is} \quad a \vee (b \wedge a) = a \wedge a$$

Notice that the dual of each axiom of a lattice is also an axiom. Accordingly, the principle of duality holds; that is:

Theorem 14.2 (Principle of Duality): The dual of any theorem in a lattice is also a theorem.

This follows from the fact that the dual theorem can be proven by using the dual of each step of the proof of the original theorem.

An important property of lattices follows directly from the absorption laws.

Theorem 14.3 (Idempotent Law): (i) $a \wedge a = a$;   (ii) $a \vee a = a$.

The proof of (i) requires only two lines:

$$a \wedge a = a \wedge (a \vee (a \wedge b)) \quad \text{(using (3b))}$$
$$= a \quad \text{(using (3a))}$$

The proof of (ii) follows from the above principle of duality (or can be proved in a similar manner).

Lattices and Order

Given a lattice L, we can define a partial order on L as follows:

$$a \precsim b \quad \text{if} \quad a \wedge b = a$$

Analogously, we could define

$$a \precsim b \quad \text{if} \quad a \vee b = b$$

We state these results in a theorem.

Theorem 14.4: Let L be a lattice. Then:

(i) $a \wedge b = a$ if and only if $a \vee b = b$.

(ii) The relation $a \precsim b$ (defined by $a \wedge b = a$ or $a \vee b = b$) is a partial order on L.

Now that we have a partial order on any lattice L, we can picture L by a diagram as was done for partially ordered sets in general.

EXAMPLE 14.10 Let C be a collection of sets closed under intersection and union. Then $(C, \cap, \cup)$ is a lattice. In this lattice, the partial order relation is the same as the set inclusion relation. Figure 14-6 shows the diagram of the lattice L of all subsets of $\{a, b, c\}$.

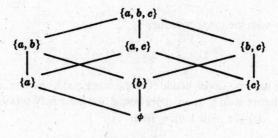

Fig. 14-6

We have shown how to define a partial order on a lattice L. The next theorem tells us when we can define a lattice on a partially ordered set P such that the lattice will give back the original order on P.

Theorem 14.5: Let P be a poset such that the $\inf(a, b)$ and $\sup(a, b)$ exist for any a, b in P. Letting

$$a \wedge b = \inf(a, b) \quad \text{and} \quad a \vee b = \sup(a, b)$$

we have that $(P, \wedge, \vee)$ is a lattice. Furthermore, the partial order on P induced by the lattice is the same as the original partial order on P.

The converse of the above theorem is also true. That is, let L be a lattice and let $\precsim$ be the induced partial order on L. Then $\inf(a, b)$ and $\sup(a, b)$ exist for any pair a, b in L and the lattice obtained from the poset $(L, \precsim)$ is the original lattice. Accordingly, we have the following:

Alternate Definition: A lattice is a partially ordered set in which

$$a \wedge b = \inf(a, b) \quad \text{and} \quad a \vee b = \sup(a, b)$$

exist for any pair of elements a and b.

We note first that any linearly ordered set is a lattice since $\inf(a, b) = a$ and $\sup(a, b) = b$ whenever $a \precsim b$. By Example 14.7, the positive integers $\mathbf{N}$ and the set $\mathbf{D}_m$ of divisors of m are lattices under the relation of divisibility.

Sublattices, Isomorphic Lattices

Suppose M is a nonempty subset of a lattice L. We say M is *a sublattice* of L if M itself is a lattice (with respect to the operations of L). We note that M is a sublattice of L if and only if M is closed under the operations of $\wedge$ and $\vee$ of L. For example, the set $\mathbf{D}_m$ of divisors of m is a sublattice of the positive integers $\mathbf{N}$ under divisibility.

Two lattices L and L' are said to be *isomorphic* if there is a one-to-one correspondence $f: L \to L'$ such that

$$f(a \wedge b) = f(a) \wedge f(b) \quad \text{and} \quad f(a \vee b) = f(a) \vee f(b)$$

for any elements a, b in L.

14.9   BOUNDED LATTICES

A lattice L is said to have a *lower bound* 0 if for any element x in L we have $0 \precsim x$. Analogously, L is said to have an *upper bound* I if for any x in L we have $x \precsim I$. We say L is *bounded* if L has both a lower bound 0 and an upper bound I. In such a lattice we have the identities

$$a \vee I = I, \quad a \wedge I = a, \quad a \vee 0 = a, \quad a \wedge 0 = 0$$

for any element a in L.

The nonnegative integers with the usual ordering,

$$0 < 1 < 2 < 3 < 4 < \cdots$$

have 0 as a lower bound but have no upper bound. On the other hand, the lattice $P(U)$ of all subsets of any universal set $\mathbf{U}$ is a bounded lattice with $\mathbf{U}$ as an upper bound and the empty set $\varnothing$ as a lower bound.

Suppose $L = \{a_1, a_2, \ldots, a_n\}$ is a finite lattice. Then

$$a_1 \vee a_2 \vee \cdots \vee a_n \quad \text{and} \quad a_1 \wedge a_2 \wedge \cdots \wedge a_n$$

are upper and lower bounds for L, respectively. Thus we have

Theorem 14.6: Every finite lattice L is bounded.

14.10 DISTRIBUTIVE LATTICES

A lattice L is said to be *distributive* if for any elements a, b, c in L we have the following:

[L_4] Distributive law:

$$(4a)\ a \wedge (b \vee c) = (a \wedge b) \vee (a \wedge c) \qquad (4b)\ a \vee (b \wedge c) = (a \vee b) \wedge (a \vee c)$$

Otherwise, L is said to be *nondistributive*. We note that by the principle of duality the condition ($4a$) holds if and only if ($4b$) holds.

Figure 14-7(a) is a nondistributive lattice since

$$a \vee (b \wedge c) = a \vee 0 = a \quad \text{but} \quad (a \vee b) \wedge (a \vee c) = I \wedge c = c$$

Figure 14-7(b) is also a nondistributive lattice. In fact, we have the following characterization of such lattices.

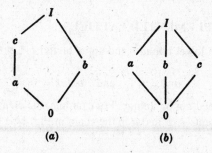

Fig. 14-7

Theorem 14.7: A lattice L is nondistributive if and only if it contains a sublattice isomorphic to Fig. 14-7(a) or to Fig. 14-7(b).

The proof of this theorem lies beyond the scope of this text.

Join Irreducible Elements, Atoms

Let L be a lattice with a lower bound 0. An element a in L is said to be *join irreducible* if $a = x \vee y$ implies $a = x$ or $a = y$. (Prime numbers under multiplication have this property, i.e., if $p = ab$ then $p = a$ or $p = b$ where p is prime.) Clearly 0 is join irreducible. If a has at least two immediate predecessors, say, b_1 and b_2 as in Fig. 14-8(a), then $a = b_1 \vee b_2$, and so a is not join irreducible. On the other hand, if a has a unique immediate predecessor c, then $a \neq \sup(b_1, b_2) = b_1 \vee b_2$ for any other elements b_1 and b_2 because c would lie between the b's and a as in Fig. 14-8(b). In other words, $a \neq 0$, is join irreducible if and only if a has a unique immediate predecessor. Those elements which immediately succeed 0, called *atoms*, are join irreducible. However, lattices can have other join irreducible elements. For example, the element c in Fig. 14-7(a) is not an atom but is join irreducible since a is its only immediate predecessor.

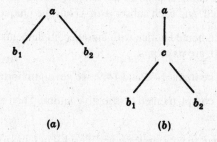

(a)   (b)

Fig. 14-8

If an element a in a finite lattice L is not join irreducible, then we can write $a = b_1 \vee b_2$. Then we can write b_1 and b_2 as the join of other elements if they are not join irreducible; and so on. Since L is finite we finally have

$$a = d_1 \vee d_2 \vee \cdots \vee d_n$$

where the d's are join irreducible. If d_i precedes d_j then $d_i \vee d_j = d_j$; so we can delete the d_i from the expression. In other words, we can assume that the d's are *irredundant*, i.e., no d precedes any other d. We emphasize that such an expression need not be unique, e.g., $I = a \vee b$ and $I = b \vee c$ in both lattices in Fig. 14-7. We now state the main theorem of this section (proved in Problem 14.28.)

Theorem 14.8: Let L be a finite distributive lattice. Then every a in L can be written uniquely (except for order) as the join of irredundant join irreducible elements.

Actually this theorem can be generalized to lattices with *finite length*, i.e., where all linearly ordered subsets are finite. (Problem 14.30 gives an infinite lattice with finite length.)

14.11 COMPLEMENTS, COMPLEMENTED LATTICES

Let L be a bounded lattice with lower bound 0 and upper bound I. Let a be an element of L. An element x in L is called a *complement* of a if

$$a \vee x = I \quad \text{and} \quad a \wedge x = 0$$

Complements need not exist and need not be unique. For example, the elements a and c are both complements of b in Fig. 14-7(a). Also, the elements y, z, and u in the chain in Fig. 14-1 have no complements. We have the following result.

Theorem 14.9: Let L be a bounded distributive lattice. Then complements are unique if they exist.

Proof: Suppose x and y are complements of any element a in L. Then

$$a \vee x = I, \quad a \vee y = I, \quad a \wedge x = 0, \quad a \wedge y = 0$$

Using distributivity,

$$x = x \vee 0 = x \vee (a \wedge y) = (x \vee a) \wedge (x \vee y) = I \wedge (x \vee y) = x \vee y$$

Similarly,

$$y = y \vee 0 = y \vee (a \wedge x) = (y \vee a) \wedge (y \vee x) = I \wedge (y \vee x) = y \vee x$$

Thus

$$x = x \vee y = y \vee x = y$$

and the theorem is proved.

Complemented Lattices

A lattice L is said to be *complemented* if L is bounded and every element in L has a complement. Figure 14-7(b) shows a complemented lattice where complements are not unique. On the other hand, the lattice $P(\mathbf{U})$ of all subsets of a universal set $\mathbf{U}$ is complemented, and each subset A of $\mathbf{U}$ has the unique complement $A^c = \mathbf{U} \setminus A$.

Theorem 14.10: Let L be a complemented lattice with unique complements. Then the join irreducible elements of L, other than 0, are its atoms.

Combining this theorem and Theorems 14.8 and 14.9, we get an important result.

Theorem 14.11: Let L be a finite complemented distributive lattice. Then every element a in L is the join of a unique set of atoms.

Remark: Some texts define a lattice L to be complemented if each a in L has a unique complement. Theorem 14.10 is then stated differently.

Solved Problems

ORDERED SETS AND SUBSETS

14.1. Let $\mathbf{N} = \{1, 2, 3, \ldots\}$ be ordered by divisibility. State whether each of the following subsets of $\mathbf{N}$ are linearly (totally) ordered.

   (a) $\{24, 2, 6\}$;   (c) $\mathbf{N} = \{1, 2, 3 \ldots\}$;   (e) $\{7\}$;

   (b) $\{3, 15, 5\}$;   (d) $\{2, 8, 32, 4\}$;   (f) $\{15, 5, 30\}$.

   (a) Since 2 divides 6 which divides 24, the set is linearly ordered.

   (b) Since 3 and 5 are not comparable, the set is not linearly ordered.

   (c) Since 2 and 3 are not comparable, the set is not linearly ordered.

   (d) This set is linearly ordered since $2 \prec 4 \prec 8 \prec 32$.

   (e) Any set consisting of one element is linearly ordered.

   (f) Since 5 divides 15 which divides 30, the set is linearly ordered.

14.2. Let $A = \{1, 2, 3, 4, 5\}$ be ordered by the Hasse diagram in Fig. 14-9(a).

   (a) Insert the correct symbol, $\prec$, $\succ$, or $\|$ (not comparable), between each pair of elements:

   (i) 1 ___ 5; (ii) 2 ___ 3; (iii) 4 ___ 1; (iv) 3 ___ 4.

   (b) Find all minimal and maximal elements of A.

   (c) Does A have a first element or a last element?

   (d) Let $L(A)$ denote the collection of all linearly ordered subsets of A with 2 or more elements, and let $L(A)$ be ordered by set inclusion. Draw the Hasse diagram of $L(A)$.

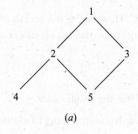

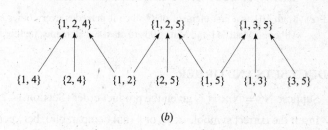

(a)   (b)

Fig. 14-9

   (a)   (i) Since there is a "path" (edges slanting upward) from 5 to 3 to 1, 5 precedes 1; hence $1 \succ 5$.

   (ii) There is no path from 2 to 3, or vice versa; hence $2 \| 3$.

   (iii) There is a path from 4 to 2 to 1; hence $4 \prec 1$.

   (iv) Neither $3 \prec 4$ nor $4 \prec 3$; hence $3 \| 4$.

   (b) No element strictly precedes 4 or 5, so 4 and 5 are minimal elements of A. No element strictly succeeds 1, so 1 is a maximal element of A.

   (c) A has no first element. Although 4 and 5 are minimal elements of A, neither precedes the other. However, 1 is a last element of A since 1 succeeds every element of A.

   (d) The elements of $L(A)$ are as follows:

   $\{1, 2, 4\}$, $\{1, 2, 5\}$, $\{1, 3, 5\}$, $\{1, 2\}$, $\{1, 4\}$, $\{1, 3\}$, $\{1, 5\}$, $\{2, 4\}$, $\{2, 5\}$, $\{3, 5\}$

   (Note $\{2, 5\}$ and $\{3, 4\}$ are not linearly ordered.) The diagram of $L(A)$ appears in Fig. 14-9(b).

14.3. Prerequisites in college is a familiar partial ordering of available classes. We write $A \prec B$ if course A is a prerequisite for course B. Let C be the ordered set consisting of the mathematics courses and their prerequisites appearing in Fig. 14-10(a).

(a) Draw the Hasse diagram for the partial ordering C of these classes.

(b) Find all minimal and maximal elements of C.

(c) Does C have a first element or a last element?

Class	Prerequisites
Math 101	None
Math 201	Math 101
Math 250	Math 101
Math 251	Math 250
Math 340	Math 201
Math 341	Math 340
Math 450	Math 201, Math 250
Math 500	Math 450, Math 251

(a)

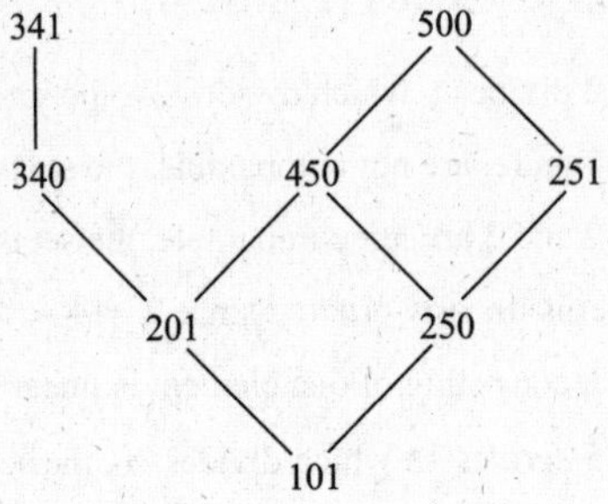

(b)

Fig. 14-10

(a) Math 101 must be on the bottom of the diagram since it is the only course with no prerequisites. Since Math 201 and Math 250 only require Math 101, we have Math 101 $\ll$ Math 201and Math 101 $\ll$ Math 250; hence draw a line slanting upward from Math 101 to Math 201 and one from Math 101 to Math 250. Continuing this process, we obtain the Hasse diagram in Fig. 14-10(b).

(b) No element strictly precedes Math 101 so Math 101 is a minimal element of C. No element strictly succeeds Math 341 or Math 500, so each is a maximal element of C.

(c) Math 101 is a first element of C since it precedes every other element of C. However, C has no last element. Although Math 341 and Math 500 are maximal elements, neither is a last element since neither precedes the other.

PRODUCT SETS AND ORDER

14.4. Suppose $\mathbf{N}^2 = \mathbf{N} \times \mathbf{N}$ is given the product order (Section 14.2) where $\mathbf{N}$ has the usual order $\leq$.

Insert the correct symbol, $\prec$, $\succ$, or $\parallel$ (not comparable), between each of the following pairs of elements of $\mathbf{N} \times \mathbf{N}$:

(a) $(5, 7)$ ___ $(7, 1)$; (c) $(5, 5)$ ___ $(4, 8)$; (e) $(7, 9)$ ___ $(4, 1)$;

(b) $(4, 6)$ ___ $(4, 2)$; (d) $(1, 3)$ ___ $(1, 7)$; (f) $(7, 9)$ ___ $(8, 2)$.

Here $(a, b) \prec (a', b')$ if $a < a'$ and $b \leq b'$ or if $a \leq a'$ and $b < b'$. Thus:

(a) $\parallel$ since $5 < 7$ but $7 > 1$. (c) $\parallel$ since $5 > 4$ and $5 < 8$. (e) $\succ$ since $7 > 4$ and $9 > 1$.
(b) $\succ$ since $4 = 4$ and $6 > 2$. (d) $\prec$ since $1 = 1$ and $3 < 7$. (f) $\parallel$ since $7 < 8$ and $9 > 2$.

14.5. Repeat Problem 14.4 using the lexicographical ordering of $\mathbf{N}^2 = \mathbf{N} \times \mathbf{N}$.

Here $(a, b) \prec (a', b')$ if $a < a'$ or if $a = a'$ but $b < b'$. Thus:

(a) $\prec$ since $5 < 7$. (c) $\succ$ since $5 > 4$. (e) $\succ$ since $7 > 4$.
(b) $\succ$ since $4 = 4$ and $6 > 2$. (d) $\prec$ since $1 = 1$ but $3 < 7$. (f) $\prec$ since $7 < 8$.

14.6. Consider the English alphabet $\mathbf{A} = \{a, b, c, \ldots, y, z\}$ with the usual (alphabetical) order. (Recall that $\mathbf{A}^*$ consisting of all words in $\mathbf{A}$.) Consider the following list of words in $\mathbf{A}^*$:

went,   forget,   to,   medicine,   me,   toast,   melt,   for,   we,   arm

(a) Sort the list of words using the short-lex (free semigroup) order.

(b) Sort the list of words using the usual (alphabetical) order of $\mathbf{A}^*$.

(a) First order the elements by length and then order them lexicographically (alphabetically):

me, to, we, arm, for, melt, went, toast, forget, medicine

(b) The usual (alphabetical) ordering yields:

arm, for, forget, me, medicine, melt, to, toast, we, went

CONSISTENT ENUMERATIONS

14.7. Suppose a student wants to take all eight mathematics courses in Problem 14.3, but only one per semester.

(a) Which choice or choices does she have for her first and for her last (eighth) semester?

(b) Suppose she wants to take Math 250 in her first year (first or second semester) and Math 340 in her senior year (seventh or eighth semester). Find all the ways that she can take the eight courses.

(a) By Fig. 14-10, Math 101 is the only minimal element and hence must be taken in the first semester, and Math 341 and 500 are the maximal elements and hence one of them must be taken in the last semester.

(b) Math 250 is not a minimal element and hence must be taken in the second semester, and Math 340 is not a maximal element so it must be taken in the seventh semester and Math 341 in the eighth semester. Also Math 500 must be taken in the sixth semester. The following gives the three possible ways to take the eight courses:

$$101, 250, 251, 201, 450, 500, 340, 341, \quad 101, 250, 201, 251, 450, 500, 340, 341,$$
$$101, 250, 201, 450, 251, 500, 340, 341$$

14.8. Prove Theorem 14.1: Suppose S is a finite poset with n elements. Then there exists a consistent enumeration $f: S \rightarrow \{1, 2, \ldots, n\}$.

The proof is by induction on the number n of elements in S. Suppose $n = 1$, say $S = \{s\}$. Then $f(s) = 1$ is a consistent enumeration of S. Now suppose $n > 1$ and the theorem holds for posets with fewer than n elements. Let $a \in S$ be a minimal element. (Such an element a exists since S is finite.) Let $T = S \backslash \{a\}$. Then T is a finite poset with $n - 1$ elements and hence, by induction, T admits a consistent enumeration; say $g: T \rightarrow \{1, 2, \ldots, n - 1\}$. Define $f: S \rightarrow \{1, 2, \ldots, n\}$ by:

$$f(x) = \begin{cases} 1, & \text{if } x = a \\ g(x) + 1 & \text{if } x \neq a \end{cases}$$

Then f is the required consistent enumeration.

UPPER AND LOWER BOUNDS, SUPREMUM AND INFIMUM

14.9. Let $S = \{a, b, c, d, e, f, g\}$ be ordered as in Fig. 14-11(a), and let $X = \{c, d, e\}$.

(a) Find the upper and lower bounds of X.

(b) Identify sup(X), the supremum of X, and inf(X), the infimum of X, if either exists.

(a) The elements e, f, and g succeed every element of X; hence e, f, and g are the upper bounds of X. The element a precedes every element of X; hence a is the lower bound of X. Note that b is not a lower bound since b does not precede c; in fact, b and c are not comparable.

(b) Since e precedes both f and g, we have $e = \text{sup}(X)$. Likewise, since a precedes (trivially) every lower bound of X, we have $a = \text{inf}(X)$. Note that sup(X) belongs to X but inf(X) does not belong to X.

14.10. Let $S = \{1, 2, 3, \ldots, 8\}$ be ordered as in Fig. 14-11(b), and let $A = (2, 3, 6)$.
(a) Find the upper and lower bounds of A.   (b) Identify sup(A) and inf(A) if either exists.

(a) The upper bound is 2, and the lower bounds are 6 and 8.

(b) Here sup(A) $= 2$ and inf(A) $= 6$.

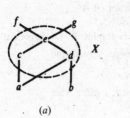

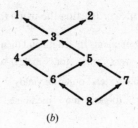

(a)   (b)

Fig. 14-11

14.11. Repeat Problem 14.10 for the subset $B = \{1, 2, 5\}$.

   (a) There is no upper bound for B since no element succeeds both 1 and 2. The lower bounds are 6, 7, 8.

   (b) Trivially, $\sup(A)$ does not exist since there are no upper bounds. Although A has three lower bounds, $\inf(A)$ does not exist since no lower bound succeeds both 6 and 7.

14.12. Consider the set $\mathbf{Q}$ of rational numbers with the usual order $\leq$. Consider the subset D of $\mathbf{Q}$ defined by

$$D = \{x \mid x \in \mathbf{Q} \text{ and } 8 < x^3 < 15\}$$

   (a) Is D bounded above or below? (b) Does $\sup(D)$ or $\inf(D)$ exist?

   (a) The subset D is bounded both above and below. For example, 1 is a lower bound and 100 an upper bound.

   (b) We claim that $\sup(D)$ does not exist. Suppose, on the contrary, $\sup(D) = x$. Since $\sqrt[3]{15}$ is irrational, $x > \sqrt[3]{15}$. However, there exists a rational number y such that $\sqrt[3]{15} < y < x$. Thus y is also an upper bound for D. This contradicts the assumption that $x = \sup(D)$. On the other hand, $\inf(D)$ does exist. Specifically, $\inf(D) = 2$.

ISOMORPHIC (SIMILAR) SETS, SIMILARITY MAPPINGS

14.13. Suppose a poset A is isomorphic (similar) to a poset B and $f: A \to B$ is a similarity mapping. Are the following statements true or false?

   (a) An element $a \in A$ is a first (last, minimal, or maximal) element of A if and only if $f(a)$ is a first (last, minimal, or maximal) element of B.

   (b) An element $a \in A$ immediately precedes an element $a' \in A$, that is, $a \ll a'$, if and only if $f(a) \ll f(a')$.

   (c) An element $a \in A$ has r immediate successors in A if and only if $f(a)$ has r immediate successors in B.

   All the statements are true; the order structure of A is the same as the order structure of B.

14.14. Let $S = \{a, b, c, d, e\}$ be the ordered set in Fig. 14-12(a). Suppose $A = \{1, 2, 3, 4, 5\}$ is isomorphic to S. Draw the Hasse diagram of A if the following is a similarity mapping from S to A:

$$f = \{(a, 1), (b, 3), (c, 5), (d, 2), (e, 4)\}$$

   The similarity mapping f preserves the order structure of S and hence f may be viewed simply as a relabeling of the vertices in the diagram of S. Thus Fig. 14-12(b) shows the Hasse diagram of A.

14.15. Let $A = \{1, 2, 3, 4, 5\}$ is ordered as in Fig. 14-12(b). Find the number n of similarity mappings $f: A \to A$.

   Since 1 is the only minimal element of A and 4 is the only maximal element, we must have $f(1) = 1$ and $f(4) = 4$. Also, $f(3) = 3$ since 3 is the only immediate successor of 1. On the other hand, there are two possibilities for $f(2)$ and $f(5)$, that is, we can have $f(2) = 2$ and $f(5) = 5$, or $f(2) = 5$ and $f(5) = 2$. Accordingly, $n = 2$.

14.16. Give an example of a finite nonlinearly ordered set $X = (A, R)$ which is isomorphic to $Y = (A, R^{-1})$, the set A with the inverse order.

   Let R be the partial ordering of $A = \{a, b, c, d, e\}$ pictured in Fig. 14-13(a).

   Then Fig. 14-13(b) shows A with the inverse order R. (The diagram of R is simply turned upside down to obtain R^{-1}.) Notice that the two diagrams are identical except for the labeling. Thus X is isomorphic to Y.

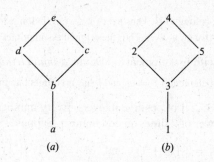

Fig. 14-12

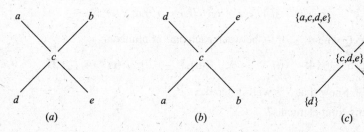

Fig. 14-13

14.17. Let A be an ordered set and, for each $a \in A$, let $p(a)$ denote the set of predecessors of a:

$$p(a) = \{x \mid x \precsim a\}$$

(called the *predecessor set* of a). Let $p(A)$ denote the collection of all predecessor sets of the elements in A ordered by set inclusion.

(a) Show that A and $p(A)$ are isomorphic by showing that the map $f: A \to p(A)$, defined by $f(a) = p(a)$, is a similarity mapping of A onto $p(A)$.

(b) Find the Hasse diagram of $p(A)$ for the set A in Fig. 14-13(a).

(a) First we show that f preserves the order relation of A. Suppose $a \precsim b$. Let $x \in p(a)$. Then $x \precsim a$, and hence $a \precsim b$; so $x \in p(b)$. Thus $p(a) \subseteq p(b)$. Suppose $a \parallel b$ (noncomparable). Then $a \in p(a)$ but $a \notin p(b)$; hence $p(a) \nsubseteq p(b)$. Similarly, $b \in p(b)$ but $b \notin p(a)$; hence $p(b) \nsubseteq p(a)$. Therefore, $p(a) \parallel p(b)$. Thus f preserves order.

We now need only show that f is one-to-one and onto. Suppose $y \in p(A)$. Then $y = p(a)$ for some $a \in A$. Thus $f(a) = p(a) = y$ so f is onto $p(A)$. Suppose $a \neq b$. Then $a \prec b$, $b \prec a$ or $a \parallel b$. In the first and third cases, $b \in p(b)$ but $b \notin p(a)$, and in the second case $a \in p(a)$ but $a \notin p(b)$. Accordingly, in all three cases, we have $p(a) \neq p(b)$. Therefore f is one-to-one.

Consequently, f is a similarity mapping of A onto $p(A)$ and so $A \simeq p(A)$.

(b) The elements of $p(A)$ follow:

$$p(a) = \{a, c, d, e\}, \quad p(b) = \{b, c, d, e\}, \quad p(c) = \{c, d, e\}, \quad p(d) = \{d\}, \quad p(e) = \{e\}$$

Figure 14-13(c) gives the diagram of $p(A)$ ordered by set inclusion. Observe that the diagrams in Fig. 14-13(a) and (c) are identical except for the labeling of the vertices.

WELL-ORDERED SETS

14.18. Prove the Principle of Transfinite Induction: Let A be a subset of a well-ordered set S with the following two properties: (i) $a_0 \in A$. (ii) If $s(a) \subseteq A$ then $a \in A$. Then $A = S$.

(Here a_0 is the first element of A, and $s(a)$ is the initial segment of a, i.e., the set of all elements strictly preceding a.)

Suppose $A \neq S$. Let $B = S \setminus A$. Then $B \neq \emptyset$. Since S is well-ordered, B has a first element b_0. Each element $x \in s(b_0)$

precedes b_0 and hence does not belong to B. Thus every $x \in s(b_0)$ belongs to A; hence $s(b_0) \subseteq A$. By (ii), $b_0 \in A$. This contradicts the assumption that $b_0 \in S \backslash A$. Thus the original assumption that $A \neq S$ is not true. Therefore $A = S$.

14.19. Let S be a well-ordered set with first element a_0. Define *a limit element* of S.

An element $b \in S$ is a limit element if $b \neq a_0$ and b has no immediate predecessor.

14.20. Consider the set $\mathbf{N} = \{1, 2, 3, \ldots\}$ of positive integers. Every number in $\mathbf{N}$ can be written uniquely as a product of a nonnegative power of 2 times an odd number. Suppose $a, a' \in \mathbf{N}$ and

$$a = 2^r(2s + 1) \quad \text{and} \quad a' = 2^{r'}(2s' + 1)$$

where r, r' and s, s' are nonnegative integers. Define:

$$a \prec a' \quad \text{if } r < r' \quad \text{or} \quad \text{if } r = r' \text{ but } s < s'.$$

(a) Insert the correct symbol, $\prec$ or $\succ$, between each pair of numbers:

(i) 5 ___ 14;   (ii) 6 ___ 9;   (iii) 3 ___ 20;   (iv) 14 ___ 21

(b) Let $S = (\mathbf{N}, \prec)$. Show that S is well-ordered.

(c) Does S have any limit elements?

(a) The elements of $\mathbf{N}$ can be listed as in Fig. 14-14. The first row consists of the odd numbers, the second row of 2 times the odd numbers, the third row of $2^2 = 4$ times the odd numbers, and so on. Then $a \prec a'$ if a is in higher row then, a' or if a and a' are in the same row but a comes before a' in the row. Accordingly:

(i) $5 \prec 14$;   (ii) $6 \succ 9$;   (iii) $3 \succ 20$;   (iv) $14 \succ 20$.

	s								
	0	1	2	3	4	5	6	7	
0	1	3	5	7	9	11	13	15	...
r 1	2	6	10	14	18	22	26	30	...
2	4	12	20	28	36	44	52	60	...
⋮	⋮	⋮	⋮	⋮	⋮	⋮	⋮	⋮	

Fig. 14-14

(b) Let A be a subset of S. The rows are well-ordered. Let r_0 denote the minimum row of elements in A. In r_0 there may be many elements of A. The columns are well-ordered, so let s_0 denote the minimum column of the elements of A in row r_0. Then $x = (r_0, s_0)$ is the first element of A. Thus S is well-ordered.

(c) As indicated by Fig. 14-14, every power of 2, that is, 1, 2, 4, 8, ..., has no immediate predecessor. Thus each, other than 1, is a limit element of S.

14.21. Let S be a well-ordered set. Let $f: S \rightarrow S$ be a similarity mapping of S into S. Prove that, for every $a \in S$, we have $a \precsim f(a)$.

Let $D = \{x \mid f(x) \prec x\}$. If D is empty, then the statement is true. Suppose $D \neq \varnothing$. Since D is well-ordered, D has a first element, say d_0. Since $d_0 \in D$, we have $f(d_0) \prec d_0$. Since f is a similarity mapping:

$$f(d_0) \prec d_0 \quad \text{implies} \quad f(f(d_0)) \prec f(d_0)$$

Thus $f(d_0)$ also belongs to D. But $f(d_0) \prec d_0$ and $f(d_0) \in D$ contradicts the fact that d_0 is the first element of D. Hence the original assumption that $D \neq \varnothing$ leads to a contradiction. Therefore D is empty and the statement is true.

14.22. Let A be a well-ordered set. Let $s(A)$ denote the collection of all initial segments $s(a)$ of elements $a \in A$ ordered by set inclusion. Prove A is isomorphic to $s(A)$ by showing that the map $f: A \to s(A)$, defined by $f(a) = s(a)$, is a similarity mapping of A onto $s(A)$. (Compare with Problem 14.17.)

First we show that f is one-to-one and onto. Suppose $y \in s(A)$. Then $y = s(a)$ for some $a \in A$. Thus $f(a) = s(a) = y$, so f is onto $s(A)$. Suppose $x \neq y$. Then one precedes the other, say, $x \prec y$. Then $x \in s(y)$. But $x \notin s(x)$. Thus $s(x) \neq s(y)$. Therefore, f is also one-to-one.

We now need only show that f preserves order, that is,

$$x \precsim y \quad \text{if and only if} \quad s(x) \subseteq s(y)$$

Suppose $x \precsim y$. If $a \in s(x)$, then $a \prec x$ and hence $a \prec y$; thus $a \in s(y)$. Thus $s(x) \subseteq s(y)$. On the other hand, suppose $x \not\precsim y$, that is, $x \succ y$. Then $y \in s(x)$. But $y \notin s(y)$; hence $s(x) \not\subseteq s(y)$. In other words, $x \precsim y$ if and only if $s(x) \subseteq s(y)$. Accordingly, f is a similarity mapping of A onto $S(A)$, and so $A \cong s(A)$.

LATTICES

14.23. Write the dual of each statement:

(a) $(a \wedge b) \vee c = (b \vee c) \wedge (c \vee a)$;   (b) $(a \wedge b) \vee a = a \wedge (b \vee a)$.

Replace $\vee$ by $\wedge$ and replace $\wedge$ by $\vee$ in each statement to obtain the dual statement:

(a) $(a \vee b) \wedge c = (b \wedge c) \vee (c \wedge a)$;   (b) $(a \vee b) \wedge a = a \vee (b \wedge a)$

14.24. Prove Theorem 14.4: Let L be a lattice. Then:

(i) $a \wedge b = a$ if and only if $a \vee b = b$.

(ii) The relation $a \precsim b$ (defined by $a \wedge b = a$ or $a \vee b = b$) is a partial order on L.

(i) Suppose $a \wedge b = a$. Using the absorption law in the first step we have:

$$b = b \vee (b \wedge a) = b \vee (a \wedge b) = b \vee a = a \vee b$$

Now suppose $a \vee b = b$. Again using the absorption law in the first step we have:

$$a = a \wedge (a \vee b) = a \wedge b$$

Thus $a \wedge b = a$ if and only if $a \vee b = b$.

(ii) For any a in L, we have $a \wedge a = a$ by idempotency. Hence $a \precsim a$, and so $\precsim$ is reflexive.

Suppose $a \precsim b$ and $b \precsim a$. Then $a \wedge b = a$ and $b \wedge a = b$. Therefore, $a = a \wedge b = b \wedge a = b$, and so $\precsim$ is antisymmetric.

Lastly, suppose $a \precsim b$ and $b \precsim c$. Then $a \wedge b = a$ and $b \wedge c = b$. Thus

$$d \wedge c = (a \wedge b) \wedge c = a \wedge (b \wedge c) = a \wedge b = a$$

Therefore $a \precsim c$, and so $\precsim$ is transitive. Accordingly, $\precsim$ is a partial order on L.

14.25. Which of the partially ordered sets in Fig. 14-15 are lattices?

A partially ordered set is a lattice if and only if $\sup(x, y)$ and $\inf(x, y)$ exist for each pair x, y in the set. Only (c) is not a lattice since $\{a, b\}$ has three upper bounds, c, d and I, and no one of them precedes the other two, that is, $\sup(a, b)$ does not exist.

14.26. Consider the lattice L in Fig. 14-15(a).

(a) Which nonzero elements are join irreducible?

(b) Which elements are atoms?

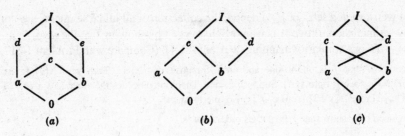

Fig. 14-15

(c) Which of the following are sublattices of L:

$$L_1 = \{0, a, b, I\}, \quad L_2 = \{0, a, e, I\}, \quad L_3 = \{a, c, d, I\}, \quad L_4 = \{0, c, d, I\}$$

(d) Is L distributive?

(e) Find complements, if they exist, for the elements, a, b and c.

(f) Is L a complemented lattice?

(a) Those nonzero elements with a unique immediate predecessor are join irreducible. Hence a, b, d, and e are join irreducible.

(b) Those elements which immediately succeed 0 are atoms, hence a and b are the atoms.

(c) A subset L' is a sublattice if it is closed under $\wedge$ and $\vee$. L_1 is not a sublattice since $a \vee b = c$, which does not belong to L_1. The set L_4 is not a sublattice since $c \wedge d = a$ does not belong to L_4. The other two sets, L_2 and L_3, are sublattices.

(d) L is not distributive since $M = \{0, a, d, e, I\}$ is a sublattice which is isomorphic to the nondistributive lattice in Fig. 14-7(a).

(e) We have $a \wedge e = 0$ and $a \vee e = I$, so a and e are complements. Similarly, b and d are complements. However, c has no complement.

(f) L is not a complemented lattice since c has no complement.

14.27. Consider the lattice M in Fig. 14-15(b).

(a) Find the nonzero join irreducible elements and atoms of M.

(b) Is M (i) distributive? (ii) complemented?

(a) The nonzero elements with a unique predecessor are a, b, and d, and of these three only a and b are atoms since their unique predecessor is 0.

(b) (i) M is distributive since M does not have a sublattice isomorphic to one of the lattices in Fig. 14-7. (ii) M is not complemented since b has no complement. Note a is the only solution to $b \wedge x = 0$ but $b \wedge a = c \neq I$.

14.28. Prove Theorem 14.8: Let L be a finite distributive lattice. Then every $a \in L$ can be written uniquely (except for order) as the join of irredundant join irreducible elements.

Since L is finite we can write a as the join of irredundant join irreducible elements as we discussed in Section 14.9. Thus we need to prove uniqueness. Suppose

$$a = b_1 \vee b_2 \vee \cdots \vee b_r = c_1 \vee c_2 \vee \cdots \vee c_s$$

where the b's are irredundant and join irreducible and the c's are irredundant and irreducible. For any given i we have

$$b_i \precsim (b_1 \vee b_2 \vee \cdots \vee b_r) = (c_1 \vee c_2 \vee \cdots \vee c_s)$$

Hence

$$b_i = b_i \wedge (c_1 \vee c_2 \vee \cdots \vee c_s) = (b_i \wedge c_1) \vee (b_i \wedge c_2) \vee \cdots \vee (b_i \wedge c_s)$$

Since b_i is join irreducible, there exists a j such that $b_i = b_i \wedge c_j$, and so $b_i \precsim c_j$. By a similar argument, for c_j there exists a b_k such that $c_j \precsim b_k$. Therefore

$$b_i \precsim c_j \precsim b_k$$

which gives $b_i = c_j = b_k$ since the b's are irredundant. Accordingly, the b's and c's may be paired off. Thus the representation for a is unique except for order.

14.29. Prove Theorem 14.10: Let L be a complemented lattice with unique complements. Then the join irreducible elements of L, other than 0, are its atoms.

Suppose a is join irreducible and a is not an atom. Then a has a unique immediate predecessor $b \neq 0$. Let b' be the complement of b. Since $b \neq 0$ we have $b' \neq I$. If a precedes b', then $b \precsim a \precsim b'$, and so $b \wedge b' = b'$, which is impossible since $b \wedge b' = I$. Thus a does not precede b', and so $a \wedge b'$ must strictly precede a. Since b is the unique immediate predecessor of a, we also have that $a \wedge b'$ precedes b as in Fig. 14-16(a). But $a \wedge b'$ precedes b'. Hence

$$a \wedge b' \precsim \inf(b, b') = b \wedge b' = 0$$

Thus $a \wedge b' = 0$. Since $a \vee b = a$, we also have that

$$a \vee b' = (a \vee b) \vee b' = a \vee (b \vee b') = a \vee I = I$$

Therefore b' is a complement of a. Since complements are unique, $a = b$. This contradicts the assumption that b is an immediate predecessor of a. Thus the only join irreducible elements of L are its atoms.

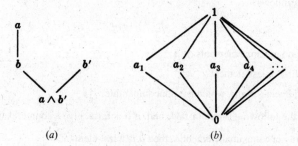

(a)   (b)

Fig. 14-16

14.30. Give an example of an infinite lattice L with finite length.

Let $L = \{0, 1, a_1, a_2, a_3, \ldots\}$ and let L be ordered as in Fig. 14-16(b). Accordingly, for each $n \in \mathbf{N}$, we have $0 < a_n < 1$. Then L has finite length since L has no infinite linearly ordered subset.

Supplementary Problems

ORDERED SETS AND SUBSETS

14.31. Let $A = \{1, 2, 3, 4, 5, 6\}$ be ordered as in Fig. 14-17(a).

    (a) Find all minimal and maximal elements of A.

    (b) Does A have a first or last element?

    (c) Find all linearly ordered subsets of A, each of which contains at least three elements.

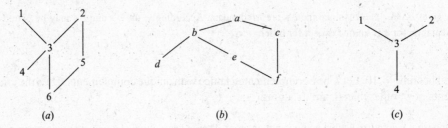

(a) (b) (c)

Fig. 14-17

14.32. Let $B = \{a, b, c, d, e, f\}$ be ordered as in Fig. 14-17(b).

    (a) Find all minimal and maximal elements of B.

    (b) Does B have a first or last element?

    (c) List two and find the number of consistent enumerations of B into the set $\{1, 2, 3, 4, 5, 6\}$.

14.33. Let $C = \{1, 2, 3, 4\}$ be ordered as in Fig. 14-17(c). Let $L(C)$ denote the collection of all nonempty linearly ordered subsets of C ordered by set inclusion. Draw a diagram of $L(C)$.

14.34. Draw the diagrams of the partitions of m (see Example 14.4) where: (a) $m = 4$; (b) $m = 6$.

14.35. Let $\mathbf{D}_m$ denote the positive divisors of m ordered by divisibility. Draw the Hasse diagrams of:
    (a) $\mathbf{D}_{12}$; (b) $\mathbf{D}_{15}$; (c) $\mathbf{D}_{16}$; (d) $\mathbf{D}_{17}$.

14.36. Let $S = \{a, b, c, d, e, f\}$ be a poset. Suppose there are exactly six pairs of elements where the first immediately precedes the second as follows:

$$f \ll a, \quad f \ll d, \quad e \ll b, \quad c \ll f, \quad e \ll c, \quad b \ll f$$

    (a) Find all minimal and maximal elements of S.

    (b) Does S have any first or last element?

    (c) Find all pairs of elements, if any, which are noncomparable.

14.37. State whether each of the following is true or false and, if it is false, give a counterexample.

    (a) If a poset S has only one maximal element a, then a is a last element.

    (b) If a finite poset S has only one maximal element a, then a is a last element.

    (c) If a linearly ordered set S has only one maximal element a, then a is a last element.

14.38. Let $S = \{a, b, c, d, e\}$ be ordered as in Fig. 14-18(a).

    (a) Find all minimal and maximal elements of S.

    (b) Does S have any first or last element?

    (c) Find all subsets of S in which c is a minimal element.

    (d) Find all subsets of S in which c is a first element.

    (e) List all linearly ordered subsets with three or more elements

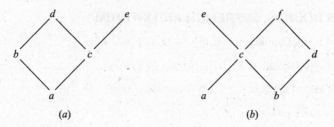

Fig. 14-18

14.39. Let $S = \{a, b, c, d, e, f\}$ be ordered as in Fig. 14-18(b).

   (a) Find all minimal and maximal elements of S.

   (b) Does S have any first or last element?

   (c) List all linearly ordered subsets with three or more elements.

14.40. Let $S = \{a, b, c, d, e, f, g\}$ be ordered as in Fig. 14-11(a). Find the number n of linearly ordered subsets of S with: (a) four elements; (b) five elements.

14.41. Let $S = \{1, 2, \ldots, 7, 8\}$ be ordered as in Fig. 14-11(b). Find the number n of linearly ordered subsets of S with: (a) five elements; (b) six elements.

CONSISTENT ENUMERATIONS

14.42. Let $S = \{a, b, c, d, e\}$ be ordered as in Fig. 14-18(a). List all consistent enumerations of S into $\{1, 2, 3, 4, 5\}$.

14.43. Let $S = \{a, b, c, d, e, f\}$ be ordered as in Fig. 14-18(b). Find the number n of consistent enumerations of S into $\{1, 2, 3, 4, 5, 6\}$.

14.44. Suppose the following are three consistent enumerations of an ordered set $A = \{a, b, c, d\}$:

$$[(a, 1), (b, 2), (c, 3), (d, 4)], \quad [(a, 1), (b, 3), (c, 2), (d, 4)], \quad [(a, 1), (b, 4), (c, 2), (d, 3)]$$

Assuming the Hasse diagram D of A is connected, draw D.

ORDER AND PRODUCT SETS

14.45. Let $M = \{2, 3, 4, \ldots\}$ and let $M^2 = M \times M$ be ordered as follows:

$$(a, b) \prec (c, d) \quad \text{if} \quad a \mid c \text{ and } b < d$$

Find all minimal and maximal elements of $M \times M$.

14.46. Consider the English alphabet $A = \{a, b, c, \ldots, y, z\}$ with the usual (alphabetical) order. Recall A^* consists of all words in A. Let L consist of the following list of elements in A^*:

   gone,   or,   arm,   go,   an,   about,   gate,   one,   at,   occur

   (a) Sort L according to the short-lex order, i.e., first by length and then alphabetically.

   (b) Sort L alphabetically.

14.47. Consider the ordered sets A and B appearing in Fig. 14-17(a) and (b), respectively. Suppose $S = A \times B$ is given the product order. Insert the correct symbol, $\prec$, $\succ$ or $\parallel$, between each pair of elements of S:

   (a) $(4, b)$___$(2, e)$;   (b) $(3, a)$___$(6, f)$;   (c) $(5, d)$___$(1, a)$;   (d) $(6, e)$___$(2, b)$.

14.48. Suppose $\mathbf{N} = \{1, 2, 3, \ldots\}$ and $\mathbf{A} = \{a, b, c, \ldots, y, z\}$ are given the usual orders, and $S = \mathbf{N} \times \mathbf{A}$ is ordered lexicographically. Sort the following elements of S:

   $(2, z)$, $(1, c)$, $(2, c)$, $(1, y)$, $(4, b)$, $(4, z)$, $(3, b)$, $(2, a)$

UPPER AND LOWER BOUNDS, SUPREMUM AND INFIMUM

14.49. Let $S = \{a, b, c, d, e, f, g\}$ be ordered as in Fig. 14-11(a). Let $A = \{a, c, d\}$.

    (a) Find the set of upper bounds of A.   (c) Does sup(A) exist?

    (b) Find the set of lower bounds of A.   (d) Does inf(A) exist?

14.50. Repeat Problem 14.49 for subset $B = \{b, c, e\}$ of S.

14.51. Let $S = \{1, 2, \ldots, 7, 8\}$ be ordered as in Fig. 14-11(b). Consider the subset $A = \{3, 6, 7\}$ of S.

    (a) Find the set of upper bounds of A.   (c) Does sup(A) exist?

    (b) Find the set of lower bounds of A.   (d) Does inf(A) exist?

14.52. Repeat Problem 14.51 for the subset $B = \{1, 2, 4, 7\}$ of S.

14.53. Consider the rational numbers $\mathbf{Q}$ with the usual order $\leq$. Let $A = \{x \mid x \in \mathbf{Q} \text{ and } 5 < x^3 < 27\}$.

    (a) Is A bounded above or below?

    (b) Does sup(A) or inf(A) exist?

14.54. Consider the real numbers $\mathbf{R}$ with the usual order $\leq$. Let $A = \{x \mid x \in \mathbf{Q} \text{ and } 5 < x^3 < 27\}$.

    (a) Is A bounded above or below?   (b) Does sup(A) or inf(A) exist?

ISOMORPHIC (SIMILAR) SETS, SIMILARITY MAPPINGS

14.55. Find the number of non-isomorphic posets with three elements a, b, c, and draw their diagrams.

14.56. Find the number of connected non-isomorphic posets with four elements a, b, c, d, and draw their diagrams.

14.57. Find the number of similarity mapings $f: S \to S$ where S is the ordered set in:

    (a) Fig. 14-17(a);   (b) Fig. 14-17(b);   (c) Fig. 14-17(c).

14.58. Show that the isomorphism relation $A \cong B$ for ordered sets is an equivalence relation, that is:

    (a) $A \cong A$ for any ordered set A.   (b) If $A \cong B$, then $B \cong A$.   (c) If $A \cong B$ and $B \cong C$, then $A \cong C$.

WELL-ORDERED SETS

14.59. Let the union S of sets $A = \{a_1, a_2, a_3, \ldots\}$, $B = \{b_1, b_2, b_3, \ldots\}$, $C = \{c_1, c_2, c_3, \ldots\}$ be ordered by:

$$S = \{A; B; C\} = \{a_1, a_2, \ldots, b_1, b_2, \ldots, c_1, c_2, \ldots\}$$

    (a) Show that S is well-ordered.

    (b) Find all limit elements of S.

    (c) Show that S is not isomorphic to $\mathbf{N} = \{1, 2, \ldots\}$ with the usual order $\leq$.

14.60. Let $A = \{a, b, c\}$ be linearly ordered by $a < b < c$, and let $\mathbf{N}$ have the usual order $\leq$.

    (a) Show that $S = \{A; \mathbf{N}\}$ is isomorphic to $\mathbf{N}$.

    (b) Show that $S' = \{\mathbf{N}; A\}$ is not isomorphic to $\mathbf{N}$.

14.61. Suppose A is a well-ordered set under the relation $\precsim$, and suppose A is also well-ordered under the inverse relation $\succsim$. Describe A.

14.62. Suppose A and B are well-ordered isomorphic sets. Show that there is only one similarity mapping $f: A \to B$.

14.63. Let S be a well-ordered set. For any $a \in S$, the set $s(a) = \{x \mid x \prec a\}$ is called an *initial segment* of a. Show that S cannot be isomorphic to one of its *initial segments*. (*Hint*: Use Problem 14.21.)

14.64. Suppose $s(a)$ and $s(b)$ are distinct initial segments of a well-ordered set S. Show that $s(a)$ and $s(b)$ cannot be isomorphic. (*Hint*: Use Problem 14.63.)

LATTICES

14.65. Consider the lattice L in Fig. 14-19(a).

    (a) Find all sublattices with five elements.   (c) Find complements of a and b, if they exist.

    (b) Find all join-irreducible elements and atoms.   (d) Is L distributive? Complemented?

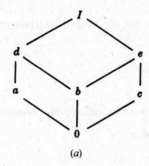   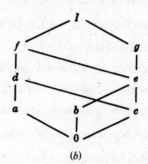

              (a)   (b)

Fig. 14-19

14.66. Consider the lattice M in Fig. 14-19(b).

    (a) Find all join-irreducible elements.

    (b) Find the atoms.

    (c) Find complements of a and b, if they exist.

    (d) Express each x in M as the join of irredundant join-irreducible elements.

    (e) Is M distributive? Complemented?

14.67. Consider the bounded lattice L in Fig. 14-20(a).

    (a) Find the complements, if they exist, of e and f.

    (b) Express I in an irredundant join-irreducible decomposition in as many ways as possible.

    (c) Is L distributive?

    (d) Describe the isomorphisms of L with itself.

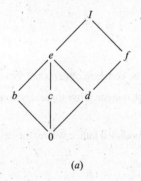 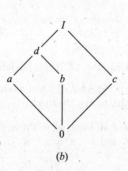 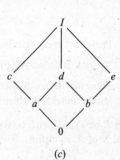

            (a)   (b)   (c)

Fig. 14-20

14.68. Consider the bounded lattice L in Fig. 14-20(b).

    (a) Find the complements, if they exist, of a and c.

    (b) Express I in an irredundant join-irreducible decomposition in as many ways as possible.

    (c) Is L distributive?

    (d) Describe the isomorphisms of L with itself.

14.69. Consider the bounded lattice L in Fig. 14-20(c).

(a) Find the complements, if they exist, of a and c.

(b) Express I in an irredundant join-irreducible decomposition in as many ways as possible.

(c) Is L distributive?

(d) Describe the isomorphisms of L with itself.

14.70. Consider the lattice $\mathbf{D}_{60} = \{1, 2, 3, 4, 5, 6, 10, 12, 15, 20, 30, 60\}$, the divisors of 60 ordered by divisibility.

(a) Draw the diagram of $\mathbf{D}_{60}$.

(b) Which elements are join-irreducible and which are atoms?

(c) Find complements of 2 and 10, if they exist.

(d) Express each number x as the join of a minimum number of irredundant join irreducible elements.

14.71. Consider the lattice $\mathbf{N}$ of positive integers ordered by divisibility.

(a) Which elements are join-irreducible ?

(b) Which elements are atoms?

14.72. Show that the following "weak" distributive laws hold for any lattice L:

(a) $a \vee (b \wedge c) \le (a \vee b) \wedge (a \vee c)$;   (b) $a \wedge (b \vee c) \ge (a \wedge b) \vee (a \wedge c)$.

14.73. Let $S = \{1, 2, 3, 4\}$. We use the notation $[12, 3, 4] \equiv [\{1, 2\}, \{3\}, \{4\}]$. Three partitions of S follow:

$$P_1 = [12, 3, 4], \quad P_2 = [12, 34], \quad P_3 = [13, 2, 4]$$

(a) Find the other twelve partitions of S.

(b) Let L be the collection of the 12 partitions of S ordered by *refinement*, that is, $P_i \prec P_j$ if each cell of P_i is a subset of a cell of P_j. For example $P_1 \prec P_2$, but P_2 and P_3 are noncomparable. Show that L is a bounded lattice and draw its diagram.

14.74. An element a in a lattice L is said to be meet-irreducible if $a = x \wedge y$ implies $a = x$ or $a = y$. Find all meet-irreducible elements in: (a) Fig. 14-19(a); (b) Fig. 14-19(b); (c) $\mathbf{D}_{60}$ (see Problem 14.70.)

14.75. A lattice M is said to be *modular* if whenever $a \le c$ we have the law

$$a \vee (b \wedge c) = (a \vee b) \wedge c$$

(a) Prove that every distributive lattice is modular.

(b) Verify that the non-distributive lattice in Fig. 14-7(b) is modular; hence the converse of (a) is not true.

(c) Show that the nondistributive lattice in Fig. 14-7(a) is non-modular. (In fact, one can prove that every non-modular lattice contains a sublattice isomorphic to Fig. 14-7(a).)

14.76. Let R be a ring. Let L be the collection of all ideals of R. Prove that L is a bounded lattice where, for any ideals J and K of R, we define: $J \vee K = J + K$ and $J \wedge K = J \cap K$.

Answers to Supplementary Problems

14.31. (a) Minimal, 4 and 6; maximal, 1 and 2. (b) First, none; last, none. (c) $\{1, 3, 4\}$, $\{1, 3, 6\}$, $\{2, 3, 4\}$, $\{2, 3, 6\}$, $\{2, 5, 6\}$.

14.32. (a) Minimal, d and f; maximal, a. (b) First, none; last, a. (c) There are eleven: *dfebca, dfecba, dfceba, fdebca, fdecba, fdceba, fedbca, fedcba, fcdeba, fecdba, fcedba*.

14.33. See Fig. 14-21.

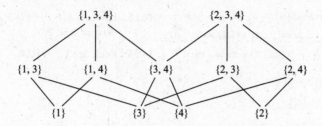

Fig. 14-21

14.34. See Fig. 14-22.

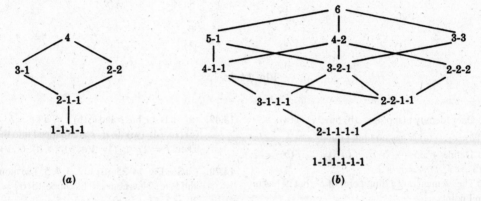

(a)   (b)

Fig. 14-22

14.35. See Fig. 14-23.

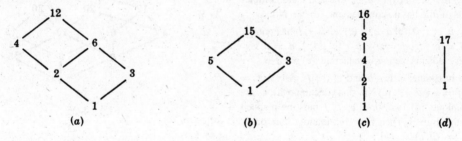

(a)   (b)   (c)   (d)

Fig. 14-23

14.36. *Hint*: Draw the diagram of *S*.

    (a) Minimal, *e*; maximal, *a*, *d*.

    (b) First, *e*; Last, none.

    (c) {*a*, *d*}, {*b*, *c*}.

14.37. (a) False. Example: **N** ∪ {*a*} where 1 ≪ *a*, and **N** ordered by ≤. (b) True. (c) True.

14.38. (a) Minimal, *a*; maximal, *d* and *e*. (b) First, *a*; last, none. (c) Any subset which contains *c* and omits *a*; that is: *c*, *cb*, *cd*, *ce*, *cbd*, *cbe*, *cde*, *cbde*. (d) *c*, *cd*, *ce*, *cde*. (e) *abd*, *acd*, *ace*.

14.39. (a) Minimal, *a* and *b*; maximal, *e* and *f*. (b) First, none; last, none. (c) *ace*, *acf*, *bce*, *bcf*, *bdf*.

14.40. (a) Four. (b) None.

14.41. (a) Six. (b) None.

14.42. *abcde*, *abced*, *acbde*, *acbed*, *acebd*.

14.43. Eleven.

14.44. $a \ll b$, $a \ll c$, $c \ll d$.

14.45. Minimal, (*p*, 2) where *p* is a prime. Maximal, none.

14.46. (a) an, at, go, or, arm, one, gate, gone, about, occur. (b) an, about, arm, at, gate, go, gone, occur, one, or.

14.47. (a) ∥; (b) >; (c) ∥; (d) <.

14.48. 1*c*, 1*y*, 2*a*, 2*c*, 2*z*, 3*b*, 4*b*, 4*z*

14.49. (a) *e*, *f*, *g*; (b) *a*; (c) sup(*A*) = *e*; (d) inf(*A*) = *a*.

14.50. (a) e, f, g;   (b) none;   (c) $\sup(B) = e$;   (d) none.

14.51. (a) 1, 2, 3;   (b) 8;   (c) $\sup(A) = 3$;   (d) $\inf(A) = 8$.

14.52. (a) None;   (b) 8;   (c) none;   (d) $\inf(B) = 8$.

14.53. (a) Both;   (b) $\sup(A) = 3$; $\inf(A)$ does not exist.

14.54. (a) Both;   (b) $\sup(A) = 3$; $\inf(A) = \sqrt[3]{5}$

14.55. Four: (1) a, b, c;   (2) $a, b \ll c$;   (3) $a \ll b, a \ll c$.
(4) $a \ll b \ll c$.

14.56. Four: See Fig. 14-24.

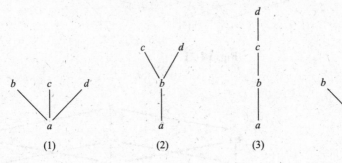

Fig. 14-24

14.57. (a) One: Identity mapping;   (b) one;   (c) two.

14.59. (b) b_1, c_1;   (c) **N** has no limit points.

14.60. (a) Define $f : S \to \mathbf{N}$ by $f(a) = 1$, $f(b) = 2$, $f(3) = 3$, $f(n) = n + 3$.
(b) The element a is a limit point of S', but **N** has no limit points.

14.61. A is a finite linearly ordered set.

14.65. (a) Six: $0abdI$, $0acdI$, $0adeI$, $0bceI$, $0aceI$, $0cdeI$;
(b) (i) $a, b, c, 0$; (ii) a, b, c. (c) c and e are complements of a. b has no complement. (d) No. No.

14.66. (a) $a, b, c, g, 0$. (b) a, b, c. (c) a has g; b has none.
(d) $I = a \vee g$, $f = a \vee b$, $e = b \vee c$, $d = a \vee c$.
Other elements are join-irreducible. (e) No. No.

14.67. (a) e has none; f has b and c. (b) $I = c \vee f = b \vee f = b \vee d \vee f$. (c) No, since decompositions are not unique. (d) Two: $0, d, e, f, I$ must be mapped into themselves. Then $F = 1_L$, identity map on L, or $F = \{(b, c), (c, b)\}$.

14.68. (a) a has c; c has a and b. (b) $I = a \vee c = b \vee c$. (c) No.
(d) Two: $0, c, d$, I must be mapped into themselves.
Then $f = 1_L$ or $f = \{(a, b), (b, a)\}$.

14.69. (a) a has e, c has b and e. (b) $I = a \vee e = b \vee c = c \vee e$.
(c) No. (d) Two: $0, d, I$ are mapped into themselves.
Then $f = 1_L$ or $f = \{(a, b), (b, a), (c, d), (d, c)\}$.

14.70. (a) See Fig. 14-25. (b) 1, 2, 3, 4, 5. The atoms are 2, 3 and 5. (c) 2 has none, 10 has none. (d) $60 = 4 \vee 3 \vee 5$; $30 = 2 \vee 3 \vee 5$; $20 = 4 \vee 5$; $15 = 3 \vee 5$; $12 = 3 \vee 4$; $10 = 2 \vee 5$; $6 = 2 \vee 3$.

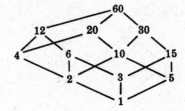

Fig. 14-25

14.73. (a) [1, 2, 3, 4], [14, 2, 3], [13, 24], [14, 23], [123, 4], [124, 3], [134, 2], [234, 1], [1234], [23, 1, 4] [24, 1, 3], [34, 1, 2]. (b) See Fig. 14-26.

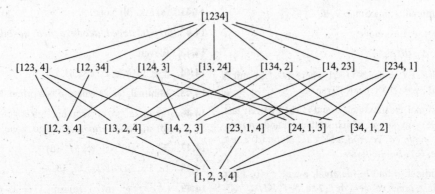

Fig. 14-26

14.74. Geometrically, an element $a \neq I$ is meet-irreducible if and only if a has only one immediate successor. (a) a, c, d, e, I; (b) a, b, d, f, g, I; (c) 4, 6, 12, 15, 60.

14.75. (a) If $a \leq c$ then $a \vee c = c$. Hence $a \vee (b \wedge c) = (a \vee b) \wedge (a \vee c) = (a \vee b) \wedge c$; (b) Here $a \leq c$. But $a \vee (b \wedge c) = a \vee 0 = a$ and $(a \vee b) \wedge c = I \wedge c = c$; hence $a \vee (b \wedge c) \neq (a \vee b) \wedge c$.

CHAPTER 15

Boolean Algebra

15.1 INTRODUCTION

Both sets and propositions satisfy similar laws, which are listed in Tables 1-1 and 4-1 (in Chapters 1 and 4, respectively). These laws are used to define an abstract mathematical structure called a *Boolean algebra*, which is named after the mathematician George Boole (1815–1864).

15.2 BASIC DEFINITIONS

Let B be a nonempty set with two binary operations $+$ and $*$, a unary operation $'$, and two distinct elements 0 and 1. Then B is called a *Boolean algebra* if the following axioms hold where a, b, c are any elements in B:

[$\mathbf{B_1}$] Commutative laws:
    (1a) $a + b = b + a$ (1b) $a * b = b * a$

[$\mathbf{B_2}$] Distributive laws:
    (2a) $a + (b * c) = (a + b) * (a + c)$ (2b) $a * (b + c) = (a * b) + (a * c)$

[$\mathbf{B_3}$] Identity laws:
    (3a) $a + 0 = a$ (3b) $a * 1 = a$

[$\mathbf{B_4}$] Complement laws:
    (4a) $a + a' = 1$ (4b) $a * a' = 0$

We will sometimes designate a Boolean algebra by $\langle B, +, *, {}', 0, 1 \rangle$ when we want to emphasize its six parts. We say 0 is the *zero* element, 1, is the *unit* element, and a' is the *complement* of a. We will usually drop the symbol $*$ and use juxtaposition instead. Then (2b) is written $a(b + c) = ab + ac$ which is the familiar algebraic identity of rings and fields. However, (2a) becomes $a + bc = (a + b)(a + c)$, which is certainly not a usual identity in algebra.

The operations $+$, $*$, and $'$ are called sum, product, and complement, respectively. We adopt the usual convention that, unless we are guided by parentheses, $'$ has precedence over $*$, and $*$ has precedence over $+$. For example,

$$a + b * c \text{ means } a + (b * c) \text{ and not } (a + b) * c; \qquad a * b' \text{ means } a * (b') \text{ and not } (a * b)'$$

Of course when $a + b * c$ is written $a + bc$ then the meaning is clear.

EXAMPLE 15.1

(a) Let $\mathbf{B} = \{0, 1\}$, the set of *bits* (binary digits), with the binary operations of $+$ and $*$ and the unary operation $'$ defined by Fig. 15-1. Then $\mathbf{B}$ is a Boolean algebra. (Note $'$ simply changes the bit, i.e., $1' = 0$ and $0' = 1$.)

+	1	0		*	1	0		'	1	0
1	1	1		1	1	0			0	1
0	1	0		0	0	0				

Fig. 15-1

(b) Let $\mathbf{B}^n = \mathbf{B} \times \mathbf{B} \times \cdots \times \mathbf{B}$ (n factors) where the operations of $+$, $*$, and $'$ are defined componentwise using Fig. 15-1. For notational convenience, we write the elements of $\mathbf{B}^n$ as n-bit sequences without commas, e.g., $x = 110011$ and $y = 111000$ belong to $\mathbf{B}^n$. Hence

$$x + y = 111011, \quad x * y = 110000, \quad x' = 001100$$

Then $\mathbf{B}^n$ is a Boolean algebra. Here $0 = 000 \cdots 0$ is the zero element, and $1 = 111 \cdots 1$ is the unit element. We note that $\mathbf{B}^n$ has 2^n elements.

(c) Let $\mathbf{D}_{70} = \{1, 2, 5, 7, 10, 14, 35, 70\}$, the divisors of 70. Define $+$, $*$, and $'$ on $\mathbf{D}_{70}$ by

$$a + b = \operatorname{lcm}(a, b), \quad a * b = \gcd(a, b), \quad a' = \frac{70}{a}$$

Then $\mathbf{D}_{70}$ is a Boolean algebra with 1 the zero element and 70 the unit element.

(d) Let C be a collection of sets closed under the set operations of union, intersection, and complement. Then C is a Boolean algebra with the empty set $\varnothing$ as the zero element and the universal set $\mathbf{U}$ as the unit element.

Subalgebras, Isomorphic Boolean Algebras

Suppose C is a nonempty subset of a Boolean algebra B. We say C is a *subalgebra* of B if C itself is a Boolean algebra (with respect to the operations of B). We note that C is a subalgebra of B if and only if C is closed under the three operations of B, i.e., $+$, $*$, and $'$. For example, $\{1, 2, 35, 70\}$ is a subalgebra of $\mathbf{D}_{70}$ in Example 15.1(c).

Two Boolean algebras B and B' are said to be *isomorphic* if there is a one-to-one correspondence $f \colon B \to B'$ which preserves the three operations, i.e., such that, for any elements, a, b in B,

$$f(a + b) = f(a) + f(b), \quad f(a * b) = f(a) * f(b) \quad \text{and} \quad f(a') = f(a)'$$

15.3 DUALITY

The *dual* of any statement in a Boolean algebra B is the statement obtained by interchanging the operations $+$ and $*$, and interchanging their identity elements 0 and 1 in the original statement. For example, the dual of

$$(1 + a) * (b + 0) = b \quad \text{is} \quad (0 * a) + (b * 1) = b$$

Observe the symmetry in the axioms of a Boolean algebra B. That is, the dual of the set of axioms of B is the same as the original set of axioms. Accordingly, the important principle of duality holds in B. Namely,

Theorem 15.1 (Principle of Duality): The dual of any theorem in a Boolean algebra is also a theorem.

In other words, if any statement is a consequence of the axioms of a Boolean algebra, then the dual is also a consequence of those axioms since the dual statement can be proven by using the dual of each step of the proof of the original statement.

15.4 BASIC THEOREMS

Using the axioms [B_1] through [B_4], we prove (Problem 15.5) the following theorem.

Theorem 15.2: Let a, b, c be any elements in a Boolean algebra B.

    (i)   Idempotent laws:

        (5a) $a + a = a$ (5b) $a * a = a$

    (ii) Boundedness laws:

        (6a) $a + 1 = 1$ (6b) $a * 0 = 0$

    (iii) Absorption laws:

        (7a) $a + (a * b) = a$ (7b) $a * (a + b) = a$

    (iv) Associative laws:

        (8a) $(a + b) + c = a + (b + c)$ (8b) $(a * b) * c = a * (b * c)$

Theorem 15.2 and our axioms still do not contain all the properties of sets listed in Table 1-1. The next two theorems give us the remaining properties.

Theorem 15.3: Let a be any element of a Boolean algebra B.

    (i) (Uniqueness of Complement) If $a + x = 1$ and $a * x = 0$, then $x = a'$.

    (ii) (Involution law) $(a')' = a$.

    (iii) (9a) $0' = 1$.   (9b) $1' = 0$.

Theorem 15.4 (DeMorgan's laws): (10a) $(a + b)' = a' * b'$. (10b) $(a * b)' = a' + b'$.

We prove these theorems in Problems 15.6 and 15.7.

15.5 BOOLEAN ALGEBRAS AS LATTICES

By Theorem 15.2 and axiom [B_1], every Boolean algebra B satisfies the associative, commutative, and absorption laws and hence is a lattice where $+$ and $*$ are the join and meet operations, respectively. With respect to this lattice, $a + 1 = 1$ implies $a \leq 1$ and $a * 0 = 0$ implies $0 \leq a$, for any element $a \in B$. Thus B is a bounded lattice. Furthermore, axioms [B_2] and [B_4] show that B is also distributive and complemented. Conversely, every bounded, distributive, and complemented lattice L satisfies the axioms [B_1] through [B_4]. Accordingly, we have the following

Alternate Definition: A Boolean algebra B is a bounded, distributive and complemented lattice.

Since a Boolean algebra B is a lattice, it has a natural partial ordering (and so its diagram can be drawn). Recall (Chapter 14) that we define $a \leq b$ when the equivalent conditions $a + b = b$ and $a * b = a$ hold. Since we are in a Boolean algebra, we can actually say much more.

Theorem 15.5: The following are equivalent in a Boolean algebra:

    (1) $a + b = b$,   (2) $a * b = a$,   (3) $a' + b = 1$,   (4) $a * b' = 0$

Thus in a Boolean alegbra we can write $a \leq b$ whenever any of the above four conditions is known to be true.

EXAMPLE 15.2

(a) Consider a Boolean algebra of sets. Then set A precedes set B if A is a subset of B. Theorem 15.4 states that if $A \subseteq B$ then the following conditions hold:

    (1) $A \cup B = B$ (2) $A \cap B = A$ (3) $A^c \cup B = \mathbf{U}$ (4) $A \cap B^c = \varnothing$

(b) Consider the Boolean algebra $\mathbf{D}_{70}$. Then a precedes b if a divides b. In such a case, $\text{lcm}(a, b) = b$ and $\gcd(a, b) = a$. For example, let $a = 2$ and $b = 14$. Then the following conditions hold:

(1)	$\text{lcm}(2, 14) = 14$.	(3)	$\text{lcm}(2', 14) = \text{lcm}(35, 14) = 70$.
(2)	$\gcd(2, 14) = 2$.	(4)	$\gcd(2, 14') = \gcd(2, 5) = 1$.

15.6 REPRESENTATION THEOREM

Let B be a finite Boolean algebra. Recall (Section 14.10) that an element a in B is an atom if a immediately succeeds 0, that is if $0 \ll a$. Let A be the set of atoms of B and let $P(A)$ be the Boolean algebra of all subsets of the set A of atoms. By Theorem 14.8, each $x \neq 0$ in B can be expressed uniquely (except for order) as the sum (join) of atoms, i.e., elements of A. Say,

$$x = a_1 + a_2 + \cdots + a_r$$

is such a representation. Consider the function $f: B \rightarrow P(A)$ defined by

$$f(x) = \{a_1, a_2, \ldots, a_r\}$$

The mapping is well defined since the representation is unique.

Theorem 15.6: The above mapping $f: B \rightarrow P(A)$ is an isomorphism.

Thus we see the intimate relationship between set theory and abstract Boolean algebras in the sense that every finite Boolean algebra is structurally the same as a Boolean algebra of sets.

If a set A has n elements, then its power set $P(A)$ has 2^n elements. Thus the above theorem gives us our next result.

Corollary 15.7: A finite Boolean algebra has 2^n elements for some positive integer n.

EXAMPLE 15.3 Consider the Boolean algebra $\mathbf{D}_{70} = \{1, 2, 5, \ldots, 70\}$ whose diagram is given in Fig. 15-2(a). Note that $A = \{2, 5, 7\}$ is the set of atoms of $\mathbf{D}_{70}$. The following is the unique representation of each nonatom by atoms:

$$10 = 2 \vee 5, \quad 14 = 2 \vee 7, \quad 35 = 5 \vee 7, \quad 70 = 2 \vee 5 \vee 7$$

Figure 15-2(b) gives the diagram of the Boolean algebra of the power set $P(A)$ of the set A of atoms. Observe that the two diagrams are structurally the same.

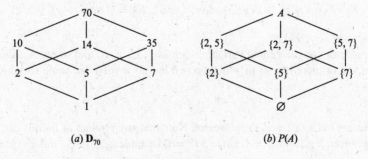

$(a)\ \mathbf{D}_{70}$ $(b)\ P(A)$

Fig. 15-2

15.7 SUM-OF-PRODUCTS FORM FOR SETS

This section motivates the concept of the sum-of-products form in Boolean algebra by an example of set theory. Consider the Venn diagram in Fig. 15-3 of three sets A, B, and C. Observe that these sets partition the

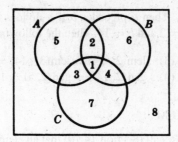

Fig. 15-3

rectangle (universal set) into eight numbered sets which can be represented as follows:

(1) $A \cap B \cap C$ (3) $A \cap B^c \cap C$ (5) $A \cap B^c \cap C^c$ (7) $A^c \cap B^c \cap C$

(2) $A \cap B \cap C^c$ (4) $A^c \cap B \cap C$ (6) $A^c \cap B \cap C^c$ (8) $A^c \cap B^c \cap C^c$

Each of these eight sets is of the form $A^* \cap B^* \cap C^*$ where:

$$A^* = A \text{ or } A^c, \quad B^* = B \text{ or } B^c, \quad C^* = C \text{ or } C^c$$

Consider any nonempty set expression E involving the sets A, B, and C, say,

$$E = [(A \cap B^c)^c \cup (A^c \cap C^c)] \cap [(B^c \cup C)^c \cap (A \cup C^c)]$$

Then E will represent some area in Fig. 15-3 and hence will uniquely equal the union of one or more of the eight sets.

Suppose we now interpret a union as a sum and an intersection as a product. Then the above eight sets are products, and the unique representation of E will be a sum (union) of products. This unique representation of E is the same as the complete sum-of-products expansion in Boolean algebras which we discuss below.

15.8 SUM-OF-PRODUCTS FORM FOR BOOLEAN ALGEBRAS

Consider a set of variables (or letters or symbols), say $x_1, x_2, \ldots, x_n$. A *Boolean expression* E in these variables, sometimes written $E(x_1, \ldots, x_n)$, is any variable or any expression built up from the variables using the Boolean operations $+$, $*$, and $'$. (Naturally, the expression E must be *well-formed*, that is, where $+$ and $*$ are used as binary operations, and $'$ is used as a unary operation.) For example,

$$E_1 = (x + y'z)' + (xyz' + x'y)' \quad \text{and} \quad E_2 = ((xy'z' + y)' + x'z)'$$

are Boolean expressions in x, y, and z.

A *literal* is a variable or complemented variable, such as x, x', y, y', and so on. A *fundamental product* is a literal or a product of two or more literals in which no two literals involve the same variable. Thus

$$xz', \quad xy'z, \quad x, \quad y', \quad x'yz$$

are fundamental products, but $xyx'z$ and $xyzy$ are not. Note that any product of literals can be reduced to either 0 or a fundamental product, e.g., $xyx'z = 0$ since $xx' = 0$ (complement law), and $xyzy = xyz$ since $yy = y$ (idempotent law).

A fundamental product P_1 is said to be *contained in* (or *included in*) another fundamental product P_2 if the literals of P_1 are also literals of P_2. For example, $x'z$ is contained in $x'yz$, but $x'z$ is not contained in $xy'z$ since x' is not a literal of $xy'z$. Observe that if P_1 is contained in P_2, say $P_2 = P_1 * Q$, then, by the absorption law,

$$P_1 + P_2 = P_1 + P_1 * Q = P_1$$

Thus, for instance, $x'z + x'yz = x'z$.

Definition 15.1: A Boolean expression E is called a *sum-of-products* expression if E is a fundamental product or the sum of two or more fundamental products none of which is contained in another.

Definition 15.2: Let E be any Boolean expression. A *sum-of-products form* of E is an equivalent Boolean sum-of-products expression.

EXAMPLE 15.4 Consider the expressions

$$E_1 = xz' + y'z + xyz' \quad \text{and} \quad E_2 = xz' + x'yz' + xy'z$$

Although the first expression E_1 is a sum of products, it is not a sum-of-products expression. Specifically, the product xz' is contained in the product xyz'. However, by the absorption law, E_1 can be expressed as

$$E_1 = xz' + y'z + xyz' = xz' + xyz' + y'z = xz' + y'z$$

This yields a sum-of-products form for E_1. The second expression E_2 is already a sum-of-products expression.

Algorithm for Finding Sum-of-Products Forms

Figure 15-4 gives a four-step algorithm which uses the Boolean algebra laws to transform any Boolean expression into an equivalent sum-of-products expression.

Algorithm 15.1: The input is a Boolean expression E. The output is a sum-of-products expression equivalent to E.

Step 1. Use DeMorgan's laws and involution to move the complement operation into any parenthesis until finally the complement operation only applies to variables. Then E will consist only of sums and products of literals.

Step 2. Use the distributive operation to next transform E into a sum of products.

Step 3. Use the commutative, idempotent, and complement laws to transform each product in E into 0 or a fundamental product.

Step 4. Use the absorption and identity laws to finally transform E into a sum-of-products expression.

Fig. 15-4

EXAMPLE 15.5 Suppose Algorithm 15.1 is applied to the following Boolean expression:

$$E = ((xy)'z)'((x' + z)(y' + z'))'$$

Step 1. Using DeMorgan's laws and involution, we obtain

$$E = (xy'' + z')((x' + z)' + (y' + z')') = (xy + z')(xz' + yz)$$

E now consists only of sums and products of literals.

Step 2. Using the distributive laws, we obtain

$$E = xyxz' + xyyz + xz'z' + yzz'$$

E now is a sum of products.

Step 3. Using the commutative, idempotent, and complement laws, we obtain

$$E = xyz' + xyz + xz' + 0$$

Each term in E is a fundamental product or 0.

Step 4. The product xz' is contained in xyz'; hence, by the absorption law,

$$xz' + (xz'y) = xz'$$

Thus we may delete xyz' from the sum. Also, by the identity law for 0, we may delete 0 from the sum. Accordingly,

$$E = xyz + xz'$$

E is now represented by a sum-of-products expression.

Complete Sum-of-Products Forms

A Boolean expression $E = E(x_1, x_2, \ldots, x_n)$ is said to be a *complete sum-of-products* expression if E is a sum-of-products expression where each product P involves all the n variables. Such a fundamental product P which involves all the variables is called a *minterm*, and there is a maximum of 2^n such products for n variables. The following theorem applies.

Theorem 15.8: Every nonzero Boolean expression $E = E(x_1, x_2, \ldots, x_n)$ is equivalent to a complete sum-of-products expression and such a representation is unique.

The above unique representation of E is called the *complete sum-of-products form* of E. Algorithm 15-1 in Fig. 15-4 tells us how to transform E into a sum-of-products form. Figure 15-5 contains an algorithm which transforms a sum-of-products form into a complete sum-of-products form.

Algorithm 15.2: The input is a Boolean sum-of-products expression $E = E(x_1, x_2, \ldots, x_n)$. The output is a complete sum-of-products expression equivalent to E.

Step 1.   Find a product P in E which does not involve the variable x_i, and then multiply P by $x_i + x_i'$, deleting any repeated products. (This is possible since $x_i + x_i' = 1$, and $P + P = P$.)

Step 2.   Repeat Step 1 until every product P in E is a minterm, i.e., every product P involves all the variables.

Fig. 15-5

EXAMPLE 15.6   Express $E(x, y, z) = x(y'z)'$ into its complete sum-of-products form.

(a) Apply Algorithm 15.1 to E so E is represented by a sum-of-products expression:

$$E = x(y'z)' = x(y + z') = xy + xz'$$

(b) Now apply Algorithm 15.2 to obtain:

$$E = xy(z + z') + xz'(y + y') = xyz + xyz' + xyz' + xy'z'$$
$$= xyz + xyz' + xy'z'$$

Now E is reprsented by its complete sum-of-products form.

Warning: The terminology in this section has not been standardized. The sum-of-products form for a Boolean expression E is also called the *disjunctive normal form* or DNF of E. The complete sum-of-products form for E is also called the *full disjunctive normal form*, or the *disjunctive canonical form*, or the *minterm canonical form* of E.

15.9 MINIMAL BOOLEAN EXPRESSIONS, PRIME IMPLICANTS

There are many ways of representing the same Boolean expression E. Here we define and investigate a minimal sum-of-products form for E. We must also define and investigate prime implicants of E since the minimal sum-of-products involves such prime implicants. Other minimal forms exist, but their investigation lies beyond the scope of this text.

Minimal Sum-of-Products

Consider a Boolean sum-of-products expression E. Let E_L denote the number of literals in E (counted according to multiplicity), and let E_S denote the number of summands in E. For instance, suppose

$$E = xyz' + x'y't + xy'z't + x'yzt$$

Then $E_L = 3 + 3 + 4 + 4 = 14$ and $E_S = 4$.

Suppose E and F are equivalent Boolean sum-of-products expressions. We say E is *simpler* than F if:

(i) $E_L < F_L$ and $E_S \leq F_L$,   or   (ii) $E_L \leq F_L$ and $E_S < F_L$

We say E is *minimal* if there is no equivalent sum-of-products expression which is simpler than E. We note that there can be more than one equivalent minimal sum-of-products expressions.

Prime Implicants

A fundamental product P is called a *prime implicant* of a Boolean expression E if

$$P + E = E$$

but no other fundamental product contained in P has this property. For instance, suppose

$$E = xy' + xyz' + x'yz'$$

One can show (Problem 15.15) that:

$$xz' + E = E \quad \text{but} \quad x + E \neq E \quad \text{and} \quad z' + E \neq E$$

Thus xz' is a prime implicant of E.

The following theorem applies.

Theorem 15.9: A minimal sum-of-products form for a Boolean expression E is a sum of prime implicants of E.

The following subsections give a method for finding the prime implicants of E based on the notion of the consensus of fundamental products. This method can then be used to find a minimal sum-of-products form for E. Section 15.12 gives a geometric method for finding such prime implicants.

Consensus of Fundamental Products

Let P_1 and P_2 be fundamental products such that exactly one variable, say x_k, appears uncomplemented in one of P_1 and P_2 and complemented in the other. Then the *consensus* of P_1 and P_2 is the product (without repetitions) of the literals of P_1 and the literals of P_2 after x_k and x_k' are deleted. (We do not define the consensus of $P_1 = x$ and $P_2 = x'$.)

The following lemma (proved in Problem 15.19) applies.

Lemma 15.10: Suppose Q is the consensus of P_1 and P_2. Then $P_1 + P_2 + Q = P_1 + P_2$.

EXAMPLE 15.7 Find the consensus Q of P_1 and P_2 where:

(a) $P_1 = xyz's$ and $P_2 = xy't$.

Delete y and y' and then multiply the literals of P_1 and P_2 (without repetition) to obtain $Q = xz'st$.

(b) $P_1 = xy'$ and $P_2 = y$.

Deleting y and y' yields $Q = x$.

(c) $P_1 = x'yz$ and $P_2 = x'yt$.

No variable appears uncomplemented in one of the products and complemented in the other. Hence P_1 and P_2 have no consensus.

(d) $P_1 = x'yz$ and $P_2 = xyz'$.

Each of x and z appear complemented in one of the products and uncomplemented in the other. Hence P_1 and P_2 have no consensus.

Consensus Method for Finding Prime Implicants

Figure 15-6 contains an algorithm, called the *consensus method*, which is used to find the prime implicants of a Boolean expression E. The following theorem gives the basic property of this algorithm.

Theorem 15.11: The consensus method will eventually stop, and then E will be the sum of its prime implicants.

Algorithm 15.3 (Consensus Method): The input is a Boolean expression
$E = P_1 + P_2 + \cdots + P_m$ where the P's are fundamental products. The output expresses E as a sum of its prime implicants (Theorem 15.11).

Step 1. Delete any fundamental product P_i which includes any other fundamental product P_j. (Permissible by the absorption law.)

Step 2. Add the consensus of any P_i and P_j providing Q does not include any of the P's. (Permissible by Lemma 15.10.)

Step 3. Repeat Step 1 and/or Step 2 until neither can be applied.

Fig. 15-6

EXAMPLE 15.8 Let $E = xyz + x'z' + xyz' + x'y'z + x'yz'$. Then:

$$
\begin{aligned}
E &= xyz + x'z' + xyz' + x'y'z && (x'yz' \text{ includes } x'z') \\
  &= xyz + x'y' + xyz' + x'y'z + xy && (\text{consensus of } xyz \text{ and } xyz') \\
  &= x'z' + x'y'z + xy && (xyz \text{ and } xyz' \text{ include } xy) \\
  &= x'z' + x'y'z + xy + x'y' && (\text{consensus of } x'z' \text{ and } x'y'z) \\
  &= x'z' + xy + x'y' && (x'y'z \text{ includes } x'y') \\
  &= x'z' + xy + x'y' + yz' && (\text{consensus of } x'z' \text{ and } xy)
\end{aligned}
$$

Now neither step in the consensus method will change E. Thus E is the sum of its prime implicants, which appear in the last line, that is, $x'z'$, xy, $x'y'$, and yz'.

Finding a Minimal Sum-of-Products Form

The consensus method (Algorithm 15.3) is used to express a Boolean expression E as a sum of all its prime implicants. Figure 15-7 contains an algorithm which uses such a sum to find a minimal sum-of-products form for E.

Algorithm 15.4: The input is a Boolean expression $E = P_1 + P_2 + \cdots + P_m$ where the Ps are all the prime implicants of E. The output expresses E as a minimal sum-of-products.

Step 1.   Express each prime implicant P as a complete sum-of-products.

Step 2.   Delete one by one those prime implicants whose summands appear among the summands of the remaining prime implicants.

Fig. 15-7

EXAMPLE 15.9 We apply Algorithm 15.4 to the following expression E which (by Example 15.8) is now expressed as the sum of all its prime implicants:

$$E = x'z' + xy + x'y' + yz'$$

Step 1. Express each prime implicant of E as a complete sum-of-products to obtain:

$$x'z' = x'z'(y + y') = x'yz' + x'y'z'$$
$$xy = xy(z + z') = xyz + xyz'$$
$$x'y' = x'y'(z + z') = x'y'z + x'y'z'$$
$$yz' = yz'(x + x') = xyz' + x'yz'$$

Step 2. The summands of $x'z'$ are $x'yz$ and $x'y'z'$ which appear among the other summands. Thus delete $x'z'$ to obtain

$$E = xy + x'y' + yz'$$

The summands of no other prime implicant appear among the summands of the remaining prime implicants, and hence this is a minimal sum-of-products form for E. In other words, none of the remaining prime implicants is *superfluous*, that is, none can be deleted without changing E.

15.10 LOGIC GATES AND CIRCUITS

Logic circuits (also called *logic networks*) are structures which are built up from certain elementary circuits called *logic gates*. Each logic circuit may be viewed as a machine L which contains one or more input devices and exactly one output device. Each input device in L sends a signal, specifically, a *bit* (binary digit),

$$0 \quad \text{or} \quad 1$$

to the circuit L, and L processes the set of bits to yield an output bit. Accordingly, an n-bit sequence may be assigned to each input device, and L processes the input sequences one bit at a time to produce an n-bit output sequence. First we define the logic gates, and then we investigate the logic circuits.

Logic Gates

There are three basic logic gates which are described below. We adopt the convention that the lines entering the gate symbol from the left are input lines and the single line on the right is the output line.

(a) *OR Gate:* Figure 15-8(a) shows an OR gate with inputs A and B and output $Y = A + B$ where "addition" is defined by the "truth table" in Fig. 15-8(b). Thus the output $Y = 0$ only when inputs $A = 0$ and $B = 0$. Such an OR gate may, have more than two inputs. Figure 15-8(c) shows an OR gate with four inputs, A, B, C, D, and output $Y = A + B + C + D$. The output $Y = 0$ if and only if all the inputs are 0.

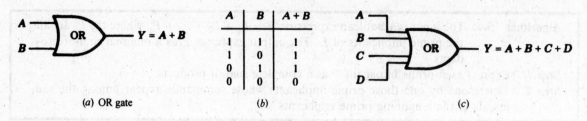

Fig. 15-8

Suppose, for instance, the input data for the OR gate in Fig. 15-15(c) are the following 8-bit sequences:

$$A = 10000101, \quad B = 10100001, \quad C = 00100100, \quad 10010101$$

The OR gate only yields 0 when all input bits are 0. This occurs only in the 2nd, 5th, and 7th positions (reading from left to right). Thus the output is the sequence $Y = 10110101$.

(b) **AND Gate:** Figure 15-9(a) shows an AND gate with inputs A and B and output $Y = A \cdot B$ (or simply $Y = AB$) where "multiplication" is defined by the "truth table" in Fig. 15-9(b). Thus the output $Y = 1$ when inputs $A = 1$ and $B = 1$; otherwise $Y = 0$. Such an AND gate may have more than two inputs. Figure 15-9(c) shows an AND gate with four inputs, A, B, C, D, and output $Y = A \cdot B \cdot C \cdot D$. The output $Y = 1$ if and only if all the inputs are 1.

Suppose, for instance, the input data for the AND gate in Fig. 15-9(c) are the following 8-bit sequences:

$$A = 11100111, \quad B = 01111011, \quad C = 01110011, \quad D = 11101110$$

The AND gate only yields 1 when all input bits are 1. This occurs only in the 2nd, 3rd, and 7th positions. Thus the output is the sequence $Y = 01100010$.

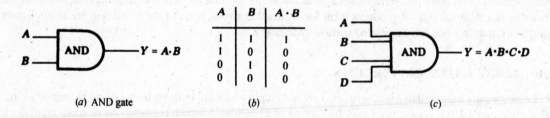

Fig. 15-9

(c) **NOT Gate:** Figure 15-10(a) shows a NOT gate, also called an *inverter*, with input A and output $Y = A'$ where "inversion," denoted by the prime, is defined by the "truth table" in Fig. 15-10(b). The value of the output $Y = A'$ is the opposite of the input A; that is, $A' = 1$ when $A = 0$ and $A' = 0$ when $A = 1$. We emphasize that a NOT gate can have only one input, whereas the OR and AND gates may have two or more inputs.

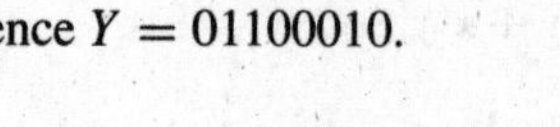

A	A'
1	0
0	1

$$A \longrightarrow \boxed{NOT} \longrightarrow y = A'$$

(a) NOT gate   (b)

Fig. 15-10

Suppose, for instance, a NOT gate is asked to process the following three sequences:

$$A_1 = 110001, \quad A_2 = 10001111, \quad A_3 = 101100111000$$

The NOT gate changes 0 to 1 and 1 to 0. Thus

$$A_1' = 001110, \quad A_2' = 01110000, \quad A_3' = 010011000111$$

are the three corresponding outputs.

Logic Circuits

A logic circuit L is a well-formed structure whose elementary components are the above OR, AND, and NOT gates. Figure 15-11 is an example of a logic circuit with inputs A, B, C and output Y. A dot indicates a place where the input line splits so that its bit signal is sent in more than one direction. (Frequently, for notational convenience, we may omit the word from the interior of the gate symbol.) Working from left to right, we express Y in terms of the inputs A, B, C as follows. The output of the AND gate is $A \cdot B$, which is then negated to yield $(A \cdot B)'$. The output of the lower OR gate is $A' + C$, which is then negated to yield $(A' + C)'$. The output of the OR gate on the right, with inputs $(A \cdot B)'$ and $(A' + C)'$, gives us our desired representation, that is,

$$Y = (A \cdot B)' + (A' + C)'$$

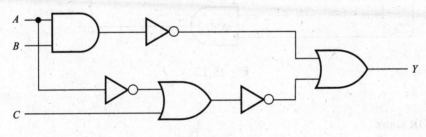

Fig. 15-11

Logic Circuits as a Boolean Algebra

Observe that the truth tables for the OR, AND, and NOT gates are respectively identical to the truth tables for the propositions $p \vee q$ (disjunction, "p or q"), $p \wedge q$ (conjunction, "p and q"), and $\neg p$ (negation, "not p"), which appear in Section 4.3. The only difference is that 1 and 0 are used instead of T and F. Thus the logic circuits satisfy the same laws as do propositions and hence they form a Boolean algebra. We state this result formally.

Theorem 15.12: Logic circuits form a Boolean Algebra.

Accordingly, all terms used with Boolean algebras, such as, complements, literals, fundamental products, minterms, sum-of-products, and complete sum-of-products, may also be used with our logic circuits.

AND-OR Circuits

The logic circuit L which corresponds to a Boolean sum-of-products expression is called an AND-OR circuit. Such a circuit L has several inputs, where:

(1) Some of the inputs or their complements are fed into each AND gate.

(2) The outputs of all the AND gates are fed into a single OR gate.

(3) The output of the OR gate is the output for the circuit L.

The following illustrates this type of a logic circuit.

EXAMPLE 15.10 Figure 15-12 is a typical AND-OR circuit with three inputs, A, B, C and output Y. We can easily express Y as a Boolean expression in the inputs A, B, C as follows. First we find the output of each AND gate:

(a) The inputs of the first AND gate are A, B, C; hence $A \cdot B \cdot C$ is the output.

(b) The inputs of the second AND gate are A, B', C; hence $A \cdot B' \cdot C$ is the output.

(c) The inputs of the third AND gate are A' and B; hence $A' \cdot B$ is the output.

Then the sum of the outputs of the AND gates is the output of the OR gate, which is the output Y of the circuit. Thus:

$$Y = A \cdot B \cdot C + A \cdot B' \cdot C + A' \cdot B$$

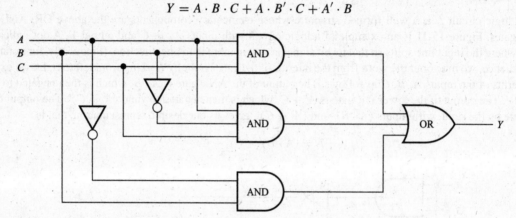

Fig. 15-12

NAND and NOR Gates

There are two additional gates which are equivalent to combinations of the above basic gates.

(a) A NAND gate, pictured in Fig. 15-13(a), is equivalent to an AND gate followed by a NOT gate.

(b) A NOR gate, pictured in Fig. 15-13(b), is equivalent to an OR gate followed by a NOT gate.

The truth tables for these gates (using two inputs A and B) appear in Fig. 15-13(c). The NAND and NOR gates can actually have two or more inputs just like the corresponding AND and OR gates. Furthermore, the output of a NAND gate is 0 if and only if all the inputs are 1, and the output of a NOR gate is 1 if and only if all the inputs are 0.

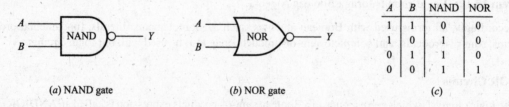

A	B	NAND	NOR
1	1	0	0
1	0	1	0
0	1	1	0
0	0	1	1

(a) NAND gate   (b) NOR gate   (c)

Fig. 15-13

Observe that the only difference between the AND and NAND gates between the OR and NOR gates is that the NAND and NOR gates are each followed by a circle. Some texts also use such a small circle to indicate a complement before a gate. For example, the Boolean expressions corresponding to two logic circuits in Fig. 15-14 are as follows:

$$\text{(a)} \quad Y = (A'B)', \quad \text{(b)} \quad Y = (A' + B' + C)'$$

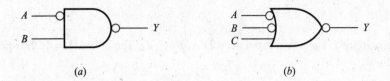

Fig. 15-14

15.11   TRUTH TABLES, BOOLEAN FUNCTIONS

Consider a logic circuit L with $n = 3$ input devices A, B, C and output Y, say

$$Y = A \cdot B \cdot C + A \cdot B' \cdot C + A' \cdot B$$

Each assignment of a set of three bits to the inputs A, B, C yields an output bit for Y. All together there are $2^n = 2^3 = 8$ possible ways to assign bits to the inputs as follows:

$$000, \quad 001, \quad 010, \quad 011, \quad 100, \quad 101, \quad 110, \quad 111$$

The assumption is that the sequence of first bits is assigned to A, the sequence of second bits to B, and the sequence of third bits to C. Thus the above set of inputs may be rewritten in the form

$$A = 00001111, \quad B = 00110011, \quad C = 01010101$$

We emphasize that these three $2^n = 8$-bit sequences contain the eight possible combinations of the input bits.

The *truth table* $T = T(L)$ of the above circuit L consists of the output sequence Y that corresponds to the input sequences A, B, C. This truth table T may be expressed using fractional or relational notation, that is, T may be written in the form

$$T(A, B, C) = Y \quad \text{or} \quad T(L) = [A, B, C; Y]$$

This form for the truth table for L is essentially the same as the truth table for a proposition discussed in Section 4.4. The only difference is that here the values for A, B, C, and Y are written horizontally, whereas in Section 4.4 they are written vertically.

Consider a logic circuit L with n input devices. There are many ways to form n input sequences $A_1, A_2, \ldots, A_n$ so that they contain the 2^n different possible combinations of the input bits. (Note that each sequence must contain 2^n bits.) One assignment scheme is as follows:

A_1: Assign 2^{n-1} bits which are 0's followed by 2^{n-1} bits which are 1's.

A_2: Repeatedly assign 2^{n-2} bits which are 0's followed by 2^{n-2} bits which are 1's.

A_3: Repeatedly assign 2^{n-3} bits which are 0's followed by 2^{n-3} bits which are 1's.

And so on. The sequences obtained in this way will be called *special sequences*. Replacing 0 by 1 and 1 by 0 in the special sequences yields the complements of the special sequences.

Remark: Assuming the input are the special sequences, we frequently do not need to distinguish between the truth table

$$T(L) = [A_1, A_2, \ldots, A_n; Y]$$

and the output Y itself.

EXAMPLE 15.11

(a) Suppose a logic circuit L has $n = 4$ input devices A, B, C, D. The $2^n = 2^4 = 16$-bit special sequences for A, B, C, D follow:

$$A = 0000000011111111, \quad C = 0011001100110011$$
$$B = 0000111100001111, \quad D = 0101010101010101$$

That is:

(1) A begins with eight 0's followed by eight 1's. (Here $2^{n-1} = 2^3 = 8$.)

(2) B begins with four 0's followed by four 1's, and so on. (Here $2^{n-2} = 2^2 = 4$.)

(3) C begins with two 0's followed by two 1's, and so on. (Here $2^{n-3} = 2^1 = 2$.)

(4) D begins with one 0 followed by one 1, and so on. (Here $2^{n-4} = 2^0 = 1$.)

(b) Suppose a logic circuit L has $n = 3$ input devices A, B, C. The $2^n = 2^3 = 8$-bit special sequences for A, B, C and their complements A', B', C' are as follows:

$$A = 00001111, \quad B = 00110011, \quad C = 01010101$$
$$A' = 11110000, \quad B' = 11001100, \quad C' = 10101010$$

Figure 15-15 contains a three-step algorithm for finding the truth table for a logic circuit L where the output Y is given by a Boolean sum-of-products expression in the inputs.

Algorithm 15.5: The input is a Boolean sum-of-products expression $Y = Y(A_1, A_2, \ldots)$.

Step 1. Write down the special sequences for the inputs $A_1, A_2, \ldots$ and their complements.

Step 2. Find each product appearing in Y. (Recall that a product $X_1 \cdot X_2 \cdots = 1$ in a position if and only if all the $X_1, X_2, \ldots$ have 1 in the position.)

Step 3. Find the sum Y of the products. (Recall that a sum $X_1 + X_2 + \cdots = 0$ in a position if and only if all the $X_1, X_2, \ldots$ have 0 in the position.)

Fig. 15-15

EXAMPLE 15.12 Algorithm 15.5 is used to find the truth table $T = T(L)$ of the logic circuit L in Fig. 15-12 or, equivalently, of the above Boolean sum-of-products expression

$$Y = A \cdot B \cdot C + A \cdot B' \cdot C + A' \cdot B$$

(1) The special sequences and their complements appear in Example 15.14(b).

(2) The products are as follows:

$$A \cdot B \cdot C = 00000001, \quad A \cdot B' \cdot C = 00000100, \quad A' \cdot B = 00110000$$

(3) The sum is $Y = 00110101$.

Accordingly,

$$T(00001111, 00110011, 01010101) = 00110101$$

or simply $T(L) = 00110101$ where we assume the input consists of the special sequences.

Boolean Functions

Let E be a Boolean expression with n variables $x_1, x_2, \ldots, x_n$. The entire discussion above can also be applied to E where now the special sequences are assigned to the variables $x_1, x_2, \ldots, x_n$ instead of the input devices $A_1, A_2, \ldots, A_n$. The truth table $T = T(E)$ of E is defined in the same way as the truth table $T = T(L)$ for a logic circuit L. For example, the Boolean expression

$$E = xyz + xy'z + x'y$$

which is analogous to the logic circuit L in Example 15.12, yields the truth table

$$T(00001111, 00110011, 01010101) = 00110101$$

or simply $T(E) = 00110101$, where we assume the input consists of the special sequences.

Remark: The truth table for a Boolean expression $E = E(x_1, x_2, \ldots, x_n)$ with n variables may also be viewed as a "Boolean" function from $\mathbf{B}^n$ into $\mathbf{B}$. (The Boolean algebras $\mathbf{B}^n$ and $\mathbf{B} = \{0, 1\}$ are defined in Example 15.1.) That is, each element in $\mathbf{B}^n$ is a list of n bits which when assigned to the list of variables in E produces an element in $\mathbf{B}$. The truth table $T(E)$ of E is simply the graph of the function.

EXAMPLE 15.13

(a) Consider Boolean expressions $E = E(x, y, z)$ with three variables. The eight minterms (fundamental products involving all three variables) are as follows:

$$xyz, \quad xyz', \quad xy'z, \quad x'yz, \quad xy'z', \quad x'yz', \quad x'y'z'$$

The truth tables for these minterms (using the special sequences for x, y, z) follow:

$$xyz = 00000001, \quad xyz' = 00000010, \quad xy'z = 00000100, \quad x'yz = 00001000$$
$$xy'z' = 00010000, \quad x'yz' = 00100000, \quad x'y'z = 01000000, \quad x'y'z' = 10000000$$

Observe that each minterm assumes the value 1 in only one of the eight positions.

(b) Consider the Boolean expression $E = xyz' + x'yz + x'y'z$. Note that E is a complete sum-of-products expression containing three minterms. Accordingly, the truth table $T = T(E)$ for E, using the special sequences for x, y, z, can be easily obtained from the sequences in part (a). Specifically, the truth table $T(E)$ will contain exactly three 1's in the same positions as the 1's in the three minterms in E. Thus

$$T(00001111, \ 00110011, \ 01010101) = 01001010$$

or simply $T(E) = 01001010$.

15.12 KARNAUGH MAPS

Karnaugh maps, where minterms involving the same variables are represented by squares, are pictorial devices for finding prime implicants and minimal forms for Boolean expressions involving at most six variables. We will only treat the cases of two, three, and four variables. In the context of Karnaugh maps, we will sometimes use the terms "squares" and "minterm" interchangeably. Recall that a minterm is a fundamental product which involves all the variables, and that a complete sum-of-products expression is a sum of minterms.

First we need to define the notion of adjacent products. Two fundamental products P_1 and P_2 are said to be *adjacent* if P_1 and P_2 have the same variables and if they differ in exactly one literal. Thus there must be an uncomplemented variable in one product and complemented in the other. In particular, the sum of two such adjacent products will be a fundamental product with one less literal.

EXAMPLE 15.14 Find the sum of adjacent products P_1 and P_2 where:

(a) $P = xyz'$ and $P_2 = xy'z'$.

$$P_1 + P_2 = xyz' + xy'z' = xz'(y + y') = xz'(1) = xz'$$

(b) $P_1 = x'yzt$ and $P_2 = x'yz't$.

$$P_1 + P_2 = x'yzt + x'yz't = x'yt(z + z') = x'yt(1) = x'yt$$

(c) $P_1 = x'yzt$ and $P_2 = xyz't$.

Here P_1 and P_2 are not adjacent since they differ in two literals. In particular,

$$P_1 + P_2 = x'yzt + xyz't = (x' + x)y(z + z')t = (1)y(1)t = yt$$

(d) $P_1 = xyz'$ and $P_2 = xyzt$.

Here P_1 and P_2 are not adjacent since they have different variables. Thus, in particular, they will not appear as squares in the same Karnaugh map.

Case of Two Variables

The Karnaugh map corresponding to Boolean expressions $E = E(x, y)$ with two variables x and y is shown in Fig. 15-16(a). The Karnaugh map may be viewed as a Venn diagram where x is represented by the points in the upper half of the map, shaded in Fig. 15-16(b), and y is represented by the points in the left half of the map, shaded in Fig. 15-16(c). Thus x' is represented by the points in the lower half of the map, and y' is represented by the points in the right half of the map. Accordingly, the four possible minterms with two literals,

$$xy, \quad xy', \quad x'y, \quad x'y'$$

are represented by the four squares in the map, as labeled in Fig. 15-16(d). Note that two such squares are adjacent, as defined above, if and only if the squares are geometrically adjacent (have a side in common).

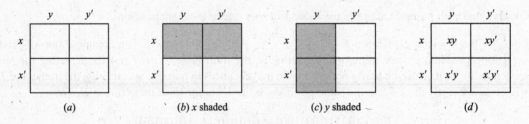

Fig. 15-16

Any complete sum-of-products Boolean expression $E(x, y)$ is a sum of minterms and hence can be represented in the Karnaugh map by placing checks in the appropriate squares. A prime implicant of $E(x, y)$ will be either a pair of adjacent squares in E or an *isolated* square, i.e., a square which is not adjacent to any other square of $E(x, y)$. A minimal sum-of-products form for $E(x, y)$ will consist of a minimal number of prime implicants which cover all the squares of $E(x, y)$ as illustrated in the next example.

EXAMPLE 15.15 Find the prime implicants and a minimal sum-of-products form for each of the following complete sum-of-products Boolean expressions:

(a) $E_1 = xy + xy'$;　(b) $E_2 = xy + x'y + x'y'$;　(c) $E_3 = xy + x'y'$

This can be solved by using Karnaugh maps as follows:

(a) Check the squares corresponding to xy and xy' as in Fig. 15-17(a). Note that E_1 consists of one prime implicant, the two adjacent squares designated by the loop in Fig. 15-17(a). This pair of adjacent squares represents the variable x, so x is a (the only) prime implicant of E_1. Consequently, $E_1 = x$ is its minimal sum.

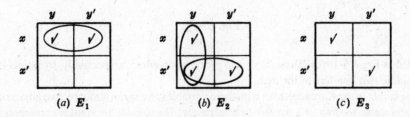

(a) E_1 (b) E_2 (c) E_3

Fig. 15-17

(b) Check the squares corresponding to xy, $x'y$, and $x'y'$ as in Fig. 15-17(b). Note that E_2 contains two pairs of adjacent squares (designated by the two loops) which include all the squares of E_2. The vertical pair represents y and the horizontal pair represents x'; hence y and x' are the prime implicants of E_2. Thus $E_2 = x' + y$ is its minimal sum.

(c) Check the squares corresponding to xy and $x'y'$ as in Fig. 15-17(c). Note that E_3 consists of two isolated squares which represent xy and $x'y'$; hence xy and $x'y'$ are the prime implicants of E_3 and $E_3 = xy + x'y'$ is its minimal sum.

Case of Three Variables

The Karnaugh map corresponding to Boolean expressions $E = E(x, y, z)$ with three variables x, y, z is shown in Fig. 15-18(a). Recall that there are exactly eight minterms with three variables:

$$xyz, \quad xyz', \quad xy'z', \quad xy'z, \quad x'yz, \quad x'yz', \quad x'y'z', \quad x'y'z$$

These minterms are listed so that they correspond to the eight squares in the Karnaugh map in the obvious way.

Furthermore, in order that every pair of adjacent products in Fig. 15-18(a) are geometrically adjacent, the right and left edges of the map must be identified. This is equivalent to cutting out, bending, and gluing the map along the identified edges to obtain the cylinder pictured in Fig. 15-18(b) where adjacent products are now represented by squares with one edge in common.

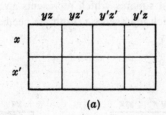   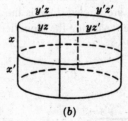

(a)   (b)

Fig. 15-18

Viewing the Karnaugh map in Fig. 15-18(a) as a Venn diagram, the areas represented by the variables x, y, and z are shown in Fig. 15-19. Specifically, the variable x is still represented by the points in the upper half of the map, as shaded in Fig. 15-19(a), and the variable y is still represented by the points in the left half of the map, as shaded in Fig. 15-19(b). The new variable z is represented by the points in the left and right quarters of

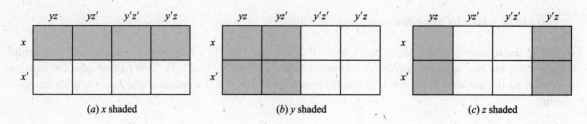

Fig. 15-19

the map, as shaded in Fig. 15-19(c). Thus x', y', and z' are represented, respectively, by points in the lower half, right half, and middle two quarters of the map.

By a *basic rectangle* in the Karnaugh map with three variables, we mean a square, two adjacent squares, or four squares which form a one-by-four or a two-by-two rectangle. These basic rectangles correspond to fundamental products of three, two, and one literal, respectively. Moreover, the fundamental product represented by a basic rectangle is the product of just those literals that appear in every square of the rectangle.

Suppose a complete sum-of-products Boolean expression $E = E(x, y, z)$ is represented in the Karnaugh map by placing checks in the appropriate squares. A prime implicant of E will be a *maximal basic rectangle* of E, i.e., a basic rectangle contained in E which is not contained in any larger basic rectangle in E. A minimal sum-of-products form for E will consist of a *minimal cover* of E, that is, a minimal number of maximal basic rectangles of E which together include all the squares of E.

EXAMPLE 15.16 Find the prime implicants and a minimal sum-of-products form for each of the following complete sum-of-products Boolean expressions:

(a) $E_1 = xyz + xyz' + x'yz' + x'y'z$.

(b) $E_2 = xyz + xyz' + xy'z + x'yz + x'y'z$.

(c) $E_3 = xyz + xyz' + x'yz' + x'y'z' + x'y'z$.

This can be solved by using Karnaugh maps as follows:

(a) Check the squares corresponding to the four summands as in Fig. 15-20(a). Observe that E_1 has three prime implicants (maximal basic rectangles), which are circled; these are xy, yz', and $x'y'z$. All three are needed to cover E_1; hence the minimal sum for E_1 is

$$E_1 = xy + yz' + x'y'z$$

(b) Check the squares corresponding to the five summands as in Fig. 15-20(b). Note that E_2 has two prime implicants, which are circled. One is the two adjacent squares which represents xy, and the other is the two-by-two square (spanning the identified edges) which represents z. Both are needed to cover E_2, so the minimal sum for E_2 is

$$E_2 = xy + z$$

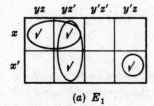

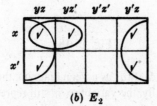

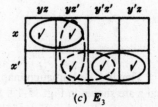

(a) E_1 (b) E_2 (c) E_3

Fig. 15-20

(c) Check the squares corresponding to the five summands as in Fig. 15-20(c). As indicated by the loops, E_3 has
     four prime implicants, xy, yz', $x'z'$, and $x'y'$. However, only one of the two dashed ones, i.e., one of yz' or
     $x'z'$, is needed in a minimal cover of E_3. Thus E_3 has two minimal sums:

$$E_3 = xy + yz' + x'y' = xy + x'z' + x'y'$$

EXAMPLE 15.17 Design a three-input minimal AND-OR circuit L with the following truth table:

$$T = [A, B, C; L] = [00001111, 00110011, 01010101; 11001101]$$

From the truth table we can read off the complete sum-of-products form for L (as in Example 15.10):

$$L = A'B'C' + A'B'C + AB'C' + AB'C + ABC$$

The associated Karnaugh map is shown in Fig. 15-21(a). Observe that L has two prime implicants, B' and AC,
in its minimal cover; hence $L = B' + AC$ is a minimal sum for L. Figure 15-21(b) gives the corresponding
minimal AND-OR circuit for L.

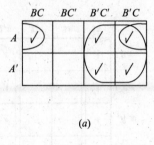

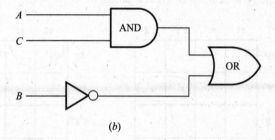

(a)   (b)

Fig. 15-21

Case of Four Variables

The Karnaugh map corresponding to Boolean expressions $E = E(x, y, z, t)$ with four variables x, y, z, t
is shown in Fig. 15-22. Each of the 16 squares corresponds to one of the 16 minterms with four variables,

$$xyzt, \quad xyzt', \quad xyz't', \quad xyz't, \quad \ldots, x'yz't$$

as indicated by the labels of the row and column of the square. Observe that the top line and the left side are
labeled so that adjacent products differ in precisely one literal. Again we must identify the left edge with the
right edge (as we did with three, variables) but we must also identify the top edge with the bottom edge. (These
identifications give rise to a donut-shaped surface called a *torus*, and we may view our map as really being a
torus.)

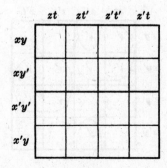

Fig. 15-22

A *basic rectangle* in a four-variable Karnaugh map is a square, two adjacent squares, four squares which form a one-by-four or two-by-two rectangle, or eight squares which form a two-by-four rectangle. These rectangles correspond to fundamental products with four, three, two, and one literal, respectively. Again, maximal basic rectangles are the prime implicants. The minimization technique for a Boolean expression $E(x, y, z, t)$ is the same as before.

EXAMPLE 15.18 Find the fundamental product P represented by the basic rectangle in the Karnaugh maps shown in Fig. 15-23.

In each case, find the literals which appear in all the squares of the basic rectangle; P is the product of such literals.

(a) xy, and z' appear in both squares; hence $P = xy'z'$.

(b) Only y and z appear in all four squares; hence $P = yz$.

(c) Only t appears in all eight squares; hence $P = t$.

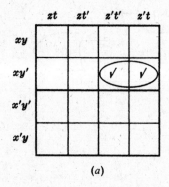 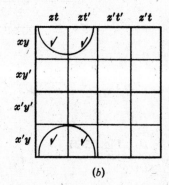 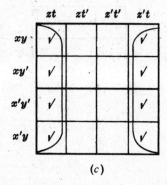

Fig. 15-23

EXAMPLE 15.19 Use a Karnaugh map to find a minimal sum-of-products form for

$$E = xy' + xyz + x'y'z' + x'yzt'$$

Check all the squares representing each fundamental product. That is, check all four squares representing xy', the two squares representing xyz, the two squares representing $x'y'z'$, and the one square representing $x'yzt'$, as in Fig. 15-24. A minimal cover of the map consists of the three designated maximal basic rectangles.

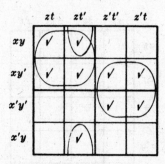

Fig. 15-24

The two-by-two squares represent the fundamental products xz and $y'z'$, and the two adjacent squares (on top and bottom) represents yzt'. Hence

$$E = xz + y'z' + yzt'$$

is a minimal sum for E.

Solved Problems

BOOLEAN ALGEBRAS

15.1. Write the dual of each Boolean equation: (a) $(a * 1) * (0 + a') = 0$; (b) $a + a'b = a + b$.

(a) To obtain the dual equation, interchange $+$ and $*$, and interchange 0 and 1. Thus

$$(a + 0) + (1 * a') = 1$$

(b) First write the equation using $*$ to obtain $a + (a' * b) = a + b$. Then the dual is $a * (a' + b) = a * b$, which can be written as

$$a(a' + b) = ab$$

15.2. Recall (Chapter 14) that the set $\mathbf{D}_m$ of divisors of m is a bounded, distributive lattice with

$$a + b = a \vee b = \mathrm{lcm}(a, b) \quad \text{and} \quad a * b = a \wedge b = \gcd(a, b).$$

(a) Show that $\mathbf{D}_m$ is a Boolean algebra if m is square free, i.e., if m is a product of distinct primes.

(b) Find the atoms of $\mathbf{D}_m$.

(a) We need only show that $\mathbf{D}_m$ is complemented. Let x be in $\mathbf{D}_m$ and let $x' = m/x$. Since m is a product of distinct primes, x and x' have different prime divisors. Hence $x * x' = \gcd(x, x') = 1$ and $x + x' = \mathrm{lcm}(x, x') = m$. Recall that 1 is the zero element (lower bound) of $\mathbf{D}_m$ and that m is the identity element (upper bound) of $\mathbf{D}_m$. Thus x' is a complement of x, and so $\mathbf{D}_m$ is a Boolean algebra.

(b) The atoms of $\mathbf{D}_m$ are the prime divisors of m.

15.3. Consider the Boolean algebra $\mathbf{D}_{210}$.

(a) List its elements and draw its diagram.

(b) Find the set A of atoms.

(c) Find two subalgebras with eight elements.

(d) Is $X = \{1, 2, 6, 210\}$ a sublattice of $\mathbf{D}_{210}$? A subalgebra?

(e) Is $Y = \{1, 2, 3, 6\}$ a sublattice of $\mathbf{D}_{210}$? A subalgebra?

(a) The divisors of 210 are 1, 2, 3, 5, 6, 7, 10, 14, 15, 21, 30, 35, 42, 70, 105, and 210. The diagram of $\mathbf{D}_{210}$ appears in Fig. 15-25.

(b) $A = \{2, 3, 5, 7\}$, the set of prime divisors of 210.

(c) $B = \{1, 2, 3, 35, 6, 70, 105, 210\}$ and $C = \{1, 5, 6, 7, 30, 35, 42, 210\}$ are subalgebras of $\mathbf{D}_{210}$.

(d) X is a sublattice since it is linearly ordered. However, X is not a subalgebra since 35 is the complement of 2 in $\mathbf{D}_{210}$ but 35 does not belong to X. (In fact, no Boolean algebra with more than two elements is linearly ordered.)

(e) Y is a sublattice of $\mathbf{D}_{210}$ since it is closed under $+$ and $*$. However, Y is not a subalgebra of $\mathbf{D}_{210}$ since it is not closed under complements in $\mathbf{D}_{210}$, e.g., $35 = 2'$ does not belong to Y. (We note that Y itself is a Boolean algebra; in fact, $Y = \mathbf{D}_6$.)

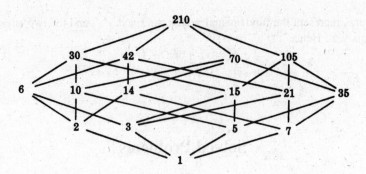

Fig. 15-25

15.4. Find the number of subalgebras of $\mathbf{D}_{210}$.

A subalgebra of $\mathbf{D}_{210}$ must contain two, four, eight or sixteen elements.

(i) There can be only one two-element subalgebra which consists of the upper bound 210 and lower bound 1, i.e., $\{1, 210\}$.

(ii) Since $\mathbf{D}_{210}$ contains sixteen elements, the only sixteen-element subalgebra is $\mathbf{D}_{210}$ itself.

(iii) Any four-element subalgebra is of the form $\{1, x, x', 210\}$, i.e., consists of the upper and lower bounds and a nonbound element and its complement. There are fourteen nonbound elements in $\mathbf{D}_{210}$ and so there are $14/2 = 7$ pairs $\{x, x'\}$. Thus $\mathbf{D}_{210}$ has seven four-element subalgebras.

(iv) Any eight-element subalgebra S will itself contain three atoms s_1, s_2, s_3. We can choose s_1 and s_2 to be any two of the four atoms of $\mathbf{D}_{210}$ and then s_3 must be the product of the other two atoms, e.g., we can let $s_1 = 2, s_2 = 3$, $s_3 = 5 \cdot 7 = 35$ (which determines the subalgebra B above), or we can let $s_1 = 5, s_2 = 7, s_3 = 2 \cdot 3 = 6$ (which determines the subalgebra C above). There are $\binom{4}{2} = 6$ ways to choose s_1 and s_2 from the four atoms of $\mathbf{D}_{210}$ and so $\mathbf{D}_{210}$ has six eight-element subalgebras.

Accordingly, $\mathbf{D}_{210}$ has $1 + 1 + 7 + 6 = 15$ subalgebras.

15.5. Prove Theorem 15.2: Let a, b, c be any element in a Boolean algebra B.

(i) Idempotent laws:
$$(5a) \quad a + a = a \qquad (5b) \quad a * a = a$$

(ii) Boundedness laws:
$$(6a) \quad a + 1 = 1 \qquad (6b) \quad a * 0 = 0$$

(iii) Absorption Laws:
$$(7a) \quad a + (a * b) = a \qquad (7b) \quad a * (a + b) = a$$

(iv) Associative Laws:
$$(8a) \quad (a + b) + c = a + (b + c) \qquad (8b) \quad (a * b) * c = a * (b * c)$$

($5b$) $a = a * 1 = a * (a + a') = (a * a) + (a * a') = (a * a) + 0 = a * a$

($5a$)   Follows from ($5b$) and duality.

($6b$) $a * 0 = (a * 0) + 0 = (a * 0) + (a * a') = a * (0 + a') = a * (a' + 0) = a * a' = 0$

($6a$)   Follows from ($6b$) and duality.

($7b$) $a * (a + b) = (a + 0) * (a + b) = a + (0 * b) = a + (b * 0) = a + 0 = a$

($7a$)   Follows from ($7b$) and duality.

($8b$)   Let $L = (a * b) * c$ and $R = a * (b * c)$. We need to prove that $L = R$. We first prove that $a + L = a + R$. Using the absorption laws in the last two steps,

$$a + L = a + ((a * b) * c) = (a + (a * b)) * (a + c) = a * (a + c) = a$$

Also, using the absorption law in the last step,

$$a + R = a + (a * (b * c)) = (a + a) * (a + (b * c)) = a * (a + (b * c)) = a$$

Thus $a + L = a + R$. Next we show that $a' + L = a' + R$. We have,

$$a' + L = a' + ((a * b) * c) = (a' + (a * b)) * (a' + c)$$
$$= ((a' + a) * (a' + b)) * (a' + c) = (1 * (a' + b)) * (a' + c)$$
$$= (a' + b) * (a' + c) = a' + (b * c)$$

Also,

$$a' + R = a' + (a * (b * c)) = (a' + a) * (a' + (b * c))$$
$$= 1 * (a' + (b * c)) = a' + (b * c)$$

Thus $a' + L = a' + R$. Consequently,

$$L = 0 + L = (a * a') + L = (a + L) * (a' + L) = (a + R) * (a' + R)$$
$$= (a * a') + R = 0 + R = R$$

(8a) Follows from (8b) and duality.

15.6. Prove Theorem 15.3: Let a be any element of a Boolean algebra B.

   (i) (Uniqueness of Complement) If $a + x = 1$ and $a * x = 0$, then $x = a'$.

   (ii) (Involution Law) $(a')' = a$

   (iii) (9a) $0' = 1$; (9b) $1' = 0$.

   (i) We have:

$$a' = a' + 0 = a' + (a * x) = (a' + a) * (a' + x) = 1 * (a' + x) = a' + x$$

   Also,

$$x = x + 0 = x + (a * a') = (x + a) * (x + a') = 1 * (x + a') = x + a'$$

   Hence $x = x + a' = a' + x = a'$.

   (ii) By definition of complement, $a + a' = 1$ and $a * a' = 0$. By commutativity, $a' + a = 1$ and $a' * a = 0$. By uniqueness of complement, a is the complement of a', that is, $a = (a')'$.

   (iii) By boundedness law (6a), $0 + 1 = 1$, and by identity axiom (3b), $0 * 1 = 0$. By uniqueness of complement, 1 is the complement of 0, that is, $1 = 0'$. By duality, $0 = 1'$.

15.7. Prove Theorem 15.4: (DeMorgan's laws): (10a) $(a + b)' = a' * b'$. (10b) $(a * b)' = a' + b'$.

   (10a) We need to show that $(a + b) + (a' * b') = 1$ and $(a + b) * (a' * b') = 0$; then by uniqueness of complement, $a' * b' = (a + b')$. We have:

$$(a + b) + (a' * b') = b + a + (a' * b') = b + (a + a') * (a + b')$$
$$= b + 1 * (a + b') = b + a + b' = b + b' + a = 1 + a = 1$$

   Also,

$$(a + b) * (a' * b') = ((a + b) * a') * b'$$
$$= ((a * a') + (b * a')) * b' = (0 + (b * a')) * b'$$
$$= (b * a') * b' = (b * b') * a' = 0 * a' = 0$$

   Thus $a' * b' = (a + b)'$.

   (10b) Principle of duality (Theorem 15.1).

15.8. Prove Theorem 15.5: The following are equivalent in a Boolean algebra:

$$(1)\ a + b = b; \quad (2)\ a * b = a; \quad (3)\ a' + b = 1; \quad (4)\ a * b' = 0.$$

By Theorem 14.4, (1) and (2) are equivalent. We show that (1) and (3) are equivalent. Suppose (1) holds. Then

$$a' + b = a' + (a + b) = (a' + a) + b = 1 + b = 1$$

Now suppose (3) holds. Then

$$a + b = 1 * (a + b) = (a' + b) * (a + b) = (a' * a) + b = 0 + b = b$$

Thus (1) and (3) are equivalent.

We next show that (3) and (4) are equivalent. Suppose (3) holds. By DeMorgan's law and involution,

$$0 = 1' = (a' + b')' = a'' * b' = a * b'$$

Conversely, if (4) holds then

$$1 = 0' = (a * b')' = a' + b'' = a' + b$$

Thus (3) and (4) are equivalent. Accordingly, all four are equivalent.

15.9. Prove Theorem 15.6: The mapping $f : B \to P(A)$ is an isomorphism where B is a Boolean algebra, $P(A)$ is the power set of the set A of atoms, and

$$f(x) = \{a_1, a_2, \ldots, a_n\}$$

where $x = a_1 + \cdots + a_n$ is the unique representation of a as a sum of atoms.

Recall (Chapter 14) that if the a's are atoms then $a_i^2 = a_i$ but $a_i a_j = 0$ for $a_i \neq a_j$. Suppose x, y are in B and suppose

$$x = a_1 + \cdots + a_r + b_1 + \cdots + b_s$$
$$y = b_1 + \cdots + b_s + c_1 + \cdots + c_t$$

where

$$A = \{a_1, \ldots, a_r, b_1, \ldots, b_s, c_1, \ldots, c_t, d_1, \ldots, d_k\}$$

is the set of atoms of B. Then

$$x + y = a_1 + \cdots + a_r + b_1 + \cdots + b_s + c_1 + \cdots + c_t$$
$$xy = b_1 + \cdots b_s$$

Hence

$$
\begin{aligned}
f(x + y) &= \{a_1, \ldots, a_r, b_1, \ldots, b_s, c_1, \ldots, c_t\} \\
&= \{a_1, \ldots, a_r, b_1, \ldots, b_s\} \cup \{b_1, \ldots, b_s, c_1, \ldots, c_t\} \\
&= f(x) \cup f(y) \\
f(xy) &= \{b_1, \ldots, b_s\} \\
&= \{a_1, \ldots, a_r, b_1, \ldots, b_s\} \cap \{b_1, \ldots, b_s, c_1, \ldots, c_t\} \\
&= f(x) \cap f(y)
\end{aligned}
$$

Let

$$y = c_1 + \cdots + c_t + d_1 + \cdots + d_k. \text{ Then } x + y = 1 \text{ and } xy = 0, \text{ and so } y = x'$$

Thus

$$f(x') = \{c_1, \ldots, c_t, d_1, \ldots, d_k\} = \{a_1, \ldots, a_r, b_1, \ldots, b_s\}^c = (f(x))^c$$

Since the representation is unique, f is one-to-one and onto. Hence f is a Boolean algebra isomorphism.

BOOLEAN EXPRESSIONS

15.10. Reduce the following Boolean products to either 0 or a fundamental product:

$$(a) \quad xyx'z; \quad (b) \quad xyzy; \quad (c) \quad xyz'yx; \quad (d) \quad xyz'yx'z'$$

Use the commutative law $x * y = y * x$, the complement law $x * x' = 0$, and the idempotent law $x * x = x$:

(a) $xyx'z = xx'yz = 0yz = 0$

(b) $xyzy = xyyz = xyz$

(c) $xyz'yx = xxyyz' = xyz'$

(d) $xyz'yx'z' = xx'yyz'z' = 0yz' = 0$

15.11. Express each Boolean expression $E(x, y, z)$ as a sum-of-products and then in its complete sum-of-products form: (a) $E = x(xy' + x'y + y'z)$; (b) $E = z(x' + y) + y'$.

First use Algorithm 15.1 to express E as a sum-of-products, and then use Algorithm 15.2 to express E as a complete sum-of-products.

(a) First we have $E = xxy' + xx'y + xy'z = xy' + xy'z$. Then

$$E = xy'(z + z') + xy'z = xy'z + xy'z' + xy'z = xy'z + xy'z'$$

(b) First we have

$$E = z(x' + y) + y' = x'z + yz + y'$$

Then

$$E = x'z + yz + y' = x'z(y + y') + yz(x + x') + y'(x + x')(z + z')$$
$$= x'yz + x'y'z + xyz + x'yz + xy'z + xy'z' + x'y'z + x'y'z'$$
$$= xyz + xy'z + xy'z' + x'yz + x'y'z + x'y'z'$$

15.12. Express $E(x, y, z) = (x' + y)' + x'y$ in its complete sum-of-products form.

We have $E = (x' + y)' + x'y = xy' + x'y$, which would be the complete sum-of-products form of E if E were a Boolean expression in x and y. However, it is specified that E is a Boolean expression in the three variables x, y, and z. Hence,

$$E = xy' + x'y = xy'(z + z') + x'y(z + z') = xy'z + xy'z' + x'yz + x'yz'$$

is the complete sum-of-products form of E.

15.13. Express each Boolean expression $E(x, y, z)$ as a sum-of-products and then in its complete sum-of-products form: (a) $E = y(x + yz)'$; (b) $E = x(xy + y' + x'y)$.

(a) $E = y(x'(yz)') = yx'(y' + z') = yx'y' + x'yz' = x'yz'$
which already is in its complete sum-of-products form.

(b) First we have $E = xxy + xy' + xx'y = xy + xy'$. Then

$$E = xy(z + z') + xy'(z + z') = xyz + xyz' + xy'z + xy'z'$$

15.14. Express each set expression $E(A, B, C)$ involving sets A, B, C as a union of intersections:

$$(a) \quad E = (A \cup B)^c \cap (C^c \cup B); \quad (b) \quad E = (B \cap C)^c \cap (A^c \cap C)^c$$

Use Boolean notation, ′ for complement, + for union, and * (or juxtaposition) for intersection, and then express E as a sum of products (union of intersections).

(a) $E = (A + B)'(C' + B) = A'B'(C' + B) = A'B'C' + A'B'B = A'B'C'$ or $E = A^c \cap B^c \cap C^c$

(b) $E = (BC)'(A' + C)' = (B' + C')(AC') = AB'C' + AC'$ or $E = (A \cap B^c \cap C^c) \cap (A \cap C^c)$

15.15. Let $E = xy' + xyz' + x'yz'$. Prove that (a) $xz' + E = E$; (b) $x + E \neq E$; (c) $z' + E \neq E$.

Since the complete sum-of-products form is unique, $A + E = E$, where $A \neq 0$, if and only if the summands in the complete sum-of-products form for A are among the summands in the complete sum-of products form for E. Hence, first find the complete sum-of-products form for E:

$$E = xy'(z + z') + xyz' + x'yz' = xy'z + xy'z' + xyz' + x'yz'$$

(a) Express xz' in complete sum-of-products form:

$$xz' = xz'(y + y') = xyz' + xy'z'$$

Since the summands of xz' are among those of E, we have $xz' + E = E$.

(b) Express x in complete sum-of-products form:

$$x = x(y + y')(z + z') = xyz + xyz' + xy'z + xy'z'$$

The summand xyz of x is not a summand of E; hence $x + E \neq E$.

(c) Express z' in complete sum-of-products form:

$$z' = z'(x + x')(y + y') = xyz' + xy'z' + x'yz' + x'y'z'$$

The summand $x'y'z'$ of z' is not a summand of E; hence $z' + E \neq E$.

MINIMAL BOOLEAN EXPRESSIONS, PRIME IMPLICANTS

15.16. For any Boolean sum-of-products expression E, we let E_L denote the number of literals in E (counting multiplicity) and E_S denote the number of summands in E. Find E_L and E_S for each of the following:

(a) $E = xy'z + x'z' + yz' + x$ (c) $E = xyt' + x'y'zt + xz't$
(b) $E = x'y'z + xyz + y + yz' + x'z$ (d) $E = (xy' + z)' + xy'$

Simply add up the number of literals and the number of summands in each expression:

(a) $E_L = 3 + 2 + 2 + 1 = 8$, $E_S = 4$.

(b) $E_L = 3 + 3 + 1 + 2 + 2 = 11$, $E_S = 5$.

(c) $E_L = 3 + 4 + 3 = 10$, $E_S = 3$.

(d) Because E is not written as a sum of products, E_L and E_S are not defined.

15.17. Given that E and F are equivalent Boolean sum-of-products, define:

(a) E is simpler than F; (b) E is minimal.

(a) E is simpler than F if $E_L < F_L$ and $E_S \leq F_S$, or if $E_L \leq F_L$ and $E_S < F_S$.

(b) E is minimal if there is no equivalent sum-of-products expression which is simpler than E.

15.18. Find the consensus Q of the fundamental products P_1 and P_2 where:

(a) $P_1 = xy'z', P_2 = xyt$ (c) $P_1 = xy'z', P_2 = x'y'zt$
(b) $P_1 = xyz't, P_2 = xzt$ (d) $P_1 = xyz', P_2 = xz't$

The consensus Q of P_1 and P_2 exists if there is exactly one variable, say x_k, which is complemented in one of P_1 and P_2 and uncomplemented in the other. Then Q is the product (without repetition) of the literals in P_1 and P_2 after x_k and x_k' have been deleted:

(a) Delete y' and y and then multiply the literals of P_1 and P_2 (without repetition) to obtain $Q = xz't$.

(b) Deleting z' and z yields $Q = xyt$.

(c) They have no consensus since both x and z appear complemented in one of the products and uncomplemented in the other.

(d) They have no consensus since no variable appears complemented in one of the products and uncomplemented in the other.

15.19. Prove Lemma 15.10: Suppose Q is the consensus of P_1 and P_2. Then $P_1 + P_2 + Q = P_1 + P_2$.

Since the literals commute, we can assume without loss of generality that

$$P_1 = a_1 a_2 \cdots a_r, t, \quad P_2 = b_1 b_2 \cdots b_s t', \quad Q = a_1 a_2 \cdots a_r b_1 b_2 \cdots b_s$$

Now, $Q = Q(t + t') = Qt + Qt'$. Because Qt contains P_1, $P_1 + Qt = P_1$; and because Qt' contains P_2, $P_2 + Qt' = P_2$. Hence

$$P_1 + P_2 + Q = P_1 + P_2 + Qt + Qt' = (P_1 + Qt) + (P_2 + Qt') = P_1 + P_2$$

15.20. Let $E = xy' + xyz' + x'yz'$. Find: (a) the prime implicants of E; (b) a minimal sum for E.

(a) Apply Algorithm 15.3 (consensus method) as follows:

$$
\begin{array}{ll}
E = xy' + xyz' + x'yz' + xz' & \text{(consensus of } xy' \text{ and } xyz') \\
= xy' + x'yz' + xz' & \text{(}xyz' \text{ includes } xz') \\
= xy' + x'yz' + xz' + yz' & \text{(consensus of } x'yz' \text{ and } xz') \\
= xy' + xz' + yz' & \text{(}x'yz' \text{ includes } yz')
\end{array}
$$

Neither step in the consensus method can now be applied. Hence xy', xz', and yz' are the prime implicants of E.

(b) Apply Algorithm 15.4 write each prime implicant of E in complete sum-of-products form obtaining:

$$xy' = xy'(z + z') = xy'z + xy'z'$$
$$xz' = xz'(y + y') = xyz' + xy'z'$$
$$yz' = yz'(x + x') = xyz' + x'yz'$$

Only the summands xyz' and $xy'z'$ of xz' appear among the other summands and hence xz' can be eliminated as superfluous. Thus $E = xy' + yz'$ is a minimal sum for E.

15.21. Let $E = xy + y't + x'yz' + xy'zt'$. Find: (a) prime implicants of E; (b) minimal sum for E.

(a) Apply Algorithm 15.3 (consensus method) as follows:

$$
\begin{array}{ll}
E = xy + y't + x'yz' + xy'zt' + xzt' & \text{(consensus of } xy \text{ and } xy'zt') \\
= xy + y't + x'yz' + xzt' & \text{(}xy'zt' \text{ includes } xzt') \\
= xy + y't + x'yz' + xzt' + yz' & \text{(consensus of } xy \text{ and } x'yz') \\
= xy + y't + xzt' + yz' & \text{(}x'yz' \text{ includes } yz') \\
= xy + y't + xzt' + yz' + xt & \text{(consensus of } xy \text{ and } y't) \\
= xy + y't + xzt' + yz' + xt + xz & \text{(consensus of } xzt' \text{ and } xt) \\
= xy + y't + yz' + xt + xz & \text{(}xzt' \text{ includes } xz) \\
= xy + y't + yz' + xt + xz + z' & \text{(consensus of } y't \text{ and } yz')
\end{array}
$$

Neither step in the consensus method can now be applied. Hence the prime implicants of E are xy, $y't$, yz', xt, xz, and $z't$.

(b) Apply Algorithm 15.4 that is, write each prime implicant in complete sum-of-products form and then delete one by one those which are superfluous, i.e. those whose summands appear among the other summands. This finally yields

$$E = y't + xz + yz'$$

as a minimal sum for E.

LOGIC GATES

15.22. Express the output Y as a Boolean expression in the inputs A, B, C for the logic circuit in:

(a) Fig. 15-26(a);   (b) Fig. 15-26(b).

(a) The inputs to the first AND gate are A and B' and to the second AND gate are B' and C. Thus $Y = AB' + B'C$.

(b) The inputs to the first AND gate are A and B' and to the second AND gate are A' and C. Thus $Y = AB' + A'C$.

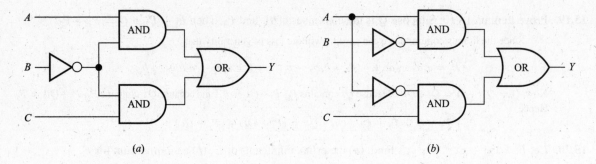

Fig. 15-26

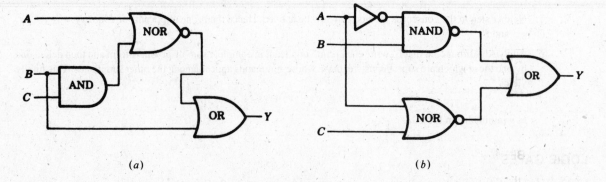

Fig. 15-27

15.23. Express the output Y as a Boolean expression in the inputs A, B, C for the logic circuit in Fig. 15-27.

The output of the first AND gate is $A'BC$, of the second AND gate is $AB'C'$, and of the last AND gate is AB'. Thus

$$Y = A'BC + AB'C' + AB'$$

15.24. Express the output Y as a Boolean expression in the inputs A, B, C for the logic circuit in:

(a) Fig. 15-28(a);   (b) Fig. 15-28(b).

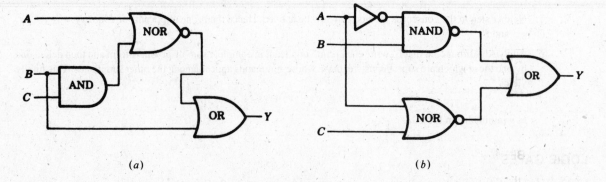

Fig. 15-28

(a) The output of the AND gate is BC and so the inputs to the NOR gate are A and BC. Thus $(A + BC)'$ is the output of the NOR gate. Therefore, the inputs to the OR gate are $(A + BC)'$ and B; hence $Y = (A + BC)' + B$.

(b) The output of the NAND gate is $(A'B)'$ and the output of the NOR gate is $(A+C)'$. Thus $Y = (A'B)' + (A+C)'$.

15.25. Express the output Y as a Boolean expression in the inputs A and B for the logic circuit in Fig. 15-29.

Here a small circle in the circuit means complement. Thus the output of the three gates on the left are AB', $(A + B)'$, and $(A'B)'$. Hence

$$Y = AB' + (A + B)' + (A'B)'$$

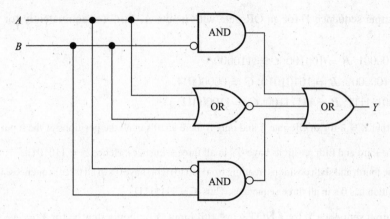

Fig. 15-29

15.26. Draw the logic circuit L with inputs A, B, C and output Y which corresponds to each Boolean expression:

(a) $Y = ABC + A'C' + B'C'$;   (b) $Y = AB'C + ABC' + AB'C'$.

These are sum-of-products expressions. Thus L will be an AND-OR circuit which has an AND gate for each product and one OR gate for the sum. The required circuits appear in Fig. 15-30(a) and (b).

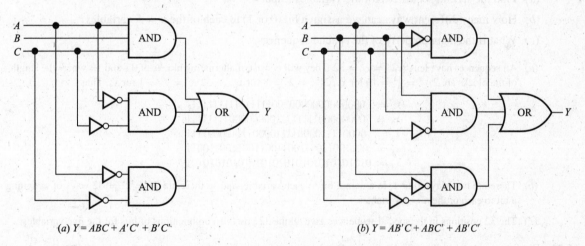

(a) $Y = ABC + A'C' + B'C'$ (b) $Y = AB'C + ABC' + AB'C'$

Fig. 15-30

TRUTH TABLES

15.27. Find the output sequence Y for an AND gate with inputs A, B, C (or equivalently for $Y = ABC$) where:

(a) $A = 111001$; $B = 100101$; $C = 110011$.

(b) $A = 11111100$; $B = 10101010$; $C = 00111100$.

(c) $A = 00111111$; $B = 11111100$; $C = 11000011$.

The output $Y = 1$ for an AND gate if and only if there are 1's in all the positions of the input sequences. Thus:

(a) Only the first and last positions have 1's in all three sequences. Hence $Y = 100001$.

(b) Only the third and fifth positions (reading from left to right) have 1's in all three sequences. Thus $Y = 00101000$.

(c) No position has 1's in all three sequences. Thus $Y = 00000000$.

15.28. Find the output sequence Y for an OR gate with inputs A, B, C (or equivalently for $Y = A + B + C$) where:

(a) $A = 100001$; $B = 100100$; $C = 110000$.

(b) $A = 11000000$; $B = 10101010$; $C = 00000011$.

(c) $A = 00111111$; $B = 11111100$; $C = 11000011$.

The output $Y = 0$ for an OR gate if and only if there are 0's in all the positions of the input sequences. Thus:

(a) Only the third and fifth positions have 0's in all three sequences. Hence $Y = 110101$.

(b) Only the fourth and sixth positions (reading from left to right) have 0's in all three sequences. Thus $Y = 11101011$.

(c) No position has 0's in all three sequences. Thus $Y = 11111111$.

15.29. Find the output sequence Y for a NOT gate with input A or, equivalently, for $Y = A'$, where:

(a) $A = 00111111$;   (b) $A = 11111100$;   (c) $A = 11000011$.

The NOT gate changes 0 to 1 and 1 to 0. Hence:

(a) $A' = 11000000$;   (b) $A' = 00000011$;   (c) $A' = 00111100$.

15.30. Consider a logic circuit L with $n = 5$ inputs A, B, C, D, E or, equivalently, consider a Boolean expression E with five variables x_1, x_2, x_3, x_4, x_5.

(a) Find the special sequences for the variables (inputs).

(b) How many different ways can we assign a bit (0 or 1) to each of the $n = 5$ variables?

(c) What is the main property of the special sequences?

(a) All sequences have length $2^n = 2^5 = 32$. They will consist of alternating blocks of 0's and 1's where the lengths of the blocks are $2^{n-1} = 2^4 = 16$ for x_1, $2^{n-2} = 2^3 = 8$ for $x_2, \ldots, 2^{n-5} = 2^0 = 1$ for x_5, Thus:

$$x_1 = \text{00000000000000001111111111111111}$$
$$x_2 = \text{00000000111111110000000011111111}$$
$$x_3 = \text{00001111000011110000111100001111}$$
$$x_4 = \text{00110011001100110011001100110011}$$
$$x_5 = \text{01010101010101010101010101010101}$$

(b) There are two ways, 0 or 1, to assign a bit to each variable, and so there are $2^n = 2^5 = 32$ ways of assigning a bit to each of the $n = 5$ variables.

(c) The 32 positions in the special sequences give all the 32 possible combinations of bits for the five variables.

15.31. Find the truth table $T = T(E)$ for the Boolean expression $E = E(x, y, z)$ where:

(a) $E = xz + x'y$; (b) $E = xy'z + xy + z'$.

The special sequences for the variables x, y, z and their complements follow:

$$x = 00001111, \quad y = 00110011, \quad z = 01010101$$
$$x' = 11110000, \quad y' = 11001100, \quad z' = 10101010$$

(a) Here $xz = 00000101$ and $x'y = 00110000$. Then $E = xz + x'y = 00110101$. Thus

$$T(00001111, 00110011, 01010101) = 00110101$$

or simply $T(E) = 00110101$ where we assume the input consists of the special sequences.

(b) Here $xy'z = 00000100$, $xy = 00000011$, and $z' = 01010101$. Then $E = xy'z + xy + z' = 01010111$. Thus

$$T(00001111, 00110011, 01010101) = 01010111$$

15.32. Find the truth table $T = T(E)$ for the Boolean expression $E = E(x, y, z)$ where:

(a) $E = xyz' + x'yz$;   (b) $E = xyz + xy'z + x'y'z$.

Here E is a complete sum-of-products expression which is the sum of minterms. Example 15.13 gives the truth tables for the minterms (using the special sequences). Each minterm contains a single 1 in its truth table; hence the truth table of E will have 1's in the same positions as the 1's in the minterms in E. Thus:

(a) $T(E) = 00001010$;   (b) $T(E) = 01000101$

15.33. Find the truth table $T = T(E)$ for the Boolean expression

$$E = E(x, y, z) = (x'y)'yz' + x'(yz + z')$$

First express E as a sum-of-products:

$$E = (x + y')yz' + x'yz + x'z' = xyz' + y'yz' + x'yz + x'z'$$
$$= xyz' + x'yz + x'z'$$

Now express E as a complete sum-of-products:

$$E = xyz' + x'yz + x'z'(y + y')$$
$$= xyz' + x'yz + x'yz' + x'y'z'$$

As in Problem 15.32, use the truth tables for the minterms appearing in Example 15.13 to obtain $T(E) = 10101010$.

15.34. Find the Boolean expression $E = E(x, y, z)$ corresponding to the truth table:

(a) $T(E) = 01001001$;   (b) $T(E) = 00010001$.

Each 1 in $T(E)$ corresponds to the minterm with the 1 in the same position (using the truth tables for the minterms appearing in Example 15.13). For example, the 1 in the second position corresponds to $x'y'z$ whose truth table has a single 1 in the second position. Then E is the sum of these minterms. Thus:

(a) $E = x'y'z + x'yz + xyz'$;   (b) $E = xy'z' + xyz$

(Again we assume the input consists of the special sequences.)

KARNAUGH MAPS

15.35. Find the fundamental product P represented by each basic rectangle in the Karnaugh map in Fig. 15-31.

In each case find those literals which appear in all the squares of the basic rectangle; then P is the product of such literals.

(a) x' and z' appear in both squares; hence $P = x'z'$.

(b) x and z appear in both squares; hence $P = xz$.

(c) Only z appears in all four squares; hence $P = z$.

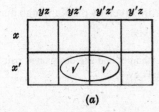   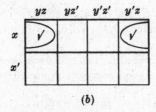   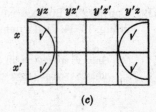

| (a) | (b) | (c) |

Fig. 15-31

15.36. Let R be a basic rectangle in a Karnaugh map for four variables x, y, z, t. State the number of literals in the fundamental product P corresponding to R in terms of the number of squares in R.

P will have one, two, three, or four literals according as R has eight, four, two, or one squares.

15.37. Find the fundamental product P represented by each basic rectangle R in the Karnaugh map in Fig. 15-32.

In each case find those literals which appear in all the squares of the basic rectangle; then P is the product of such literals. (Problem 15.36 indicates the number of such literals in P.)

(a) There are two squares in R, so P has three literals. Specifically, x', y', t' appear in both squares; hence $P = x'y't'$.

(b) There are four squares in R, so P has two literals. Specifically, only y' and t appear in all four squares; hence $P = y't$.

(c) There are eight squares in R, so P has only one literal. Specifically, only y appears in all eight squares; hence $P = y$.

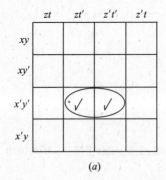 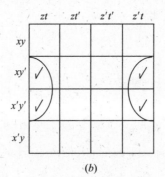 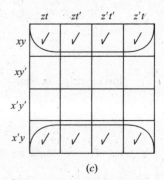

$\qquad$ (a) $\qquad\qquad\qquad\qquad$ (b) $\qquad\qquad\qquad\qquad$ (c)

Fig. 15-32

15.38. Let E be the Boolean expression given in the Karnaugh map in Fig. 15-33.

(a) Write E in its complete sum-of-products form. (b) Find a minimal form for E.

(a) List the seven fundamental products checked to obtain

$$E = xyz't' + xyz't + xy'zt + xy'zt' + x'y'zt + x'y'zt' + x'yz't'$$

(b) The two-by-two maximal basic rectangle represents $y'z$ since only y' and z appear in all four squares. The horizontal pair of adjacent squares represents xyz', and the adjacent squares overlapping the top and bottom edges represent $yz't'$. As all three rectangles are needed for a minimal cover,

$$E = y'z + xyz' + yz't'$$

is the minimal sum for E.

15.39. Consider the Boolean expressions E_1 and E_2 in variables x, y, z, t which are given by the Karnaugh maps in Fig. 15-34. Find a minimal sum for (a) E_1; (b) E_2.

(a) Only y' appears in all eight squares of the two-by-four maximal basic rectangle, and the designated pair of adjacent squares represents xzt'. As both rectangles are needed for a minimal cover,

$$E_1 = y' + xzt'$$

is the minimal sum for E_1.

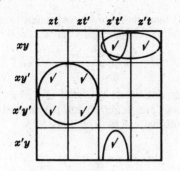 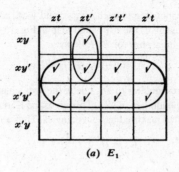 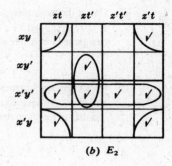

Fig. 15-33   Fig. 15-34

(b) The four corner squares form a two-by-two maximal basic rectangle which represents yt, since only y and t appear in all the four squares. The four-by-one maximal basic rectangle represents $x'y'$, and the two adjacent squares represent $y'zt'$. As all three rectangles are needed for a minimal cover,

$$E_2 = yt + x'y' + y'zt'$$

is the minimal sum for E_2.

15.40. Consider the Boolean expressions E_1 and E_2 in variables x, y, z, t which are given by the Karnaugh maps in Fig. 15-35. Find a minimal sum for (a) E_1; (b) E_2.

(a) There are five prime implicants, designated by the four loops and the dashed circle. However, the dashed circle is not needed to cover all the squares, whereas the four loops are required. Thus the four loops give the minimal sum for E_1; that is,
$$E_1 = xzt' + xy'z' + x'y'z + x'z't'$$

(b) There are five prime implicants, designated by the five loops of which two are dashed. Only one of the two dashed loops is needed to cover the square $x'y'z't'$. Thus there are two minimal sums for E_2 as follows:

$$E_2 = x'y + yt + xy't' + y'z't' = x'y + yt + xy't' + x'z't'$$

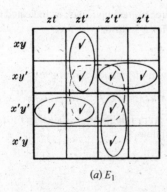 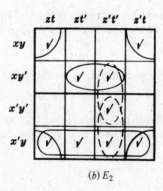

(a) E_1 (b) E_2

Fig. 15-35

15.41. Use a Karnaugh map to find a minimal sum for:

(a) $E_1 = x'yz + x'yz't + y'zt' + xyzt' + xy'z't'$.

(b) $E_2 = y't' + y'z't + x'y'zt + yzt'$.

(a) Check the two squares corresponding to each of $x'yz$ and $y'zt'$, and check the square corresponding to each of $x'yz't$, $xyzt'$, and $xy'z't'$. This gives the Karnaugh map in Fig. 15-36(a). A minimal cover consists of the three designated loops. Thus a minimal sum for E_1 follows:

$$E_1 = zt' + xy't' + x'yt$$

(b) Check the four squares corresponding to zt', check the two squares corresponding to each of $y'z't$ and yzt', and check the square corresponding to $x'y'zt$. This gives the Karnaugh map in Fig. 15-36(b). A minimal cover consists of the three designated maximal basic rectangles. Thus a minimal sum for E_2 follows:

$$E_2 = zt' + xy't' + x'yt$$

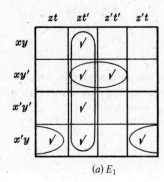
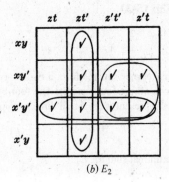

(a) E_1 (b) E_2

Fig. 15-36

15.42. Find a minimal sum-of-products form for the Boolean expression E with the following truth tables:

(a) $T(00001111, 00110011, 01010101) = 10100110$.

(b) $T(00001111, 00110011, 01010101) = 00101111$.

(a) From the given truth table T (and the truth tables in Example 15.13 for the minterms in variables x, y, z) we can read off the complete sum-of-products form for E:

$$E = x'y'z' + x'yz' + xy'z + xyz'$$

Its Karnaugh map appears in Fig. 15-37(a). There are three prime implicants, as indicated by the three loops, which form a minimal cover of E. Thus a minimal form for E follows:

$$E = yz' + x'z' + xy'z$$

(b) From the given truth table we can read off the complete sum-of-products form for E:

$$E = x'yz' + x'y'z + xy'z + xyz' + xyz$$

Its Karnaugh map appears in Fig. 15-37(b). There are two prime implicants, as indicated by the two loops, which form a minimal cover of E. Thus a minimal form for E follows:

$$E = xz + y$$

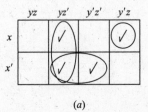
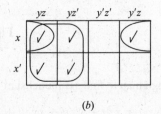

(a)   (b)

Fig. 15-37

Supplementary Problems

BOOLEAN ALGEBRAS

15.43. Write the dual of each Boolean expression:

  (a) $a(a' + b) = ab$ (b) $(a + 1)(a + 0) = a$ (c) $(a + b)(b + c) = ac + b$

15.44. Consider the lattices D_m of divisors of m (where $m > 1$).

  (a) Show that D_m is a Boolean algebra if and only if m is *square-free*, that is, if m is a product of distinct primes.

  (b) If D_m is a Boolean algebra, show that the atoms are the distinct prime divisors of m.

15.45. Consider the following lattices: (a) D_{20}; (b) D_{55}; (c) D_{99}; (d) D_{130}. Which of them are Boolean algebras, and what are their atoms?

15.46. Consider the Boolean algebra D_{110}

  (a) List its elements and draw its diagram.

  (b) Find all its subalgebras.

  (c) Find the number of sublattices with four elements.

  (d) Find the set A of atoms of D_{110}.

  (e) Give the isomorphic mapping $f: D_{110} \to P(A)$ as defined in Theorem 15.6.

15.47. Let B be a Boolean algebra. Show that:

  (a) For any x in B, $0 \leq x \leq 1$. (b) $a < b$ if and only if $b' < a'$.

15.48. An element x in a Boolean algebra is called a *maxterm* if the identity 1 is its only successor. Find the maxterms in the Boolean algebra D_{210} pictured in Fig. 15-25.

15.49. Let B be a Boolean algebra.

  (a) Show that complements of the atoms of B are the maxterms.

  (b) Show that any element x in B can be expressed uniquely as a product of maxterms.

15.50. Let B be a 16-element Boolean algebra and let S be an 8-element subalgebra of B. Show that two of the atoms of S must be atoms of B.

15.51. Let $B = (B, +, *, ', 0, 1)$ be a Boolean algebra. Define an operation Δ on B (called the *symmetric difference*) by

$$x \Delta y = (x * y') + (x' * y)$$

Prove that $R = (B, \Delta, *)$ is a commutative Boolean ring. (See Section B.6 and Problem B.72.)

15.52. Let $R = (R, \oplus, \cdot)$ be a Boolean ring with identity $1 \neq 0$. Define

$$x' = 1 \oplus x, \quad x + y = x \otimes y \oplus x \cdot y, \quad x * y = x \cdot y$$

Prove that $B = (R, +, *, ', 0, 1)$ is a Boolean algebra.

BOOLEAN EXPRESSIONS, PRIME IMPLICANTS

15.53. Reduce the following Boolean products to either 0 or a fundamental product:

  (a) $xy'zxy'$; (b) $xyz'sy'ts$; (c) $xy'xz'ty'$; (d) $xyz'ty't$.

15.54. Express each Boolean expression $E(x, y, z)$ as a sum-of-products and then in its complete sum-of-products form:

  (a) $E = x(xy' + x'y + y'z)$; (b) $E = (x + y'z)(y + z')$; (c) $E = (x' + y)' + y'z$.

15.55. Express each Boolean expression $E(x, y, z)$ as a sum-of-products and then in its complete sum-of-products form:

  (a) $E = (x'y)'(x' + xyz')$; (b) $E = (x + y)'(xy')'$; (c) $E = y(x + yz)'$.

15.56. Find the consensus Q of the fundamental products P_1 and P_2 where:

    (a) $P_1 = xy'z, P_2 = xyt$;   (c) $P_1 = xy'zt, P_2 = xyz'$;

    (b) $P_1 = xyz't', P_2 = xzt'$;   (d) $P_1 = xy't, P_2 = xzt$.

15.57. For any Boolean sum-of-products expression E, we let E_L denote the number of literals in E (counting multiplicity) and E_S denote the number of summands in E. Find E_L and E_S for each of the following:

    (a) $E = xyz't + x'yt + xy'zt$;   (b) $E = xyzt + xt' + x'y't + yt$.

15.58. Apply the consensus method (Algorithm 15.3) to find the prime implicants of each Boolean expression:

    (a) $E_1 = xy'z' + x'y + x'y'z' + x'yz$;

    (b) $E_2 = xy' + x'z't + xyzt' + x'y'zt'$;

    (c) $E_3 = xyzt + xyz't' + xz't' + x'y'z' + x'yz't$.

15.59. Find a minimal sum-of-products form for each of the Boolean expressions in Problem 15.58.

LOGIC GATES, TRUTH TABLES

15.60. Express the output Y as a Boolean expression in the inputs A, B, C for the logic circuit in:

    (a) Fig. 15-38(a);   (b) Fig. 15-38(b).

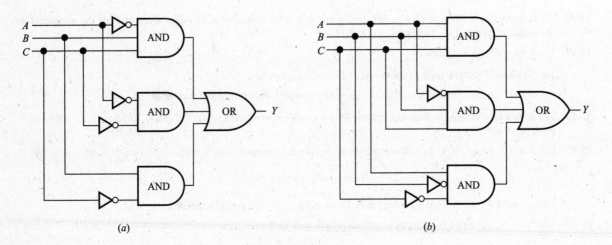

                (a)   (b)

Fig. 15-38

15.61. Express the output Y as a Boolean expression in the inputs A, B, C for the logic circuit in:

    (a) Fig. 15-39(a);   (b) Fig. 15-39(b).

15.62. Draw the logic circuit L with inputs A, B, C and output Y which corresponds to each Boolean expression:

    (a) $Y = AB'C + AC' + A'C$;   (b) $Y = A'BC + A'BC' + ABC'$.

15.63. Find the output sequence Y for an AND gate with inputs A, B, C (or equivalently for $Y = ABC$) where:

    (a) $A = 110001; B = 101101; C = 110011$.

    (b) $A = 01111100; B = 10111010; C = 00111100$.

    (c) $A = 00111110; B = 01111100; C = 11110011$.

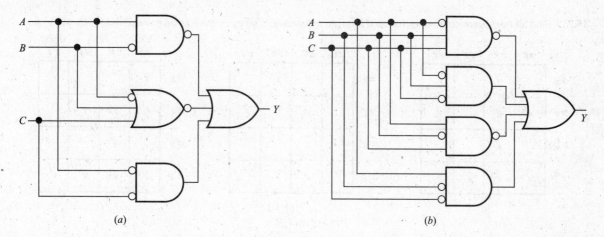

Fig. 15-39

15.64. Find the output sequence Y for an OR gate with inputs A, B, C (or equivalently for $Y = A + B + C$) where:

(a) $A = 100011$; $B = 100101$; $C = 1000001$.

(b) $A = 10000001$; $B = 00100100$; $C = 00000011$.

(c) $A = 00111100$; $B = 11110000$; $C = 10000001$.

15.65. Find the output sequence Y for a NOT gate with input A or, equivalently, for $Y = A'$, where:

(a) $A = 11100111$;   (b) $A = 10001000$;   (c) $A = 11111000$.

15.66. Consider a logic circuit L with $n = 6$ inputs A, B, C, D, E, F, or, equivalently, consider a Boolean expression E with six variables $x_1, x_2, x_3, x_4, x_5, x_6$.

(a) How many different ways can we assign a bit (0 or 1) to each of the $n = 6$ variables?

(b) Find the first three special sequences for the variables (inputs).

15.67. Find the truth table $T = T(E)$ for the Boolean expression $E = E(x, y, z)$ where:

(a) $E = xy + x'z$;   (b) $E = xyz' + y + xy'$.

15.68. Find the truth table $T = T(E)$ for the Boolean expression $E = E(x, y, z)$ where:

(a) $E = x'yz' + x'y'z$;   (b) $E = xyz' + xy'z' + x'y'z'$.

15.69. Find the Boolean expression $E = E(x, y, z)$ corresponding to the truth tables:

(a) $T(E) = 10001010$;   (b) $T(E) = 00010001$;   (c) $T(E) = 00110000$.

15.70. Find all possible minimal sums for the each Boolean expression E given by the Karnaugh maps in Fig. 15-40.

	yz	yz'	$y'z'$	$y'z$
x	√		√	√
x'	√	√		

(a)

	yz	yz'	$y'z'$	$y'z$
x	√		√	√
x'	√	√		√

(b)

	yz	yz'	$y'z'$	$y'z$
x	√			√
x'	√	√	√	√

(c)

Fig. 15-40

15.71. Find all possible minimal sums for each Boolean expression E given by the Karnaugh maps in Fig. 15-41.

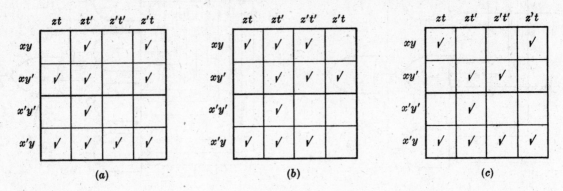

Fig. 15-41

15.72. Use a Karnaugh map to find a minimal sum for the Boolean expression:

(a) $E = xy + x'y + x'y'$;   (b) $E = x + x'yz + xy'z'$.

15.73. Find the minimal sum for each Boolean expression:

(a) $E = y'z + y'z't' + z't$;   (b) $E = y'zt + xzt' + xy'z'$.

15.74. Use Karnaugh maps to redesign each circuit in Fig. 15-42 so that it becomes a minimal AND-OR circuit.

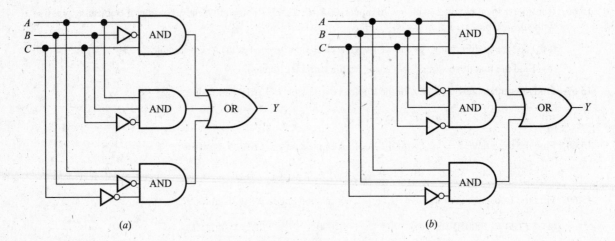

Fig. 15-42

15.75. Suppose three switches A, B, C are connected to the same hall light. At any moment a switch may be "up" denoted by 1 or "down" denoted by 0. A change in any switch will change the parity (odd or even) of the number of 1's. The switches will be able to control the light if it associates, say, an odd parity with the light being "on" (represented by 1), and an even parity with the light being "off" (represented by 0).

(a) Show that the following truth table satisfies these conditions:

$$T(A, B, C) = T(00001111, 00110011, 01010101) = 01101001$$

(b) Design a minimal AND-OR circuit L with the above truth table.

Answers to Supplementary Problems

15.43. (a) $a + a'b = a + b$.
   (b) $a \cdot 0 + a \cdot 1 = a$.
   (c) $ab + bc = (a + c)b$.

15.45. (b) $\mathbf{D}_{55}$; atoms 5 and 11. (d) $\mathbf{D}_{130}$; atoms 2, 5 and 13.

15.46. (a) There are eight elements 1, 2, 5, 10, 11, 22, 55, 110. See Fig. 15-43(a).

(b) There are five subalgebras: $\{1, 110\}$, $\{1, 2, 55, 110\}$, $\{1, 5, 22, 110\}$, $\{1, 10, 11, 110\}$, $\mathbf{D}_{110}$.

(c) There are 15 sublattices which include the above three subalgebras.

(d) $A = \{2, 5, 11\}$.

(e) See Fig. 15-43(b).

15.48. Maxterms: 30, 42, 70, 105

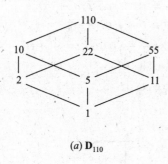

(a) $\mathbf{D}_{110}$

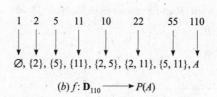

(b) $f : \mathbf{D}_{110} \longrightarrow P(A)$

Fig. 15-43

15.49. (b) *Hint*: Use duality.

15.53. (a) $xy'z$; (b) 0; (c) $xy'z't$; (d) 0.

15.54. (a) $E = xy' + xy'z = xy'z' + xy'z$.
   (b) $E = xy + xz' = xyz + xyz' + xy'z'$.
   (c) $E = xy' + y'z = xy'z + xy'z' + x'y'z$.

15.55. (a) $E = xyz' + x'y' = xyz' + x'y'z + x'y'z'$.
   (b) $E = x'y' = x'y'z + x'y'z'$.
   (c) $E = x'yz'$.

15.56. (a) $Q = xzt$. (b) $Q = xyt'$. (c) and (d) Does not exist.

15.57. (a) $E_L = 11, E_S = 3$; (b) $E_L = 11, E_S = 4$.

15.58. (a) $x'y, x'z', y'z'$.
   (b) $xy', xzt', y'zt', x'z't, y'z't$.
   (c) $xyzt, xz't', y'z't', x'y'z', x'z't$.

15.59. (a) $E = x'y + x'z'$.
   (b) $E = xy' + xzt' + x'z't + y'z't$.
   (c) $E = xyzt + xz't' + x'y'z' + x'z't$.

15.60. (a) $Y = A'BC + A'C' + BC'$;
   (b) $ABC + A'BC + AB'C'$.

15.61. (a) $Y = (AB')' + (A' + B + C)' + AC$
   (b) $Y = (A'BC)' + A'BC' + (AB'C)' + AB'C'$

15.62. See Fig. 15-44.

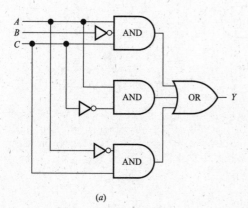

(a)

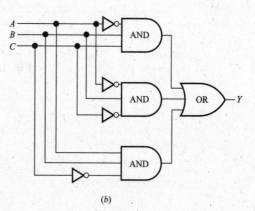

(b)

Fig. 15-44

15.63. (a) $Y = 100001$; (b) $Y = 00111000$;
(c) $Y = 00110000$.

15.64. (a) $Y = 100111$; (b) $Y = 10100111$;
(c) $Y = 11111101$.

15.65. (a) $A' = 00011000$; (b) $A' = 01110111$;
(c) $A' = 00000111$.

15.66. (a) $2^n = 2^6 = 64$.
(b) $x_1 = 000 \cdots 00111 \ldots 11$ (32 zeros) (32 ones).
$x_2 = (0000000000000000011111111111111111)_2$.
$x_3 = (0000000011111111)^4$.

15.67. (a) $T(E) = 01010011$; (b) $T(E) = 00111111$.

15.68. (a) $T(E) = 01000000$; (b) $T(E) = 10001010$.

15.69. Use truth tables for minterms in Example 15.13.

(a) $E = x'y'z' + x'yz + xyz'$.

(b) $E = xy'z' + xyz$.

(c) $E = x'yz' + xy'z'$.

15.70. (a) $E = xy' + x'y + yz = xy' + x'y + xz'$.
(b) $E = xy' + x'y + z$.
(c) $E = x' + z$.

15.71. (a) $E = x'y + zt' + xz't + xy'z$
$= x'y + zt' + xz't + xy't$.
(b) $E = yz + yt' + zt' + xy'z'$.
(c) $E = x'y + yt + xy't + x'zt$
$= x'y + yt + xy't + y'zt$.

15.72. (a) $E = x' + y$; (b) $E = xz' + yz$.

15.73. (a) $E = y' + z't$; (b) $E = xy' + zt' + y'zt$.

15.74. (a) See Fig. 15-45.

15.75. (b) See Fig. 15-46.

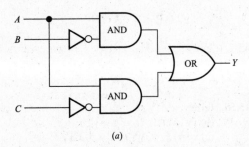
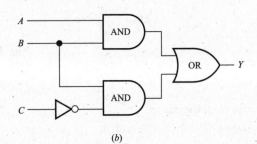

(a)   (b)

Fig. 15-45

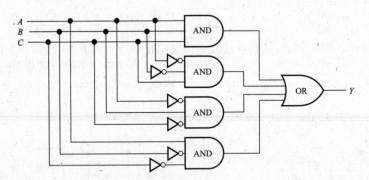

Fig. 15-46

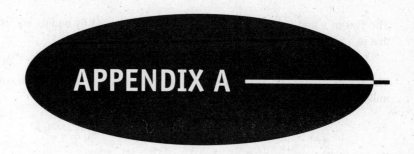

APPENDIX A

Vectors and Matrices

A.1 INTRODUCTION

Data is frequently arranged in *arrays*, that is, sets whose elements are indexed by one or more subscripts. If the data consists of numbers, then a one-dimensional array is called a *vector* and a two-dimensional array is called a *matrix* (where the dimension denotes the number of subscripts). This appendix investigates these vectors and matrices, and certain algebraic operations involving them. In this context, the numbers themselves are called *scalars*.

A.2 VECTORS

By a *vector u*, we mean a list of numbers, say, $a_1, a_2, \ldots, a_n$. Such a vector is denoted by

$$u = (a_1, a_2, \ldots, a_n)$$

The numbers a_i are called the *components* or *entries* of u. If all the $a_i = 0$, then u is called the *zero vector*. Two such vectors, u and v, are *equal*, written $u = v$, if they have the same number of components *and* corresponding components are equal.

EXAMPLE A.1

(a) The following are vectors where the first two have two components and the last two have three components:

$$(3, -4), \quad (6, 8), \quad (0, 0, 0), \quad (2, 3, 4)$$

The third vector is the zero vector with three components.

(b) Although the vectors $(1, 2, 3)$ and $(2, 3, 1)$ contain the same numbers, they are not equal since corresponding components are not equal.

Vector Operations

Consider two arbitrary vectors u and v with the same number of components, say

$$u = (a_1, a_2, \ldots, a_n) \quad \text{and} \quad v = (b_1, b_2, \ldots, b_n)$$

The *sum* of u and v, written $u + v$, is the vector obtained by adding corresponding components from u and v; that is,

$$u + v = (a_1 + b_1, a_2 + b_2, \ldots, a_n + b_n)$$

The *scalar product* or, simply, *product*, of a scalar k and the vector u, written ku, is the vector obtained by multiplying each component of u by k; that is,

$$ku = (ka_1, ka_2, \ldots, ka_n)$$

We also define

$$-u = -1(u) \quad \text{and} \quad u - v = u + (-v)$$

and we let 0 denote the zero vector. The vector $-u$ is called the *negative* of the vector u.

The *dot product* or *inner product* of the above vectors u and v is denoted and defined by

$$u \cdot v = a_1 b_1 + a_2 b_2 + \cdots + a_n b_n$$

The *norm* or *length* of the vector u is denoted and defined by

$$\|u\| = \sqrt{u \cdot u} = \sqrt{a_1^2 + a_2^2 + \cdots + a_n^2}$$

We note that $\|u\| = 0$ if and only if $u = 0$; otherwise $\|u\| > 0$.

EXAMPLE A.2 Let $u = (2, 3, -4)$ and $v = (1, -5, 8)$. Then

$$
\begin{aligned}
u + v &= (2 + 1, \ 3 - 5, \ -4 + 8) = (3, -2, 4) \\
5u &= (5 \cdot 2, \ 5 \cdot 3, \ 5 \cdot (-4)) = (10, 15, -20) \\
-v &= -1 \cdot (1, -5, 8) = (-1, 5, -8) \\
2u - 3v &= (4, 6, -8) + (-3, 15, -24) = (1, 21, -32) \\
u \cdot v &= 2 \cdot 1 + 3 \cdot (-5) + (-4) \cdot 8 = 2 - 15 - 32 = -45 \\
\|u\| &= \sqrt{2^2 + 3^2 + (-4)^2} = \sqrt{4 + 9 + 16} = \sqrt{29}
\end{aligned}
$$

Vectors under the operations of vector addition and scalar multiplication have various properties, e.g.,

$$k(u + v) = ku + kv$$

where k is a scalar and u and v are vectors. Many such properties appear in Theorem A.1, which also holds for vectors since vectors may be viewed as a special case of matrices.

Column Vectors

Sometimes a list of numbers is written vertically rather than horizontally, and the list is called a *column vector*. In this context, the above horizontally written vectors are called *row vectors*. The above operations for row vectors are defined analogously for column vectors.

A.3 MATRICES

A *matrix A* is a rectangular array of numbers usually presented in the form

$$
A = \begin{bmatrix}
a_{11} & a_{12} & \cdots & a_{1n} \\
a_{21} & a_{22} & \cdots & a_{2n} \\
\hline
\multicolumn{4}{c}{\cdots\cdots\cdots\cdots\cdots\cdots\cdots\cdots} \\
a_{m1} & a_{m2} & \cdots & a_{mn}
\end{bmatrix}
$$

The m horizontal lists of numbers are called the *rows* of A, and the n vertical lists of numbers its *columns*. Thus the element a_{ij}, called the *ij entry*, appears in row i and column j. We frequently denote such a matrix by simply writing $A = [a_{ij}]$.

A matrix with m rows and n columns is called an m by n matrix, written $m \times n$. The pair of numbers m and n is called the *size* of the matrix. Two matrices A and B are equal, written $A = B$, if they have the same size and if corresponding elements are equal. Thus, the equality of two $m \times n$ matrices is equivalent to a system of mn equalities, one for each corresponding pair of elements.

A matrix with only one row is called a *row matrix* or *row vector*, and a matrix with only one column is called a *column matrix* or *column vector*. A matrix whose entries are all zero is called a *zero matrix* and will usually be denoted by 0.

EXAMPLE A.3

(a) The rectangular array $A = \begin{bmatrix} 1 & -4 & 5 \\ 0 & 3 & -2 \end{bmatrix}$ is a 2×3 matrix. Its rows are $[1, -4, 5]$ and $[0, 3, -2]$, and its

columns are $\begin{bmatrix} 1 \\ 0 \end{bmatrix}, \begin{bmatrix} -4 \\ 3 \end{bmatrix}, \begin{bmatrix} 5 \\ -2 \end{bmatrix}$.

(b) The 2×4 zero matrix is the matrix $0 = \begin{bmatrix} 0 & 0 & 0 & 0 \\ 0 & 0 & 0 & 0 \end{bmatrix}$.

(c) Suppose

$$\begin{bmatrix} x+y & 2z+t \\ x-y & z-t \end{bmatrix} = \begin{bmatrix} 3 & 7 \\ 1 & 5 \end{bmatrix}$$

Then the four corresponding entries must be equal. That is,

$$x + y = 3, \quad x - y = 1, \quad 2z + t = 7, \quad z - t = 5$$

The solution of the system of equations is

$$x = 2, \quad y = 1, \quad z = 4, \quad t = -1$$

A.4   MATRIX ADDITION AND SCALAR MULTIPLICATION

Let $A = [a_{ij}]$ and $B = [b_{ij}]$ be two matrices of the same size, say, $m \times n$ matrices. The *sum* of A and B, written $A + B$, is the matrix obtained by adding corresponding elements from A and B. The (*scalar*) *product* of the matrix A by a scalar k, written kA, is the matrix obtained by multiplying each element of A by k. These operations are pictured in Fig. A-1.

$$A + B = \begin{bmatrix} a_{11}+b_{11} & a_{12}+b_{12} & \cdots & a_{1n}+b_{1n} \\ a_{21}+b_{21} & a_{22}+b_{22} & \cdots & a_{2n}+b_{2n} \\ \hline a_{m1}+b_{m1} & a_{m2}+b_{m2} & \cdots & a_{mn}+b_{mn} \end{bmatrix} \quad \text{and} \quad kA = \begin{bmatrix} ka_{11} & ka_{12} & \cdots & ka_{1n} \\ ka_{21} & ka_{22} & \cdots & ka_{2n} \\ \hline ka_{m1} & ka_{m2} & \cdots & ka_{mn} \end{bmatrix}$$

Fig. A-1

Observe that $A + B$ and kA are also $m \times n$ matrices. We also define

$$-A = (-1)A \quad \text{and} \quad A - B = A + (-B)$$

The matrix $-A$ is called the *negative* of A. The sum of matrices with different sizes is not defined.

EXAMPLE A.4 Let $A = \begin{bmatrix} 1 & -2 & 3 \\ 0 & 4 & 5 \end{bmatrix}$ and $B = \begin{bmatrix} 4 & 6 & 8 \\ 1 & -3 & -7 \end{bmatrix}$. Then

$$A + B = \begin{bmatrix} 1+4 & -2+6 & 3+8 \\ 0+1 & 4+(-3) & 5+(-7) \end{bmatrix} = \begin{bmatrix} 5 & 4 & 11 \\ 1 & 1 & -2 \end{bmatrix}$$

$$3A = \begin{bmatrix} 3(1) & 3(-2) & 3(3) \\ 3(0) & 3(4) & 3(5) \end{bmatrix} = \begin{bmatrix} 3 & -6 & 9 \\ 0 & 12 & 15 \end{bmatrix}$$

$$2A - 3B = \begin{bmatrix} 2 & -4 & 6 \\ 0 & 8 & 10 \end{bmatrix} + \begin{bmatrix} -12 & -18 & -24 \\ -3 & 9 & 21 \end{bmatrix} = \begin{bmatrix} -10 & -22 & -18 \\ -3 & 17 & 31 \end{bmatrix}$$

Matrices under matrix addition and scalar multiplication have the following properties.

Theorem A.1: Let A, B, C be matrices with the same size, and let k and k' be scalars. Then:

(i) $(A + B) + C = A + (B + C)$ (v) $k(A + B) = kA + kB$

(ii) $A + 0 = 0 + A$ (vi) $(k + k')A = kA + k'A$

(iii) $A + (-A) = (-A) + 0 = A$ (vii) $(kk')A = k(k'A)$

(iv) $A + B = B + A$ (viii) $1A = A$

Note first that the 0 in (ii) and (iii) refers to the zero matrix. Also, by (i) and (iv), any sum of matrices

$$A_1 + A_2 + \cdots + A_n$$

requires no parentheses, and the sum does not depend on the order of the matrices. Furthermore, using (vi) and (viii), we also have

$$A + A = 2A, \quad A + A + A = 3A, \quad \cdots$$

Lastly, since n-component vectors may be identified with either $1 \times n$ or $n \times 1$ matrices, Theorem A.1 also holds for vectors under vector addition and scalar multiplication.

The proof of Theorem A.1 reduces to showing that the ij entries on both sides of each matrix equation are equal.

A.5 MATRIX MULTIPLICATION

The product of matrices A and B, written AB, is somewhat complicated. For this reason, we first begin with a special case. (The reader is referred to Section 3.5 for a discussion of the summation symbol Σ, the Greek capital letter sigma.)

The product AB of a row matrix $A = [a_i]$ and a column matrix $B = [b_i]$ with the same number of elements is defined as follows:

$$AB = [a_1, a_2, \ldots, a_n] \begin{bmatrix} b_1 \\ b_2 \\ .. \\ b_n \end{bmatrix} = a_1 b_1 + a_2 b_2 + \cdots + a_n b_n = \sum_{k=1}^{n} a_k b_k$$

That is, AB is obtained by multiplying corresponding entries in A and B and then adding all the products. We emphasize that AB is a scalar (or a 1×1 matrix). The product AB is not defined when A and B have different numbers of elements.

EXAMPLE A.5

(a) $[7, -4, 5] \begin{bmatrix} 3 \\ 2 \\ -1 \end{bmatrix} = 7(3) + (-4)(2) + 5(-1) = 21 - 8 - 5 = 8$

(b) $[6, -1, 8, 3] \begin{bmatrix} 4 \\ -9 \\ -2 \\ 5 \end{bmatrix} = 24 + 9 - 16 + 15 = 32$

We are now ready to define matrix multiplication in general.

Definition A.1: Let $A = [a_{ik}]$ and $B = [b_{kj}]$ be matrices such that the number of columns of A is equal to the number of rows of B, say, A is an $m \times p$ matrix and B is a $p \times n$ matrix. Then the product AB is the $m \times n$ matrix $C = [c_{ij}]$ whose ij-entry is obtained by multiplying the ith row of A by the jth column of B, that is,

$$c_{ij} = a_{i1}b_{1j} + a_{i2}b_{2j} + \cdots + a_{ip}b_{pj} = \sum_{k=1}^{p} a_{ik}b_{kj}$$

The product AB is pictured in Fig. A-2.

$$\begin{bmatrix} a_{11} & \cdots & a_{1p} \\ \cdot & \cdots & \cdot \\ a_{i1} & \cdots & a_{ip} \\ \cdot & \cdots & \cdot \\ a_{m1} & \cdots & a_{mp} \end{bmatrix} \begin{bmatrix} b_{11} & \cdots & b_{1j} & \cdots & b_{1n} \\ \cdot & \cdots & \cdot & \cdots & \cdot \\ \cdot & \cdots & \cdot & \cdots & \cdot \\ \cdot & \cdots & \cdot & \cdots & \cdot \\ b_{p1} & \cdots & b_{pj} & \cdots & b_{pn} \end{bmatrix} = \begin{bmatrix} c_{11} & \cdots & c_{1n} \\ \cdot & \cdots & \cdot \\ \cdot & c_{ij} & \cdot \\ \cdot & \cdots & \cdot \\ c_{m1} & \cdots & c_{mn} \end{bmatrix}$$

Fig. A-2

We emphasize that the product AB is not defined if A is an $m \times p$ matrix and B is a $q \times n$ matrix where $p \neq q$.

EXAMPLE A.6

(a) Find AB where $A = \begin{bmatrix} 1 & 3 \\ 2 & -1 \end{bmatrix}$ and $B = \begin{bmatrix} 2 & 0 & -4 \\ 5 & -2 & 6 \end{bmatrix}$.

Since A is 2×2 and B is 2×3, the product AB is defined and AB is a 2×3 matrix. To obtain the first row of the product matrix AB, multiply the first row $(1, 3)$ of A times each column of B,

$$\begin{bmatrix} 2 \\ 5 \end{bmatrix}, \quad \begin{bmatrix} 0 \\ -2 \end{bmatrix}, \quad \begin{bmatrix} -4 \\ 6 \end{bmatrix}$$

respectively. That is,

$$AB = \begin{bmatrix} 2 + 15 & 0 - 6 & -4 + 18 \end{bmatrix} = \begin{bmatrix} 17 & -6 & 14 \end{bmatrix}$$

To obtain the second row of the product AB, multiply the second row $(2, -1)$ of A times each column of B, respectively. Thus

$$AB = \begin{bmatrix} 17 & -6 & 14 \\ 4 - 5 & 0 + 2 & -8 - 6 \end{bmatrix} = \begin{bmatrix} 17 & -6 & 14 \\ -1 & 2 & -14 \end{bmatrix}$$

(b) Suppose $A = \begin{bmatrix} 1 & 2 \\ 3 & 4 \end{bmatrix}$ and $B = \begin{bmatrix} 5 & 6 \\ 0 & -2 \end{bmatrix}$. Then

$$AB = \begin{bmatrix} 5 + 0 & 6 - 4 \\ 15 + 0 & 18 - 8 \end{bmatrix} = \begin{bmatrix} 5 & 2 \\ 15 & 10 \end{bmatrix} \quad \text{and} \quad BA = \begin{bmatrix} 5 + 18 & 10 + 24 \\ 0 - 6 & 0 - 8 \end{bmatrix} = \begin{bmatrix} 23 & 34 \\ -6 & -8 \end{bmatrix}$$

The above Example A.6(b) shows that matrix multiplication is not commutative, that is, that the products AB and BA of matrices need not be equal.

Matrix multiplication does, however, satisfy the following properties:

Theorem A.2: Let A, B, C be matrices. Then, whenever the products and sums are defined:

(i) $(AB)C = A(BC)$ (Associative Law).
(ii) $A(B + C) = AB + AC$ (Left Distributive Law).
(iii) $(B + C)A = BA + CA$ (Right Distributive Law).
(iv) $k(AB) = (kA)B = A(kB)$ where k is a scalar.

Matrix Multiplication and Systems of Linear Equations

Any system S of linear equations is equivalent to the matrix equation

$$AX = B$$

where A is the matrix consisting of the coefficients, X is the column vector of unknowns, and B is the column vector of constants. (Here *equivalent* means that any solution of the system S is a solution to the matrix equation $AX = B$, and vice versa.) For example, the system

$$\begin{array}{l} x + 2y - 3z = 4 \\ 5x - 6y + 8z = 9 \end{array} \quad \text{is equivalent to} \quad \begin{bmatrix} 1 & 2 & -3 \\ 5 & -6 & 8 \end{bmatrix} \begin{bmatrix} x \\ y \\ z \end{bmatrix} = \begin{bmatrix} 4 \\ 9 \end{bmatrix}$$

Observe that the system is completely determined by the matrix

$$M = [A, B] = \begin{bmatrix} 1 & 2 & -3 & 4 \\ 5 & -6 & 8 & 9 \end{bmatrix}$$

which is called the *augmented matrix* of the system.

A.6   TRANSPOSE

The *transpose* of a matrix A, written A^T, is the matrix obtained by writing the rows of A, in order, as columns. For example,

$$\begin{bmatrix} 1 & 2 & 3 \\ 4 & 5 & 6 \end{bmatrix}^T = \begin{bmatrix} 1 & 4 \\ 2 & 5 \\ 3 & 6 \end{bmatrix} \quad \text{and} \quad [1, -3, -5]^T = \begin{bmatrix} 1 \\ -3 \\ -5 \end{bmatrix}$$

Note that if A is an $m \times n$ matrix, then A^T is an $n \times m$ matrix. In particular, the transpose of a row vector is a column vector, and vice versa. Furthermore, if $B = [b_{ij}]$ is the transpose of $A = [a_{ij}]$, then $b_{ij} = a_{ji}$ for all i and j.

A.7   SQUARE MATRICES

A matrix with the same number of rows as columns is called a *square matrix*. A square matrix with n rows and n columns is said to be of *order n*, and is called an *n-square matrix*.

The *main diagonal*, or simply *diagonal*, of an n-square matrix $A = [a_{ij}]$ consists of the elements $a_{11}, a_{22}, \ldots, a_{nn}$, that is, the elements from the upper left corner to the lower right corner of the matrix. The *trace* of A, written $\text{tr}(A)$, is the sum of the diagonal elements, that is, $\text{tr}(A) = a_{11} + a_{22} + \cdots + a_{nn}$.

The n-square *unit matrix*, denoted by I_n, or simply I, is the square matrix with 1's along the diagonal and 0's elsewhere. The unit matrix I plays the same role in matrix multiplication as the number 1 does in the usual multiplication of numbers. Specifically, for any matrix A,

$$AI = IA = A$$

Consider, for example, the matrices

$$\begin{bmatrix} 1 & -2 & 0 \\ 0 & -4 & -6 \\ 5 & 3 & 2 \end{bmatrix} \quad \text{and} \quad \begin{bmatrix} 1 & 0 & 0 & 0 \\ 0 & 1 & 0 & 0 \\ 0 & 0 & 1 & 0 \\ 0 & 0 & 0 & 1 \end{bmatrix}$$

Both are square matrices. The first is of order 3, and its diagonal consists of the elements 1, −4, 2 so its trace equals $1 − 4 + 2 = −1$. The second matrix is of order 4; its diagonal consists only of 1's, and there are only 0's elsewhere. Thus the second matrix is the unit matrix of order 4.

Algebra of Square Matrices

Let A be any square matrix. Then we can multiply A by itself. In fact, we can form all nonnegative *powers* of A as follows:

$$A^2 = AA, \quad A^3 = A^2A, \ldots, \quad A^{n+1} = A^nA, \ldots, \quad \text{and} \quad A^0 = I \text{ (when } A \neq 0)$$

Polynomials in the matrix A are also defined. Specifically, for any polynomial

$$f(x) = a_0 + a_1x + a_2x^2 + \cdots + a_nx^n$$

where the a_i are scalars, we define $f(A)$ to be the matrix

$$f(A) = a_0I + a_1A + a_2A^2 + \cdots + a_nA^n$$

Note that $f(A)$ is obtained from $f(x)$ by substituting the matrix A for the variable x and substituting the scalar matrix a_0I for the scalar term a_0. In the case that $f(A)$ is the zero matrix, the matrix A is then called a *zero* or *root* of the polynomial $f(x)$.

EXAMPLE A.7 Suppose $A = \begin{bmatrix} 1 & 2 \\ 3 & -4 \end{bmatrix}$. Then

$$A^2 = \begin{bmatrix} 1 & 2 \\ 3 & -4 \end{bmatrix} \begin{bmatrix} 1 & 2 \\ 3 & -4 \end{bmatrix} = \begin{bmatrix} 7 & -6 \\ -9 & 22 \end{bmatrix} \quad \text{and}$$

$$A^3 = A^2A = \begin{bmatrix} 7 & -6 \\ -9 & 22 \end{bmatrix} \begin{bmatrix} 1 & 2 \\ 3 & -4 \end{bmatrix} = \begin{bmatrix} -11 & 38 \\ 57 & -106 \end{bmatrix}$$

Suppose $f(x) = 2x^2 − 3x + 5$. Then

$$f(A) = 2\begin{bmatrix} 7 & -6 \\ -9 & -22 \end{bmatrix} - 3\begin{bmatrix} 1 & 2 \\ 3 & -4 \end{bmatrix} + 5\begin{bmatrix} 1 & 0 \\ 0 & 1 \end{bmatrix} = \begin{bmatrix} 16 & -18 \\ -27 & 61 \end{bmatrix}$$

Suppose $g(x) = x^2 + 3x − 10$. Then

$$g(A) = \begin{bmatrix} 7 & -6 \\ -9 & 22 \end{bmatrix} + 3\begin{bmatrix} 1 & 2 \\ 3 & -4 \end{bmatrix} - 10\begin{bmatrix} 1 & 0 \\ 0 & 1 \end{bmatrix} = \begin{bmatrix} 0 & 0 \\ 0 & 0 \end{bmatrix}$$

Thus A is a zero of the polynomial $g(x)$.

A.8   INVERTIBLE (NONSINGULAR) MATRICES, INVERSES

A square matrix A is said to be *invertible* (or *nonsingular*) if there exists a matrix B such that

$$AB = BA = I, \quad \text{(the identity matrix)}.$$

Such a matrix B is unique; it is called the *inverse* of A and is denoted by A^{-1}. Observe that B is the inverse of A if and only if A is the inverse of B. For example, suppose

$$A = \begin{bmatrix} 2 & 5 \\ 1 & 3 \end{bmatrix} \quad \text{and} \quad B = \begin{bmatrix} 3 & -5 \\ -1 & 2 \end{bmatrix}$$

Then

$$AB = \begin{bmatrix} 6-5 & -10+10 \\ 3-3 & -5+6 \end{bmatrix} = \begin{bmatrix} 1 & 0 \\ 0 & 1 \end{bmatrix} \quad \text{and} \quad BA = \begin{bmatrix} 6-5 & 15-15 \\ -2+2 & -5+6 \end{bmatrix} = \begin{bmatrix} 1 & 0 \\ 0 & 1 \end{bmatrix}$$

Thus A and B are inverses.

It is known that $AB = I$ if and only if $BA = I$; hence it is only necessary to test one product to determine whether two matrices are inverses. For example,

$$\begin{bmatrix} 1 & 0 & 2 \\ 2 & -1 & 3 \\ 4 & 1 & 8 \end{bmatrix} \begin{bmatrix} -11 & 2 & 2 \\ -4 & 0 & 1 \\ 6 & -1 & -1 \end{bmatrix} = \begin{bmatrix} -11+0+12 & 2+0-2 & 2+0-2 \\ -22+4+18 & 4+0-3 & 4-1-3 \\ -44-4+48 & 8+0-8 & 8+1-8 \end{bmatrix} = \begin{bmatrix} 1 & 0 & 0 \\ 0 & 1 & 0 \\ 0 & 0 & 1 \end{bmatrix}$$

Thus the two matrices are invertible and are inverses of each other.

A.9 DETERMINANTS

To each n-square matrix $A = [a_{ij}]$ we assign a specific number called the *determinant* of A and denoted by $\det(A)$ or $|A|$ or

$$\begin{vmatrix} a_{11} & a_{12} & \cdots & a_{1n} \\ a_{21} & a_{22} & \cdots & a_{2n} \\ \hdotsfor{4} \\ a_{n1} & a_{n2} & \cdots & a_{nn} \end{vmatrix}$$

We emphasize that a square array of numbers enclosed by straight lines, called a *determinant of order n*, is not a matrix but denotes the number that the determinant function assigns to the enclosed array of numbers, i.e., the enclosed square matrix.

The determinants of order 1, 2, and 3 are defined as follows:

$$|a_{11}| = a_{11} \qquad \begin{vmatrix} a_{11} & a_{12} \\ a_{21} & a_{22} \end{vmatrix} = a_{11}a_{22} - a_{12}a_{21}$$

$$\begin{vmatrix} a_{11} & a_{12} & a_{13} \\ a_{21} & a_{22} & a_{23} \\ a_{31} & a_{32} & a_{33} \end{vmatrix} = a_{11}a_{22}a_{33} + a_{12}a_{23}a_{31} + a_{13}a_{21}a_{32} - a_{13}a_{22}a_{31} - a_{12}a_{21}a_{33} - a_{11}a_{23}a_{32}$$

The diagram in Fig. A-3(a) may help the reader remember the determinant of order 2. That is, the determinant equals the product of the elements along the plus-labeled arrow minus the product of the elements along the minus-labeled arrow. There is an analogous diagram to remember a determinant of order 3 which appears in Fig. A-3(b). For notational convenience, we have separated the three plus-labeled arrows and the three minus-labeled arrows. We emphasize that there are no such diagrammatic tricks to remember determinants of higher order.

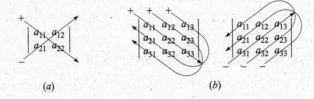

$$(a) \qquad\qquad\qquad\qquad\qquad (b)$$

Fig. A-3

EXAMPLE A.8

(a) $\begin{vmatrix} 5 & 4 \\ 2 & 3 \end{vmatrix} = 5(3) - 4(2) = 15 - 8 = 7, \qquad \begin{vmatrix} 2 & 1 \\ -4 & 6 \end{vmatrix} = 2(6) - 1(-4) = 12 + 4 = 16.$

(b) $\begin{vmatrix} 2 & 1 & 3 \\ 4 & 6 & -1 \\ 5 & 1 & 0 \end{vmatrix} = 2(6)(0) + 1(-1)(5) + 3(1)(4) - 5(6)(3) - 1(-1)(2) - 0(1)(4)$

$\qquad\qquad\qquad = 0 - 5 + 12 - 90 + 2 - 0 = 81$

General Definition of Determinants

The general definition of a determinant of order n is as follows:

$$\det(A) = \sum \operatorname{sgn}(\sigma) a_{1j_1} a_{2j_2} \ldots a_{nj_n}$$

where the sum is taken over all permutations $\sigma = \{j_1, j_2, \ldots, j_n\}$ of $\{1, 2, \ldots, n\}$. Here $\operatorname{sgn}(\sigma)$ equals $+1$ or -1 according as an even or an odd number of interchanges are required to change σ so that its numbers are in the usual order. We have included the general definition of the determinant function for completeness. The reader is referred to texts in matrix theory or linear algebra for techniques for computing determinants of order greater than 3. Permutations are studied in Chapter 5.

An important property of the determinant function is that it is multiplicative. That is:

Theorem A.3: Let A and B be any n-square matrices. Then

$$\det(AB) = \det(A) \cdot \det(B)$$

The proof of the above theorem lies beyond the scope of this text.

Determinants and Inverses of 2 × 2 Matrices

Consider an arbitrary 2×2 matrix $A = \begin{bmatrix} a & b \\ c & d \end{bmatrix}$. Suppose $|A| = ad - bc \neq 0$. Then one can prove that

$$A^{-1} = \begin{bmatrix} a & b \\ c & d \end{bmatrix}^{-1} = \begin{bmatrix} d/|A| & -b/|A| \\ -c/|A| & a/|A| \end{bmatrix} = \frac{1}{|A|} \begin{bmatrix} d & -b \\ -c & a \end{bmatrix}$$

In other words, when $|A| \neq 0$, the inverse of a 2×2 matrix A is obtained as follows:

(1) Interchange the elements on the main diagonal.

(2) Take the negatives of the other elements.

(3) Multiply the matrix by $1/|A|$ or, equivalently, divide each element by $|A|$.

For example, if $A = \begin{bmatrix} 2 & 3 \\ 4 & 5 \end{bmatrix}$, then $|A| = -2$ and so

$$A^{-1} = \frac{1}{-2} \begin{bmatrix} 5 & -3 \\ -4 & 2 \end{bmatrix} = \begin{bmatrix} -\frac{5}{2} & \frac{3}{2} \\ 2 & -1 \end{bmatrix}$$

On the other hand, if $|A| = 0$, then A^{-1} does not exist. Although there is no simple formula for matrices of higher order, this result does hold in general. Namely:

Theorem A.4: A matrix A is invertible if and only if it has a nonzero determinant.

A.10   ELEMENTARY ROW OPERATIONS, GAUSSIAN ELIMINATION (OPTIONAL)

This section discusses the Gaussian elimination algorithm in the context of elementary row operations.

Elementary Row Operations

Consider a matrix $A = [a_{ij}]$ whose rows will be denoted, respectively, by $R_1, R_2, \ldots, R_m$. The first nonzero element in a row R_i is called the *leading* nonzero element. A row with all zeros is called a *zero row*. Thus a zero row has no leading nonzero element.

The following three operations on A are called the *elementary row operations*:

[E₁]   Interchange row R_i and row R_j. This operation will be indicated by writing: "Interchange R_i and R_j."

[E₂]   Multiply each element in a row R_i by a nonzero constant k. This operation will be indicated by writing: "Multiply R_i by k."

[E₃]   Add a multiple of one row R_i to another row R_j or, in other words, replace R_j by the sum $kR_i + R_j$. This operation will be indicated by writing: "Add kR_i to R_j."

To avoid fractions, we may perform **[E₂]** and **[E₃]** in one step; that is, we may apply the following operation:

[E]   Add a multiple of one row R_i to a nonzero multiple of another row R_j or, in other words, replace R_j by the sum $kR_i + k'R_j$ where $k' \neq 0$. We indicate this operation by writing: "Add kR_i to $k'R_j$."

We emphasize that, in the row operations **[E₃]** and **[E]**, only row R_j is actually changed.

Notation: Matrices A and B are said to be *row equivalent*, written $A \sim B$, if matrix B can be obtained from matrix A by using elementary row operations.

Echelon Matrices

A matrix A is called an *echelon matrix*, or is said to be in *echelon form*, if the following two conditions hold:

(i)   All zero rows, if any, are on the bottom of the matrix.

(ii) Each leading nonzero entry is to the right of the leading nonzero entry in the preceding row.

The matrix is said to be in *row canonical form* if it has the following two additional properties:

(iii) Each leading nonzero entry is 1.

(vi) Each leading nonzero entry is the only nonzero entry in its column.

The zero matrix 0, for any number of rows or columns, is a special example of a matrix in row canonical form. The n-square identity matrix I_n is another example of a matrix in row canonical form.

A square matrix A is said to be in *triangular form* if its diagonal entries $a_{11}, a_{22}, \ldots, a_{nn}$ are the leading nonzero entries. Thus a square matrix in triangular form is a special case of an echelon matrix. The identity matrix I is the only example of a square matrix which is in triangular form and in row canonical form.

EXAMPLE A.9   Consider the echelon matrices in Fig. A-4 whose leading nonzero entries have been circled. (The zeros preceding and below the leading nonzero entries in an echelon matrix form a "staircase" pattern as indicated above by the shading.) The third matrix is in row canonical form. The second matrix is not in row canonical form since the third column contains a leading nonzero entry and another nonzero entry. The first matrix is not in row canonical form since some leading nonzero entries are not 1. The last matrix is in triangular form.

$$
\begin{bmatrix} ②&3&2&0&4&5&-6 \\ 0&0&①&1&-3&2&0 \\ 0&0&0&0&0&⑥&2 \\ 0&0&0&0&0&0&0 \end{bmatrix}, \quad
\begin{bmatrix} ①&2&3 \\ 0&0&① \\ 0&0&0 \end{bmatrix}, \quad
\begin{bmatrix} 0&①&3&0&0&4 \\ 0&0&0&①&0&-3 \\ 0&0&0&0&①&2 \end{bmatrix}, \quad
\begin{bmatrix} ②&4&7 \\ 0&⑤&8 \\ 0&0&⑥ \end{bmatrix}
$$

Fig. A-4

Gaussian Elimination in Matrix Form

Consider any matrix A. Two algorithms, Algorithms A-1 and A-2, are given in Fig. A-5 and Fig. A-6, respectively. The first algorithm transforms the matrix A into an echelon form (using only elementary row operations), and the second algorithm transforms the echelon matrix into a matrix in row canonical form. (The two algorithms together are called *Gaussian elimination*.)

At the end of the Algorithm A-1, the *pivot* (leading nonzero) entries will be

$$a_{1j_1}, a_{2j_2}, \ldots, a_{rj_r}$$

where r denotes the number of nonzero rows in the matrix in echelon form.

Remark 1: The number $m = -\dfrac{a_{ij_1}}{a_{1j_1}} = -\dfrac{\text{coefficient to be deleted}}{\text{pivot}}$ is called the *multiplier*.

Remark 2: One could replace the operation in Step 1 (*b*) by

$$\text{``Add } -a_{ij_1} R_1 \text{ to } a_{1j_1} R_i \text{''}$$

This would avoid fractions if all the scalars were originally integers.

EXAMPLE A.10 Find the row canonical form of $A = \begin{bmatrix} 1 & 2 & -3 & 1 & 2 \\ 2 & 4 & -4 & 6 & 10 \\ 3 & 6 & -6 & 9 & 13 \end{bmatrix}$.

First reduce A to echelon form using Algorithm A-1. Specifically, use $a_{11} = 1$ as a pivot to obtains zeros below a_{11}, that is, apply the row operations "Add $-2R_1$ to R_2" and "Add $-3R_1$ to R_3." Then use $a_{23} = 2$ as a pivot to obtain 0 below a_{23}, that is, apply the row operation "Add $-\frac{3}{2}R_2$ to R_3." This yields

$$A \sim \begin{bmatrix} 1 & 2 & -3 & 1 & 2 \\ 0 & 0 & 2 & 4 & 6 \\ 0 & 0 & 3 & 6 & 7 \end{bmatrix} \sim \begin{bmatrix} 1 & 2 & -3 & 1 & 2 \\ 0 & 0 & 2 & 4 & 6 \\ 0 & 0 & 0 & 0 & -2 \end{bmatrix}$$

The matrix A is now in echelon form.

Now use Algorithm A-2 to further reduce A to row canonical form. Specifically, multiply R_3 by $-1/2$ so the pivot entry $a_{35} = 1$, and then use $a_{35} = 1$ as a pivot to obtain zeros above it by the operations "Add $-6R_3$ to R_2" and "Add $-2R_3$ to R_1." This yields:

$$A \sim \begin{bmatrix} 1 & 2 & -3 & 1 & 2 \\ 0 & 0 & 2 & 4 & 6 \\ 0 & 0 & 0 & 0 & 1 \end{bmatrix} \sim \begin{bmatrix} 1 & 2 & -3 & 1 & 0 \\ 0 & 0 & 2 & 4 & 0 \\ 0 & 0 & 0 & 0 & 1 \end{bmatrix}$$

Multiply R_2 by $\frac{1}{2}$ so the pivot entry $a_{23} = 1$, and then use $a_{23} = 1$ as a pivot to obtain 0 above it by the operation "Add $3R_1$ to R_1". This yields:

$$A \sim \begin{bmatrix} 1 & 2 & -3 & 1 & 0 \\ 0 & 0 & 1 & 2 & 0 \\ 0 & 0 & 0 & 0 & 1 \end{bmatrix} \sim \begin{bmatrix} 1 & 2 & 0 & 7 & 0 \\ 0 & 0 & 1 & 2 & 0 \\ 0 & 0 & 0 & 0 & 1 \end{bmatrix}$$

The last matrix is the row canonical form of A.

Algorithm A-1: (Forward Elimination): The input is an arbitrary matrix $A = [a_{ij}]$.

Step 1. Find the first column with a nonzero entry. If no such column exists, then EXIT. (We have the zero matrix.) Otherwise, let j_1 denote the number of this column.

(a)   Arrange so that $a_{1j_1} \neq 0$. That is, if necessary, interchange rows so that a nonzero entry appears in the first row in column j_1.

(b)   Use a_{1j_1} as a *pivot* to obtain zeros below a_{1j_1}. That is, for $i > 1$:
   (1) Set $m = -a_{ij_1}/a_{1j_1}$.
   (2) Add aL_1 to L_i.

   (This replaces row R_i by $-(a_{ij_1}/a_{1j_1})\,R_1 + R_j$.)

Step 2. Repeat Step1 with the submatrix formed by all the rows, excluding the first row. Here we let j_2 denote the first column in the submatrix with a nonzero entry. Hence, at the end of Step 2, we have $a_{2j_2} \neq 0$.

Step 3 to r + 1. Continue the above process until the submatrix has no nonzero entry.

Fig. A-5

The final Step r in Algorithm A-2 in Fig. A-6 changes the first pivot to 1.

Algorithm A-2: (Backward Elimination): The input is a matrix $A = [a_{ij}]$ in echelon form with pivot entries $a_{1j_1}, a_{2j_2}, ..., a_{rj_r}$.

Step 1. (a)   Multiply the last nonzero row R_r by $1/a_{rj_r}$ so that the pivot entry is equal to 1.

(b)   Use $a_{rj_r} = 1$ to obtain zeros above the pivot. That is, for $i = r - 1, r - 2, ..., 1$:
   (1) Set $m = -a_{ir_i}$.
   (2) Add mR_r to R_i.

   In other words, apply the elementry row operations

$$\text{“Add } -a_{ir_1} R_r \text{ to } R_i\text{”}$$

   (This replaces row R_i by $-a_{ir_i} R_r + R_i$.)

Step 2 to r − 1. Repeat Step 1 for rows $R_{r-1}, R_{r-2}, ..., R_2$.

Step r. Multiply R_1 by $1/a_{1j_1}$.

Fig. A-6

Algorithms A-1 and A-2 show that any matrix is row equivalent to at least one matrix in row canonical form. Actually, one proves in linear algebra that such a matrix is unique; it is called the *row canonical form* of A.

Theorem A.5: Any matrix A is row equivalent to a unique matrix in row canonical form.

Matrix Solution of a System of Linear Equations

Consider a system S of linear equations or, equivalently, a matrix equation $AX = B$ with augmented matrix $M = [A, B]$. The system is solved by applying the above Gaussian elimination algorithm to M as follows.

Part A (*Reduction*): Reduce the augmented matrix M to echelon form. If a row of the form $(0, 0, \ldots, 0, b)$, with $b \neq 0$, appears, then *stop*. The system does not have a solution.

Part B (*Back-Substitution*): Further reduce the augmented matrix M to its row canonical form.

The unique solution of the system or, when the solution is not unique, the free variable form of the solution is easily obtained from the row canonical form of M.

The following example applies the above algorithm to a system S with a unique solution. The cases where S has no solution and where S has an infinite number of solutions are shown in Problem A.23.

EXAMPLE A.11

Solve the system: $\begin{cases} x + 2y + z = 3 \\ 2x + 5y - z = -4 \\ 3x - 2y - z = 5 \end{cases}$

Reduce its augmented matrix M to echelon form and then to row canonical form as follows:

$$M = \begin{bmatrix} 1 & 2 & 1 & 3 \\ 2 & 5 & -1 & -4 \\ 3 & -2 & -1 & 5 \end{bmatrix} \sim \begin{bmatrix} 1 & 2 & 1 & 3 \\ 0 & 1 & -3 & -10 \\ 0 & -8 & -4 & -4 \end{bmatrix} \sim \begin{bmatrix} 1 & 2 & 1 & 3 \\ 0 & 1 & -3 & -10 \\ 0 & 0 & -28 & -84 \end{bmatrix}$$

$$\sim \begin{bmatrix} 1 & 2 & 1 & 3 \\ 0 & 1 & -3 & -10 \\ 0 & 0 & 1 & 3 \end{bmatrix} \sim \begin{bmatrix} 1 & 2 & 0 & 0 \\ 0 & 1 & 0 & -1 \\ 0 & 0 & 1 & 3 \end{bmatrix} \sim \begin{bmatrix} 1 & 0 & 0 & 2 \\ 0 & 1 & 0 & -1 \\ 0 & 0 & 1 & 3 \end{bmatrix}$$

Thus the system has the unique solution $x = 2$, $y = -1$, $z = 3$ or, equivalently, the vector $u = (2, -1, 3)$. We note that the echelon form of M already indicated that the solution was unique since it corresponded to a triangular system.

Inverse of an $n \times n$ Matrix

Figure A-7 contains Algorithm A-3 which finds the inverse A^{-1} of any arbitrary $n \times n$ matrix.

Algorithm A-3: Find the inverse of an $n \times n$ matrix A.

Step 1. Form the $n \times 2n$ matrix $M = [A, I]$; that is, A is in the left half of M and the identity matrix I is in the right half of M.

Step 2. Row reduce M to an echelon form. If the process generates a zero row in the A-half of M, then *stop* (A has no inverse). Otherwise the A-half is now in triangular form.

Step 3. Further row reduce M to the row canonical form

$$M \sim [I, B]$$

where I has replaced A in the left half of M.

Step 4. Set $A^{-1} = B$, where B is the matrix that is now in the right half of M.

Fig. A-7

EXAMPLE A.12

Find the inverse of $A = \begin{bmatrix} 1 & 0 & 2 \\ 2 & -1 & 3 \\ 4 & 1 & 8 \end{bmatrix}$.

Form the matrix $M = (A, I)$ and reduce M to echelon form:

$$M = \left[\begin{array}{ccc:ccc} 1 & 0 & 2 & 1 & 0 & 0 \\ 2 & -1 & 3 & 0 & 1 & 0 \\ 4 & 1 & 8 & 0 & 0 & 1 \end{array}\right] \sim \left[\begin{array}{ccc:ccc} 1 & 0 & 2 & 1 & 0 & 0 \\ 0 & -1 & -1 & 2 & 1 & 0 \\ 0 & 1 & 0 & -4 & 0 & 1 \end{array}\right] \sim \left[\begin{array}{ccc:ccc} 1 & 0 & 2 & 1 & 0 & 0 \\ 0 & -1 & -1 & -2 & 1 & 0 \\ 0 & 0 & -1 & -6 & 1 & 1 \end{array}\right]$$

In echelon form, the left half of M is in triangular form; hence A is invertible. Further row reduce M to row canonical form:

$$M \sim \left[\begin{array}{ccc:ccc} 1 & 0 & 0 & -11 & 2 & 2 \\ 0 & -1 & 0 & 4 & 0 & -1 \\ 0 & 0 & 1 & 6 & -1 & -1 \end{array}\right] \sim \left[\begin{array}{ccc:ccc} 1 & 0 & 0 & -11 & 2 & 2 \\ 0 & 1 & 0 & -4 & 0 & 1 \\ 0 & 0 & 1 & 6 & -1 & -1 \end{array}\right]$$

The identity matrix is in the left half of the final matrix; hence the right half is A^{-1}. In other words,

$$A^{-1} = \begin{bmatrix} -11 & 2 & 2 \\ -4 & 0 & 1 \\ 6 & -1 & -1 \end{bmatrix}$$

A.11 BOOLEAN (ZERO-ONE) MATRICES

The *binary digits* or *bits* are the symbols 0 and 1. Consider the following operations on these digits:

+	0	1
0	0	1
1	1	1

×	0	1
0	0	0
1	0	1

Viewing these bits as logical values (0 representing FALSE and 1 representing TRUE), the above operations correspond, respectively, to the logical operations of OR ($\vee$) and AND ($\wedge$); that is,

$\vee$	F	T
F	F	T
T	T	T

$\wedge$	F	T
F	F	F
T	F	T

(The above operations on 0 and 1 are called *Boolean operations* since they also correspond to the operations of a Boolean algebra discussed in Chapter 15.)

Now let $A = \begin{bmatrix} a_{ij} \end{bmatrix}$ be a matrix whose entries are the bits 0 and 1 subject to the above Boolean operations. Then A is called a *Boolean matrix*. The *Boolean product* of two such matrices is the usual product except that now we use the Boolean operations of addition and multiplication. For example, if

$$A = \begin{bmatrix} 1 & 1 \\ 1 & 0 \end{bmatrix} \text{ and } B = \begin{bmatrix} 0 & 1 \\ 0 & 1 \end{bmatrix}, \text{ then } AB = \begin{bmatrix} 0+0 & 1+1 \\ 0+0 & 1+0 \end{bmatrix} = \begin{bmatrix} 0 & 1 \\ 0 & 1 \end{bmatrix}$$

One can easily show that if A and B are Boolean matrices, then the Boolean product AB can be obtained by finding the usual product of A and B and then replacing any nonzero digit by 1.

Solved Problems

VECTORS

A.1. Let $u = (2, -7, 1)$, $v = (-3, 0, 4)$, and $w = (0, 5, -8)$. Find: (a) $3u - 4v$; (b) $2u + 3v - 5w$.

First perform the scalar multiplication and then the vector addition.

(a) $3u - 4v = 3(2, -7, 1) - 4(-3, 0, 4) = (6, -21, 3) + (12, 0, -16) = (18, -21, -13)$.

(b) $2u + 3v - 6w = 2(2, -7, 1) + 3(-3, 0, 4) - 5(0, 5, -8) = (4, -14, 2) + (-9, 0, 12) + (0, -25, 40)$
$= (-5, -39, 54)$.

A.2. For the vector u, v, w in Problem A.1, find: (a) $u \cdot v$; (b) $u \cdot w$; (c) $v \cdot w$.

Multiply corresponding components and then add:

(a) $u \cdot v = 2(-3) - 7(0) + 1(4) = -6 + 0 + 4 = -2$.

(b) $u \cdot w = 2(0) - 7(5) + 1(-8) = 0 - 35 - 8 = -43$.

(c) $v \cdot w = -3(0) + 0(5) + 4(-8) = 0 + 0 - 32 = -32$.

A.3. Find $\|u\|$ where: (a) $u = (3, -12, -4)$; (b) $u = (2, -3, 8, -7)$.

First find $\|u\|^2 = u \cdot u$ by squaring the components and adding. Then $\|u\| = \sqrt{\|u\|^2}$.

(a) $\|u\|^2 = (3)^2 + (-12)^2 + (-4)^2 = 9 + 144 + 16 = 169$. Hence $\|u\| = \sqrt{169} = 13$.

(b) $\|u\|^2 = 4 + 9 + 64 + 49 = 126$. Hence $\|u\| = \sqrt{126}$.

A.4. Find x and y if $x(1, 1) + y(2, 1) = (1, 4)$.

First multiply by the scalars x and y and then add:

$$x(1, 1) + y(2, -1) = (x, x) + (2y, -y) = (x + 2y, x - y) = (1, 4)$$

Two vectors are equal only when their corresponding components are equal; hence set the corresponding components equal to each other to obtain $x + 2y = 1$ and $x - y = 4$. Finally, solve the system of equations to obtain $x = 3$ and $y = -1$.

A.5. Suppose $u = \begin{bmatrix} 5 \\ 3 \\ -4 \end{bmatrix}$, $v = \begin{bmatrix} -1 \\ 5 \\ 2 \end{bmatrix}$, $w = \begin{bmatrix} 3 \\ -1 \\ -2 \end{bmatrix}$. Find: (a) $5u - 2v$; (b) $-2u + 4v - 3w$.

(a) $5u - 2v = 5\begin{bmatrix} 5 \\ 3 \\ -4 \end{bmatrix} - 2\begin{bmatrix} -1 \\ 5 \\ 2 \end{bmatrix} = \begin{bmatrix} 25 \\ 15 \\ -20 \end{bmatrix} + \begin{bmatrix} 2 \\ -10 \\ -4 \end{bmatrix} = \begin{bmatrix} 27 \\ 5 \\ -24 \end{bmatrix}$.

(b) $-2u + 4v - 3w = \begin{bmatrix} -10 \\ -6 \\ 8 \end{bmatrix} + \begin{bmatrix} -4 \\ 20 \\ 8 \end{bmatrix} + \begin{bmatrix} -9 \\ 3 \\ 6 \end{bmatrix} = \begin{bmatrix} -23 \\ 17 \\ 22 \end{bmatrix}$.

MATRIX ADDITION AND SCALAR MULTIPLICATION

A.6. Find $2A - 3B$, where $A = \begin{bmatrix} 1 & -2 & 3 \\ 4 & 5 & -6 \end{bmatrix}$ and $B = \begin{bmatrix} 3 & 0 & 2 \\ -7 & 1 & 8 \end{bmatrix}$.

First perform the scalar multiplications, and then a matrix addition:

$$2A - 3B = \begin{bmatrix} 2 & -4 & 6 \\ 8 & 10 & -12 \end{bmatrix} + \begin{bmatrix} -9 & 0 & -6 \\ 21 & -3 & -24 \end{bmatrix} = \begin{bmatrix} -7 & -4 & 0 \\ 29 & 7 & -36 \end{bmatrix}$$

(Note that we multiply B by -3 and then add, rather than multiplying B by 3 and subtracting. This usually avoids errors.)

A.7. Find x, y, z, t, where $3 \begin{bmatrix} x & y \\ z & t \end{bmatrix} = \begin{bmatrix} x & 6 \\ -1 & 2t \end{bmatrix} + \begin{bmatrix} 4 & x+y \\ z+t & 3 \end{bmatrix}$.

First write each side as a single matrix:

$$\begin{bmatrix} 3x & 3y \\ 3z & 3t \end{bmatrix} = \begin{bmatrix} x+4 & x+y+6 \\ z+t-1 & 2t+3 \end{bmatrix}$$

Set corresponding entries equal to each other to obtain the system of four equations.

$$3x = x+4, \quad 3y = x+y+6, \quad 3z = z+t = 1, \quad 3t = 2t+3$$

or

$$2x = 4, \quad 2y = 6+x, \quad 2z = t-1, \quad t = 3$$

The solution is $x = 2, y = 4, z = 1, t = 3$.

A.8. Prove Theorem A.1(v): $k(A + B) = kA + kB$.

Let $A = [a_{ij}]$ and $B = [b_{ij}]$. Then the ij entry of $A + B$ is $a_{ij} + b_{ij}$. Hence $k(a_{ij} + b_{ij})$ is the ij entry of $k(A + B)$. On the other hand, the ij entries of kA and kB are ka_{ij} and kb_{ij}, respectively. Thus $ka_{ij} + kb_{ij}$ is the ij entry of $kA + kB$. However, for scalars, $k(a_{ij} + b_{ij}) = ka_{ij} + kb_{ij}$. Thus $k(A + B)$ and $kA + kB$ have the same ij entries. Therefore, $k(A + B) = kA + kB$.

MATRIX MULTIPLICATION AND TRANSPOSE

A.9. Calculate: (a) $[3, \ -2, \ 5] \begin{bmatrix} 6 \\ 1 \\ -4 \end{bmatrix}$;   (b) $[2, \ -1, \ 7, \ 4] \begin{bmatrix} 5 \\ -3 \\ -6 \\ 9 \end{bmatrix}$.

Multiply corresponding entries and then add:

(a) $[3, \ -2, \ 5] \begin{bmatrix} 6 \\ 1 \\ -4 \end{bmatrix} = 18 - 2 - 20 = -4$.   (b) $[2, \ -1, \ 7, \ 4] \begin{bmatrix} 5 \\ -3 \\ -6 \\ 9 \end{bmatrix} = 10 + 3 - 42 + 36 = 7$.

A.10. Let $A = \begin{bmatrix} 1 & 3 \\ 2 & -1 \end{bmatrix}$ and $B = \begin{bmatrix} 2 & 0 & -4 \\ 3 & -2 & 6 \end{bmatrix}$. Find: (a) AB; (b) BA.

(a) Since A is 2×2 and B is 2×3, the product AB is defined and is a 2×3 matrix. To obtain the first row of AB, multiply the first row $[1, 3]$ of A by the columns $\begin{bmatrix} 2 \\ 3 \end{bmatrix}, \begin{bmatrix} 0 \\ -2 \end{bmatrix}, \begin{bmatrix} -4 \\ 6 \end{bmatrix}$ of B, respectively:

$$\begin{bmatrix} 1 & 3 \\ 2 & -1 \end{bmatrix}\begin{bmatrix} 2 & 0 & -4 \\ 3 & -2 & 6 \end{bmatrix} = \begin{bmatrix} 1(6) + 3(3) & 1(0) + 3(-2) & 1(-4) + 3(6) \end{bmatrix}$$
$$= \begin{bmatrix} 2+9 & 0-6 & -4+18 \end{bmatrix} = \begin{bmatrix} 11 & -6 & 14 \end{bmatrix}$$

To obtain the entries in the second row of AB, multiply the second row $[2, -1]$ of A by the columns of B, respectively:

$$\begin{bmatrix} 1 & 3 \\ 2 & -1 \end{bmatrix}\begin{bmatrix} 2 & 0 & -4 \\ 3 & -2 & 6 \end{bmatrix} = \begin{bmatrix} 11 & -6 & 14 \\ 4-3 & 0+2 & -8-6 \end{bmatrix}$$

Thus

$$AB = \begin{bmatrix} 11 & -6 & 14 \\ 1 & 2 & -14 \end{bmatrix}$$

(b) Note that B is 2×3 and A is 2×2. Since the inner numbers, 3 and 2, are not equal, the product BA is not defined.

A.11. Find the transpose of each matrix:

$$A = \begin{bmatrix} 1 & -2 & 3 \\ 7 & 8 & -9 \end{bmatrix}; \quad B = \begin{bmatrix} 1 & 2 & 3 \\ 2 & 4 & 5 \\ 3 & 5 & 6 \end{bmatrix}; \quad C = [1, \; -3, \; 5, \; -7]; \quad D = \begin{bmatrix} 2 \\ -4 \\ 6 \end{bmatrix}$$

Rewrite the rows of each matrix as columns to obtain the transposes of the matrices:

$$A^T = \begin{bmatrix} 1 & 7 \\ -2 & 8 \\ 3 & -9 \end{bmatrix}, \quad B^T = \begin{bmatrix} 1 & 2 & 3 \\ 2 & 4 & 5 \\ 3 & 5 & 6 \end{bmatrix}, \quad C^T = \begin{bmatrix} 1 \\ -3 \\ 5 \\ -7 \end{bmatrix}, \quad D^T = [2, \; -4, \; 6]$$

(Note that $B^T = B$; such a matrix is said to be *symmetric*. Note also that the transpose of the row vector C is a column vector, and the transpose of the column vector D is a row vector.)

A.12. Prove Theorem A.2(i): $A(BC) = A(BC)$.

Let $A = [a_{ij}]$, $B = [b_{jk}]$, and $C = [c_{kl}]$. Furthermore, let $AB = S = [s_{ik}]$ and $BC = T = [t_{jl}]$. Then

$$s_{ik} = a_{i1}b_{1k} + a_{i2}b_{2k} + \cdots + a_{im}b_{mk} = \sum_{j=1}^{m} a_{ij}b_{jk}$$

$$t_{jl} = b_{j1}c_{1i} + b_{j2}c_{2i} + \cdots + b_{jn}c_{nl} = \sum_{k=l}^{n} b_{jk}c_{kl}$$

Now, multiplying S by C, i.e., (AB) by C, the element in the ith row and lth column of the matrix $(AB)C$ is

$$s_{i1}c_{1l} + s_{i2}c_{2l} + \cdots + s_{in}c_{nl} = \sum_{k=1}^{n} s_{ik}c_{kl} = \sum_{k=1}^{n}\sum_{j=1}^{m} (a_{ij}b_{jk})c_{kl}$$

On the other hand, multiplying A by T, i.e., A by BC, the element in the ith row and lth column of the matrix $A(BC)$ is

$$a_{i1}t_{1l} + a_{i2}t_{2l} + \cdots + a_{im}t_{ml} = \sum_{j=1}^{m} a_{ij}t_{jl} = \sum_{k=1}^{m}\sum_{j=1}^{n} a_{ij}(b_{jk}c_{kl})$$

Since the above sums are equal, the theorem is proven.

SQUARE MATRICES, DETERMINANTS, INVERSES

A.13. Find the diagonal and trace of each matrix:

$$(a) \; A = \begin{bmatrix} 1 & 3 & 6 \\ 2 & -5 & 8 \\ 4 & -2 & 7 \end{bmatrix}; \quad (b) \; B = \begin{bmatrix} t-2 & 3 \\ -4 & t+5 \end{bmatrix}; \quad (c) \; C = \begin{bmatrix} 1 & 2 & -3 \\ 4 & -5 & 6 \end{bmatrix}.$$

(a) The diagonal consists of the elements a_{11}, a_{22}, a_{33}, that is, the scalars $1, -5, 7$. The trace is the sum of the diagonal elements; hence $\text{tr}(A) = 1 - 5 + 7 = 3$.

(b) The diagonal consists of the pair $\{t - 2, t + 5\}$. Thus $\text{tr}(B) = t - 2 + t + 5 = 2t + 3$.

(c) The diagonal and trace are defined only for square matrices.

A.14. Let $A = \begin{bmatrix} 1 & 2 \\ 4 & -3 \end{bmatrix}$. Find: (a) A^2; (b) A^3; (c) $f(A)$ where $f(x) = 2x^3 - 4x + 5$; (d) $g(A)$ where $g(x) = x^2 + 2x - 11$.

(a) $A^2 = AA = \begin{bmatrix} 1 & 2 \\ 4 & -3 \end{bmatrix}\begin{bmatrix} 1 & 2 \\ 4 & -3 \end{bmatrix} = \begin{bmatrix} 1+8 & 2-6 \\ 4-12 & 8+9 \end{bmatrix} = \begin{bmatrix} 9 & -4 \\ -8 & 17 \end{bmatrix}.$

(b) $A^3 = AA^2 = \begin{bmatrix} 1 & 2 \\ 4 & -3 \end{bmatrix}\begin{bmatrix} 9 & -4 \\ -8 & 17 \end{bmatrix} = \begin{bmatrix} 9-16 & -4+34 \\ 36+24 & -16-51 \end{bmatrix} = \begin{bmatrix} -7 & 30 \\ 60 & -67 \end{bmatrix}.$

(c) Compute $f(A)$ by first substituting A for x and $5I$ for the constant term 5 in $f(x) = 2x^3 - 4x + 5$:

$$f(A) = 2A^3 - 4A + 5I = 2\begin{bmatrix} -7 & 30 \\ 60 & -67 \end{bmatrix} - 4\begin{bmatrix} 1 & 2 \\ 4 & -3 \end{bmatrix} + 5\begin{bmatrix} 1 & 0 \\ 0 & 1 \end{bmatrix}.$$

Then multiply each matrix by its respective scalar:

$$f(A) = \begin{bmatrix} -14 & 60 \\ 120 & -134 \end{bmatrix} + \begin{bmatrix} -4 & -8 \\ -16 & 12 \end{bmatrix} + \begin{bmatrix} 5 & 0 \\ 0 & 5 \end{bmatrix}.$$

Lastly, add the corresponding elements in the matrices:

$$f(A) = \begin{bmatrix} -14 - 4 + 5 & 60 - 8 + 0 \\ 120 - 16 + 0 & -134 + 12 + 5 \end{bmatrix} = \begin{bmatrix} -13 & 52 \\ 104 & -117 \end{bmatrix}.$$

(d) Compute $g(A)$ by first substituting A for x and $11I$ for the constant term 11 in $g(x) = x^2 + 2x - 11$:

$$g(A) = A^2 + 2A - 11I = \begin{bmatrix} 9 & -4 \\ -8 & 17 \end{bmatrix} + 2\begin{bmatrix} 1 & 2 \\ 4 & -3 \end{bmatrix} - 11\begin{bmatrix} 1 & 0 \\ 0 & 1 \end{bmatrix}$$

$$= \begin{bmatrix} 9 & -4 \\ -8 & 17 \end{bmatrix} + \begin{bmatrix} 2 & 4 \\ 8 & -6 \end{bmatrix} + \begin{bmatrix} -11 & 0 \\ 0 & -11 \end{bmatrix} = \begin{bmatrix} 0 & 0 \\ 0 & 0 \end{bmatrix}.$$

(Since $g(A) = 0$, the matrix A is a zero of the polynomial $g(x)$.)

A.15. Compute each determinant: (a) $\begin{vmatrix} 4 & 5 \\ -3 & -2 \end{vmatrix}$;   (b) $\begin{vmatrix} a-b & b \\ b & a+b \end{vmatrix}$.

(a) $\begin{vmatrix} 4 & 5 \\ -3 & -2 \end{vmatrix} = 4(-2) - (-3)(5) = -8 + 15 = 7.$

(b) $\begin{vmatrix} a-b & b \\ b & a+b \end{vmatrix} = (a-b)(a+b) - b^2 = a^2 - b^2 - b^2 = a^2 - 2b^2.$

A.16. Find the determinant of each matrix:

(a) $A = \begin{bmatrix} 1 & 2 & 3 \\ 4 & -2 & 3 \\ 0 & 5 & -1 \end{bmatrix}$;   (b) $B = \begin{bmatrix} 4 & -1 & -2 \\ 0 & 2 & -3 \\ 5 & 2 & 1 \end{bmatrix}$;   (c) $C = \begin{bmatrix} 2 & -3 & 4 \\ 1 & 2 & -3 \\ -1 & -2 & 5 \end{bmatrix}$

(Hint: Use the diagram in Fig. A-3 (b)):

(a) $|A| = 2 + 0 + 60 - 0 - 15 + 8 = 55$

(b) $|B| = 8 + 15 + 0 + 20 + 24 + 0 = 67$

(c) $|C| = 20 - 9 - 8 + 8 - 12 + 15 = 14$

A.17. Find the inverse of: (a) $A = \begin{bmatrix} 5 & 3 \\ 4 & 2 \end{bmatrix}$; (b) $B = \begin{bmatrix} -2 & 6 \\ 3 & -9 \end{bmatrix}$.

Use the formula in Section A.9.

(a) First find $|A| = 5(2) - 3(4) = 10 - 12 = -2$. Next, interchange the diagonal elements, take the negatives of the nondiagonal elements, and multiply by $1/|A|$:

$$A^{-1} = -\frac{1}{2}\begin{bmatrix} 2 & -3 \\ -4 & 5 \end{bmatrix} = \begin{bmatrix} -1 & \frac{3}{2} \\ 2 & -\frac{5}{2} \end{bmatrix}$$

(b) First find $|B| = -2(-9) - 6(3) = 18 - 18 = 0$. Since $|B| = 0$, B has no inverse.

A.18. Find the inverse of: (a) $A = \begin{bmatrix} 1 & -2 & 2 \\ 2 & -3 & 6 \\ 1 & 1 & 7 \end{bmatrix}$; (b) $B = \begin{bmatrix} 1 & 3 & -4 \\ 1 & 5 & -1 \\ 3 & 13 & -6 \end{bmatrix}$.

(a) Form the matrix $M = [A, I]$ and row reduce M to echelon form:

$$M = \begin{bmatrix} 1 & -2 & 2 & \vdots & 1 & 0 & 0 \\ 2 & -3 & 6 & \vdots & 0 & 1 & 0 \\ 1 & 1 & 7 & \vdots & 0 & 0 & 1 \end{bmatrix} \sim \begin{bmatrix} 1 & -2 & 2 & \vdots & 1 & 0 & 0 \\ 0 & 1 & 2 & \vdots & -2 & 1 & 0 \\ 0 & 3 & 5 & \vdots & -1 & 0 & 1 \end{bmatrix} \sim \begin{bmatrix} 1 & -2 & 2 & \vdots & 1 & 0 & 0 \\ 0 & 1 & 2 & \vdots & -2 & 1 & 0 \\ 0 & 0 & -1 & \vdots & 5 & -3 & 1 \end{bmatrix}$$

In echelon form, the left half of M is in triangular form; hence A has an inverse. Further reduce M to row canonical form:

$$M = \begin{bmatrix} 1 & -2 & 0 & \vdots & 11 & -6 & 2 \\ 0 & 1 & 0 & \vdots & 8 & -5 & 2 \\ 0 & 0 & 1 & \vdots & -5 & 3 & -1 \end{bmatrix} \sim \begin{bmatrix} 1 & 0 & 0 & \vdots & 27 & -16 & 6 \\ 0 & 1 & 0 & \vdots & 8 & -5 & 2 \\ 0 & 0 & 1 & \vdots & -5 & 3 & -1 \end{bmatrix}$$

The final matrix has the form $[I, A^{-1}]$; that is, A^{-1} is the right half of the last matrix. Thus

$$A^{-1} = \begin{bmatrix} 27 & -16 & 6 \\ 8 & -5 & 2 \\ -5 & 3 & -1 \end{bmatrix}$$

(b) Form the matrix $M = [B, I]$ and row reduce M to echelom form:

$$M = \begin{bmatrix} 1 & 3 & -4 & \vdots & 1 & 0 & 0 \\ 1 & 5 & -1 & \vdots & 0 & 1 & 0 \\ 3 & 13 & -6 & \vdots & 0 & 0 & 1 \end{bmatrix} \sim \begin{bmatrix} 1 & 3 & -4 & \vdots & 1 & 0 & 0 \\ 0 & 2 & 3 & \vdots & -1 & 1 & 0 \\ 0 & 4 & 6 & \vdots & -3 & 0 & 1 \end{bmatrix} \sim \begin{bmatrix} 1 & 3 & -4 & \vdots & 1 & 0 & 0 \\ 0 & 2 & 3 & \vdots & -1 & 1 & 0 \\ 0 & 0 & 0 & \vdots & -1 & -2 & 1 \end{bmatrix}$$

In echelon form, M has a zero row in its left half; that is, B is not row reducible to triangular form. Accordingly, B has no inverse.

ECHELON MATRICES, ROW REDUCTION, GAUSSIAN ELIMINATION

A.19. Interchange the rows in each matrix to obtain an echelon matrix:

$$(a) \begin{bmatrix} 0 & 1 & -3 & 4 & 6 \\ 4 & 0 & 2 & 5 & -3 \\ 0 & 0 & 7 & -2 & 8 \end{bmatrix}; \quad (b) \begin{bmatrix} 0 & 0 & 0 & 0 & 0 \\ 1 & 2 & 3 & 4 & 5 \\ 0 & 0 & 5 & -4 & 7 \end{bmatrix}; \quad (c) \begin{bmatrix} 0 & 2 & 2 & 2 & 2 \\ 0 & 3 & 1 & 0 & 0 \\ 0 & 0 & 0 & 0 & 0 \end{bmatrix}$$

(a) Interchange the first and second rows.

(b) Bring the zero row to the bottom of the matrix.

(c) No number of row interchange can produce an echelon matrix.

A.20. Row reduce the matrix $A = \begin{bmatrix} 1 & 2 & -3 & 0 \\ 2 & 4 & -2 & 2 \\ 3 & 6 & -4 & 3 \end{bmatrix}$ to echelon form.

Use a_{11} as a pivot to obtain zeros below a_{11}, that is, apply the row operations "Add $-2R_1$ to R_2" and "Add $-3R_1$ to R_3;" and then use $a_{23} = 4$ as a pivot to obtain a zero below a_{23}, that is, by applying the row operation "Add $-5R_2$ to $4R_3$." These operations yield the following where the last matrix is in echelon form:

$$A \sim \begin{bmatrix} 1 & 2 & -3 & 0 \\ 0 & 0 & 4 & 2 \\ 0 & 0 & 5 & 3 \end{bmatrix} \sim \begin{bmatrix} 1 & 2 & -3 & 0 \\ 0 & 0 & 4 & 2 \\ 0 & 0 & 0 & 2 \end{bmatrix}$$

A.21. Which of the following matrices are in row canonical form?

$$\begin{bmatrix} 1 & 2 & -3 & 0 & 1 \\ 0 & 0 & 5 & 2 & -4 \\ 0 & 0 & 0 & 7 & 3 \end{bmatrix}, \quad \begin{bmatrix} 0 & 1 & 7 & -5 & 0 \\ 0 & 0 & 0 & 0 & 1 \\ 0 & 0 & 0 & 0 & 0 \end{bmatrix}, \quad \begin{bmatrix} 1 & 0 & 5 & 0 & 2 \\ 0 & 1 & 2 & 0 & 4 \\ 0 & 0 & 0 & 1 & 7 \end{bmatrix}$$

The first matrix is not in row canonical form since, for example, two leading nonzero entries are 5 and 7, not 1. Also, there are nonzero entries above the leading nonzero entries 5 and 7. The second and third matrices are in row canonical form.

A.22. Reduce the matrix $A = \begin{bmatrix} 1 & -2 & 3 & 1 & 2 \\ 1 & 1 & 4 & -1 & 3 \\ 2 & 5 & 9 & -2 & 8 \end{bmatrix}$ to row canonical form.

First reduce A to echelon form by applying the operations "Add $-R_1$ to R_2" and "Add $-2R_1$ to R_3," and then the operation "Add $-3R_2$ to R_3." These operations yield

$$A \sim \begin{bmatrix} 1 & -2 & 3 & 1 & 2 \\ 0 & 3 & 1 & -2 & 1 \\ 0 & 9 & 3 & -4 & 4 \end{bmatrix} \sim \begin{bmatrix} 1 & -2 & 3 & 1 & 2 \\ 0 & 3 & 1 & -2 & 1 \\ 0 & 0 & 0 & 2 & 1 \end{bmatrix}$$

Now use back-substitution on the echelon matrix to obtain the row canonical form of A. Specifically, first multiply R_3 by $\frac{1}{2}$ to obtain the pivot $a_{34} = 1$, and then apply the operations "Add $2R_3$ to R_2" and "Add $-R_3$ to R_t." These operations yield

$$A \sim \begin{bmatrix} 1 & -2 & 3 & 1 & 2 \\ 0 & 3 & 1 & -2 & 1 \\ 0 & 0 & 0 & 1 & \frac{1}{2} \end{bmatrix} \sim \begin{bmatrix} 1 & -2 & 3 & 0 & \frac{3}{2} \\ 0 & 3 & 1 & 0 & 2 \\ 0 & 0 & 0 & 1 & \frac{1}{2} \end{bmatrix}$$

Now multiply R_2 by $\frac{1}{3}$ making the pivot $a_{22} = 1$, and then apply the operation "Add $2R_2$ to R_1." We obtain

$$A \sim \begin{bmatrix} 1 & -2 & 3 & 0 & \frac{3}{2} \\ 0 & 1 & \frac{1}{3} & 0 & \frac{2}{3} \\ 0 & 0 & 0 & 1 & \frac{1}{2} \end{bmatrix} \sim \begin{bmatrix} 1 & 0 & \frac{11}{3} & 0 & \frac{17}{6} \\ 0 & 1 & \frac{1}{3} & 0 & \frac{2}{3} \\ 0 & 0 & 0 & 1 & \frac{1}{2} \end{bmatrix}$$

Since $a_{11} = 1$, the last matrix is the desired row canonical form of A.

A.23. Solve each system using its augmented matrix M:

$$(a) \quad \begin{aligned} x + y - 2z + 4t &= 5 \\ 2x + 2y - 3z + t &= 4 \\ 3x + 3y - 4z - 2t &= 3 \end{aligned} \qquad (b) \quad \begin{aligned} x - 2y + 4z &= 2 \\ 2x - 3y + 5z &= 3 \\ 3x - 4y + 6z &= 7 \end{aligned}$$

(a) Reduce its augmented matrix M to echelon form and then to row canonical form:

$$M = \begin{bmatrix} 1 & 1 & -2 & 4 & 5 \\ 2 & 2 & -3 & 1 & 4 \\ 3 & 3 & -4 & -2 & 3 \end{bmatrix} \sim \begin{bmatrix} 1 & 1 & -2 & 4 & 5 \\ 0 & 0 & 1 & -7 & -6 \\ 0 & 0 & 2 & 14 & 12 \end{bmatrix} \sim \begin{bmatrix} 1 & 1 & 0 & -10 & -7 \\ 0 & 0 & 1 & -7 & -6 \end{bmatrix}$$

(The third row of the second matrix is deleted since it is a multiple of the second row and will result in a zero row.)

Write down the system corresponding to the row canonical form of M and then transfer the free variables to the other side to obtain the free variable form of the solution:

$$\begin{aligned} x + y - 10t &= -7 \\ z - 7t &= -6 \end{aligned} \quad \text{and then} \quad \begin{aligned} x &= -7 - y + 10t \\ z &= -6 + 7t \end{aligned}$$

Here x and z are the basic variables and y and t are the free variables.

The *parametric* form of the solution can be obtained by setting the free variables equal to *parameters*, say $y = a$ and $t = b$. This process yields $x = -7 - a + 10b$, $y = a$, $z = -6 + 7b$, $t = b$ or $u = (-7 - a + 10b, \ a, \ -6 + 7b, \ b)$ (which is another form of the solution).

A *particular solution* can be obtained by assigning any values to the free variables (or parameters) and solving for the basic variables using either form of the general solution. For example, setting $y = 2, t = 3$, we obtain $x = 21, z = 15$. Thus the following is a particular solution of the system:

$$x = 21, \quad y = 2, \quad z = 15, \quad t = 3 \quad \text{or} \quad u = (21, \ 2, \ 15, \ 3)$$

(b) First row reduce its augmented matrix M to echelon form:

$$M = \begin{bmatrix} 1 & -2 & 4 & 2 \\ 2 & -3 & 5 & 3 \\ 3 & -4 & 6 & 7 \end{bmatrix} \sim \begin{bmatrix} 1 & -2 & 4 & 2 \\ 0 & 1 & -3 & -1 \\ 0 & 2 & -6 & 1 \end{bmatrix} \sim \begin{bmatrix} 1 & -2 & 4 & 2 \\ 0 & 1 & -3 & -1 \\ 0 & 0 & 0 & 3 \end{bmatrix}$$

In echelon form, the third row corresponds to the degenerate equation $0x + 0y + 0z = 3$.

Thus the system has no solution. (Note that the echelon form indicates whether or not the system has a solution.)

MISCELLANEOUS PROBLEMS

A.24. Let $A = \begin{bmatrix} 1 & 0 & 0 \\ 0 & 0 & 1 \\ 1 & 1 & 0 \end{bmatrix}$ and $B = \begin{bmatrix} 0 & 1 & 1 \\ 1 & 0 & 0 \\ 0 & 1 & 0 \end{bmatrix}$ be Boolean matrices.

Find the Boolean products AB, BA, and A^2.

Find the usual matrix product and then substitute 1 for any nonzero scalar. Thus:

$$AB = \begin{bmatrix} 0 & 1 & 1 \\ 0 & 1 & 0 \\ 1 & 1 & 1 \end{bmatrix}; \quad BA = \begin{bmatrix} 1 & 1 & 1 \\ 1 & 0 & 0 \\ 0 & 0 & 1 \end{bmatrix}; \quad A^2 = \begin{bmatrix} 1 & 0 & 0 \\ 1 & 1 & 0 \\ 1 & 0 & 1 \end{bmatrix}$$

A.25. Let $A = \begin{bmatrix} 1 & 3 \\ 4 & -3 \end{bmatrix}$. (a) Find a nonzero column vector $u = \begin{bmatrix} x \\ y \end{bmatrix}$ such that $Au = 3u$. (b) Describe all such vectors.

(a) First set up the matrix equation $Au = 3u$ and then write each side as a single matrix (column vector):

$$\begin{bmatrix} 1 & 3 \\ 4 & -3 \end{bmatrix}\begin{bmatrix} x \\ y \end{bmatrix} = 3\begin{bmatrix} x \\ y \end{bmatrix} \quad \text{and} \quad \begin{bmatrix} x + 3y \\ 4x - 3y \end{bmatrix} = \begin{bmatrix} 3x \\ 3y \end{bmatrix}$$

Set corresponding elements equal to each other to obtain a system of equations, and reduce the system to echelon form:

$$\begin{array}{l} x + 3y = 3x \\ 4x - 3y = 3y \end{array} \quad \text{or} \quad \begin{array}{l} 2x - 3y = 0 \\ 4x - 6y = 0 \end{array} \quad \text{to} \quad \begin{array}{l} 2x - 3y = 0 \\ 0 = 0 \end{array} \quad \text{or} \quad 2x - 3y = 0$$

The system reduces to one (nondegenerate) linear equation in two unknowns, and so it has an infinite number of solutions. To obtain a nonzero solution, set $y = 2$, say; then $x = 3$. Thus $u = [3, 2]^T$ is a desired nonzero solution.

(b) To find the general solution, set $y = a$, where a is a parameter. Substitute $y = a$ into $2x - 3y = 0$ to obtain $x = 3a/2$. Thus $u = [3a/2, a]^T$ represents all such solutions. Alternatively, let $y = 2b$ so $v = [3b, 2b]$ represents all such solutions.

Supplementary Problems

VECTORS

A.26. Let $u = (2, -1, 0, -3)$, $v = (1, -1, -1, 3)$, $w = (1, 3, -2, 2)$. Find: (a) $2u - 3v$; (b) $5u - 3v - 4w$; (c) $-u + 2v - 2w$; (d) $u \cdot v, u \cdot w, v \cdot w$; (e) $\|u\|, \|v\|, \|w\|$.

A.27. Let $u = \begin{bmatrix} 1 \\ 3 \\ -4 \end{bmatrix}$, $v = \begin{bmatrix} 2 \\ 1 \\ 5 \end{bmatrix}$, $w = \begin{bmatrix} 3 \\ -2 \\ 6 \end{bmatrix}$. Find: (a) $5u - 3v$; (b) $2u + 4v - 6w$;

(c) $u \cdot v, u \cdot w, v \cdot w$; (d) $\|u\|, \|v\|, \|w\|$.

A.28. Find x and y where: (a) $x(2, 5) + y(4, -3) = (8, 33)$; (b) $x(1, 4) + y(2, -5) = (7, 2)$.

MATRIX OPERATIONS

A.29. Let $A = \begin{bmatrix} 1 & 2 \\ 3 & -4 \end{bmatrix}$, $B = \begin{bmatrix} 5 & 0 \\ -6 & 7 \end{bmatrix}$, $C = \begin{bmatrix} 1 & -3 & 4 \\ 2 & 6 & -5 \end{bmatrix}$, $D = \begin{bmatrix} 3 & 7 & -1 \\ 4 & -8 & 9 \end{bmatrix}$. Find:

(a) $5A - 2B$ and $2C - 3D$;　　　(c) AC and AD;　　　(e) A^T and C^T;

(b) AB and BA;　　　　　　　　(d) BC and BD;　　　(f) A^2, B^2, C^2.

A.30. Let $A = \begin{bmatrix} 1 & -1 & 2 \\ 0 & 3 & 4 \end{bmatrix}$, $B = \begin{bmatrix} 4 & 0 & -3 \\ -1 & -2 & 3 \end{bmatrix}$, $C = \begin{bmatrix} 2 & -3 & 0 & 1 \\ 5 & -1 & -4 & 2 \\ -1 & 0 & 0 & 3 \end{bmatrix}$, $D = \begin{bmatrix} 2 \\ -1 \\ 3 \end{bmatrix}$.

Find: (a) $3A - 4B$;   (b) AB, AC, AD;   (c) BC, BD, CD;   (d) A^T and $A^T B$.

A.31. Let $A = \begin{bmatrix} 1 & 2 \\ 3 & 6 \end{bmatrix}$. Find a 2×3 matrix B with distinct entries such that $AB = 0$.

SQUARE MATRICES

A.32. Find the diagonal and trace of: (a) $A = \begin{bmatrix} 2 & -7 & 8 \\ 3 & -6 & -5 \\ 4 & 0 & -1 \end{bmatrix}$; (b) $B = \begin{bmatrix} 1 & 2 & -9 \\ -3 & 2 & 8 \\ 5 & -6 & -1 \end{bmatrix}$.

A.33. Let $A = \begin{bmatrix} 2 & -5 \\ 3 & 1 \end{bmatrix}$. Find: (a) A^2 and A^3; (b) $f(A)$ where $f(x) = x^3 - 2x^2 - 5$.

A.34. Let $B = \begin{bmatrix} 4 & -2 \\ 1 & -6 \end{bmatrix}$. Find: (a) B^2 and B^3; (b) $f(B)$ where $f(x) = x^2 + 2x - 22$.

A.35. Let $A = \begin{bmatrix} 6 & -4 \\ 3 & -2 \end{bmatrix}$. Find a nonzero vector $u = \begin{bmatrix} x \\ y \end{bmatrix}$ such that $Au = 4u$.

DETERMINANTS AND INVERSES

A.36. Find each determinant: (a) $\begin{vmatrix} 2 & 5 \\ 4 & 1 \end{vmatrix}$; (b) $\begin{vmatrix} 6 & 1 \\ 3 & -2 \end{vmatrix}$; (c) $\begin{vmatrix} -2 & 8 \\ -5 & -2 \end{vmatrix}$; (d) $\begin{vmatrix} a-b & a \\ a & a+b \end{vmatrix}$.

A.37. Compute the determinant of each matrix in Problem A.32.

A.38. Find the inverse of: (a) $A = \begin{bmatrix} 7 & 4 \\ 5 & 3 \end{bmatrix}$; (b) $B = \begin{bmatrix} 5 & -2 \\ 6 & -3 \end{bmatrix}$; (c) $C = \begin{bmatrix} 4 & -6 \\ -2 & 3 \end{bmatrix}$.

A.39. Find the inverse of each matrix (if it exists):

$$A = \begin{bmatrix} 1 & 2 & -4 \\ -1 & -1 & 5 \\ 2 & 7 & -3 \end{bmatrix}; \quad B = \begin{bmatrix} 1 & -1 & 1 \\ 0 & 2 & -2 \\ 1 & 3 & -1 \end{bmatrix}; \quad C = \begin{bmatrix} 1 & 2 & 3 \\ 2 & 5 & -1 \\ 5 & 12 & 1 \end{bmatrix}.$$

ECHELON MATRICES, ROW REDUCIONS, GAUSSIAN ELIMINATION

A.40. Reduce A to echelon form and then to row canonical form, where:

(a) $A = \begin{bmatrix} 1 & 2 & -1 & 2 & 1 \\ 2 & 4 & 1 & -2 & 3 \\ 3 & 6 & 2 & -6 & 5 \end{bmatrix}$;   (b) $A = \begin{bmatrix} 2 & 3 & -2 & 5 & 1 \\ 3 & -1 & 2 & 0 & 4 \\ 4 & -5 & 6 & -5 & 7 \end{bmatrix}$.

A.41. Using only 0's and 1's, list all 2×2 matrices in echelon form.

A.42. Using only 0's and 1's find the number of 3×3 matrices in row canonical form.

A.43. Solve each system:
$$\begin{array}{lll} & x + 2y - 4z = -3 & x + 2y - 4z = 3 \\ \text{(a)} & 2x + 6y - 5z = 2 \quad \text{(b)} & 2x + 6y - 5z = 10 \\ & 3x + 11y - 4z = 12 & 3x + 10y - 6z = 14 \end{array}$$

A.44. Solve each system:
$$\begin{array}{ll} & x + 2y + 3z = 7 \\ x - 3y + 2z - t = 2 & x + 3y + z = 6 \\ \text{(a)} \quad 3x - 9y + 7z - t = 7 \quad \text{(b)} & 2x + 6y + 5z = 15 \\ 2x - 6y + 7z + 4t = 7 & 3x + 10y + 7z = 23 \end{array}$$

MISCELLANEOUS PROBLEMS

A.45. Let $A = \begin{bmatrix} 1 & 2 \\ 0 & 1 \end{bmatrix}$. Find: (a) A^n; (b) A^{-1}; (c) matrix B such that $B^2 = A$.

A.46. Matrices A and B are said to commute if $AB = BA$. Find all matrices $\begin{bmatrix} x & y \\ z & t \end{bmatrix}$ which commute with $\begin{bmatrix} 1 & 1 \\ 0 & 1 \end{bmatrix}$.

A.47. Let $A = \begin{bmatrix} 0 & 1 & 0 \\ 1 & 0 & 1 \\ 1 & 0 & 0 \end{bmatrix}$ and $B = \begin{bmatrix} 1 & 0 & 0 \\ 1 & 0 & 0 \\ 0 & 1 & 1 \end{bmatrix}$ be Boolean matrices.

Find the Boolean matrices: (a) $A + B$; (b) AB; (c) BA; (d) A^2; (e) B^2.

Answers to Supplementary Problems

Notation: $M = [R_1; R_2; \ldots; R_n]$ denotes a matrix with rows $R_1, \ldots, R_n$.

A.26. (a) $(1, 1, 3, -15)$; (b) $(3, -14, 11, -32)$;
(c) $(-2, -7, 2, 5)$; (d) $-6, -7, 6$;
(e) $\sqrt{14}$, $\sqrt{12} = 2\sqrt{3}$, $\sqrt{18} = 3\sqrt{2}$.

A.27. (a) $[-1, 12, -35]^T$; (b) $[-8, 22, -24]^T$;
(c) $-15, -27, 34$; (d) $\sqrt{26}$, $\sqrt{30}$, 7.

A.28. (a) $x = 6$, $y = -1$; (b) $x = 3$, $y = 2$.

A.29. (a) $[-5, 10; 27, -34]$, $[-7, 27, 11; -8, 36, -37]$;
(b) $[-7, 14; 39, -28]$, $[5, 10; 15, -40]$;
(c) $[5, 9, -6; -5, -33, 32]$, $[11, -9, 17; -7, 53, 39]$;
(d) $[5, -15, 20; 8, 60, -59]$, $[15, 35, -5; 10, -98, 69]$;
(e) $[1, 3; 2, -4]$, $[1, 2; -3, 6; 4, -5]$;
(f) $[7, -6; -9, 22]$, $[25, 0; -72, 49]$, C^2 not defined.

A.30. (a) $[-13, -3, 18; 4, 17, 0]$; (b) AB not defined,
$[-5, -2, 4, 5; 11, -3, -12, 18]$, $[9; 9]$;
(c) $[11, -12, 0, -5; -15, 5, 8, 4]$, $[-1; 9]$, CD not
defined; (d) $[1, 0; -1, 3; 2, 4]$, $[4, 0, -3; -7, -6, 12; 4, -8, 6]$.

A.31. $[2, 4, 6; -1, -2, -3]$

A.32. (a) $[2, -6, -1]$, -5; (b) $[1, 2, -1]$, 2

A.33. (a) $[-11, -15; 9, -14]$, $[-67, 40; -24, -59]$;
(b) $[-50, 70; -42, -36]$.

A.34. (a) $[14, 4; -2, 34]$, $[60, -52; 26, -200]$
(b) $f(B) = 0$.

A.35. $[2a; a]$, for any nonzero a.

A.36. (a) -18; (b) -15; (c) 44; (d) $-b^2$.

A.37. (a) 323; (b) 48.

A.38. (a) $[3, -4; -5, 7]$; (b) $[1, -2/3; 2, -5/3]$;
(c) Not defined.

A.39. (a) $[-16, -11, 3; 7/2, 5/2, -1/2; -5/2, -3/2, 1/2]$; (b) $[1, 1/2, 0; -1/2, -1/2, 1/2; -1/2, -1, 1/2]$; (c) Not defined.

A.40. (a) $[1, 2, -1, 2, 1; 0, 0, 3, -6, 1; 0, 0, 0, -6, 1]$, $[1, 2, 0, 0, 4/3; 0, 0, 1, 0, 0; 0, 0, 0, 1, -1/6]$;
(b) $[2, 3, -2, 5, 1; 0, -11, 10, -15, 5; 0, \ldots, 0]$, $[1, 0, 4/11, 5/11, 13/11; 0, 1, -10/11, 15/11, -5/11; 0, \ldots, 0]$

A.41. $[1, 1; 0, 1]$, $[1, 1; 0, 0]$, $[1, 0; 0, 0]$, $[0, 1; 0, 0]$, $[0, 0; 0, 0]$, $[1, 0; 0, 1]$

A.42. There are 13.

A.43. (a) $x = 3$, $y = 1$, $z = 2$; (b) No solution.

A.44. (a) $x = 3y + 5t$, $z = 1 - 2t$; (b) $x = 2$, $y = 1$, $z = 1$.

A.45. (a) $[1, 2n; 0, 1]$; (b) $[1, -2; 0, 1]$; (c) $[1, 1; 0, 1]$.

A.46. $[a, b; 0, a]$

A.47. (a) $[110; 101; 111]$; (b) $[100; 111; 100]$;
(c) $[010; 010; 101]$; (d) $[101; 110; 010]$;
(e) $[100; 100; 111]$.

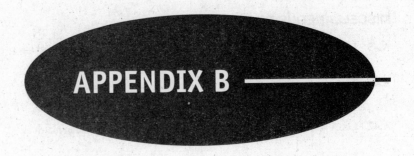

APPENDIX B

Algebraic Systems

B.1 INTRODUCTION

This Appendix investigates some of the major algebraic systems in mathematics: semigroups, groups, rings, and fields. We also define the notion of a homomorphism and the notion of a quotient structure. We begin with the formal definition of an operation, and discuss various types of operations.

B.2 OPERATIONS

The reader is familiar with the operations of addition and multiplication of numbers, union and intersection of sets, and the composition of functions. These operations are denoted as follows:

$$a + b = c, \quad a \cdot b = c, \quad A \cup B = C, \quad A \cap B = C, \quad g \circ f = h.$$

In each situation, an element (c, C, or h) is assigned to an original pair of elements. We make this notion precise.

Definition B.1: Let S be a nonempty set. An *operation* on S is a function $*$ from $S \times S$ into S. In such a case, instead of $*(a, b)$, we usually write

$$a * b \quad \text{or sometimes} \quad ab$$

The set S and an operation $*$ on S is denoted by $(S, *)$ or simply S when the operation is understood.

Remark: An operation $*$ from $S \times S$ into S is sometimes called a *binary operation*. A *unary* operation is a function from S into S. For example, the absolute value $|n|$ of an integer n is a unary operation on $\mathbf{Z}$, and the complement A^C of a set A is a unary operation on the power set $P(X)$ of a set X. A *ternary* (3-ary) operation is a function from $S \times S \times S$ into S. More generally, an *n*-ary operation is a function from $S \times S \times \cdots \times S$ (n factors) into S. Unless otherwise stated, the word operation shall mean binary operation. We will also assume that our underlying set S is nonempty.

Suppose S is a finite set. Then an operation $*$ on S can be presented by its operation (multiplication) table where the entry in the row labeled a and the column labeled b is $a * b$.

Suppose S is a set with an operation $*$, and suppose A is a subset of S. Then A is said to be *closed under* $*$ if $a * b$ belongs to A for any elements a and b in A.

EXAMPLE B.1 Consider the set **N** of positive integers.

(a) Addition ($+$) and multiplication ($\times$) are operations on **N**. However, subtraction ($-$) and division ($/$) are not operations on **N** since the difference and the quotient of positive integers need not be positive integers. For example, $2 - 9$, and $7/3$ are not positive integers.

(b) Let A and B denote, respectively, the set of even and odd positive integers. Then A is closed under addition and multiplication since the sum and product of any even numbers are even. On the other hand, B is closed under multiplication but not addition since, for example, $3 + 5 = 8$ is even.

EXAMPLE B.2 Let $S = \{a, b, c, d\}$. The tables in Fig. B-1 define operations $*$ and $\cdot$ on S. Note that $*$ can be defined by the following operation where x and y are any elements of S:

$$x * y = x$$

$*$	a	b	c	d
a	a	a	a	a
b	b	b	b	b
c	c	c	c	c
d	d	d	d	d

$\cdot$	a	b	c	d
a	a	b	c	d
b	b	a	a	b
c	c	b	a	a
d	d	a	a	a

(a)   (b)

Fig. B-1

Next we list a number of important properties of our operations.

Associative Law:

An operation $*$ on a set S is said to be *associative* or to satisfy the *Associative Law* if, for any elements a, b, c in S, we have

$$(a * b) * c = a * (b * c)$$

Generally speaking, if an operation is not associative, then there may be many ways to form a product. For example, the following shows five ways to form the product $abcd$:

$$((ab)c)d, \quad (ab)(cd), \quad (a(bc))d, \quad a((bc)d), \quad a(b(cd))$$

If the operation is associative, then the following theorem (proved in Problem B.4) applies.

Theorem B.1: Suppose $*$ is an associative operation on a set S. Then any product $a_1 * a_2 * \cdots * a_n$ requires no parentheses, that is, all possible products are equal.

Commutative Law:

An operation $*$ on a set S is said to be *commutative* or satisfy the *Commutative Law* if, for any elements a, b in S,

$$a * b = b * a$$

EXAMPLE B.3

(a) Consider the set **Z** of integers. Addition and multiplication of integers are associative and commutative. On the other hand, subtraction is nonassociative. For example,

$$(8 - 4) - 3 = 1 \quad \text{but} \quad 8 - (4 - 3) = 7$$

Moreover, subtraction is not commutative since, for example, $3 - 7 \neq 7 - 3$.

(b) Consider the operation of matrix multiplication on the set M of n-square matrices. One can prove that matrix multiplication is associative. On the other hand, matrix multiplication is not commutative. For example,

$$\begin{bmatrix} 1 & 2 \\ 3 & 4 \end{bmatrix}\begin{bmatrix} 5 & 6 \\ 0 & -2 \end{bmatrix} = \begin{bmatrix} 5 & 2 \\ 15 & 10 \end{bmatrix} \quad \text{but} \quad \begin{bmatrix} 5 & 6 \\ 0 & -2 \end{bmatrix}\begin{bmatrix} 1 & 2 \\ 3 & 4 \end{bmatrix} = \begin{bmatrix} 23 & 34 \\ -6 & -8 \end{bmatrix}$$

Identity Element:

Consider an operation $*$ on a set S. An element e in S is called an *identity* element for $*$ if, for any element a in S,

$$a * e = e * a = a$$

More generally, an element e is called a *left identity* or a *right identity* according as $e * a = a$ or $a * e = a$ where a is any element in S. The following theorem applies.

Theorem B.2: Suppose e is a left identity and f is a right identity for an operation on a set S. Then $e = f$

The proof is very simple. Since e is a left identity, $ef = f$; but since f is a right identity, $ef = e$. Thus $e = f$. This theorem tells us, in particular, that an identity element is unique, and that if an operation has more than one left identity then it has no right identity, and vice versa.

Inverses:

Suppose an operation $*$ on a set S does have an identity element e. The *inverse* of an element a in S is an element b such that

$$a * b = b * a = e$$

If the operation is associative, then the inverse of a, if it exists, is unique (Problem B.2). Observe that if b is the inverse of a, then a is the inverse of b. Thus the inverse is a symmetric relation, and we can say that the elements a and b are inverses.

Notation: If the operation on S is denoted by $a * b$, $a \times b$, $a \cdot b$, or ab, then S is said to be written *multiplicatively* and the inverse of an element $a \in S$ is usually denoted by a^{-1}. Sometimes, when S is commutative, the operation is denoted by $+$ and then S is said to be written *additively*. In such a case, the identity element is usually denoted by 0 and it is called the *zero* element; and the inverse is denoted by $-a$ and it is called the *negative* of a.

EXAMPLE B.4 Consider the rational numbers $\mathbf{Q}$. Under addition, 0 is the identity element, and -3 and 3 are (additive) inverses since

$$(-3) + 3 = 3 + (-3) = 0$$

On the other hand, under multiplication, 1 is the identity element, and -3 and $-1/3$ are (multiplicative) inverses since

$$(-3)(-1/3) = (-1/3)(-3) = 1$$

Note 0 has no multiplicative inverse.

Cancellation Laws:

An operation $*$ on a set S is said to satisfy the *left cancellation law* or the *right cancellation law* according as:

$$a * b = a * c \text{ implies } b = c \quad \text{or} \quad b * a = c * a \text{ implies } b = c$$

Addition and subtraction of integers in $\mathbf{Z}$ and multiplication of nonzero integers in $\mathbf{Z}$ do satisfy both the left and right cancellation laws. On the other hand, matrix multiplication does not satisfy the cancellation laws. For example, suppose

$$A = \begin{bmatrix} 1 & 1 \\ 0 & 0 \end{bmatrix}, \quad B = \begin{bmatrix} 1 & 1 \\ 0 & 1 \end{bmatrix}, \quad C = \begin{bmatrix} 0 & -3 \\ 1 & 5 \end{bmatrix}, \quad D = \begin{bmatrix} 1 & 2 \\ 0 & 0 \end{bmatrix}$$

Then $AB = AC = D$, but $B \neq C$.

B.3　SEMIGROUPS

Let S be a nonempty set with an operation. Then S is called a *semigroup* if the operation is associative. If the operation also has an identity element, then S is called a *monoid*.

EXAMPLE B.5

(a) Consider the positive integers N. Then $(N, +)$ and $(N, \times)$ are semigroups since addition and multiplication on N are associative. In particular, $(N, \times)$ is a monoid since it has the identity element 1. However, $(N, +)$ is not a monoid since addition in N has no zero element.

(b) Let S be a finite set, and let $F(S)$ be the collection of all functions $f: S \to S$ under the operation of composition of functions. Since the composition of functions is associative, $F(S)$ is a semigroup. In fact, $F(S)$ is a monoid since the identity function is an identity element for $F(S)$.

(c) Let $S = \{a, b, c, d\}$. The multiplication tables in Fig. B-1 define operations $*$ and $\cdot$ on S. Note that $*$ can be defined by the formula $x * y = x$ for any x and y in S. Hence

$$(x * y) * z = x * z = x \quad \text{and} \quad x * (y * z) = x * y = x$$

Therefore, $*$ is associative and hence $(S, *)$ is a semigroup. On the other hand, $\cdot$ is not associative since, for example,

$$(b \cdot c) \cdot c = a \cdot c = c \quad \text{but} \quad b \cdot (c \cdot c) = b \cdot a = b$$

Thus $(S, \cdot)$ is not a semigroup.

Free Semigroup, Free Monoid

Let A be a nonempty set. A *word* w on A is a finite sequence of its elements. For example, the following are words on $A = \{a, b, c\}$:

$$u = ababbbb = abab^4 \quad \text{and} \quad v = baccaaaa = bac^2a^4$$

(We write a^2 for aa, a^3 for aaa, and so on.) The *length* of a word w, denoted by $l(w)$, is the number of elements in w. Thus $l(u) = 7$ and $l(v) = 8$.

The concatenation of words u and v on a set A, written $u * v$ or uv, is the word obtained by writing down the elements of u followed by the elements of v. For example,

$$uv = (abab^4)(bac^2a^4) = abab^5c^2a^4$$

Now let $F = F(A)$ denote the collection of all words on A under the operation of concatenation. Clearly, for any words u, v, w, the words $(uv)w$ and $u(vw)$ are identical; they simply consist of the elements of u, v, w written down one after the other. Thus F is a semigroup; it is called the *free semigroup* on A, and the elements of A are called the *generators* of F.

The empty sequence, denoted by λ, is also considered as a word on A. However, we do not assume that λ belongs to the free semigroup $F = F(A)$. The set of all words on A including λ is frequently denoted by A^*. Thus A^* is a monoid under concatenation; it is called the *free monoid* on A.

Subsemigroups

Let A be a nonempty subset of a semigroup S. Then A is called a *subsemigroup* of S if A itself is a semigroup with respect to the operation on S. Since the elements of A are also elements of S, the Associative Law automatically holds for the elements of A. Therefore, A is a subsemigroup of S if and only if A is closed under the operation on S.

EXAMPLE B.6

(a) Let A and B denote, respectively, the set of even and odd positive integers. Then $(A, \times)$ and $(B, \times)$ are subsemigroups of $(\mathbf{N}, \times)$ since A and B are closed under multiplication. On the other hand, $(A, +)$ is a subsemigroup of $(\mathbf{N}, +)$ since A is closed under addition, but $(B, +)$ is not a subsemigroup of $(\mathbf{N}, +)$ since B is not closed under addition.

(b) Let F be the free semigroup on the set $A = \{a, b\}$. Let H consist of all even words, that is, words with even length. The concatenation of two such words is also even. Thus H is a subsemigroup of F.

Congruence Relations and Quotient Structures

Let S be a semigroup and let $\sim$ be an equivalence relation on S. Recall that the equivalence relation $\sim$ induces a partition of S into equivalence classes. Also, $[a]$ denotes the equivalence class containing the element $a \in S$, and that the collection of equivalence classes is denoted by $S/\sim$.

Suppose that the equivalence relation $\sim$ on S has the following property:

$$\boxed{\text{If } a \sim a' \text{ and } b \sim b', \text{ then } ab \sim a'b'.}$$

Then $\sim$ is called a *congruence relation* on S. Furthermore, we can now define an operation on the equivalence classes by

$$[a] * [b] = [a * b] \quad \text{or, simply,} \quad [a][b] = [ab]$$

Furthermore, this operation on $S/\sim$ is associative; hence $S/\sim$ is a semigroup. We state this result formally.

Theorem B.3: Let $\sim$ be a congruence relation on a semigroup S. Then $S/\sim$, the equivalence classes under $\sim$, form a semigroup under the operation $[a][b] = [ab]$.

This semigroup $S/\sim$ is called the quotient of S by $\sim$.

EXAMPLE B.7

(a) Let F be the free semigroup on a set A. Define $u \sim u'$ if u and u' have the same length. Then $\sim$ is an equivalence relation on F. Furthermore, suppose $u \sim u'$ and $v \sim v'$, say,

$$l(u) = l(u') = m \quad \text{and} \quad l(v) = l(v') = n$$

Then $l(uv) = l(u'v') = m + n$, and so $uv \sim u'v'$. Thus $\sim$ is a congruence relation on F.

(b) Consider the integers $\mathbf{Z}$ and a positive integer $m > 1$. Recall (Section 11.8) that we say that a is congruent to b modulo m, written

$$a \equiv b \pmod{m}$$

if m divides the difference $a - b$. Theorem 11.21 states that this relation is an equivalence relation on $\mathbf{Z}$. Furthermore, Theorem 11.22 tells us that if $a \equiv c \pmod{m}$ and $b \equiv d \pmod{m}$ then:

$$a + b \equiv c + d \pmod{m} \quad \text{and} \quad ab \equiv cd \pmod{m}$$

In other words, this relation is a congruence relation on $\mathbf{Z}$.

Homomorphism of Semigroups

Consider two semigroups $(S, *)$ and $(S', *')$. A function $f: S \to S'$ is called a *semigroup homomorphism* or, simply, a *homomorphism* if

$$f(a * b) = f(a) *' f(b) \quad \text{or, simply} \quad f(ab) = f(a)f(b)$$

Suppose f is also one-to-one and onto. Then f is called an *isomorphism* between S and S', and S and S' are said to be *isomorphic* semigroups, written $S \cong S$.

EXAMPLE B.8

(a) Let M be the set of all 2×2 matrices with integer entries. The determinant of any matrix $A = \begin{bmatrix} a & b \\ c & d \end{bmatrix}$ is denoted and defined by $\det(A) = |A| = ad - bc$. One proves in Linear Algebra that the determinant is a *multiplicative function*, that is, for any matrices A and B,

$$\det(AB) = \det(A) \cdot \det(B)$$

Thus the determinant function is a semigroup homomorphism on $(M, \times)$, the matrices under matrix multiplication. On the other hand, the determinant function is not additive, that is, for some matrices,

$$\det(A + B) \neq \det(A) + \det(B)$$

Thus the determinant function is not a semigroup homomorphism on $(M, +)$.

(b) Figure B-2(a) gives the addition table for $\mathbf{Z_4}$, the integers modulo 4 under addition; and Fig. B-2(b) gives the multiplication table for $S = \{1, 3, 7, 9\}$ in $\mathbf{Z_{10}}$. (We note that S is a reduced residue system for the integers $\mathbf{Z}$ modulo 10.) Let $f: \mathbf{Z_4} \to S$ be defined by

$$f(0) = 1, \quad f(1) = 3, \quad f(2) = 9, \quad f(3) = 7$$

+	0	1	2	3
0	0	1	2	3
1	1	2	3	0
2	2	3	0	1
3	3	0	1	2

×	1	3	7	9
1	1	3	7	9
3	3	9	1	7
7	7	1	9	3
9	9	7	3	1

(a)   (b)

Fig. B-2

One can show that f is a homomorphism. Since f is also one-to-one and onto, f is an isomorphism. Thus $\mathbf{Z_4}$ and S are isomorphic semigroups.

(c) Let $\sim$ be a congruence relation on a semigroup S. Let $\phi: S \to S/\sim$ be the *natural mapping* from S into the factor semigroup $S/\sim$ defined by

$$\phi(a) = [a]$$

That is, each element a in S is assigned its equivalence class $[a]$. Then ϕ is a homomorphism since

$$\phi(ab) = [ab] = [a][b] = \phi(a)\phi(b)$$

Fundamental Theorem of Semigroup Homomorphisms

Recall that the image of a function $f: S \to S'$, written $f(S)$ of Im f, consists of the images of the elements of S under f. Namely:

$$\text{Im } f = \{b \in S' \mid \text{there exists } a \in S \text{ for which } f(a) = b\}$$

The following theorem (proved in Problem B.5) is fundamental to semigroup theory.

Theorem B.4: Let $f: S \to S'$ be a semigroup homomorphism. Let $a \sim b$ if $f(a) = f(b)$. Then:
(i) $\sim$ is a congruence relation on S. (ii) $S/\sim$ is isomorphic to $f(S)$.

EXAMPLE B.9

(a) Let F be the free semigroup on $A = \{a, b\}$. The function $f: F \to \mathbf{Z}$ defined by

$$f(u) = l(u)$$

is a homomorphism. Note $f(F) = \mathbf{N}$. Thus $F/\sim$ is isomorphic to $\mathbf{N}$.

(b) Let M be the set of 2×2 matrices with integer entries. Consider the determinant function det: $M \to \mathbf{Z}$. We note that the image of det is $\mathbf{Z}$. By Theorem B.4, $M/\sim$ is isomorphic to $\mathbf{Z}$.

Semigroup Products

Let $(S_1, *_1)$ and $(S_2, *_2)$ be semigroups. We form a new semigroup $S = S_1 \otimes S_2$, called the *direct product* of S_1 and S_2, as follows.

(1) The elements of S come from $S_1 \times S_2$, that is, are ordered pairs (a, b) where $a \in S_1$ and $b \in S_2$

(2) The operation $*$ in S is defined componentwise, that is,

$$(a, b) * (a', b') = (a *_1 a', \; b *_2 b') \quad \text{or simply} \quad (a, b)(a', b') = (aa', bb')$$

One can easily show (Problem B.3) that the above operation is associative.

B.4   GROUPS

Let G be a nonempty set with a binary operation (denoted by juxtaposition). Then G is called a *group* if the following axioms hold:

[$\mathbf{G_1}$] Associative Law: For any a, b, c in G, we have $(ab)c = a(bc)$.
[$\mathbf{G_2}$] Identity element: There exists an element e in G such that $ae = ea = a$ for every a in G.
[$\mathbf{G_3}$] Inverses: For each a in G, there exists an element a^{-1} in G (the *inverse* of a) such that

$$aa^{-1} = a^{-1}a = e$$

A group G is said to be *abelian* (or *commutative*) if $ab = ba$ for every $a, b \in G$, that is, if G satisfies the Commutative Law.

When the binary operation is denoted by juxtaposition as above, the group G is said to be written *multiplicatively*. Sometimes, when G is abelian, the binary operation is denoted by $+$ and G is said to be written *additively*. In such a case the identity element is denoted by 0 and it is called the *zero* element; and the inverse is denoted by $-a$ and it is called the *negative* of a.

The number of elements in a group G, denoted by $|G|$, is called the *order* of G. In particular, G is called a *finite group* if its order is finite.

Suppose A and B are subsets of a group G. Then we write:

$$AB = \{ab \mid a \in A, \; b \in B\} \quad \text{or} \quad A + B = \{a + b \mid a \in A, \; b \in B\}$$

EXAMPLE B.10

(a) The nonzero rational numbers $\mathbf{Q} \backslash \{0\}$ form an abelian group under multiplication. The number 1 is the identity element and q/p is the multiplicative inverse of the rational number p/q.

(b) Let S be the set of 2×2 matrices with rational entries under the operation of matrix multiplication. Then S is not a group since inverses do not always exist. However, let G be the subset of 2×2 matrices with a nonzero determinant. Then G is a group under matrix multiplication. The identity element is

$$I = \begin{bmatrix} 1 & 0 \\ 0 & 1 \end{bmatrix} \text{ and the inverse of } A = \begin{bmatrix} a & b \\ c & d \end{bmatrix} \text{ is } A^{-1} = \begin{bmatrix} d/|A| & -b/|A| \\ -c/|A| & a/|A| \end{bmatrix}$$

This is an example of a nonabelian group since matrix multiplication is noncommutative.

(c) Recall that $\mathbf{Z}_m$ denotes the integers modulo m. $\mathbf{Z}_m$ is a group under addition, but it is not a group under multiplication. However, let $\mathbf{U}_m$ denote a reduced residue system modulo m which consists of those integers relatively prime to m. Then $\mathbf{U}_m$ is a group under multiplication (modulo m). Figure B-3 gives the multiplication table for $\mathbf{U}_{12} = \{1, 5, 7, 11\}$.

$\times$	1	5	7	11
1	1	5	7	11
5	5	1	11	7
7	7	11	1	5
11	11	7	5	1

Fig. B-3

	ε	σ_1	σ_2	σ_3	ϕ_1	ϕ_2
ε	ε	ϕ_1	σ_3	σ_3	ϕ_1	ϕ_2
σ_1	σ_1	ε	ϕ_1	ϕ_2	σ_2	σ_3
σ_2	σ_2	ϕ_2	ε	ϕ_1	σ_3	σ_1
σ_3	σ_3	ϕ_1	ϕ_2	ε	σ_1	σ_2
ϕ_1	ϕ_1	σ_3	σ_1	σ_2	ϕ_2	ε
ϕ_2	ϕ_2	σ_2	σ_3	σ_1	ε	ϕ_1

Fig. B-4

Symmetric Group S_n

A one-to-one mapping σ of the set $\{1, 2, \ldots, n\}$ onto itself is called a *permutation*. Such a permutation may be denoted as follows where $j_i = \sigma(i)$:

$$\sigma = \begin{pmatrix} 1 & 2 & 3 & \cdots & n \\ j_1 & j_2 & j_3 & \cdots & j_n \end{pmatrix}$$

The set of all such permutations is denoted by S_n, and there are $n! = n(n-1) \cdot \ldots \cdot 2 \cdot 1$ of them. The composition and inverses of permutations in S_n belong to S_n, and the identity function ε belongs to S_n. Thus S_n forms a group under composition of functions called the *symmetric group of degree n*.

The symmetric group S_3 has $3! = 6$ elements as follows:

$$\varepsilon = \begin{pmatrix} 1 & 2 & 3 \\ 1 & 2 & 3 \end{pmatrix}, \quad \sigma_2 = \begin{pmatrix} 1 & 2 & 3 \\ 3 & 2 & 1 \end{pmatrix}, \quad \phi_1 = \begin{pmatrix} 1 & 2 & 3 \\ 2 & 3 & 1 \end{pmatrix}$$

$$\sigma_1 = \begin{pmatrix} 1 & 2 & 3 \\ 1 & 3 & 2 \end{pmatrix}, \quad \sigma_3 = \begin{pmatrix} 1 & 2 & 3 \\ 2 & 1 & 3 \end{pmatrix}, \quad \phi_2 = \begin{pmatrix} 1 & 2 & 3 \\ 3 & 1 & 2 \end{pmatrix}$$

The multiplication table of S_3 appears in Fig. B-4.

MAP(A), PERM(A), and AUT(A)

Let A be a nonempty set. The collection MAP(A) of all functions (mappings) $f: A \rightarrow A$ is a semigroup under composition of functions; it is not a group since some functions may have no inverses. However, the subsemigroup PERM(A) of all one-to-one correspondences of A with itself (called *permutations* of A) is a group under composition of functions.

Furthermore, suppose A contains some type of geometric or algebraic structure; for example, A may be the set of vertices of a graph, or A may be an ordered set or a semigroup. Then the set AUT(A) of all isomorphisms of A with itself (called *automorphisms* of A) is also a group under compositions of functions.

B.5 SUBGROUPS, NORMAL SUBGROUPS, AND HOMOMORPHISMS

Let H be a subset of a group G. Then H is called a *subgroup* of G if H itself is a group under the operation of G. Simple criteria to determine subgroups follow.

Proposition B.5: A subset H of a group G is a subgroup of G if:

> (i) The identity element $e \in H$.
> (ii) H is closed under the operation of G, i.e. if $a, b \in H$, then $ab \in H$.
> (iii) H is closed under inverses, that is, if $a \in H$, then $a^{-1} \in H$.

Every group G has the subgroups $\{e\}$ and G itself. Any other subgroup of G is called a *nontrivial subgroup*.

Cosets

Suppose H is a subgroup of G and $a \in G$. Then the set

$$Ha = \{ha \mid h \in H\}$$

is called a *right coset* of H. (Analogously, aH is called a *left coset* of H.) We have the following important results (proved in Problems B.13 and B.15).

Theorem B.6: Let H be a subgroup of a group G. Then the right cosets Ha form a partition of G.

Theorem B.7 (Lagrange): Let H be a subgroup of a finite group G. Then the order of H divides the order of G.

The number of right cosets of H in G, called the index of H in G, is equal to the number of left cosets of H in G; and both numbers are equal to $|G|$ divided by $|H|$.

Normal Subgroups

The following definition applies.

Definition B.2: A subgroup H of G is a *normal* subgroup if $a^{-1}Ha \subseteq H$, for every $a \in G$, or, equivalently, if $aH = Ha$, i.e., if the right and left cosets coincide.

Note that every subgroup of an abelian group is normal.

The importance of normal subgroups comes from the following result (proved in Problem B.17).

Theorem B.8: Let H be a normal subgroup of a group G. Then the cosets of H form a group under coset multiplication:

$$(aH)(bH) = abH$$

This group is called the *quotient group* and is denoted by G/H.

Suppose the operation in G is addition or, in other words, G is written additively. Then the cosets of a subgroup H of G are of the form $a + H$. Moreover, if H is a normal subgroup of G, then the cosets form a group under coset addition, that is,

$$(a + H) + (b + H) = (a + b) + H$$

EXAMPLE B.11

(a) Consider the permutation group S_3 of degree 3 which is investigated above. The set $H = \{\varepsilon, \sigma_1\}$ is a subgroup of S_3. Its right and left cosets follow:

<table>
<tr><td>**Right Cosets**</td><td>**Left Cosets**</td></tr>
<tr><td>$H = \{\varepsilon, \sigma_1\}$</td><td>$H = \{\varepsilon, \sigma_1\}$</td></tr>
<tr><td>$H\phi_1 = \{\phi_1, \sigma_2\}$</td><td>$\phi_1 H = \{\phi_1, \sigma_3\}$</td></tr>
<tr><td>$H\phi_2 = \{\phi_2, \sigma_3\}$</td><td>$\phi_2 H = \{\phi_2, \sigma_2\}$</td></tr>
</table>

Observe that the right cosets and the left cosets are distinct; hence H is not a normal subgroup of S_3.

(b) Consider the group G of 2×2 matrices with rational entries and nonzero determinants. (See Example A.10.) Let H be the subset of G consisting of matrices whose upper-right entry is zero; that is, matrices of the form

$$\begin{bmatrix} a & 0 \\ c & d \end{bmatrix}$$

Then H is a subgroup of G since H is closed under multiplication and inverses and $I \in H$. However, H is not a normal subgroup since, for example, the following product does not belong to H:

$$\begin{bmatrix} 1 & 2 \\ 1 & 3 \end{bmatrix}^{-1} \begin{bmatrix} 1 & 0 \\ 1 & 1 \end{bmatrix} \begin{bmatrix} 1 & 2 \\ 1 & 3 \end{bmatrix} = \begin{bmatrix} -1 & -4 \\ 1 & 3 \end{bmatrix}$$

On the other hand, let K be the subset of G consisting of matrices with determinant 1. One can show that K is also a subgroup of G. Moreover, for any matrix X in G and any matrix A in K, we have

$$\det(X^{-1}AX) = 1$$

Hence $X^{-1}AX$ belongs to K, so K is a normal subgroup of G.

Integers Modulo m

Consider the group $\mathbf{Z}$ of integers under addition. Let H denote the multiples of 5, that is,

$$H = \{\ldots, -10, -5, 0, 5, 10, \ldots\}$$

Then H is a subgroup (necessarily normal) of $\mathbf{Z}$. The cosets of H in $\mathbf{Z}$ appear in Fig. B-5(a). By the above Theorem B.8, $\mathbf{Z}/H = \{0, 1, 2, 3, 4\}$ is a group under coset addition; its addition table appears in Fig. B-5(b).

This quotient group $\mathbf{Z}/H$ is referred to as the integers modulo 5 and it is frequently denoted by $\mathbf{Z}_5$. Analogously, for any positive integer n, there exists the quotient group $\mathbf{Z}_n$ called the *integers modulo n*.

$\bar{0} = 0 + H = H = \{\ldots, -10, -5, 0, 5, 10, \ldots\}$

$\bar{1} = 1 + H = \{\ldots, -9, -4, 1, 6, 11, \ldots\}$

$\bar{2} = 2 + H = \{\ldots, -8, -3, 2, 7, 12, \ldots\}$

$\bar{3} = 3 + H = \{\ldots, -7, -2, 3, 8, 13, \ldots\}$

$\bar{4} = 4 + H = \{\ldots, -6, -1, 4, 9, 14, \ldots\}$

$+$	$\bar{0}$	$\bar{1}$	$\bar{2}$	$\bar{3}$	$\bar{4}$
$\bar{0}$	$\bar{0}$	$\bar{1}$	$\bar{2}$	$\bar{3}$	$\bar{4}$
$\bar{1}$	$\bar{1}$	$\bar{2}$	$\bar{3}$	$\bar{4}$	$\bar{0}$
$\bar{2}$	$\bar{2}$	$\bar{3}$	$\bar{4}$	$\bar{0}$	$\bar{1}$
$\bar{3}$	$\bar{3}$	$\bar{4}$	$\bar{0}$	$\bar{1}$	$\bar{2}$
$\bar{4}$	$\bar{4}$	$\bar{0}$	$\bar{1}$	$\bar{2}$	$\bar{3}$

(a)   (b)

Fig. B-5

Cyclic Subgroups

Let G be any group and let a be any element of G. As usual, we define $a^0 = e$ and $a^{n+1} = a^n \cdot a$. Clearly, $a^m a^n = a^{m+n}$ and $(a^m)^n = a^{mn}$, for any integers m and n. Let S denote the set of all the powers of a; that is

$$S = \{\cdots, a^{-3}, a^{-2}, a^{-1}, e, a, a^2, a^3, \cdots\}$$

Then S is a subgroup of G called the cyclic group generated by a. We denote this group by $gp(a)$.

Furthermore, suppose that the powers of a are not distinct, say $a^r = a^s$ with, say, $r > s$. Then $a^{r-s} = e$ where $r, s > 0$. The smallest positive integer m such that $a^m = e$ is called the *order* of a and it will be denoted by $|a|$. If $|a| = m$, then the cyclic subgroup $gp(a)$ has m elements as follows:

$$gp(a) = \{e, a, a^2, a^3, \ldots, a^{m-1}\}$$

Consider, for example, the element ϕ_1 in the symmetric group S_3 discussed above. Then:

$$\phi_1{}^1 = \phi_1, \quad \phi_1{}^2 = \phi_2, \quad \phi_1{}^3 = \phi_2 \cdot \phi_1 = e$$

Hence $|\phi_1| = 3$ and $gp(\phi_1) = \{e, \phi_1, \phi_2\}$. Observe that $|\phi_1|$ divides the order of S_3. This is true in general; that is, for any element a in a group G, $|a|$ equals the order of $gp(a)$ and hence $|a|$ divides $|G|$ by Lagrange's Theorem B.7. We also remark that a group G is said to be *cyclic* if it has an element a such that $G = gp(a)$.

Generating Sets, Generators

Consider any subset A of a group G. Let $gp(A)$ denote the set of all elements x in G such that x is equal to a product of elements where each element comes from the set $A \cup A^{-1}$ (where A^{-1} denotes the set of inverses of elements of A). That is,

$$gp(A) = \{x \in G \mid x = b_1 b_2 \ldots b_m \text{ where each } b_i \in A \cup A^{-1}\}$$

Then $gp(A)$ is a subgroup of G with *generating set* A. In particular, A is said to generate the group G if $G = gp(A)$, that is, if every g in G is a product of elements from $A \cup A^{-1}$. We say A is a *minimal set of generators* of G if A generates G and if no set with fewer elements than A generates G. For example, the permutations $a = \sigma_1$ and $b = \phi_1$ form a minimal set of generators of the symmetric group S_3 (Fig. B-4). Specifically,

$$e = a^2, \quad \sigma_1 = a, \quad \sigma_2 = ab, \quad \sigma_3 = ab^2, \quad \phi_1 = b, \quad \phi_2 = b^2$$

and S_3 is not cyclic so it cannot be generated by one element.

Homomorphisms

A mapping f from a group G into a group G' is called a homomorphism if, for every $a, b \in G$,

$$f(ab) = f(a)f(b)$$

In addition, if f is one-to-one and onto, then f is called an *isomorphism*; and G and G' are said to be *isomorphic*, written $G \cong G'$.

If $f : G \to G'$ is a homomorphism, then the kernel of f, written Ker f, is the set of elements whose image is the identity element e' of G'; that is,

$$\text{Ker } f = \{a \in G \mid f(a) = e'\}$$

Recall that the image of f, written $f(G)$ or Im f, consists of the images of the elements under f; that is,

$$\text{Im } f = \{b \in G' \mid \text{there exists } a \in G \text{ for which } f(a) = b\}.$$

The following theorem (proved in Problem B.19) is fundamental to group theory.

Theorem B.9:   Suppose $f\colon G \to G'$ is a homomorphism with kernel K. Then K is a normal subgroup of G, and the quotient group G/K is isomorphic to $f(G)$.

EXAMPLE B.12

(a) Let G be the group of real numbers under addition, and let G' be the group of positive real numbers under multiplication. The mapping $f\colon G \to G'$ defined by $f(a) = 2^a$ is a homomorphism because

$$f(a + b) = 2^{a+b} = 2^a 2^b = f(a)f(b)$$

In fact, f is also one-to-one and onto; hence G and G' are isomorphic.

(b) Let a be any element in a group G. The function $f\colon \mathbf{Z} \to G$ defined by $f(n) = a^n$ is a homomorphism since

$$f(m + n) = a^{m+n} = a^m \cdot a^n = f(m) \cdot f(n)$$

The image of f is $gp(a)$, the cyclic subgroup generated by a. By Theorem B.9,

$$gp(a) \cong \mathbf{Z}/\mathbf{K}$$

where K is the kernel of f. If $K = \{0\}$, then $gp(a) = \mathbf{Z}$. On the other hand, if m is the order of a, then $K = \{$multiples of $m\}$, and so $gp(a) \cong \mathbf{Z}_m$. In other words, any cyclic group is isomorphic to either the integers $\mathbf{Z}$ under addition, or to $\mathbf{Z}_m$, the integers under addition modulo m.

B.6   RINGS, INTEGRAL DOMAINS, AND FIELDS

Let R be a nonempty set with two binary operations, an operation of addition (denoted by $+$) and an operation of multiplication (denoted by juxtaposition). Then R is called a *ring* if the following axioms are satisfied:

[$\mathbf{R_1}$]   For any $a, b, c \in R$, we have $(a + b) + c = a + (b + c)$.

[$\mathbf{R_2}$]   There exists an element $0 \in R$, called the *zero* element, such that, for every $a \in R$,

$$a + 0 = 0 + a = a.$$

[$\mathbf{R_3}$]   For each $a \in R$ there exists an element $-a \in R$, called the *negative* of a, such that

$$a + (-a) = (-a) + a = 0.$$

[$\mathbf{R_4}$]   For any $a, b \in R$, we have $a + b = b + a$.

[$\mathbf{R_5}$]   For any $a, b, c \in R$, we have $(ab)c = a(bc)$.

[$\mathbf{R_6}$]   For any $a, b, c \in R$, we have: (i) $a(b + c) = ab + ac$, and (ii) $(b + c)a = ba + ca$.

Observe that the axioms [$\mathbf{R_1}$] through [$\mathbf{R_4}$] may be summarized by saying that R is an abelian group under addition.

Subtraction is defined in R by $a - b = a + (-b)$.

One can prove (Problem B.21) that $a \cdot 0 = 0 \cdot a = 0$ for every $a \in R$.

A subset S of R is a *subring* of R if S itself is a ring under the operations in R. We note that S is a subring of R if: (i) $0 \in S$, and (ii) for any $a, b \in S$, we have $a - b \in S$ and $ab \in S$.

Special Kinds of Rings: Integral Domains and Fields

This subsection defines a number of different kinds of rings, including integral domains and fields.

R is called a *commutative ring* if $ab = ba$ for every $a, b \in R$.

R is called a *ring with an identity element 1* if the element 1 has the property that $a \cdot 1 = 1 \cdot a = a$ for every element $a \in R$. In such a case, an element $a \in R$ is called a *unit* if a has a multiplicative inverse, that is, an element a^{-1} in R such that $a \cdot a^{-1} = a^{-1} \cdot a = 1$.

R is called a *ring with zero divisors* if there exist nonzero elements $a, b \in R$ such that $ab = 0$. In such a case, a and b are called *zero divisors*.

Definition B.3: A commutative ring R is an *integral domain* if R has no zero divisors, that is, if $ab = 0$ implies $a = 0$ or $b = 0$.

Definition B.4: A commutative ring R with an identity element 1 (not equal to 0) is a field if every nonzero $a \in R$ is a unit, that is, has a multiplicative inverse.

A field is necessarily an integral domain; for if $ab = 0$ and $a \neq 0$, then

$$b = 1 \cdot b = a^{-1}ab = a^{-1} \cdot 0 = 0$$

We remark that a field may also be viewed as a commutative ring in which the nonzero elements form a group under multiplication.

EXAMPLE B.13

(a) The set $\mathbf{Z}$ of integers with the usual operations of addition and multiplication is the classical example of an integral domain (with an identity element). The units in $\mathbf{Z}$ are only 1 and -1, that is, no other element in $\mathbf{Z}$ has a multiplicative inverse.

(b) The set $\mathbf{Z}_m = \{0, 1, 2, \ldots, m - 1\}$ under the operation of addition and multiplication modulo m is a ring; it is called the *ring of integers modulo m*. If m is a prime, then $\mathbf{Z}_m$ is a field. On the other hand, if m is not a prime then $\mathbf{Z}_m$ has zero divisors. For instance, in the ring $\mathbf{Z}_6$,

$$2 \cdot 3 = 0 \quad \text{but} \quad 2 \not\equiv 0 \ (\text{mod } 6) \quad \text{and} \quad 3 \not\equiv 0 \ (\text{mod } 6)$$

(c) The rational numbers $\mathbf{Q}$ and the real numbers $\mathbf{R}$ each form a field with respect to the usual operations of addition and multiplication.

(d) Let M denote the set of 2×2 matrices with integer or real entries. Then M is a noncommutative ring with zero divisors under the operations of matrix addition and matrix multiplication. M does have an identity element, the identity matrix.

(e) Let R be any ring. Then the set $R[x]$ of all polynomials over R is a ring with respect to the usual operations of addition and multiplication of polynomials. Moreover, if R is an integral domain then $R[x]$ is also an integral domain.

Ideals

A subset J of a ring R is called an *ideal* in R if the following three properties hold:

    (i) $0 \in J$.

    (ii) For any $a, b \in J$, we have $a - b \in J$.

    (iii) For any $r \in R$ and $a \in J$, we have $ra, ar \in J$.

Note first that J is a subring of R. Also, J is a subgroup (necessarily normal) of the additive group of R. Thus we can form the following collection of cosets which form a partition of R:

$$\{a + J \mid a \in R\}$$

The importance of ideals comes from the following theorem which is analogous to Theorem B.7 for normal subgroups.

Theorem B.10: Let J be an ideal in a ring R. Then the cosets $\{a + J \mid a \in R\}$ form a ring under the coset operations

$$(a + J) + (b + J) = a + b + J \quad \text{and} \quad (a + J)(b + J) = ab + J$$

This ring is denoted by R/J and is called the *quotient ring*.

Now let R be a commutative ring with an identity element 1. For any $a \in R$, the following set is an ideal:

$$(a) = \{ra \mid r \in R\} = aR$$

It is called the *principal ideal generated* by a. If every ideal in R is a principal ideal, then R is called a *principal ideal ring*. In particular, if R is also an integral domain, then R is called a *principal ideal domain* (PID).

EXAMPLE B.14

(a) Consider the ring $\mathbf{Z}$ of integers. Then every ideal J in $\mathbf{Z}$ is a principal ideal, that is, $J = (m) = m\mathbf{Z}$, for some integer m. Thus $\mathbf{Z}$ is a principal ideal domain (PID). The quotient ring $\mathbf{Z_m} = \mathbf{Z}/(m)$ is simply the ring of integers modulo m. Although $\mathbf{Z}$ is an integral domain (no zero divisors), the quotient ring $\mathbf{Z_m}$ may have zero divisors, e.g., 2 and 3 are zero divisors in $\mathbf{Z_6}$.

(b) Let R be any ring. Then $\{0\}$ and R are ideals. In particular, if R is a field, then $\{0\}$ and R are the only ideals.

(c) Let K be a field. Then the ring $K[x]$ of polynomials over K is a PID (principal ideal domain). On the other hand, the ring $K[x, y]$ of polynomials in two variables is not a PID.

Ring Homomorphisms

A mapping f from a ring R into a ring R' is called a *ring homomorphism* or, simply, *homomorphism* if, for every $a, b \in R$,

$$f(a + b) = f(a) + f(b), \quad f(ab) = f(a)f(b)$$

In addition, if f is one-to-one and onto, then f is called an *isomorphism*; and R and R' are said to be *isomorphic*, written $R \cong R'$.

Suppose $f : R \to R'$ is a homomorphism. Then the kernel of f, written Ker f, is the set of elements whose image is the zero element 0 of R'; that is,

$$\text{Ker } f = \{r \in R \mid f(r) = 0\}$$

The following theorem (analogous to Theorem B.9 for groups) is fundamental to ring theory.

Theorem B.11: Let $f : R \to R'$ be a ring homomorphism with kernel K. Then K is an ideal in R, and the quotient ring R/K is isomorphic to $f(R)$.

Divisibility in Integral Domains

Now let D be an integral domain. We say that b divides a in D if $a = bc$ for some $c \in D$. An element $u \in D$ is called a *unit* if u divides 1, i.e., if u has a multiplicative inverse. An element $b \in D$ is called an *associate* of $a \in D$ if $b = ua$ for some unit $u \in D$. A nonunit $p \in D$ is said to be *irreducible* if $p = ab$ implies a or b is a unit.

An integral domain D is called a *unique factorization domain* (UFD), if every nonunit $a \in D$ can be written uniquely (up to associates and order) as a product of irreducible elements.

EXAMPLE B.15

(a) The ring $\mathbf{Z}$ of integers is the classical example of a unique factorization domain. The units of $\mathbf{Z}$ are 1 and -1. The only associates of $n \in Z$ are n and $-n$. The irreducible elements of $\mathbf{Z}$ are the prime numbers.

(b) The set $D = \{a + b\sqrt{13} \mid a, b \text{ integers}\}$ is an integral domain. The units of D follow:

$$\pm 1, \quad 18 \pm 5\sqrt{13}, \quad -18 \pm 5\sqrt{13}$$

The elements 2, $3 - \sqrt{13}$ and $-3 - \sqrt{13}$ are irreducible in D. Observe that

$$4 = 2 \cdot 2 = (3 - \sqrt{13})(-3 - \sqrt{13})$$

Thus D is not a unique factorization domain. (See Problem B.97.)

B.7 POLYNOMIALS OVER A FIELD

This section investigates polynomials whose coefficients come from some integral domain or field K. In particular, we show that polynomials over a field K have many of the same properties as the integers.

Basic Definitions

Let K be an integral domain or a field. Formally, a polynomial f over K is an infinite sequence of elements from K in which all except a finite number of them are 0; that is,

$$f = (\ldots, 0, a_n, \ldots, a_1, a_0) \quad \text{or, equivalently,} \quad f(t) = a_n t^n + \cdots + a_1 t + a_0$$

where the symbol t is used as an indeterminate. The entry a_k is called the kth coefficient of f. If n is the largest integer for which $a \neq 0$, then we say that the degree of f is n, written $\deg(f) = n$. We also call a_n the leading coefficient of f. If $a_n = 1$, we call f a *monic* polynomial. On the other hand, if every coefficient of f is 0 then f is called the *zero* polynomial, written $f \equiv 0$. The degree of the zero polynomial is not defined.

Let $K[t]$ be the collection of all polynomials $f(t)$ over K. Consider the polynomials

$$f(t) = a_n t^n + \cdots + a_1 t + a_0 \quad \text{and} \quad g(t) = b_m t^m + \cdots + b_1 t + b_0$$

Then the sum $f + g$ is the polynomial obtained by adding corresponding coefficients; that is, if $m \leq n$, then

$$f(t) + g(t) = a_n t^n + \cdots + (a_m + b_m)t^m + \cdots + (a_1 + b_1)t + (a_0 + b_0)$$

Furthermore, the product of f and g is the polynomial

$$f(t)g(t) = (a_n b_m)t^{n+m} + \cdots + (a_1 b_0 + a_0 b_1)t + (a_0 b_0)$$

That is,

$$f(t)g(t) = c_{n+m}t^{n+m} + \cdots + c_1 t + c_0 \quad \text{where} \quad c_k = \sum_{i=0}^{k} a_i b_{k-i} = a_0 b_k + a_1 b_{k-1} + \cdots + a_k b_0$$

The set K of scalars is viewed as a subset of $K[t]$. Specifically, we identify the scalar $a_0 \in K$ with the polynomial

$$f(t) = a_0 \quad \text{or} \quad a_0 = (\cdots, 0, 0, a_0)$$

Then the operators of addition and scalar multiplication are preserved by this identification. Thus, the mapping $\psi: K \to K[t]$ defined by $\psi(a_0) = a_0$ is an isomorphism which embeds K into $K[t]$.

Theorem B.12: Let K be an integral domain. Then $K[t]$ under the operations of addition and multiplication of polynomials is a commutative ring with an identity element 1.

The following simple result has important consequences.

Lemma B.13: Suppose f and g are polynomials over an integral domain K. Then

$$\deg(fg) = \deg(f) + \deg(g).$$

The proof follows directly from the definition of the product of polynomials. Namely, suppose

$$f(t) = a_n t^n + \cdots + a_1 t + a_0 \quad \text{and} \quad g(t) = b_m t^m + \cdots + b_1 t + b_0$$

where $a_n \neq 0$ and $b_m \neq 0$. Thus $\deg(f) = n$ and $\deg(g) = m$. Then

$$f(t)g(t) = a_n b_m t^{n+m} + \text{ terms of lower degree}$$

Also, since K is an integral domain with no zero divisors, $a_n b_m \neq 0$. Thus

$$\deg(fg) = m + n = \deg(f) + \deg(g)$$

and the lemma is proved.

The following proposition lists many properties of our polynomials. (Recall that a polynomial g is said to *divide* a polynomial f if there exists a polynomial h such that $f(t) = g(t)h(t)$.)

Proposition B.14: Let K be an integral domain and let f and g be polynomials over K.

   (i) $K[t]$ is an integral domain.

   (ii) The units of $K[t]$ are the units in K.

   (iii) If g divides f, then $\deg(g) \leq \deg(f)$ or $f \equiv 0$.

   (iv) If g divides f and f divides g, then $f(t) = kg(t)$ where k is a unit in K.

   (v) If d and d' are monic polynomials such that d divides d' and d' divides d, then $d = d'$.

Euclidean Algorithm, Roots of Polynomials

This subsection discusses the roots of a polynomial $f(t)$, where we now assume the coefficients of $f(t)$ come from a field K. Recall that a scalar $a \in K$ is a *root* of a polynomial $f(t)$ if $f(a) = 0$. First we begin with an important theorem which is very similar to a corresponding theorem for the integers **Z**.

Theorem B.15 (Euclidean Division Algorithm): Let $f(t)$ and $g(t)$ be polynomials over a field K with $g(t) \neq 0$. Then there exist polynomials $q(t)$ and $r(t)$ such that

$$f(t) = q(t)g(t) + r(t)$$

where either $r(t) \equiv 0$ or $\deg(r) < \deg(g)$.

The above theorem (proved in Problem B.30) formalizes the process known as "long division." The polynomial $q(t)$ is called the *quotient* and the polynomial $r(t)$ is called the *remainder* when $f(t)$ is divided by $g(t)$.

Corollary B.16 (Remainder Theorem): Suppose $f(t)$ is divided by $g(t) = t - a$. Then $f(a)$ is the remainder.

The proof follows from the Euclidean Algorithm. That is, dividing $f(t)$ by $t - a$ we get

$$f(t) = q(t)(t - a) + r(t)$$

where $\deg(r) < \deg(t - a) = 1$. Hence $r(t) = r$ is a scalar. Substituting $t = a$ in the equation for $f(t)$ yields

$$f(a) = q(a)(a - a) + r = q(t) \cdot 0 + r = r$$

Thus $f(a)$ is the remainder, as claimed.

Corollary B.16 also tells us that $f(a) = 0$ if and only if the remainder $r = r(t) \equiv 0$. Accordingly:

Corollary B.17 (Factor Theorem): The scalar $a \in K$ is a root of $f(t)$ if and only if $t - a$ is a factor of $f(t)$.

The next theorem (proved in Problem B.31) tells us the number of possible roots of a polynomial.

Theorem B.18: Suppose $f(t)$ is a polynomial over a field K, and $\deg(f) = n$. Then $f(t)$ has at most n roots.

The following theorem (proved in Problem B.32) is the main tool for finding rational roots of a polynomial with integer coefficients.

Theorem B.19: Suppose a rational number p/q (reduced to lowest terms) is a root of the polynomial

$$f(t) = a_n t^n + \cdots + a_1 t + a_0$$

where all the coefficients $a_n, \ldots, a_1, a_0$ are integers. Then p divides the constant term a_0 and q divides the leading coefficient a_n. In particular, if $c = p/q$ is an integer, then c divides the constant term a_0.

EXAMPLE B.16

(a) Suppose $f(t) = t^3 + t^2 - 8t + 4$. Assuming $f(t)$ has a rational root, find all the roots of $f(t)$.

Since the leading coefficient is 1, the rational roots of $f(t)$ must be integers from among $\pm 1, \pm 2, \pm 4$. Note $f(1) \neq 0$ and $f(-1) \neq 0$. By synthetic division, or dividing by $t - 2$, we get

$$
\begin{array}{r|rrrr}
2 & 1 & +\,1 & -\,8 & +\,4 \\
  &   & 2 & +\,6 & -\,4 \\
\hline
  & 1 & +\,3 & -\,2 & +\,0
\end{array}
$$

Therefore $t = 2$ is a root and $f(t) = (t - 2)(t^2 + 3t - 2)$. Using the quadratic formula for $t^2 + 3t - 2 = 0$, we obtain the following three roots of $f(t)$:

$$t = 2, \quad t = (-3 + \sqrt{17})/2, \quad t = (-3 - \sqrt{17})/2$$

(b) Suppose $h(t) = t^4 - 2t^3 + 11t - 10$. Find all the real roots of $h(t)$ assuming there are two integer roots.

The integer roots must be among $\pm 1, \pm 2, \pm 5, \pm 10$. By synthetic division (or dividing by $t - 1$ and then $t + 2$) we get

$$
\begin{array}{r|rrrrr}
1 & 1 & -\,2 & +\,0 & +\,11 & -\,10 \\
  &   & 1 & -\,1 & -\,1 & +\,10 \\
\hline
-2 & 1 & -\,1 & -\,1 & +\,10 & +\,0 \\
  &   & -\,2 & +\,6 & -\,10 & \\
\hline
  & 1 & -\,3 & +\,5 & +\,0 &
\end{array}
$$

Thus $t = 1$ and $t = -2$ are roots and $h(t) = (t-1)(t+2)(t^2 - 3t + 5)$. The quadratic formula with $t^2 - 3t + 5$ tells us that there are no other real roots. That is, $t = 1$ and $t = -2$ are the only real roots of $h(t)$.

$K[t]$ as a PID and UFD

The following theorems (proved in Problems B.33 and B.34) apply.

Theorem B.20: The ring $K[t]$ of polynomials over a field K is a principal ideal domain (PID). That is, if J is an ideal in $K[t]$, then there exists a unique monic polynomial d which generates J, that is, every polynomial f in J is a multiple of d.

Theorem B.21: Let f and g be polynomials in $K[t]$, not both zero. Then there exists a unique monic polynomial d such that:

      (i) d divides both f and g.   (ii) If d' divides f and g, then d' divides d.

The polynomial d in the above Theorem B.21 is called the *greatest common divisor* of f and g, written $d = gcd(f, g)$. If $d = 1$, then f and g are said to be *relatively prime*.

Corollary B.22: Let d be the greatest common divisor of f and g. Then there exist polynomials m and n such that $d = mf + ng$. In particular, if f and g are relatively prime, then there exist polynomials m and n such that $mf + ng = 1$.

A polynomial $p \in K[t]$ is said to be *irreducible* if p is not a scalar and if $p = fg$ implies f or g is a scalar. In other words, p is irreducible if its only divisors are its associates (scalar multiples). The following lemma (proved in Problem B.36) applies.

Lemma B.23: Suppose $p \in K[t]$ is irreducible. If p divides the product fg of polynomials f and g in $K[t]$, then p divides f or p divides g. More generally, if p divides the product $f_1 f_2 \cdots f_n$ of n polynomials, then p divides one of them.

The next theorem (proved in Problem B.37) states that the polynomials over a field form a *unique factorization domain* (UFD).

Theorem B.24 (Unique Factorization Theorem): Let f be a nonzero polynomial in $K[t]$. Then f can be written uniquely (except for order) as a product

$$f = kp_1 p_2 \ldots p_n$$

where $k \in K$ and the p's are monic irreducible polynomials in $K[t]$.

Fundamental Theorem of Algebra

The proof of the following theorem lies beyond the scope of this text.

Fundamental Theorem of Algebra: Any nonzero polynomial $f(t)$ over the complex field $\mathbf{C}$ has a root in $\mathbf{C}$.

Thus $f(t)$ can be written uniquely (except for order) as a product

$$f(t) = k(t - r_1)(t - r_2) \cdots (t - r_n)$$

where k and the r_i are complex numbers and $\deg(f) = n$.

The above theorem is certainly not true for the real field $\mathbf{R}$. For example, $f(t) = t^2 + 1$ is a polynomial over R, but $f(t)$ has no real root.

The following theorem (proved in Problem B.38) does apply.

Theorem B.25: Suppose $f(t)$ is a polynomial over the real field $\mathbf{R}$, and suppose the complex number $z = a + bi$, $b \neq 0$, is a root of $f(t)$. Then the complex conjugate $\bar{z} = a - bi$ is also a root of $f(t)$. Hence the following is a factor of $f(t)$:

$$c(t) = (t - z)(t - \bar{z}) = t^2 - 2at + a^2 + b^2$$

The following theorem follows from Theorem B.25 and the Fundamental Theorem of Algebra.

Theorem B.26: Let $f(t)$ be a nonzero polynomial over the real field $\mathbf{R}$. Then $f(t)$ can be written uniquely (except for order) as a product

$$f(t) = kp_1(t)p_2(t) \cdots p_n(t)$$

where $k \in \mathbf{R}$ and the $p_i(t)$ are real monic polynomials of degree 1 or 2.

EXAMPLE B.17 Let $f(t) = t^4 - 3t^3 + 6t^2 + 25t - 39$. Find all the roots of $f(t)$ given that $t = 2 + 3i$ is a root.

Since $2 + 3i$ is a root, then $2 - 3i$ is a root and $c(t) = t^2 - 4t + 13$ is a factor of $f(t)$. Dividing $f(t)$ by $c(t)$ we get

$$f(t) = (t^2 - 4t + 13)(t^2 + t - 3)$$

The quadratic formula with $t^2 + t - 3$ gives us the other roots of $f(t)$. That is, the four roots of $f(t)$ are as follows:

$$t = 2 + 3i, \quad t = 2 - 3i, \quad t = (-1 + \sqrt{13})/2, \quad t = (-1 - \sqrt{13})/2$$

Solved Problems

OPERATIONS AND SEMIGROUPS

B.1. Consider the set **Q** of rational numbers, and let $*$ be the operation on **Q** defined by

$$a * b = a + b - ab$$

(a) Find: (i) $3 * 4$; (ii) $2 * (-5)$; (iii) $7 * (1/2)$.

(b) Is $(\mathbf{Q}, *)$ a semigroup? Is it commutative?

(c) Find the identity element for $*$.

(d) Do any of the elements in **Q** have an inverse? What is it?

(a)   (i) $3 * 4 = 3 + 4 - 3(4) = 3 + 4 - 12 = -5$

    (ii) $2 * (-5) = 2 + (-5) + 2(-5) = 2 - 5 + 10 = 7$

    (iii) $7 * (1/2) = 7 + (1/2) - 7(1/2) = 4$

(b) We have:

$$\begin{aligned}
(a * b) * c &= (a + b - ab) * c = (a + b - ab) + c - (a + b - ab)c \\
&= a + b - ab + c - ac - bc + abc = a + b + c - ab - ac - bc + abc \\
a * (b * c) &= a * (b + c - bc) = a + (b + c - bc) - a(b + c - bc) \\
&= a + b + c - bc - ab - ac + abc
\end{aligned}$$

Hence $*$ is associative and $(\mathbf{Q}, *)$ is a semigroup. Also

$$a * b = a + b - ab = b + a - ba = b * a$$

Hence $(\mathbf{Q}, *)$ is a commutative semigroup.

(c) An element e is an identity element if $a * e = a$ for every $a \in \mathbf{Q}$. Compute as follows:

$$a * e = a, \quad a + e - ae = a, \quad e - ea = 0, \quad e(1 - a) = 0, \quad e = 0$$

Accordingly, 0 is the identity element.

(d) In order for a to have an inverse x, we must have $a * x = 0$ since 0 is the identity element by Part (c). Compute as follows:

$$a * x = 0, \quad a + x - ax = 0, \quad a = ax - x, \quad a = x(a - l), \quad x = a/(a - l)$$

Thus if $a \neq 1$, then a has an inverse and it is $a/(a - 1)$.

B.2. Let S be a semigroup with identity e, and let b and b' be inverses of a. Show that $b = b'$, that is, that inverses are unique if they exist.

We have:

$$b * (a * b') = b * e = b \quad \text{and} \quad (b * a) * b' = e * b' = b'$$

Since S is associative, $(b * a) * b' = b * (a * b')$; hence $b = b'$.

B.3. Let $S = \mathbf{N} \times \mathbf{N}$. Let $*$ be the operation on S defined by $(a, b) * (a', b') = (aa', bb')$.

   (*a*) Show that $*$ is associative. (Hence S is a semigroup.)

   (*b*) Define $f: (S, *) \to (\mathbf{Q}, \times)$ by $f(a, b) = a/b$. Show that f is a homomorphism.

   (*c*) Find the congruence relation $\sim$ in S determined by the homomorphism f, that is, where $x \sim y$ if $f(x) = f(y)$. (See Theorem B.4.)

   (*d*) Describe $S/\sim$. Does $S/\sim$ have an identity element? Does it have inverses?

   Suppose $x = (a, b)$, $y = (c, d)$, $z = (e, f)$.

   (*a*) We have

$$(xy)z = (ac, bd) * (e, f) = [(ac)e, (bd)f]$$
$$x(yz) = (a, b) * (ce, df) = [a(ce), b(df)]$$

   Since a, b, c, d, e, f, are positive integers, $(ac)e = a(ce)$ and $(bd)f = b(df)$. Thus $(xy)z = x(yz)$ and hence $*$ is associative. That is, $(S, *)$ is a semigroup.

   (*b*) f is a homomorphism since

$$f(x * y) = f(ac, bd) = (ac)/(bd) = (a/b)(c/d) = f(x)f(y)$$

   (*c*) Suppose $f(x) = f(y)$. Then $a/b = c/d$ and hence $ad = bc$. Thus f determines the congruence relation $\sim$ on S defined by $(a, b) \sim (c, d)$ if $ad = bc$.

   (*d*) The image of f is $\mathbf{Q}^+$, the set of positive rational numbers. By Theorem B.3, $S/\sim$ is isomorphic to $\mathbf{Q}^+$. Thus $S/\sim$ does have an identity element, and every element has an inverse.

B.4. Prove Theorem B.1. Suppose $*$ is an associative operation on a set S. Then any product $a_1 * a_2 * \ldots * a_n$ requires no parenthesis, that is, all possible products are equal.

   The proof is by induction on n. Since n is associative, the theorem holds for $n = 1, 2$, and 3. Suppose $n \geq 4$. We use the notation:

$$(a_1 a_2, \cdots a_n) = (\cdots ((a_1 a_2) a_3) \cdots) a_n \quad \text{and} \quad [a_1 a_2 \cdots a_n] = \text{any product}$$

We show $[a_1 a_2 \cdots a_n] = (a_1 a_2 \cdots a_n)$ and so all such products will be equal. Since $[a_1 a_2 \cdots a_n]$ denotes some product, there exists an $r < n$ such that $[a_1 a_2 \cdots a_n] = [a_1 a_2 \cdots a_r][a_{r+1} \cdots a_n]$. Therefore, by induction,

$$[a_1 a_2 \cdots a_n] = [a_1 a_2 \cdots a_r][a_{r+1} \cdots a_n] = [a_1 a_2 \cdots a_r](a_{r+1} \cdots a_n)$$
$$= [a_1 \cdots a_r]((a_{r+1} \cdots a_{n-1})a_n) = ([a_1 \cdots a_r](a_{r-1} \cdots a_{n-1}))a_n$$
$$= [a_1 \cdots a_{n-1}]a_n = (a_1 \cdots a_{n-1})a_n = (a_1 a_2 \cdots a_n)$$

Thus the theorem is proved.

B.5. Prove Theorem B.4: Let $f : S \to S'$ be a semigroup homomorphism. Let $a \sim b$ if $f(a) = f(b)$. Then: (i) $\sim$ is a congruence relation; (ii) $S/\sim$ is isomorphic to $f(S)$.

   (i) First we show that $\sim$ is an equivalence relation. Since $f(a) = f(a)$, we have $a \sim a$.

      If $a \sim b$, then $f(a) = f(b)$ or $f(b) = f(a)$; hence $b \sim a$. Lastly, if $a \sim b$ and $b \sim c$, then $f(a) = f(b)$ and $f(b) = f(c)$; hence $f(a) = f(c)$. Thus $a \sim c$. That is, $\sim$ is an equivalence relation. Suppose now $a \sim a'$ and $b \sim b'$. Then $f(a) = f(a')$ and $f(b) = f(b')$.

      Since f is a homomorphism,

$$f(ab) = f(a)f(b) = f(a')f(b') = f(a'b')$$

   Therefore $ab \sim a'b'$. That is, $\sim$ is a congruence relation.

   (ii) Define $\Psi: S/\sim \to f(S)$ by $\Psi([a]) = f(a)$. We need to prove: (1) Ψ is well-defined, that is, $\Psi([a]) \in f(S)$, and if $[a] = [b]$ then $f([a]) = f([b])$. (2) Ψ is an isomorphism, that is, Ψ is a homomorphism, one-to-one and onto.

(1) *Proof that* Ψ *is well-defined*: We have $\Psi([a]) = f(a)$. Since $a \in S$, we have $f(a) \in f(S)$. Hence $\Psi[a]) \in f(S)$, as required. Now suppose $[a] = [b]$. Then $a \sim b$ and hence $f(a) = f(b)$. Thus

$$\Psi([a]) = f(a) = f(b) = \Psi([b])$$

That is, Ψ is well-defined.

(2) *Proof that* Ψ *is an isomorphism*: Since f is a homomorphism,

$$\Psi([a][b]) = \Psi[ab] = f(ab) = f(a)f(b) = \Psi([a])\Psi([b])$$

Hence Ψ is a homomorphism. Suppose $\Psi([a]) = \Psi([b])$. Then $f(a) = f(b)$, and so $a \sim b$. Thus $[a] = [b]$ and Ψ is one-to-one. Lastly, let $y \in f(S)$. Then, $f(a) = y$ for some $a \in S$. Hence $\Psi([a]) = f(a) = y$. Thus Ψ is onto $f(S)$. Accordingly, Ψ is an isomorphism.

GROUPS

B.6. Consider the group $G = \{1, 2, 3, 4, 5, 6\}$ under multiplication modulo 7.

(a) Find the multiplication table of G.   (b) Find $2^{-1}, 3^{-1}, 6^{-1}$.

(c) Find the orders and subgroups generated by 2 and 3.   (d) Is G cyclic?

(a) To find $a * b$ in G, find the remainder when the product ab is divided by 7.

For example, $5 \cdot 6 = 30$ which yields a remainder of 2 when divided by 7; hence $5 * 6 = 2$ in G. The multiplication table of G appears in Fig. B-6(a).

(b) Note first that 1 is the identity element of G. Recall that a^{-1} is that element of G such that $aa^{-1} = 1$. Hence $2^{-1} = 4$, $3^{-1} = 5$ and $6^{-1} = 6$.

(c) We have $2^1 = 2$, $2^2 = 4$, but $2^3 = 1$. Hence $|2| = 3$ and $gp(2) = \{1, 2, 4\}$. We have $3^1 = 3$, $3^2 = 2$, $3^3 = 6$, $3^4 = 4$, $3^5 = 5$, $3^6 = 1$. Hence $|3| = 6$ and $gp(3) = G$.

(d) G is cyclic since $G = gp(3)$.

*	1	2	3	4	5	6
1	1	2	3	4	5	6
2	2	4	6	1	3	5
3	3	6	2	5	1	4
4	4	1	5	2	6	3
5	5	3	1	6	4	2
6	6	5	4	3	2	1

*	1	2	4	7	8	11	13	14
1	1	2	4	7	8	11	13	14
2	2	4	8	14	1	7	11	13
4	4	8	1	13	2	14	7	11
7	7	14	13	4	11	2	1	8
8	8	1	2	11	4	13	14	7
11	11	7	14	2	13	1	8	4
13	13	11	7	1	14	8	4	2
14	14	13	11	8	7	4	2	1

(a)   (b)

Fig. B-6

B.7. Let G be a reduced residue system modulo 15, say, $G = \{1, 2, 4, 7, 8, 11, 13, 14\}$ (the set of integers between 1 and 15 which are coprime to 15). Then G is a group under multiplication modulo 15.

(a) Find the multiplication table of G.   (b) Find $2^{-1}, 7^{-1}, 11^{-1}$.

(c) Find the orders and subgroups generated by 2, 7, and 11.   (d) Is G cyclic?

(a) To find $a * b$ in G, find the remainder when the product ab is divided by 15. The multiplication table appears in Fig. B-6(b).

(b) The integers r and s are inverses if $r * s = 1$. Hence: $2^{-1} = 8$, $7^{-1} = 13$, $11^{-1} = 11$.

(c) We have $2^2 = 4$, $2^3 = 8$, $2^4 = 1$. Hence $|2| = 4$ and $gp(2) = \{1, 2, 4, 8\}$. Also,

$7^2 = 4$, $7^3 = 4 * 7 = 13$, $7^4 = 13 * 7 = 1$. Hence $|7| = 4$ and $gp(7) = \{1, 4, 7, 13\}$. Lastly, $11^2 = 1$. Hence $|11| = 2$ and $gp(11) = \{1, 11\}$.

(d) No, since no element generates G.

B.8. Consider the symmetric group S_3 whose multiplication table is given in Fig. B-4.

(a) Find the order and the group generated by each element of S_3.

(b) Find the number and all subgroups of S_3.

(c) Let $A = \{\sigma_1, \sigma_2\}$ and $B = \{\phi_1, \phi_2\}$. Find AB, $\sigma_3 A$, and $A\sigma_3$.

(d) Let $H = gp(\sigma_1)$ and $K = gp(\sigma_2)$. Show that HK is not a subgroup of S_3.

(e) Is S_3 cyclic?

(a) There are six elements: (1) ε, (2) σ_1, (3) σ_2, (4) σ_3, (5) ϕ_1, (6) ϕ_2. Find the powers of each element x until $x^n = \varepsilon$. Then $|x| = n$, and $gp(x) = \{\varepsilon, x^1, x^2, \ldots, x^{n-1}\}$. Note $x^1 = x$, so we need only begin with $n = 2$ when $x \neq \varepsilon$.

   (1) $\varepsilon^1 = \varepsilon$; so $|\varepsilon| = 1$ and $g(\varepsilon) = \{\varepsilon\}$.

   (2) $\sigma_1^2 = \varepsilon$; hence $|\sigma_1| = 2$ and $gp(\sigma_1) = \{\varepsilon, \sigma_1\}$.

   (3) $\sigma_2^2 = \varepsilon$; hence $|\sigma_2| = 2$ and $gp(\sigma_2) = \{\varepsilon, \sigma_2\}$.

   (4) $\sigma_3^2 = \varepsilon$; hence $|\sigma_3| = 2$ and $gp(\sigma_3) = \{\varepsilon, \sigma_3\}$.

   (5) $\phi_1^2 = \phi_2$, $\phi_1^3 = \phi_2\phi_1 = \varepsilon$; hence $|\phi_1| = 3$ and $gp(\phi_1) = \{\varepsilon, \phi_1, \phi_2\}$.

   (6) $\phi_2^2 = \phi_1$, $\phi_2^3 = \phi_1\phi_2 = \varepsilon$; hence $|\phi_2| = 3$ and $gp(\phi_1) = \{\varepsilon, \phi_2, \phi_1\}$.

(b) First of all, $H_1 = \{\varepsilon\}$ and $H_2 = S_3$ are subgroups of S_3. Any other subgroup of S_3 must have order 2 or 3 since its order must divide $|S_3| = 6$. Since 2 and 3 are prime numbers, these subgroups must be cyclic (Problem B.61) and hence must appear in part (a). Thus the other subgroups of S_3 follow:

$$H_3 = \{\varepsilon, \sigma_1\}, \quad H_4 = \{\varepsilon, \sigma_2\}, \quad H_5 = \{\varepsilon, \sigma_3\}, \quad H_6 = \{\varepsilon, \phi_1, \phi_2\}$$

Accordingly, S_3 has six subgroups.

(c) Multiply each element of A by each element of B:

$$\sigma_1\phi_1 = \sigma_2, \quad \sigma_1\phi_2 = \sigma_3, \quad \sigma_3\phi_1 = \sigma_3, \quad \sigma_2\phi_2 = \sigma_1$$

Hence $AB = \{\sigma_1, \sigma_2, \sigma_3\}$.

   Multiply σ_3 by each element of A:

$$\sigma_3\sigma_1 = \phi_1, \quad \sigma_3\sigma_2 = \phi_2, \quad \text{hence} \quad \sigma_3 A = \{\phi_1, \phi_2\}$$

   Multiply each element of A by σ_3:

$$\sigma_1\sigma_3 = \phi_2, \quad \sigma_2\sigma_3 = \phi_1, \quad \text{hence} \quad A\sigma_3 = \{\phi_1, \phi_2\}$$

(d) $H = \{e, \sigma_1\}$, $K = \{e, \sigma_2\}$ and then $HK = \{e, \sigma_1, \sigma_2, \phi_1\}$, which is not a subgroup of S_3 since HK has four elements.

(e) S_3 is not cyclic since S_3 is not generated by any of its elements.

B.9. Let σ and τ be the following elements of the symmetric group S_6:

$$\sigma = \begin{pmatrix} 1 & 2 & 3 & 4 & 5 & 6 \\ 3 & 1 & 5 & 4 & 6 & 2 \end{pmatrix} \quad \text{and} \quad \tau = \begin{pmatrix} 1 & 2 & 3 & 4 & 5 & 6 \\ 5 & 3 & 1 & 6 & 2 & 4 \end{pmatrix}$$

Find: $\tau\sigma, \sigma\tau, \sigma^2$, and σ^{-1}. (Since σ and τ are functions, $\tau\sigma$ means apply σ and then τ.)

Figure B-7 shows the effect on 1, 2, ..., 6 of the composition of the permutations:

(*a*) σ and then τ;   (*b*) τ and then σ;   (*c*) σ and then σ, i.e. σ^2.

Thus:

$$\tau\sigma = \begin{pmatrix} 1 & 2 & 3 & 4 & 5 & 6 \\ 1 & 5 & 2 & 6 & 4 & 3 \end{pmatrix}, \quad \sigma\tau = \begin{pmatrix} 1 & 2 & 3 & 4 & 5 & 6 \\ 6 & 5 & 3 & 2 & 1 & 4 \end{pmatrix}, \quad \sigma^2 = \begin{pmatrix} 1 & 2 & 3 & 4 & 5 & 6 \\ 5 & 3 & 6 & 4 & 2 & 1 \end{pmatrix}$$

We obtain σ^{-1} by interchanging the top and bottom rows of σ and then rearranging:

$$\sigma^{-1} = \begin{pmatrix} 3 & 1 & 5 & 4 & 6 & 2 \\ 1 & 2 & 3 & 4 & 5 & 6 \end{pmatrix} = \begin{pmatrix} 1 & 2 & 3 & 4 & 5 & 6 \\ 2 & 6 & 1 & 4 & 3 & 5 \end{pmatrix}$$

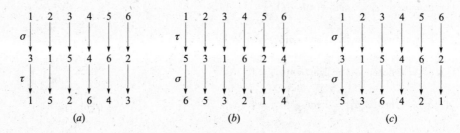

(*a*)   (*b*)   (*c*)

Fig. B-7

B.10. Let H and K be groups.

   (*a*) Define the direct product $G = H \times K$ of H and K.

   (*b*) What is the identity element and the order of $G = H \times K$?

   (*c*) Describe and find the multiplication table of the group $G = \mathbf{Z}_2 \times \mathbf{Z}_2$.

   (*a*) Let $G = H \times K$, the Cartesian product of H and K, with the operation $*$ defined componentwise by

$$(h, k) * (h', k') = (hh', kk')$$

   Then G is a group (Problem B.68), called the *direct product* of H and K.

   (*b*) The element $e = (e_H, e_K)$ is the identity element of G, and $|G| = |H| \cdot |K|$.

   (*c*) Since $\mathbf{Z}_2$ has two elements, G has four elements. Let

$$e = (0, 0), \quad a = (1, 0), \quad b = (0, 1), \quad c = (1, 1)$$

   The multiplication table of G appears in Fig. B-8(*a*). Note that G is abelian since the table is symmetric. Also, $a^2 = e, b^2 = e, c^2 = e$. Thus G is not cyclic, and hence $G \not\cong \mathbf{Z}_4$.

B.11. Let S be the square in the plane $\mathbf{R}^2$ pictured in Fig. B-8(*b*), with its center at the origin 0. Note that the vertices of S are numbered counterclockwise from 1 to 4.

   (*a*) Define the group G of symmetries of S.

   (*b*) List the elements of G.

   (*c*) Find a minimum set of generators of G.

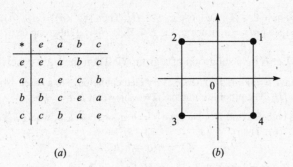

$$
\begin{array}{c|cccc}
* & e & a & b & c \\
\hline
e & e & a & b & c \\
a & a & e & c & b \\
b & b & c & e & a \\
c & c & b & a & e \\
\end{array}
$$

(a)   (b)

Fig. B-8

(a) A symmetry σ of S is a rigid one-to-one correspondence between S and itself. (Here rigid means that distances between points do not change.) The group G of symmetries of S is the set of all symmetries of S under composition of mappings.

(b) There are eight symmetries as follows. For $\alpha = 0°$, $90°$, $180°$, $270°$, let $\sigma(\alpha)$ be the symmetry obtained by rotating S about its center α degrees, and let $\tau(\alpha)$ be the symmetry obtained by reflecting S about the y-axis and then rotating S about its center α degrees. Note that any symmetry σ of S is completely determined by its effect on the vertices of S and hence σ can be represented as a permutation in S_4. Thus:

$$
\sigma(0°) = \begin{pmatrix} 1 & 2 & 3 & 4 \\ 1 & 2 & 3 & 4 \end{pmatrix}, \quad
\sigma(90°) = \begin{pmatrix} 1 & 2 & 3 & 4 \\ 2 & 3 & 4 & 1 \end{pmatrix},
$$

$$
\sigma(180°) = \begin{pmatrix} 1 & 2 & 3 & 4 \\ 3 & 4 & 1 & 4 \end{pmatrix}, \quad
\sigma(270°) = \begin{pmatrix} 1 & 2 & 3 & 4 \\ 4 & 1 & 2 & 3 \end{pmatrix}
$$

$$
\tau(0°) = \begin{pmatrix} 1 & 2 & 3 & 4 \\ 2 & 1 & 4 & 3 \end{pmatrix}, \quad
\tau(90°) = \begin{pmatrix} 1 & 2 & 3 & 4 \\ 3 & 2 & 1 & 4 \end{pmatrix},
$$

$$
\tau(180°) = \begin{pmatrix} 1 & 2 & 3 & 4 \\ 4 & 3 & 2 & 1 \end{pmatrix}, \quad
\tau(270°) = \begin{pmatrix} 1 & 2 & 3 & 4 \\ 1 & 4 & 3 & 2 \end{pmatrix}
$$

(c) Let $a = \sigma(90°)$ and $b = \tau(0°)$. Then a and b form a maximum set of generators of G. Specifically,

$$
\sigma(0°) = a^4, \quad \sigma(90°) = a, \quad \sigma(180°) = a^2, \quad \sigma(270°) = a^3
$$
$$
\tau(0°) = b, \quad \tau(90°) = ba, \quad \tau(180°) = ba^2, \quad \tau(270°) = ba^3
$$

and G is not cyclic so it is not generated by one element. (One can show that the relations $a^4 = e$, $b^2 = e$, and $bab = a^{-1}$ completely describe G.)

B.12. Let G be a group and let A be a nonempty set.

(a) Define the meaning of the statement "G acts on A."

(b) Define the stabilizer H_a of an element $a \in A$.

(c) Show that H_a is a subgroup of G.

(a) Let PERM(A) denote the group of all permutations of A. Let $\psi : G \to$ PERM(A) be any homomorphism. Then G is said to act on A where each element g in G defines a permutation $g : A \to A$ by

$$
g(a) = (\psi(g))(a)
$$

(Frequently, the permutation $g : A \to A$ is given directly and hence the homomorphism is implicitly defined.)

(b) The stabilizer H_a of $a \in A$ consists of all elements of G which "fix a," that is,

$$
H_a = \{ g \in G \mid g(a) = a \}
$$

(c) Since $e(a) = a$, we have $e \in H_a$. Suppose $g, g' \in H_a$. Then $(gg')(a) = g(g'(a)) = g(a) = a$; hence $gg' \in H_a$. Also, $g^{-1}(a) = a$ since $g(a) = a$; hence $g^{-1} \in H_a$. Thus H_a is a subgroup of G.

B.13. Prove Theorem B.6: Let H be a subgroup of a group G. Then the right cosets Ha form a partition of G.

Since $e \in H$, we have $a = ea \in Ha$; hence every element belongs to a coset. Now suppose Ha and Hb are not disjoint. Say $c \in Ha \cap Hb$. The proof is complete if we show that $Ha = Hb$.

Since c belongs to both Ha and Hb, we have $c = h_1 a$ and $c = h_2 b$, where $h_1, h_2 \in H$. Then $h_1 a = h_2 b$, and so $a = h_1^{-1} h_2 b$. Let $x \in Ha$. Then

$$x = h_3 a = h_3 h_1^{-1} h_2 b$$

where $h_3 \in H$. Since H is a subgroup, $h_3 h_1^{-1} h_2 \in H$; hence $x \in Hb$. Since x was any element of Ha, we have $Ha \subseteq Hb$. Similarly, $Hb \subseteq Ha$. Both inclusions imply $Ha = Hb$, and the theorem is proved.

B.14. Let H be a finite subgroup of G. Show that H and any coset Ha have the same number of elements.

Let $H = \{h_1, h_2, \ldots, h_k\}$, where H has k elements. Then $Ha = \{h_1 a, h_2 a, \ldots, h_k a\}$.

However, $h_i a = h_j a$ implies $h_i = h_j$; hence the k elements listed in Ha are distinct. Thus H and Ha have the same number of elements.

B.15. Prove Theorem B.7 (Lagrange): Let H be a subgroup of a finite group G. Then the order of H divides the order of G.

Suppose H has r elements and there are s right cosets; say

$$Ha_1, Ha_2, \ldots, Ha_s$$

By Theorem B.6, the cosets partition G and by Problem B.14, each coset has r elements. Therefore G has rs elements, and so the order of H divides the order of G.

B.16. Prove: Every subgroup of a cyclic group G is cyclic.

Since G is cyclic, there is an element $a \in G$ such that $G = gp(a)$. Let H be a subgroup of G. If $H = \{e\}$, then $H = gp(e)$ and H is cyclic. Otherwise, H contains a nonzero power of a. Since H is a subgroup, it must be closed under inverses and so H contains positive powers of a. Let m be the smallest positive power of a such that a^m belongs to H. We claim that $b = a^m$ generates H. Let x be any other element of H; since x belongs to G we have $x = a^n$ for some integer n. Dividing n by m we get a quotient q and a remainder r, that is,

$$n = mq + r$$

where $0 \le r < m$. Then

$$a^n = a^{mq+r} = a^{mq} \cdot a^r = b^q \cdot a^r \quad \text{so} \quad a^r = b^{-q} a^n$$

But $a^n, b \in H$. Since H is a subgroup, $b^{-q} a^n \in H$, which means $a^r \in H$. However, m is the smallest positive power of a belonging to H. Therefore, $r = 0$. Hence $x = a^n = b^q$. Thus b generates H, and H is cyclic.

B.17. Prove Theorem B.8: Let H be a normal subgroup of a group G. Then the cosets of H in G form a group under coset multiplication defined by $(aH)(bH) = abH$.

Coset multiplication is well-defined, since

$$(aH)(bH) = a(Hb)H = a(bH)H = ab(HH) = abH$$

(Here we have used the fact that H is normal, so $Hb = bH$, and, from Problem B.57, that $HH = H$.) Associativity of coset multiplication follows from the fact that associativity holds in G. H is the identity element of G/H, since

$$(aH)H = a(HH) = aH \quad \text{and} \quad H(aH) = (Ha)H = (aH)H = aH$$

Lastly, $a^{-1}H$ is the inverse of aH since

$$(a^{-1}H)(aH) = a^{-1}aHH = eH = H \quad \text{and} \quad (aH)(a^{-1}H) = aa^{-1}HH = eH = H$$

Thus G/H is a group under coset multiplication.

B.18. Suppose $F : G \to G'$ is a group homomorphism. Prove: (a) $f(e) = e'$; (b) $(fa^{-6}) = f(a)^{-1}$.

(a) Since $e = ee$ and f is a homomorphism, we have

$$f(e) = f(ee) = f(e)f(e)$$

Multiplying both sides by $f(e)^{-1}$ gives us our result.

(b) Using part (a) and that $aa^{-1} = a^{-1}a = e$, we have

$$e' = f(e) = f(aa^{-1}) = f(a)f(a^{-1}) \quad \text{and} \quad e' = f(e) = f(a^{-1}a) = f(a^{-1})f(a)$$

Hence $f(a^{-1})$ is the inverse of $f(a)$; that is, $f(a^{-1}) = (f(a))^{-1}$.

B.19. Prove Theorem B.9: Let $f : G \to G'$ be a homomorphism with kernel K. Then K is a normal subgroup of G, and G/K is isomorphic to the image of f. (Compare with Problem B.5, the analogous theorem for semigroups.)

Proof that K is normal: By Problem B.18, $f(e) = e'$, so $e \in K$. Now suppose $a, b \in K$ and $g \in G$. Then $f(a) = e'$ and $f(b) = e'$. Hence

$$f(ab) = f(a)f(b) = e'e' = e'$$
$$f(a^{-1}) = f(a)^{-1} = e'^{-1} = e'$$
$$f(gag^{-1}) = f(g)f(a)f(g^{-1}) = f(g)e'f(g)^{-1} = e'$$

Hence ab, a^{-1}, and gag^{-1} belong to K, so K is a normal subgroup.

Proof that $G/K \cong H$, where H is the image of f: Let $\varphi: G/K \to H$ be defined by

$$\varphi(Ka) = f(a)$$

We show that φ is well-defined, i.e., if $Ka = Kb$ then $\varphi(Ka) = \varphi(Kb)$. Suppose $Ka = Kb$. Then $ab^{-1} \in K$ (Problem B.57). Then $f(ab^{-1}) = e'$, and so

$$f(a)f(b)^{-1} = f(a)f(b^{-1}) = f(ab^{-1}) = e'$$

Hence $f(a) = f(b)$, and so $\varphi(Ka) = \varphi(Kb)$. Thus φ is well-defined.

We next show that φ is a homomorphism:

$$\varphi(KaKb) = \varphi(Kab) = f(ab) = f(a)f(b) = \varphi(Ka)\varphi(Kb)$$

Thus φ is a homomorphism. We next show that φ is one-to-one. Suppose $\varphi(Ka) = \varphi(Kb)$. Then

$$f(a) = f(b) \quad \text{or} \quad f(a)f(b)^{-1} = e' \quad \text{or} \quad f(a)f(b^{-1}) = e' \quad \text{or} \quad f(ab^{-1}) = e'$$

Thus $ab^{-1} \in K$, and by Problem B.57 we have $Ka = Kb$. Thus φ is one-to-one. We next show that φ is onto. Let $h \in H$. Since H is the image of f, there exists $a \in G$ such that $f(a) = h$. Thus $\varphi(Ka) = f(a) = h$, and so φ is onto. Consequently $G/K \cong H$ and the theorem is proved.

RINGS, INTEGRAL DOMAINS, FIELDS

B.20. Consider the ring $\mathbf{Z}_{10} = \{0, 1, 2, \ldots, 9\}$ of integers modulo 10. (a) Find the units of $\mathbf{Z}_{10}$. (b) Find -3, -8, and 3^{-1}. (c) Let $f(x) = 2x^2 + 4x + 4$. Find the roots of $f(x)$ over $\mathbf{Z}_{10}$.

(a) By Problem B.78 those integers relatively prime to the modulus $m = 10$ are the units in $\mathbf{Z}_{10}$. Hence the units are 1, 3, 7, and 9.

(b) Recall that $-a$ in a ring R is the element such that $a + (-a) = (-a) + a = 0$. Hence $-3 = 7$ since $3 + 7 = 7 + 3 = 0$ in $\mathbf{Z}_{10}$. Similarly $-8 = 2$. Recall that a^{-1} in a ring R is the element such that $a \cdot a^{-1} = a^{-1} \cdot a = 1$. Hence $3^{-1} = 7$ since $3 \cdot 7 = 7 \cdot 3 = 1$ in $\mathbf{Z}_{10}$.

(c) Substitute each of the ten elements of $\mathbf{Z}_{10}$ into $f(x)$ to see which elements yield 0. We have:

$$f(0) = 4, \quad f(2) = 0, \quad f(4) = 2, \quad f(6) = 0, \quad f(8) = 4$$
$$f(1) = 0, \quad f(3) = 4, \quad f(5) = 4, \quad f(7) = 0, \quad f(9) = 2$$

Thus the roots are 1, 2, 6, and 7. (This example shows that a polynomial of degree n can have more than n roots over an arbitrary ring. This cannot happen if the ring is a field.)

B.21. Prove that in a ring R: (i) $a \cdot 0 = 0 \cdot a = 0$; (ii) $a(-b) = (-a)b = -ab$; (iii) $(-1)a = -a$ (when R has an identity element 1).

(i) Since $0 = 0 + 0$, we have
$$a \cdot 0 = a(0 + 0) = a \cdot 0 + a \cdot 0$$
Adding $-(a \cdot 0)$ to both sides yields $0 = a \cdot 0$. Similarly $0 \cdot a = 0$.

(ii) Using $b + (-b) = (-b) + b = 0$, we have
$$ab + a(-b) = a(b + (-b)) = a \cdot 0 = 0$$
$$a(-b) + ab = a((-b) + b) = a \cdot 0 = 0$$
Hence $a(-b)$ is the negative of ab; that is, $a(-b) = -ab$. Similarly, $(-a)b = -ab$.

(iii) We have
$$a + (-1)a = 1 \cdot a + (-1)a = (1 + (-1))a = 0 \cdot a = 0$$
$$(-1)a + a = (-1)a + 1 \cdot a = ((-1) + 1)a = 0 \cdot a = 0$$
Hence $(-1)a$ is the negative of a; that is, $(-1)a = -a$.

B.22. Let D be an integral domain. Show that if $ab = ac$ with $a \neq 0$ then $b = c$.

Since $ab = ac$, we have
$$ab - ac = 0 \quad \text{and so} \quad a(b - c) = 0$$
Since $a \neq 0$, we must have $b - c = 0$, since D has no zero divisors. Hence $b = c$.

B.23. Suppose J and K are ideals in a ring R. Prove that $J \cap K$ is an ideal in R.

Since J and K are ideals, $0 \in J$ and $0 \in K$. Hence $0 \in J \cap K$. Now let $a, b \in J \cap K$ and let $r \in R$. Then $a, b \in J$ and $a, b \in K$. Since J and K are ideals,
$$a - b, ra, ar \in J \quad \text{and} \quad a - b, ra, ar \in K$$
Hence $a - b, ra, ar \in J \cap K$. Therefore $J \cap K$ is an ideal.

B.24. Let J be an ideal in a ring R with an identity element 1. Prove: (a) If $1 \in J$ then $J = R$; (b) If any unit $u \in J$ then $J = R$.

(a) If $1 \in J$ then for any $r \in R$ we have $r \cdot 1 \in R$ or $r \in J$. Hence $J = R$.

(b) If $u \in J$ then $u^{-1} \cdot u \in J$ or $1 \in J$. Hence $J = R$ by part (a).

B.25. Prove: (a) A finite integral domain D is a field. (b) $\mathbf{Z}_p$ is a field where p is a prime number. (c) (Fermat) If p is prime, then $a^p \equiv a \pmod{p}$ for any integer a.

(a) Suppose D has n elements, say $D = \{a_1, a_2, \ldots, a_n\}$. Let a be any nonzero element of D. Consider the n elements
$$aa_1, aa_2, \ldots, a_n$$
Since $a \neq 0$, we have $aa_i = aa_k$ implies $a_i = a_k$ (Problem B.22). Thus the above n elements are distinct, and so they must be a rearrangement of the elements of D. One of them, say aa_k, must equal the identity element 1 of D; that is, $aa_k = 1$. Thus a_k is the inverse of a. Since a was any nonzero element of D, we have that D is a field.

(b) Recall $\mathbf{Z}_p = \{0, 1, 2, \ldots, p - 1\}$. We show that $\mathbf{Z}_p$ has no zero divisors. zero divisors. Suppose $a * b = 0$ in $\mathbf{Z}_p$; that is, $0 \pmod{p}$. Then p divides ab. Since p is prime, p divides a or p divides b. Thus $a \equiv 0 \pmod{p}$ or $b \equiv 0 \pmod{p}$; that is, $a = 0$ or $b = 0$ in $\mathbf{Z}_p$. Accordingly, $\mathbf{Z}_p$ has no zero divisors and hence $\mathbf{Z}_p$ is an integral domain. By part (a), $\mathbf{Z}_p$ is a field.

(c) If p divides a, then $a \equiv 0 \pmod{p}$ and so $a^p \equiv a \equiv 0 \pmod{p}$. Suppose p does not divide a, then a may be viewed as a nonzero element of $\mathbf{Z}_p$ is a field, its nonzero elements form a group G under multiplication of order $p - 1$. By Problem B.45, $a^{p-1} = 1$ in $\mathbf{Z}_p$.

In other words, $a^{p-1} \equiv 1 \pmod{p}$. Multiplying by a gives $a^p \equiv a \pmod{p}$, and the theorem is proved.

POLYNOMIALS OVER A FIELD

B.26. Suppose $f(t) = 2t^3 - 3t^2 - 6t - 2$. Find all the roots of $f(t)$ knowing that $f(t)$ has a rational root.

The rational roots of $f(t)$ must be among $\pm 1, \pm 2, \pm 1/2$. Testing each possible root, we get, by synthetic division (or dividing by $2t + 1$),

$$
-\tfrac{1}{2} \begin{array}{|r}
2 - 3 - 6 - 2 \\
 - 1 + 2 + 2 \\
\hline
2 - 4 - 4 + 0
\end{array}
$$

Therefore $t = -1/2$ is a root and

$$f(t) = (t + 1/2)(2t^2 - 4t - 4) = (2t + 1)(t^2 - 2t - 2)$$

We can now use the quadratic formula on $t^2 - 2t - 2$ to obtain the following three roots of $f(t)$:

$$t = -1/2, \quad t = 1 + \sqrt{3}, \quad t = 1 - \sqrt{3}$$

B.27. Let $f(t) = t^4 - 3t^3 + 3t^2 + 3t - 20$. Find all the roots of $f(t)$ given that $t = 1 + 2i$ is a root.

Since $1 + 2i$ is a root, then $1 - 2i$ is a root and $c(t) = t^2 - 2t + 5$ is a factor of $f(t)$. Dividing $f(t)$ by $c(t)$ we get

$$f(t) = (t^2 - 2t + 5)(t^2 - t - 4)$$

The quadratic formula with $t^2 - t - 4$ gives us the other roots of $f(t)$. That is, the four roots of $f(t)$ follow:

$$t = 1 + 2i, \quad t = 1 - 2i, \quad t = (1 + \sqrt{17})/2, \quad t = (1 - \sqrt{17})/2$$

B.28. Let $K = \mathbf{Z}_8$. Find all roots of $f(t) = t^2 + 6t$.

Here $\mathbf{Z}_8 = \{0, 1, 2, \ldots, 7\}$. Substitute each element of $\mathbf{Z}_8$ into $f(t)$ to obtain:

$$f(0) = 0, \quad f(2) = 0, \quad f(4) = 0, \quad f(6) = 0$$

Then $f(t)$ has four roots, $t = 0, 2, 4, 6$. (Theorem B.21 does not hold here since K is not a field.)

B.29. Suppose $f(t)$ is a real polynomial with odd degree n. Show that $f(t)$ has a real root.

The complex (nonreal) roots come in pairs. Since $f(t)$ has an odd number n of roots (counting multiplicity), $f(t)$ must have at least one real root.

B.30. Prove Theorem B.15 (Euclidean Division Algorithm): Let $f(t)$ and $g(t)$ be polynomials over a field K with $g(t) \neq 0$. Then there exist polynomials $q(t)$ and $r(t)$ such that

$$f(t) = q(t)g(t) + r(t)$$

where either $r(t) \equiv 0$ or $\deg(r) < \deg(g)$.

If $f(t) = 0$ or if $\deg(f) < \deg(g)$, then we have the required representation $f(t) = 0g(t) + f(t)$. Now suppose $\deg(f) \geq \deg(g)$, say

$$f(t) = a_n t^n + \cdots + a_1 t + a_0 \quad \text{and} \quad g(t) = b_m t^m + \cdots + b_1 t + b_0$$

where $a_n, b_m \neq 0$ and $n > m$. We form the polynomial

$$f_1(t) = f(t) - \frac{a_n}{b_m} t^{n-m} g(t) \tag{1}$$

(This is the first subtraction step in "long division.") Then $\deg(f_1) < \deg(f)$. By induction, there exist polynomials $q_1(t)$ and $r(t)$ such that $f_1(t) = q_1(t)g(t) + r(t)$ where either $r(t) \equiv 0$ or $\deg(r) < \deg(g)$. Substituting this into (1) and solving for $f(t)$, we get

$$f(t) = \left[q_1(t) + \frac{a_n}{b_m} t^{n-m} \right] g(t) + r(t)$$

which is the desired representation.

B.31. Prove Theorem B.18: Suppose $f(t)$ is a polynomial over a field K, and $\deg(f) = n$. Then $f(t)$ has at most n roots.

The proof is by induction on n. If $n = 1$, then $f(t) = at + b$ and $f(t)$ has the unique root $t = -b/a$. Suppose $n > 1$. If $f(t)$ has no roots, then the theorem is true. Suppose $a \in K$ is a root of $f(t)$. Then

$$f(t) = (t - a)g(t) \tag{1}$$

where $\deg(g) = n - 1$. We claim that any other root of $f(t)$ must also be a root of g(t).

Suppose $b \neq a$ is another root of $f(t)$. Substituting $t = b$ in (1) yields $0 = f(b) = (b - a)g(b)$.

Since K has no zero divisors and $b - a \neq 0$, we must have $g(b) = 0$. By induction, $g(t)$ has at most $n - 1$ roots. Thus $f(t)$ has at most $n - 1$ roots other than a. Thus $f(t)$ has at most n roots.

B.32. Prove Theorem B.19: Suppose a rational number p/q (reduced to lowest terms) is a root of the polynomial

$$f(t) = a_n t^n + \cdots + a_1 t + a_0$$

where all the coefficients $a_n, \ldots, a_1, a_0$ are integers. Then p divides the constant term a_0 and q divides the leading coefficients a_n. In particular, if $c = p/q$ is an integer, then c divides the constant term a_0.

Substitute $t = p/q$ into $f(t) = 0$ to obtain $a_n(p/q)^n + \cdots + a_1(p/q) + a_0 = 0$. Multiply both sides of the equation by q^n to obtain

$$a_n p^n + a_{n-1} p^{n-1} q + a_{n-2} p^{n-2} q^2 + \cdots + a_1 p q^{n-1} + a_0 q^n = 0 \tag{1}$$

Since p divides all of the first n terms of (1), p must divide the last term $a_0 q^n$. Assuming p and q are relatively prime, p divides a_0. Similarly, q divides the last n terms of (1), hence q divides the first term $a_n p^n$. Since p and q are relatively prime, q divides a_n.

B.33. Prove Theorem B.20: The ring $K[t]$ of polynomials over a field K is a principal ideal domain (PID). If J is an ideal in $K[t]$, then there exists a unique monic polynomial d which generates J, that is, every polynomial f in J is a multiple of d.

Let d be a polynomial of lowest degree in J. Since we can multiply d by a nonzero scalar and still remain in J, we can assume without loss in generality that d is a monic polynomial (leading coefficient equal 1). Now suppose $f \in J$. By the division algorithm there exist polynomials q and r such that $f = qd + r$ where either $r \equiv 0$ or $\deg(r) < \deg(d)$. Now $f, d \in J$ implies $qd \in J$ and hence $r = f - qd \in J$. But d is a polynomial of lowest degree in J. Accordingly, $r \equiv 0$ and $f = qd$, that is, d divides f. It remains to show that d is unique. If d' is another monic polynomial which generates J, then d divides d' and d' divides d. This implies that $d = d'$, because d and d' are monic. Thus the theorem is proved.

B.34. Prove Theorem B.21: Let f and g be polynomials in $K[t]$, not both the zero polynomial. Then there exists a unique monic polynomial d such that: (i) d divides both f and g. (ii) If d' divides f and g, then d' divides d.

The set $I = \{mf + ng \mid m, n \in K[t]\}$ is an ideal. Let d be the monic polynomial which generates I. Note $f, g \in I$; hence d divides f and g. Now suppose d' divides f and g. Let J be the ideal generated by d'. Then $f, g \in J$ and hence $I \subseteq J$. Accordingly, $d \in J$ and so d' divides d as claimed. It remains to show that d is unique. If d_1 is another (monic) greatest common divisor of f and g, then d divides d_1 and d_1 divides d. This implies that $d = d_1$ because d and d_1 are monic. Thus the theorem is proved.

B.35. Prove Corollary B.22: Let d be the greatest common divisor of f and g. Then there exist polynomials m and n such that $d = mf + ng$. In particular, if f and g are relatively prime, then there exist polynomials m and n such that $mf + ng = 1$.

From the proof of Theorem B.21 in Problem B.34, the greatest common divisor d generates the ideal $I = \{mf + ng \mid m, n \in K[t]\}$. Thus there exist polynomials m and n such that $d = mf + ng$.

B.36. Prove Lemma B.23: Suppose $p \in K[t]$ is irreducible. If p divides the product fg of polynomials $f, g \in K[t]$, then p divides f or p divides g. More generally, if p divides the product $f_1 f_2 \cdots f_n$ of n polynomials, then p divides one of them.

Suppose p divides fg but not f. Since p is irreducible, the polynomials f and p must then be relatively prime. Thus there exist polynomials $m, n \in K[t]$ such that $mf + np = 1$. Multiplying this equation by g, we obtain $mfg + npg = g$. But p divides fg and so p divides mfg. Also, p divides npg. Therefore, p divides the sum $g = mfg + npg$.

Now suppose p divides $f_1 f_2 \cdots f_n$. If p divides f_1, then we are through. If not, then by the above result p divides the product $f_2 \cdots f_n$. By induction on n, p divides one of the polynomials in the product $f_2 \cdots f_n$. Thus the lemma is proved.

B.37. Prove Theorem B.24 (Unique Factorization Theorem): Let f be a nonzero polynomial in $K[t]$. Then f can be written uniquely (except for order) as a product $f = kp_1 p_2 \cdots p_n$ where $k \in K$ and the p's are monic irreducible polynomials in $K[t]$.

We prove the existence of such a product first. If f is irreducible or if $f \in K$, then such a product clearly exists. On the other hand, suppose $f = gh$ where g and h are nonscalars. Then g and h have degrees less than that of f. By induction, we can assume $g = k_1 g_1 g_2 \cdots g_r$ and $h = k_2 h_1 h_2 \cdots h_s$ where $k_1, k_2 \in K$ and the g_i and h_j are monic irreducible polynomials. Accordingly, our desired representation follows:

$$f = (k_1 k_2) g_1 g_2 \cdots g_r h_1 h_2 \cdots h_s$$

We next prove uniqueness (except for order) of such a product for f. Suppose

$$f = kp_1 p_2 \cdots p_n = k' q_1 q_2 \ldots q_m \quad \text{where} \quad k, k' \in K$$

and the $p_1, \ldots, p_n, q_1, \ldots, q_m$ are monic irreducible polynomials. Now p_1 divides $k' q_1 \ldots q_m$. Since p_1 is irreducible it must divide one of the q's by Lemma B.23. Say p_1 divides q_1. Since p_1 and q_1 are both irreducible and monic, $p_1 = q_1$. Accordingly, $kp_2 \ldots p_n = k' q_2 \ldots q_m$. By induction, we have that $n = m$ and $p_2 = q_2, \ldots, p_n = q_m$ for some rearrangement of the q's. We also have that $k = k'$. Thus the theorem is proved.

B.38. Prove Theorem B.25: Suppose $f(t)$ is a polynomial over the real field R, and suppose the complex number $z = a + bi, b \neq 0$, is a root of $f(t)$. Then the complex conjugate $\bar{z} = a - bi$ is also a root of $f(t)$. Hence the following is a factor of $f(t)$:

$$c(t) = (t - z)(t - \bar{z}) = t^2 - 2at + a^2 + b^2$$

Dividing $f(t)$ by $c(t)$ where $\deg(c) = 2$, there exist $q(t)$ and real numbers M and N such that

$$f(t) = c(t)q(t) + Mt + N \tag{1}$$

Since $z = a + bi$ is a root of $f(t)$ and $c(t)$, we have, by substituting $t = a + bi$ in (1),

$$f(z) = c(z)q(z) + M(z) + n \quad \text{or} \quad 0 = 0q(z) + M(z) + N \quad \text{or} \quad M(a + bi) + N = 0$$

Thus $Ma + N = 0$ and $Mb = 0$. Since $b \neq 0$, we must have $M = 0$. Then $0 + N = 0$ or $N = 0$. Accordingly, $f(t) = c(t)q(t)$ and $\bar{z} = a - bi$ is a root of $f(t)$.

Supplementary Problems

OPERATIONS AND SEMIGROUPS

B.39. Consider the set **N** of positive integers, and let $*$ denote least common multiple (lcm) operation on N.

(a) Find $4 * 6, 3 * 5, 9 * 18, 1 * 6$.

(b) Is $(\mathbf{N}, *)$ a semigroup? Is it commutative?

(c) Find the identity element of $*$.

(d) Which elements in N, if any, have inverses and what are they?

B.40. Let $*$ be the operation on the set **R** of real numbers defined by $a * b = a + b + 2ab$.

(a) Find $2 * 3, 3 * (-5)$, and $7 * (1/2)$. (b) Is $(\mathbf{R}, *)$ a semigroup? Is it commutative?

(c) Find the identity element of $*$. (d) Which elements have inverses and what are they?

B.41. Let A be a nonempty set with the operation $*$ defined by $a * b = a$, and assume A has more than one element.

(a) Is A a semigroup?   (c) Does A have an identity element?

(b) Is A commutative?   (d) Which elements, if any, have inverses and what are they?

B.42. Let $A = \{a, b\}$. (a) Find the number of operations on A. (b) Exhibit one which is neither associative nor commutative.

B.43. For each of the following sets, state which are closed under: (a) multiplication; (b) addition.

$$A = \{0, l\}, \quad B = \{1, 2\}, \quad C = \{x \mid x \text{ is prime}\}, \quad D = \{2, 4, 8, \ldots\} = \{x \mid x = 2^n\}.$$

B.44. Let $A = \{\ldots, -9, -6, -3, 0, 3, 6, 9, \ldots\}$, the multiples of 3. Is A closed under:

(a) addition; (b) multiplication; (c) subtraction; (d) division (except by 0)?

B.45. Find a set A of three integers which is closed under: (a) multiplication; (b) addition.

B.46. Let S be an infinite set. Let A be the collection of finite subsets of S and let B be the collection of infinite subsets of S

(a) Is A closed under: (i) union; (ii) intersection; (iii) complements?

(b) Is B closed under: (i) union; (ii) intersection; (iii) complements?

B.47. Let $S = \mathbf{Q} \times \mathbf{Q}$, the set of ordered pairs of rational numbers, with the operation $*$ defined by

$$(a, b) * (x, y) = (ax, ay + b)$$

(a) Find $(3, 4) * (1, 2)$ and $(-1, 3) * (5, 2)$.   (c) Find the identity element of S.

(b) Is S a semigroup? Is it commutative?   (d) Which elements, if any, have inverses and what are they?

B.48. Let $S = \mathbf{N} \times N$, the set of ordered pairs of positive integers, with the operation $*$ defined by

$$(a, b) * (c, d) = (ad + bc, bd)$$

(a) Find $(3, 4) * (1, 5)$ and $(2, 1) * (4, 7)$.

(b) Show that $*$ is associative. (Hence that S is a semigroup.)

(c) Define $f : (S, *) \to (\mathbf{Q}, +)$ by $f(a, b) = a/b$. Show that f is a homomorphism.

(d) Find the congruence relation $\sim$ in S determined by the homomorphism f, that is, $x \sim y$ if $f(x) = f(y)$.

(e) Describe $S/\sim$. Does $S/\sim$ have an identity element? Does it have inverses?

B.49. Let $S = \mathbf{N} \times \mathbf{N}$. Let $*$ be the operation on S defined by

$$(a, b) * (a', b') = (a + a', b + b')$$

(a) Find $(3, 4) * (1, 5)$ and $(2, 1) * (4, 7)$.

(b) Show that $*$ is associative. (Hence that S is a semigroup.)

(c) Define $f: (S, *) \to (\mathbf{Z}, +)$ by $f(a, b) = a - b$. Show that f is a homomorphism.

(d) Find the congruence relation $\sim$ in S determined by the homomorphism f.

(e) Describe $S/\sim$. Does $S/\sim$ have an identity element? Does it have inverses?

GROUPS

B.50. Consider $\mathbf{Z}_{20} = \{0, 1, 2, \ldots, 19\}$ under addition modulo 20. Let H be the subgroup generated by 5. (a) Find the elements and order of H. (b) Find the cosets of H in $\mathbf{Z}_{20}$.

B.51. Consider $G = \{1, 5, 7, 11\}$ under multiplication modulo 12. (a) Find the order of each element. (b) Is G cyclic? (c) Find all subgroups of G.

B.52. Consider $G = \{1, 5, 7, 11, 13, 17\}$ under multiplication modulo 18. (a) Construct the multiplication table of G. (b) Find $5^{-1}, 7^{-1}$, and 17^{-1}. (c) Find the order and group generated by: (i) 5; (ii) 13; (d) Is G cyclic?

B.53. Consider the symmetric group S_4. Let $\alpha = \begin{pmatrix} 1 & 2 & 3 & 4 \\ 3 & 4 & 2 & 1 \end{pmatrix}$ and $\beta = \begin{pmatrix} 1 & 2 & 3 & 4 \\ 2 & 4 & 3 & 1 \end{pmatrix}$.

　　(a) Find $\alpha\beta$, $\beta\alpha$, α^2, α^{-1}.　　(b) Find the orders of α, β, and $\alpha\beta$.

B.54. Prove the following results for a group G.

　　(a) The identity element e is unique.

　　(b) Each a in G has a unique inverse a^{-1}.

　　(c) $(a^{-1})^{-1} = a$, $(ab)^{-1} = b^{-1}a^{-1}$, and, more generally, $(a_r a_2 \ldots a_n) = a_n^{-1} \ldots a_2^{-1}a_1^{-1}$.

　　(d) $ab = ac$ implies $b = c$, and $ba = ca$ implies $b = c$.

　　(e) For any integers r and s, we have $a^r a^s = a^{r+s}$, $(a^r)^s = a^{rs}$.

　　(f) G is abelian if and only if $(ab)^2 = a^2b^2$ for all $a, b \in G$.

B.55. Let H be a subgroup of G. Prove: (a) $H = Ha$ if and only if $a \in H$. (b) $Ha = Hb$ if and only if $ab^{-1} \in H$, (c) $HH = H$.

B.56. Prove Proposition B.5: A subset H of a group G is a subgroup of G if: (i) $e \in H$, (ii) for all $a, b \in H$, we have $ab, a^{-1} \in H$.

B.57. Let G be a group. Prove:

　　(a) The intersection of any number of subgroups of G is a subgroup of G.

　　(b) For any $A \subseteq G$, $gp(A)$ is equal to the intersection of all subgroups of G containing A.

　　(c) The intersection of any number of normal subgroups of G is a normal subgroup of G.

B.58. Suppose G is an abelian group. Show that any factor group G/H is also abelian.

B.59. Suppose $|G| = p$, where p is a prime. Prove: (a) G has no subgroups except G and $\{e\}$. (b) G is cyclic and every element $a \neq e$ generates G.

B.60. Show that $G = \{1, -1, i, -i\}$ is a group under multiplication, and show that $G \cong \mathbf{Z}_4$ by giving an explicit isomorphism $f: G \to \mathbf{Z}_4$.

B.61. Let H be a subgroup of G with only two right cosets. Show that H is normal.

B.62. Let $S = \mathbf{R}^2$, the Cartesian plane. Find the stabilizer H_a of $a = (1, 0)$ in S where G is the following group acting on S:

　　(a) $G = \mathbf{Z} \times \mathbf{Z}$ and G acts on S by $g(x, y) = (x + m, y + n)$ where $g = (m, n)$. That is, each element g in G is a translation of S.

　　(b) $G = (\mathbf{R}, +)$ and G acts on S by $g(x, y) = (x\cos g - y\sin g, x\sin g + y\cos g)$. That is, each element in G rotates S about the origin by an angle g.

B.63. Let S be the regular polygon with n sides, and let G be the group of symmetries of S.

　　(a) Find the order of G.

　　(b) Show that G is generated by two elements a and b such that $a^n = e$, $b^2 = e$, and $b^{-1}ab = a^{-1}$. (G is called the *dihedral group*.)

B.64. Suppose a group G acts on a set S, say by the homomorphism $\psi: \to \mathrm{PERM}(S)$.

　　(a) Prove that, for any $s \in S$: (i) $e(s) = s$, and (ii) $(gg')(s) = g(g'(s))$ where $g, g' \in G$.

　　(b) The orbit G_s of any $s \in S$ is defined by $G_s = \{g(s) \mid g \in G\}$. Show that the orbits form a partition of S.

　　(c) Show that $|G_s| =$ the number of cosets of the stabilizer H_s of s in G. (Recall $H_s = \{g \in G \mid g(s) \in s\}$.)

B.65. Let G be an abelian group and let n be a fixed positive integer. Show that the function $f: G \to G$ defined by $f(a) = a^n$ is a homomorphism.

B.66. Let G be the multiplicative group of complex numbers z such that $|z| = 1$, and let $\mathbf{R}$ be the additive group of real numbers. Prove $G \cong \mathbf{R}/\mathbf{Z}$.

B.67. Suppose H and N are subgroups of G with N normal. Show that: (a) HN is a subgroup of G. (b) $H \cap N$ is a normal subgroup of H. (c) $H/(H \cap N) \cong HN/N$.

B.68. Let H and K be groups. Let G be the product set $H \times K$ with the operation

$$(h, k) * (h', k') = (hh', kk').$$

　　(a) Show that G is a group (called the *direct product* of H and K).

　　(b) Let $H' = H \times \{e\}$. Show that: (i) $H' \cong H$; (ii) H' is a normal subgroup of G; (iii) $G/H' \cong K$.

RINGS

B.69. Consider the ring $\mathbf{Z}_{12} = \{0, 1, \ldots, 11\}$ of integers modulo 12. (a) Find the units of $\mathbf{Z}_{12}$. (b) Find the roots of $f(x) = x^2 + 4x + 4$ over $\mathbf{Z}_{12}$. (c) Find the associates of 2.

B.70. Consider the ring $\mathbf{Z}_{30} = \{0, 1, \ldots, 29\}$ of integers modulo 30.

   (a) Find -2, -7, and -11. (b) Find: 7^{-1}, 11^{-1}, and 26^{-1}.

B.71. Show that in a ring R: (a) $(-a)(-b) = ab$; (b) $(-1)(-1) = 1$, if R has an identity element 1.

B.72. Suppose $a^2 = a$ for every $a \in R$. (Such a ring is called a *Boolean* ring). Prove that R is commutative.

B.73. Let R be a ring with an identity element 1. We make R into another ring R' by defining:

$$a \oplus b = a + b + 1 \quad \text{and} \quad a * b = ab + a + b$$

   (a) Verify that R' is a ring. (b) Determine the 0-element and the 1-element of R'.

B.74. Let G be any (additive) abelian group. Define a multiplication in G by $a * b = 0$ for every $a, b \in G$. Show that this makes G into a ring.

B.75. Let J and K be ideals in a ring R. Prove that $J + K$ nd $J \cap K$ are also ideals.

B.76. Let R be a ring with unity 1. Show that $(a) = \{ra \mid r \in R\}$ is the smallest ideal containing a.

B.77. Show that R and $\{0\}$ are ideals of any ring R.

B.78. Prove: (a) The units of a ring R form a group under multiplication. (b) The units in $\mathbf{Z}_m$ are those integers which are relatively prime to m.

B.79. For any positive integer m, verify that $m\mathbf{Z} = \{rm \mid r \in \mathbf{Z}\}$ is a ring. Show that $2\mathbf{Z}$ and $3\mathbf{Z}$ are not isomorphic.

B.80. Prove Theorem B.10: Let J be an ideal in a ring R. Then the cosets $\{a + J \mid a \in R\}$ form a ring under the coset operations $(a + J) + (b + J) = a + b + J$ and $(a + J)(b + J) = ab + J$.

B.81. Prove Theorem B.11: Let $f: R \to R'$ be a ring homomorphism with kernel K. Then K is an ideal in R, and the quotient ring R/K is isomorphic to $f(R)$.

B.82. Let J be an ideal in a ring R. Consider the (canonical) mapping $f: R \to R/J$ defined by $f(a) = a + J$. Show that: (a) f is a ring homomorphism; (b) f is an onto mapping.

B.83. Suppose J is an ideal in a ring R. Show that: (a) If R is commutative, then R/J is commutative. (b) If R has a unity element 1 and $1 \notin J$, then $1 + R$ is a unity element for R/J.

INTEGRAL DOMAINS AND FIELDS

B.84. Prove that if $x^2 = 1$ in an integral domain D, then $x = -1$ or $x = 1$.

B.85. Let $R \neq \{0\}$ be a finite commutative ring with no zero divisors. Show that R is an integral domain, that is, that R has an identity element 1.

B.86. Prove that $F = \{a + b\sqrt{2} \mid a, b \text{ rational}\}$ is a field.

B.87. Prove that $F = \{a + b\sqrt{2} \mid a, b \text{ integers}\}$ is an integral domain but not a field.

B.88. A complex number $a + bi$ where a, b are integers is called a *Gaussian integer*. Show that the set G of Gaussian integers is an integral domain. Also show that the units are $\pm 1, \pm i$.

B.89. Let R be an integral domain and let J be an ideal in R. Prove that the factor ring R/J is an integral domain if and only if J is a prime ideal. (An ideal J is *prime* if $J \neq R$ and if $ab \in J$ implies $a \in J$ or $b \in J$.)

B.90. Let R be a commutative ring with unity element 1, and let J be an ideal in R. Prove that the factor ring R/J is a field if and only if J is a maximal ideal. (An ideal J is maximal if $J \neq R$ and no ideal K lies strictly between J and R, that is, if $J \subseteq K \subseteq R$ then $J = K$ or $K = R$.)

B.91. Let D be the ring of real 2×2 matrices of the form $\begin{bmatrix} a & -b \\ b & a \end{bmatrix}$. Show that D is isomorphic to the complex field C, when D is a field.

B.92. Show that the only ideal in a field K is $\{0\}$ or K itself.

B.93. Suppose $f: K \to K'$ is a homomorphism from a field K to a field K'. Show that f is an *embedding*; that is, f is one-to-one. (We assume $f(1) \neq 0$.)

B.94. Consider the integral domain $D = \{a + b\sqrt{13} \mid a, b \text{ integers}\}$. (See Example B.15(b).) If $\alpha = a + \sqrt{13}$, we define $N(\alpha) = a^2 - 13b^2$. Prove:

    (i) $N(\alpha\beta) = N(\alpha)N(\beta)$.

    (iii) Among the units of D are ± 1, $18 \pm 5\sqrt{13}$; and $-18 \pm 5\sqrt{13}$.

    (ii) α is a unit if and only if $N(\alpha) = +1$.

    (iv) The numbers 2, $3 - \sqrt{13}$ and $-3 - \sqrt{13}$ are irreducible.

POLYNOMIALS OVER A FIELD

B.95. Find the roots of $f(t)$ assuming $f(t)$ has an integer root: (a) $f(t) = t^3 - 2t^2 - 6t - 3$; (b) $f(t) = t^3 - t^2 - 11t - 10$; (c) $f(t) = t^3 + 2t^2 - 13t - 6$.

B.96. Find the roots of $f(t)$ assuming $f(t)$ has a rational root: (a) $f(t) = 2t^3 - 3t^2 - 16t - 7$; (b) $f(t) = 2t^3 - t^2 - 9t + 9$.

B.97. Find the roots of $f(t) = t^4 - 5t^3 + 16t^2 - 9t - 13$, given that $t = 2 + 3i$ is a root.

B.98. Find the roots of $f(t) = t^4 - t^3 - 5t^2 + 12t - 10$, given that $t = 1 - i$ is a root.

B.99. For any scalar $a \in K$, define the *evaluation map* $\psi_a \colon K[t] \to K$ by $\psi_a(f(t)) = f(a)$. Show that ψ_a is a ring homomorphism.

B.100. Prove: (a) Proposition B.14. (b) Theorem B.26.

Answers to Supplementary Problems

B.39. (a) 12, 15, 18, 6; (b) Yes, yes; (c) 1; (d) Only 1 and it is its own inverse.

B.40. (a) 17, -32, 29/2; (b) Yes, yes; (c) Zero; (d) If $a \neq 1/2$, then a has an inverse which is $-a/(1 + 2a)$.

B.41. (a) Yes; (b) No; (c) No; (d) It is meaningless to talk about inverses when no identity element exists.

B.42. (a) Sixteen, since there are two choices, a or b, for each of the four products aa, ab, ba, and bb. (b) Let $aa = b$, $ab = a$, $ba = b$, $bb = a$. Then $ab \neq ba$. Also, $(aa)b = bb = a$, but $a(ab)$ as $= b$.

B.43. (a) A, D; (b) none.

B.44. (a) Yes; (b) yes; (c) yes; (d) no.

B.45. (a) $\{1, -1, 0\}$; (b) There is no set.

B.46. (a) Yes, yes, no; (b) Yes, no, no.

B.47. (a) (3, 10), $(-5, 1)$; (b) yes, no; (c) (1, 0); (d) The element (a, b) has an inverse if $a \neq 0$, and its inverse is $(1/a, -b/a)$.

B.48. (a) (19, 20), (18, 7). (d) $(a, b) \sim (c, d)$ if $ad = bc$. (e) $S/\sim$ is isomorphic to the positive rational numbers under addition. Thus $S/\sim$ has no identity element and no inverses.

B.49. (a) (4, 9), (6, 8); (d) $(a, b) \sim (c, d)$ if $a + d = b + c$. (e) $S/\sim$ is isomorphic $\mathbf{Z}$ since every integer is the difference of two positive integers. Thus $S/\sim$ has an identity element, and every element has an inverse.

B.50. (a) $H = l\{0, 5, 10, 15\}$ and $|H| = 4$. (b) H, $1 + H = \{1, 6, 11, 16\}$, $2 + H = \{2, 7, 12, 17\}$, $3 + H = \{3, 8, 13, 18\}$, $4 + H = \{4, 9, 14, 19\}$.

B.51. (a) $x^2 = 1$ if $x \neq 1$. (b) No. (c) $\{1\}$, $\{1, 5\}$, $\{1, 7\}$, $\{1, 11\}$, G.

B.52. (a) See Fig. B-9(a). (b) 11, 13, 17; (c) (i) $|\%| = 6$, $gp(5) = G$; (ii) $|13| = 3$, $gp(13) = \{1, 7, 13\}$; (d) Yes, since $G = gp(5)$.

B.53. (a) See Fig. B-9(b). (b) 4, 3, 4.

×	1	5	7	11	13	17
1	1	5	7	11	13	17
5	5	7	17	1	11	13
7	7	17	13	5	1	11
11	11	1	5	13	17	7
13	13	11	1	17	7	5
17	17	13	11	7	5	1

(a)

$$\alpha\beta = \begin{pmatrix} 1 & 2 & 3 & 4 \\ 3 & 1 & 4 & 2 \end{pmatrix}, \quad \beta\alpha = \begin{pmatrix} 1 & 2 & 3 & 4 \\ 4 & 1 & 2 & 3 \end{pmatrix}$$

$$\alpha^2 = \begin{pmatrix} 1 & 2 & 3 & 4 \\ 2 & 1 & 4 & 3 \end{pmatrix}, \quad \alpha^{-1} = \begin{pmatrix} 1 & 2 & 3 & 4 \\ 4 & 3 & 1 & 2 \end{pmatrix}$$

(b)

Fig. B-9

B.60. $f(1) = 0$, $f(i) = 1$, $f(-1) = 2$, $f(-i) = 3$

B.62. (a) $\{(0, 0)\}$, (b) $\{2\pi r \mid r \in Z\}$.

B.69. (a) $1, 5, 7, 11$; (b) $4, 10$; (c) $\{2, 10\}$.

B.70. (a) $28, 23, 19$; (b) $13, 11$, 26^{-1} does not exist since 26 is not a unit.

B.72. Show $-a = a$ using $a + a = (a + a)^2$. Then show $ab = -ba$ by $(a + b) = (a + b)^2$.

B.73. (b) $-1 = 0-$element, $0 = 1-$element.

B.91. Show f is an isomorphism where $f\left(\begin{bmatrix} a & -b \\ b & a \end{bmatrix}\right) = a + bi$.

B.93. *Hint*: Use Problem B.92.

B.95. (a) -1, $(3 \pm \sqrt{21})/2$; (b) -2, $(3 \pm \sqrt{29})/2$; (c) 3, $(-5 \pm \sqrt{17})/2$

B.96. (a) $-1/2$, $1 \pm 2\sqrt{2}$; (b) $3/2$, $(-1 \pm \sqrt{13})/2$

B.97. $2 \pm 3i$, $(1 \pm \sqrt{5})/2$

B.98. $1 \pm i$, $(-1 \pm \sqrt{21})/2$

INDEX